Sample Entry

(1)—**ALLOWING TO DIE / NEWBORNS / DOWN'S SYNDROME**

(3)

(2)—**Gustafson, James M.**   Mongolism, parental
desires, and the right to life.

(4)   *Perspectives in Biology and Medicine*   (7)

(5)—16(4): 529-557, Spring 1973.   20 refs.—(8)

(6)   12 fn.   BE00469.   (10)

(9)   *allowing to die; *Down's syndrome;
extraordinary treatment; intelligence;

(11)   *moral obligations; *newborns;
nurses; *parental consent; physicians;
quality of life; *treatment refusal;
*value of life; Northeastern University—(12)
Hospital

1. Subject heading
2. Author
3. Title of article
4. Name of journal
5. Volume and issue number
6. Pagination
7. Date of publication
8. Number of bibliographical references
9. Number of footnotes
10. Bioethics Accession Number
11. Descriptors: *allowing to die. . . *value of life.
The most important concepts in the document
are represented by asterisked terms.
12. Identifier: Northeastern University Hospital

BIBLIOGRAPHY
OF
BIOETHICS

BIBLIOGRAPHY
OF
BIOETHICS

Volume 5

Editor
LeRoy Walters

Associate Editors
Maureen L. Canick
Renée Johnson

Center for Bioethics, Kennedy Institute
Georgetown University, Washington, D.C. 20057

Gale Research Company
Book Tower • Detroit, Michigan 48226

This *Bibliography of Bioethics* is one of the ongoing research projects of the Kennedy Institute. New volumes of the *Bibliography of Bioethics* will be produced annually by the Kennedy Institute and together with an *Encyclopedia of Bioethics,* published in 1978, will provide basic research information in the field of bioethics.

Library of Congress Catalog Card Number 75-4140
ISBN 0-8103-0986-6

This publication was supported in part by NIH Grant LM 01702 from the National Library of Medicine.

Contents

v

Staff

LeRoy Walters, Ph.D.
Editor

Maureen Canick, M.L.S.
Senior Bibliographer

Renée Johnson, M.S.
Bibliographer

Emilio Jaksetic, B.S., J.D.
Research Assistant

Marlene Fine
Norina Moy
Student Assistants

Subject-Matter Consultants

Bioethics, in General

J. Russell Elkinton, M.D.
Former Editor
Annals of Internal Medicine

Behavior Control

Daniel N. Robinson, M.D.
Department of Psychology
Georgetown University

Transplants and Artificial Organs—Heart

Theodore Cooper, M.D.
School of Medicine
Cornell University

Genetics

Robert C. Baumiller, Ph.D.
Department of Obstetrics and Gynecology
Georgetown University

Drug Testing and Use

William T. Beaver, M.D.
Department of Pharmacology
Georgetown University

Population

Arthur J. Dyck, Ph.D.
School of Public Health
Harvard University

Human Experimentation

Jay Katz, M.D.
School of Law
Yale University

Death

William E. Flynn, M.D.
Department of Psychiatry
Georgetown University

Transplants and Artificial Organs—Kidney

George E. Schreiner, M.D.
Department of Medicine
Georgetown University

Human Reproduction

André E. Hellegers, M.D.
Kennedy Institute
Georgetown University

Health-Care Delivery

Robert R. Huntley, M.D.
Department of Community Medicine
Georgetown University

Law and Bioethics

Lawrence R. Tancredi, M.D., J.D.
Schools of Medicine and Law
New York University

INTRODUCTION

INTRODUCTION

The Field of Bioethics

Bioethics can be defined as the systematic study of value questions which arise in the biomedical and behavioral fields. Specific bioethical issues which have recently received national and international attention include euthanasia, psychosurgery, human experimentation, genetic engineering, abortion, the definition of death, medical confidentiality, and the allocation of scarce medical resources.

As this list of topics suggests, the field of bioethics includes several dimensions. The first is health care ethics, a consideration of the rights and duties of health professionals and their patients. Traditionally, the accent in discussions of health care ethics has been on the duties of health professionals—duties which, since the time of Hippocrates, have frequently been delineated in codes of professional ethics. In more recent times the rights of patients have also received considerable attention. Research ethics, the study of value problems in biomedical and behavioral research, constitutes a second dimension of bioethics. During the 20th century, as both the volume and visible achievements of such research have increased, new questions have arisen concerning the investigator-subject relationship and the potential social impact of biomedical and behavioral research and technology. Within the past decade a third dimension of bioethics has emerged—the quest to develop reasonable public-policy guidelines for both the delivery of health care and the conduct of biomedical and behavioral research.

It is obvious that no single academic discipline is adequate to discuss these various dimensions of bioethics. For this reason bioethics has been, since its inception in the late 1960s, a cross-disciplinary field. The primary participants in the interdisciplinary discussion have been physicians, biologists, psychologists, sociologists, lawyers, historians, and philosophical and religious ethicists.

During the 1970s there has been a rapid growth of academic, professional, and public interest in the field of bioethics. One evidence of this interest is the establishment of numerous research institutes and teaching programs in bioethics, both in the United States and abroad. In recent years, professional societies, federal and state legislatures, and the courts have also turned increasing attention to problems in the field. In addition, during the past several years there has been a veritable explosion of literature on bioethical issues.

Despite the increase of public interest and available literature in bioethics, no systematic effort to achieve bibliographic control of the field has been undertaken. The literature of bioethics appears in widely scattered sources and is reported, if at all, in diverse indexes which employ a bewildering variety of subject headings. This *Bibliography* represents an initial attempt to identify the central issues of bioethics, to develop an index language appropriate to the field, and to provide comprehensive, cross-disciplinary coverage of current English-language materials on bioethical topics.

3

The Scope of the Bibliography

This fifth volume of the *Bibliography of Bioethics* includes materials which discuss the ethical aspects of the following major topics and subtopics:

BIOETHICS, IN GENERAL
HISTORY OF MEDICAL ETHICS
CODES OF PROFESSIONAL ETHICS
PHYSICIAN PATIENT RELATIONSHIP
    Disclosure
    Informed Consent
    Treatment Refusal
    Confidentiality
    Malpractice
HEALTH CARE
    Patients' Rights
    Cost of Health Care
    Resource Allocation
CONTRACEPTION
    Availability of Contraceptives to Minors
    Involuntary Sterilization
ABORTION
POPULATION
    Right to Reproduce
    Population Policy
REPRODUCTIVE TECHNOLOGIES
    Artificial Insemination
    Sex Predetermination
    In Vitro Fertilization
    Embryo Transfer
    Cloning
GENETIC INTERVENTION
    Genetic Counseling
    Genetic Screening
    DNA Therapy
    Recombinant DNA Research
    Eugenics
SOCIOBIOLOGY
MENTAL HEALTH THERAPIES
    Psychotherapy

    Operant Conditioning
    Psychoactive Drugs
    Electrical Stimulation of the Brain
    Psychosurgery
    Involuntary Commitment
    Right to Treatment
HUMAN EXPERIMENTATION
    Informed Consent
    Behavioral Research
    Research on Children
    Research on Pregnant Women
    Research on Fetuses
    Research on Prisoners
    Research on the Mentally Handicapped
    Social Control
ARTIFICIAL AND TRANSPLANTED
  ORGANS OR TISSUES
    Organ or Tissue Donation
    Heart
    Kidney
    Blood
    Other Organs or Tissues
DEATH AND DYING
    Determination of Death
    Terminal Care
    Prolongation of Life
    Allowing to Die
    Euthanasia
INTERNATIONAL DIMENSIONS OF
  BIOLOGY AND MEDICINE
    Biological Warfare
    Physicians and Violence
    Resource Allocation.[1]

The *Bibliography* seeks to be comprehensive for all English-language materials—both print and nonprint—which discuss ethical aspects of the topics and subtopics listed above. It therefore incorporates a variety of media and literary forms, including journal and newspaper articles, monographs, essays in books, court decisions, bills, audiovisual materials, and unpublished documents. This fifth volume of the *Bibliography* indexes documents[2] which were published during calendar years 1973 through 1978, concentrating primarily on 1977 documents.

A cross-disciplinary monitoring system has been devised in an effort to secure documents falling within the subject-matter scope outlined above. The Research Assistant systematically examines 59 reference tools for pertinent citations:

Advance Bibliography of Contents: Political
    Science and Government
All England Law Reports (subject index)
Australian National Bibliography
Bibliographic Index

British Humanities Index
British National Bibliography
Canadian Periodical Index
Catholic Periodical and Literature Index

---

[1] Several additional topics are closely related to the field of bioethics but have not been included in the *Bibliography*. Among these are animal experimentation, human sexuality, ecology, and the world hunger problem.

[2] In this Introduction the word "document" is used in an extended sense to refer to both print and nonprint materials.

Choice
The Citation
Clearinghouse Review
Congressional Information Service
Congressional Record (subject index)
Contents of Current Legal Periodicals
Cumulative Book Index
Cumulative Index to Nursing and Allied Health
   Literature
Current Contents: Clinical Practice
Current Contents: Social and Behavioral Sciences
Current Work in the History of Medicine
Dissertation Abstracts (A): Humanities
   and Social Sciences
Dissertation Abstracts (B): Sciences and
   Engineering
Dominion Law Reports (subject index)
Essay and General Literature Index
Family Planning/Population Reporter
General Digest
Harvard Annual Legal Bibliography
Hospital Literature Index
Humanities Index
Index to Canadian Legal Periodical Literature
Index to the Christian Science Monitor
Index to Foreign Legal Periodicals
Index to Legal Periodicals
Index to Periodical Articles Related to Law

Index to Religious Periodical Literature
International Nursing Index
Library of Congress Catalog: Films and Other
   Materials for Projection
Medical Socioeconomic Research Sources
Mental Retardation and the Law
Modern Federal Practice Digest
Monthly Catalog of U.S. Government Publications
Philosopher's Index
Popular Periodical Index
Population Index
Psychological Abstracts
Public Affairs Information Service Bulletin
Publishers' Weekly
Reader's Guide to Periodical Literature
Religious and Theological Abstracts
Reporter on Human Reproduction and the Law
Science Books and Films
Selected Rand Abstracts
Social Sciences Index
Sociological Abstracts
State Health Legislation Report
Subject Guide to Books in Print
Times [of London] Index
Vertical File Index
Wall Street Journal Index
Western Weekly Reports (subject index).

Three data bases are also searched, the New York Times Information Bank and the National Library of Medicine's MEDLINE and CATLINE files.

In addition, the *Bibliography* staff directly monitors 82 journals and newspapers for articles and citations falling within the scope of bioethics:

America
American Journal of Law and Medicine
American Journal of Nursing
American Journal of Psychiatry
American Journal of Public Health
American Medical News
American Psychologist
American Scientist
Annals of Internal Medicine
Atlantic
BioScience
British Medical Journal
Bulletin of the American Academy of Psychiatry
   and the Law
Bulletin of the Atomic Scientists
Bulletin of the History of Medicine
Chemical and Engineering News
Christian Century
Chronicle of Higher Education
Commentary
Commonweal
Connecticut Medicine
Dimensions in Health Service

Ebony
Ethics
Ethics in Science and Medicine
Futurist
Hastings Center Report
Hospital Medical Staff
Hospital Progress
Hospitals
Human Life Review
Impact of Science on Society
International Digest of Health Legislation
Journal of Health, Politics, Policy and Law
Journal of Medical Ethics
Journal of Religious Ethics
Journal of the American Medical Association
Judaism
Lancet
Legal Aspects of Medical Practice
Linacre Quarterly
Man and Medicine
Medical Care
Medical World News
Medico-Legal Bulletin

Medico-Legal Journal
Medico-Moral Newsletter
Minerva
Modern Medicine
Nature
New England Journal of Medicine
New Physician
New Scientist
New York Times
New York Times Magazine
Nursing Mirror
Nursing Times
Ob. Gyn. News
Omega
Perspectives in Biology and Medicine
Pharos
Philosophy and Public Affairs
Philosophy East and West
Psychology Today

Review of Metaphysics
Saturday Review
Science
Science and Government Report
Science and Society
Science for the People
Science News
The Sciences
Scientific American
Sh'ma
Social Biology
Social Science and Medicine
Tradition
Washington Post
WHO Chronicle
World Medical Journal
Yale Journal of Biology and Medicine
Zygon.

This fifth volume of the *Bibliography* includes bibliographical data for 1,601 documents, of which 6 were published in 1973, 14 in 1974, 25 in 1975, 90 in 1976, 1,160 in 1977 and 306 in 1978.

Instructions for Use of the Bibliography

The *Bibliography of Bioethics* is divided into six parts:

1. Introduction
2. List of Journals Cited
3. Bioethics Thesaurus
4. Subject Entry Section
5. Title Index
6. Author Index.

Parts 3 and 4, the Bioethics Thesaurus and the Subject Entry Section, constitute the core of the *Bibliography;* relatively detailed information concerning their construction and use will be provided. Also included are brief instructions for the use of parts 2, 5, and 6.

List of Journals Cited

The second part of the *Bibliography,* the List of Journals Cited, records the name of each journal cited in the Subject Entry Section. When available, the International Standard Serial Number (ISSN) of the journal is also listed. The dual purpose of this list is to indicate precisely which journal is designated by each journal title and to obviate the need to include an ISSN number with each individual citation from a journal.

Bioethics Thesaurus

The Bioethics Thesaurus is an index language developed by the Bibliographers specifically for the cross-disciplinary field of bioethics. Within the *Bibliography* the primary functions of the Thesaurus are to translate the user's concepts into searchable terms and to enable the user to broaden or narrow the scope of a search.

Since only the standardized language of the Bioethics Thesaurus is employed in the subject headings of the Subject Entry Section, we encourage users to consult the Thesaurus before searching in the Subject Entry Section for documents on particular topics. This two-step method is more likely to be successful than a direct approach to the Subject Entry Section. For example, a user who searches the Subject Entry Section for documents on "the sanctity of life" will find nothing. In contrast, a user who consults the Bioethics Thesaurus will discover the following entry:

Sanctity of Life
Use VALUE OF LIFE

The further step of checking the Subject Entry Section for documents on the "Value of Life" will yield 11 citations and 17 cross-references.

The following are examples of a main entry and a cross-reference in the Bioethics Thesaurus:[3]

> **HOSPITALS**
> SN Institutions in which sick or injured
> persons are given medical or surgical
> treatment
> UF Clinics
> Hospital policies
> BT Institutions
> NT Mental institutions
> Private hospitals
> RT Intensive care units

> **Clinics**
> *Use* HOSPITALS

SN	The SN, or Scope Note, is a brief statement of the intended usage of a term. It may be used to clarify an ambiguous term or to restrict the usage of a term.
UF, *Use*	Terms following the UF (Used For) notation are synonyms or variant forms of the searchable term. They are not used for indexing or searching but indicate to the user what other terms or concepts are linguistically covered by the searchable term. *Use* references are the reciprocals of UF references.
BT, NT	The BT (Broader Term) and NT (Narrower Term) notations indicate hierarchical relationships among terms. A Broader Term (for example, Institutions) designates the class of which the entry term (**HOSPITALS**) is a member. Narrower Terms are the reciprocals of Broader Terms.
RT	Terms which do not fall into the broader-narrower relationship, but which are conceptually related, are listed as Related Terms (RT).
↓	If there are narrower terms for either a Narrower Term or a Related Term, the existence of those narrower terms is indicated by an arrow (↓). By turning to the point in the Thesaurus where the NT or RT appears as an entry term, the user can discover what the narrower terms are. The arrow thus assists the user who wishes to follow a hierarchy from a broad term to a highly specific term.
†	The remaining symbol, the dagger (†) indicates that a phrase is represented in the Bioethics Thesaurus by two terms rather than by one term. Thus, a user who consulted **"Hospital Policies"** in the appropriate section of the Thesaurus would discover that the concept **"Hospital Policies"** is represented in this index language by the two terms **"INSTITUTIONAL POLICIES"** and **"HOSPITALS."**[4]

As the foregoing examples illustrate, the word order of natural language is employed throughout the vocabulary of the Bioethics Thesaurus. Phrases like "value of life" or "private hospitals" are always expressed in the order of ordinary speech rather than in the inverted order sometimes found in traditional subject headings, e.g., "life, value of."

This edition of the Bioethics Thesaurus contains 580 searchable terms; it lists, in addition, 229 *Use* references which lead the user from non-searchable to searchable terms. The approach to thesaurus construction has been inductive rather than deductive; only concepts actually encountered in the indexed documents have been included in the Thesaurus.

---

[3] These examples have been slightly modified to exhibit all of the abbreviations and symbols employed in the Bioethics Thesaurus.

[4] These abbreviations and symbols closely follow the recommendations of the *American National Standard Guidelines for Thesaurus Structure, Construction, and Use, Z39.19-1974* (New York: American National Standards Institute, 1974).

Subject Entry Section

The Subject Entry Section, the main part of the *Bibliography,* contains entries, or bibliographic descriptions of documents, which are arranged according to subject headings. In Volume 5 of the *Bibliography,* entries for 1,601 documents have been included in the Subject Entry Section. By form, these 1,601 documents can be categorized as follows:

Journal articles	925	57.8%
Essays in books	247	15.4%
Newspaper articles	187	11.7%
Monographs	130	8.1%
Court decisions	62	3.7%
Audiovisuals	24	1.5%
Bills or laws	23	1.4%
Unpublished documents	3	0.2%

Entries

A sample subject heading and entry for a journal article follow:[5]

ALLOWING TO DIE / NEWBORNS / DOWN'S SYNDROME

Gustafson, James M. Mongolism, parental desires, and the right to life. *Perspectives in Biology and Medicine* 16(4): 529-557, Spring 1973. 20 refs. 12 fn. BE00469.
*allowing to die; *Down's syndrome; extraordinary treatment; intelligence; *moral obligations; *newborns; nurses; *parental consent; physicians; quality of life; *treatment refusal; *value of life; Northeastern University Hospital.

Twelve data elements appear in the preceding entry for a journal article:

1. Author: Gustafson, James M.
2. Title of article: Mongolism, parental desires, and the right to life.
3. Name of journal: Perspectives in Biology and Medicine[6]
4. Volume number: 16
5. Issue number: (4)
6. Pagination: 529-557
7. Date of publication: Spring 1973
8. Number of references: 20
9. Number of footnotes: 12
10. Bioethics Accession Number:[7] BE00469
11. Descriptors: *allowing to die; *Down's syndrome; etc.
12. Identifier(s): Northeastern University Hospital.

Elements 8, 9, 11, and 12 require brief explanation. Within this work, references (element 8) are defined as bibliographical citations which appear at the conclusion of a document, usually in alphabetical order. References are not repeated within a single document. Footnotes, on the other hand, are defined as explanatory comments or bibliographical citations which parallel the sequence of ideas expressed in the document. Footnotes may appear either at the foot of a page or at the conclusion of a document.[8]

---

[5] This example has been slightly modified for illustrative purposes.

[6] International Standard Serial Numbers (ISSN's) are not included in individual entries. Instead, the ISSN's of all journals cited in the Subject Entry Section are included in the part of the *Bibliography* entitled "List of Journals Cited."

[7] The Bioethics Accession Number is a notation intended only for the use of the *Bibliography* staff.

[8] When there is no mention of footnotes or references in a particular entry, none appeared in the original document.

One of the most important elements in each entry is the series of descriptors (element 11). The descriptors, all of which have been chosen from the controlled vocabulary of the Bioethics Thesaurus, summarize the content of each document. The most important concepts in the document are designated by asterisks. All descriptors are arranged in alphabetical sequence.

Approximately 10–12 descriptors have been assigned to each journal article of average length. The general indexing policy followed in the *Bibliography* has been to employ the most specific term available. Thus, in the example cited above, the term "Down's syndrome" is preferred to the more general "chromosomal disorders."

Identifiers (element 12) are proper nouns which are not part of the Bioethics Thesaurus. In most cases identifiers refer to a particular person, organization, political entity, or time. All identifiers are listed in alphabetical order following the last descriptor.

The sample entry presented above displays the format and elements which appear in a journal article entry. The *Bibliography* staff has consciously sought to devise distinctive formats for each of the major forms of material encountered, for example, essays in books, monographs, and court decisions. Letters to the editor, editorials, and specific types of audiovisual materials are also identified in the bibliographic data.

Several print and nonprint forms contain data elements which do not appear in journal-article entries. Among these additional elements the most important are the following:

1. Essays in books
   Editor
   Title of book
   International Standard Book Number (ISBN)
2. Monographs
   Imprint
   Total pagination
   Pagination of bibliography
   ISBN
3. Court decisions
   Date of decision
4. Bills
   Date of introduction
   Sponsor
   Legislative committee
5. Laws
   Date of approval
6. Audiovisual materials
   Medium designation
   Physical description

Subject Headings

The subject headings which appear in the Subject Entry Section are comprised of one to three asterisked terms from the list of descriptors in the entry itself.[9] Important identifiers may also be included in subject headings.

In the example cited above, the composite subject heading is:

ALLOWING TO DIE / NEWBORNS / DOWN'S SYNDROME.

Since the composite subject headings are a distinctive feature of the *Bibliography of Bioethics*,[10] their function will be briefly explained. Most bibliographies and indexes which use controlled vocabularies employ one of two systems: (1) they print the same citation under several simple subject headings (for example, "allowing to die," "newborns," and

---

[9] Most, but not all, of the asterisked descriptors which follow citations have been included in subject headings.

[10] A similar system of composite subject headings is employed in the *British Technology Index*, edited by E. J. Coates.

"Down's syndrome"); or (2) they print the citation only once but repeat its accession number under several simple subject headings. In either case the user who wishes to coordinate several concepts or topics must either manually compare lists of citations or accession numbers or rely on a computer for the coordination of terms. Within the present *Bibliography* the composite subject headings enable the user to coordinate important terms or concepts with a minimum of effort and without recourse to a computer.

The Subject Heading approach employed in Volumes 1-4 of the *Bibliography* has been refined to limit the total number of terms in each heading to no more than three. The terms which comprise the composite subject headings are arranged in a definite sequence according to a general formula. Since most users of the *Bibliography of Bioethics* will wish to retrieve citations which discuss either a specific *topic* (for example, human experimentation) or a specific *issue* (for example, informed consent), these elements, or facets, have been given higher priority in the formula. The complete list of facets contained in the formula follows:

Facets: Topic / Issue / Agent / Target / Context / Generalities / Space / Time

Illustrations: HUMAN EXPERIMENTATION / INFORMED CONSENT / PHYSICIANS /
ADULTS / CANCER / LEGAL ASPECTS / NEW JERSEY / 1956.

No subject heading will contain terms corresponding to all eight facets. However, the facets which are represented will be arranged in the order indicated by the formula. For instance, in the example cited earlier, the following three facets are present:

ALLOWING TO DIE / NEWBORNS / DOWN'S SYNDROME
Topic   Target   Context.

The cross-reference structure of the Subject Entry Section ensures that users who search for facets of lower priority (for example, Targets or Contexts) will be led to main subject headings. In the example cited above, a user who searched for documents on "Down's syndrome" would encounter a series of cross-references like the following:

Down's Syndrome   *See under*

ALLOWING TO DIE / NEWBORNS / DOWN'S
SYNDROME

AMNIOCENTESIS / SELECTIVE ABORTION/
DOWN'S SYNDROME

MASS SCREENING / PREGNANT WOMEN / DOWN'S
SYNDROME. . . .

Thus, it is possible to find, grouped together at a single point in the *Bibliography,* cross-references to all subject headings which contain a particular term, even if that term usually appears in a facet of lower priority.[11]

Finally, it should be noted that not every term from the Bioethics Thesaurus is included in the subject headings of the Subject Entry Section. Only terms which represent major themes or concepts in the indexed materials are employed in the subject headings.

Title Index, Author Index

The fifth and sixth parts, the Title Index and the Author Index, require little explanation. The numbers following the title in both indexes refer to page numbers in the Subject Entry Section. If a page number is repeated, the entry in question appears more than once on a single page.

An effort has been made to determine the authorship of every document indexed in the Subject Entry Section. In cases where no author could be identified, the document is cited only in the Title Index. For the convenience of users, the titles of individual documents are included with the authors' names in the Author Index.

---

[11] If an indexed document includes discussion of more than one major *topic,* multiple main subject headings have been created for that document.

Acknowledgments

It is a pleasure to acknowledge the assistance of numerous persons who played a major role in the production of this fifth volume of the *Bibliography of Bioethics.*

The Senior Bibliographer, Maureen Canick, supervised the day-to-day operations of the project. Ms. Canick and Renée Johnson continued to develop the language of the Bioethics Thesaurus and indexed the documents cited in this volume. Emilio Jaksetic, Research Assistant, monitored the indexes and journals listed in the Introduction and played a significant role in securing documents. Marlene Fine assisted in the monitoring of newspapers and in document acquisition, as well as providing secretarial support for the project, while Norina Moy was instrumental in keypunching and preparation of computer tapes.

The Librarian of the Center for Bioethics, Doris Goldstein, provided helpful advice on the retrieval of documents and, together with her staff, secured many of the documents which are included in the *Bibliography.* We also wish to thank Christine W. Nord, who prepared the final typed draft of the List of Journals Cited.

The fifth volume of the *Bibliography* has been produced with the aid of computer programs developed by Dr. Richard H. Lineback of Bowling Green State University. Competent technical assistance in the running of the programs was provided by Maria Limarzi, University Information Systems, Georgetown University. Our sincere thanks are also extended to John Vytowich of the International Computaprint Corporation, who efficiently transformed the data into camera-ready pages.

The staff owes a debt of gratitude to Professor David Batty, Information Science Consultant, for his expert advice in the production of the *Bibliography.* Professor Batty, more than any other single individual, conceptualized the information-science methodology employed in this project. We also wish to thank Dr. Clifford A. Bachrach and Edith Calhoun of the Medical Subject Headings Division at the National Library of Medicine, who generously shared their expertise with the staff. Valuable comments have also been contributed by the scholars at the Kennedy Institute and by our Subject-Matter Consultants.

The Gale Research Company has provided helpful technical advice concerning the production of the *Bibliography.* We thank James Ethridge, Vice-President of Gale Research Company, for his interest in the project since its inception and Dedria Bryfonski, Editor, for suggesting ways of producing a more useful and readable book.

Finally, the staff acknowledges with gratitude the generous financial support of two organizations. The Extramural Programs of the National Library of Medicine have provided funding for approximately 90% of the total costs of this fifth volume. We wish to thank the Library and, in particular, our Project Officer, Dr. Roger W. Dahlen. The remaining support for the project, as well as funding for the Bioethics Library which has helped to make the project possible, has been provided by the Joseph P. Kennedy, Jr. Foundation, to which we also extend our sincere thanks.

The bibliographical information published in Volumes 1-5 of the *Bibliography of Bioethics* is also available in a new National Library of Medicine data base, BIOETHICSLINE. For further information about access to BIOETHICSLINE, please contact Medlars Management, National Library of Medicine, 8600 Rockville Pike, Bethesda, Maryland 20209.

* * *

The staff welcomes suggestions for the improvement of future volumes of the *Bibliography of Bioethics.* Please send all comments to:

Editor, *Bibliography of Bioethics*
Center for Bioethics, Kennedy Institute
Georgetown University
Washington, D.C.   20057.

April 18, 1979   LeRoy Walters

LIST OF JOURNALS CITED

LIST OF JOURNALS CITED

Acta Sociologica   ISSN 0001-6993
Administration in Mental Health   ISSN 0090-1180
Advocate: The Suffolk University Law School Journal
Air Force Law Review   ISSN 0094-8381
Alabama Journal of Medical Sciences   ISSN 0002-4252
Alcoholism: Clinical and Experimental Research
America   ISSN 0002-7030
American Historical Review   ISSN 0002-8762
American Journal of Clinical Pathology   ISSN 0002-9173
American Journal of Digestive Diseases
American Journal of Epidemiology   ISSN 0002-9262
American Journal of Human Genetics   ISSN 0002-9297
American Journal of Law and Medicine   ISSN 0098-8588
American Journal of Nursing   ISSN 0002-936X
American Journal of Occupational Therapy   ISSN 0002-9386
American Journal of Orthopsychiatry   ISSN 0002-9432
American Journal of Psychiatry   ISSN 0002-953X
American Journal of Public Health   ISSN 0002-9572
American Journal of Roentgenology   ISSN 0002-9580
American Medical News   ISSN 0001-1843
American Philosophical Quarterly   ISSN 0003-0481
American Psychologist   ISSN 0003-066X
American Sociological Review   ISSN 0003-1224
Anaesthesia and Intensive Care
Analysis   ISSN 0003-2638
Anesthesiology   ISSN 0003-3022
Annals of Internal Medicine   ISSN 0003-4819
Annals of Neurology
Archives of Surgery   ISSN 0004-0010
Arizona Law Review   ISSN 0004-153X
Arizona State Law Journal
Arkansas Law Review   ISSN 0004-1831
Arkansas Statutes Annotated
Atlantic Reporter, 2d Series

Behavior Modification   ISSN 0145-4455
Behavior Therapy   ISSN 0005-7894
Behavioral Engineering

Bioethics Digest
Biological Psychiatry   ISSN 0066-3223
BioScience   ISSN 0006-3568
Brethren Life and Thought   ISSN 0006-9663
British Journal of Psychiatry   ISSN 0007-1250
British Medical Journal   ISSN 0007-1447
Brooklyn Law Review   ISSN 0007-2362
Bulletin of the American Academy of Psychiatry and
    the Law   ISSN 0091-634X
Bulletin of the Atomic Scientists
Bulletin of the History of Medicine   ISSN 0007-5140
Bulletin of the New York Academy of
    Medicine   ISSN 0028-7091

CA-A Cancer Journal for Clinicians   ISSN 0007-9235
California Law Review   ISSN 0008-1221
California Penal Code (Deering)
California Reporter
Canadian Family Physician   ISSN 0008-350X
Canadian Journal of Philosophy   ISSN 0045-5091
Canadian Medical Association Journal   ISSN 0008-4409
Canadian Nurse   ISSN 0008-4581
Canadian Psychiatric Association Journal   ISSN 0008-4824
Canadian Psychological Review
Case and Comment   ISSN 0008-7238
Catholic Hospital   ISSN 0008-8099
Catholic Mind   ISSN 0008-8242
Change   ISSN 0009-1383
Chemical and Engineering News   ISSN 0009-2347
Christian Century   ISSN 0009-5281
Christian Science Monitor (Eastern Edition)
Christianity and Crisis   ISSN 0009-5745
Chronicle of Higher Education   ISSN 0009-5982
Circulation   ISSN 0009-7322
Clinical Research   ISSN 0009-9279
Clinics in Perinatology   ISSN 0095-5108
Colorado Lawyer
Columbia Journalism Review   ISSN 0010-194X
Commentary   ISSN 0010-2601
Commonweal   ISSN 0010-3330
Congressional Record (Daily Edition)
Connecticut Medicine   ISSN 0010-6178

Corrective and Social Psychiatry and Journal of
    Behavior Technology Methods and Therapy
Crime and Delinquency   ISSN 0011-1287
Criminal Justice Journal
Criminal Law Review   ISSN 0011-135X
Critical Care Medicine   ISSN 0090-3493
Cumberland Law Review   ISSN 0360-8298

Daedalus   ISSN 0011-5266
Daily Washington Law Reporter
Dalhousie Law Journal
DePaul Law Review   ISSN 0011-7188
Dickinson Law Review   ISSN 0012-2459
Dimensions in Health Service   ISSN 0317-7645
Diseases of the Nervous System   ISSN 0012-3714
Dissent
Dominion Law Reports, 3d Series
Drake Law Review   ISSN 0012-5938
Draper World Population Fund Report
Duquesne Law Review   ISSN 0012-7213

Ebony   ISSN 0012-9011
Economist   ISSN 0013-0613
Ethics   ISSN 0014-1704
Ethics in Science and Medicine

Family Law Reporter
Family Planning Perspectives   ISSN 0014-7354
Family Planning/Population Reporter   ISSN
    0090-0923
Federal Register
Federal Reporter, 2d Series
Federal Supplement

General Statutes of North Carolina
George Washington Law Review   ISSN 0016-8076
Georgia Law Review   ISSN 0016-8300
Geriatrics   ISSN 0016-867X

Hastings Center Report   ISSN 0093-0334
Homiletic and Pastoral Review   ISSN 0018-4268
Hospital and Community Psychiatry   ISSN
    0022-1597
Hospital and Health Services Administration
Hospital Financial Management   ISSN 0018-5639
Hospital Medical Staff   ISSN 0090-0710
Hospital Physician   ISSN 0018-5795
Hospital Progress   ISSN 0018-5817
Hospitals   ISSN 0018-5973
Human Behavior   ISSN 0046-8134
Human Events   ISSN 0018-7186
Human Life Review
Humanist   ISSN 0018-7399

Indian Journal of Social Work   ISSN 0019-5634
Indiana Law Journal   ISSN 0019-6665
Intellect
International Digest of Health Legislation   ISSN

0020-6563
International Journal of Clinical Pharmacology
International Journal of Health Services   ISSN
    0020-7314
International Journal of Occupational Health and
    Safety   ISSN 0093-2205
International Journal of Social Psychiatry   ISSN
    0020-7640
International Surgery   ISSN 0020-8868
Irish Journal of Medical Science
Israel Law Review   ISSN 0021-2237

Journal of Autism and Childhood
    Schizophrenia   ISSN 0021-9185
Journal of Chronic Diseases   ISSN 0021-9681
Journal of Clinical Pharmacology   ISSN 0021-9754
Journal of Consulting and Clinical Psychology
Journal of Drug Issues   ISSN 0022-0426
Journal of Experimental Social Psychology   ISSN
    0022-1031
Journal of Family Law   ISSN 0022-1066
Journal of Family Practice   ISSN 0094-3509
Journal of Forensic Sciences   ISSN 0022-1198
Journal of Health Politics, Policy and Law   ISSN
    0361-6878
Journal of Legal Medicine   ISSN 0093-1748
Journal of Medical Education   ISSN 0022-2593
Journal of Medical Ethics   ISSN 0306-6800
Journal of Medicine and Philosophy   ISSN
    0360-5310
Journal of Occupational Medicine   ISSN 0022-3212
Journal of Pediatrics   ISSN 0022-3468
Journal of Personality and Social Psychology   ISSN
    0022-3514
Journal of Practical Nursing   ISSN 0022-3867
Journal of Religion and Health   ISSN 0022-4197
Journal of School Health   ISSN 0022-4391
Journal of Social Issues   ISSN 0022-4537
Journal of the American Association of Nurse
    Anesthetists   ISSN 0002-7448
Journal of the American Geriatrics Society   ISSN
    0002-8614
Journal of the American Medical Association   ISSN
    0002-9955
Journal of the Indian Law Institute   ISSN 0019-5731
Journal of the Irish Medical Association   ISSN
    0021-129X
Journal of the Kentucky Medical Association   ISSN
    0023-0294
Journal of the Medical Society of New Jersey   ISSN
    0025-7524
Journal of the National Association of Private
    Psychiatric Hospitals   ISSN 0027-8629
Journal of the Royal Society of Arts   ISSN
    0036-9114
Journal of the South Carolina Medical
    Association   ISSN 0038-3139
Journal of the Tennessee Medical Association   ISSN

0040-3318
Journal of Urban Law   ISSN 0041-9559
Juris Doctor   ISSN 0047-3014

Kentucky Law Journal   ISSN 0023-026X

Lancet   ISSN 0023-7507
Law and Liberty   ISSN 0094-0615
Linacre Quarterly   ISSN 0024-3639
Louvain Studies   ISSN 0024-6964
Loyola of Los Angeles Law Review
Loyola University Law Journal

Man and Medicine: The Journal of Values and Ethics
   in Health Care
Manchester School of Economics and Social
   Studies   ISSN 0025-2034
Manitoba Law Journal
Maryland Annotated Code
Maryland State Medical Journal   ISSN 0025-4363
MCN: American Journal of Maternal Child Nursing
Medical and Pediatric Oncology   ISSN 0098-1532
Medical Clinics of North America   ISSN 0025-7125
Medical Economics   ISSN 0025-7206
Medical History   ISSN 0025-7273
Medical Journal of Australia   ISSN 0025-729X
Medical Trial Technique Quarterly   ISSN 0025-7591
Medical World News   ISSN 0025-763X
Medical-Moral Newsletter   ISSN 0025-7397
Medicine, Science, and the Law   ISSN 0025-8024
Medicolegal News   ISSN 0097-0085
Memphis State University Law Review   ISSN
   0047-6714
Mental Disability Law Reporter
Mental Retardation   ISSN 0047-6765
Mental Retardation and the Law   ISSN 0098-8111
Mercer Law Review   ISSN 0025-987X
Midwife, Health Visitor and Community Nurse
Milbank Memorial Fund Quarterly   ISSN 0026-3745
Military Medicine   ISSN 0026-4075
Minnesota Law Review   ISSN 0026-5535
Missouri Law Review   ISSN 0026-6604
Missouri Medicine   ISSN 0026-6620
Modern Healthcare
Modern Law Review   ISSN 0026-7961
Montana Law Review   ISSN 0026-9972
Mount Sinai Journal of Medicine   ISSN 0027-2507

National Journal
National Observer
Natural History   ISSN 0028-0712
Nature   ISSN 0028-0836
Nevada Revised Statutes
New England Journal of Medicine   ISSN 0028-4793
New England Journal on Prison Law   ISSN
   0095-7364
New Humanist
New Law Journal   ISSN 0306-6479

New Physician   ISSN 0028-6451
New Republic   ISSN 0028-6583
New Scholasticism   ISSN 0028-6621
New Scientist   ISSN 0028-6664
New Statesman   ISSN 0028-6842
New Universities Quarterly
New York Law School Law Review
New York State Journal of Medicine   ISSN
   0028-7628
New York Supplement, 2d Series
New York Times
New Zealand Medical Journal   ISSN 0028-8446
New Zealand Nursing Journal   ISSN 0028-8535
Newsletter on Science, Technology, and Human
   Values
Newsweek   ISSN 0028-9604
North Carolina Journal of Mental Health
North Carolina Law Review   ISSN 0029-2524
North Eastern Reporter, 2d Series
North Western Reporter, 2d Series
Nursing Forum   ISSN 0029-6473
Nursing Mirror   ISSN 0029-6511
Nursing Outlook   ISSN 0029-6554
Nursing Research   ISSN 0029-6562
Nursing Times   ISSN 0029-6589

Ob. Gyn. News   ISSN 0029-7437
Occupational Health Nursing   ISSN 0029-7933
Ohio Northern University Law Review   ISSN
   0094-534X
Oklahoma Law Review   ISSN 0030-1752

Pacific Law Journal   ISSN 0030-8757
Pacific Reporter, 2d Series
Pediatrics   ISSN 0031-4005
Pennsylvania District and County Reports, 3d Series
Pennsylvania Medicine   ISSN 0031-4595
People
Personalist   ISSN 0031-5621
Perspectives in Biology and Medicine   ISSN
   0031-5982
Pharos   ISSN 0031-7179
Philosophy   ISSN 0031-8191
Philosophy and Public Affairs   ISSN 0048-3915
Phylon   ISSN 0031-8106
Population and Development Review   ISSN
   0098-7921
Postgraduate Medicine   ISSN 0032-5481
Progressive   ISSN 0033-0736
Psychiatric Annals   ISSN 0048-5713
Psychiatric Opinion   ISSN 0033-2712
Public Interest   ISSN 0033-3557

Radiology   ISSN 0033-8419
Rutgers-Camden Law Journal   ISSN 0036-0449

San Diego Law Review   ISSN 0036-4037
Saturday Review   ISSN 0036-4983

Scandanavian Journal of Social Medicine   ISSN
   0300-8037
Schizophrenia Bulletin   ISSN 0586-7614
Science   ISSN 0036-8075
Science and Government Report   ISSN 0048-9581
Science for the People   ISSN 0048-9662
Science News   ISSN 0036-8423
Sciences   ISSN 0036-861X
Scientific American   ISSN 0036-8733
Scottish Medical Journal
Sh'ma   ISSN 0049-0385
Social Biology   ISSN 0037-766X
Social Science and Medicine   ISSN 0037-7856
Social Theory and Practice   ISSN 0037-802X
Society
Sociological Review   ISSN 0038-0261
South African Medical Journal   ISSN 0038-2469
South Dakota Journal of Medicine   ISSN 0038-3317
South Western Reporter, 2d Series
Southern California Law Review   ISSN 0038-3910
Southern Reporter, 2d Series
Soviet Law and Government   ISSN 0038-5530
St. Luke's Journal of Theology
Stanford Law Review   ISSN 0038-9765
Studies in Family Planning
Suffolk University Law Review   ISSN 0039-4696
Sunday Times (London)
Supervisor Nurse   ISSN 0039-5870

Temple Law Quarterly   ISSN 0040-2974
Texas Law Review   ISSN 0040-4411
Texas Medicine   ISSN 0040-4470
Texas Revised Civil Statutes Annotated (Vernon)
Thomist   ISSN 0040-6325
Time   ISSN 0040-781X
Times (London)

Tradition: A Journal of Orthodox Thought   ISSN
   0041-0608
Transactions of the American Academy of
   Ophthalmology and Otolaryngology
Trusts and Estates   ISSN 0041-3682

U.S. Catholic   ISSN 0041-7548
U.S. Medicine   ISSN 0042-1227
United States Reports
University of Chicago Law Review   ISSN 0041-9494
University of Cincinnati Law Review   ISSN
   0009-6881
University of Colorado Law Review   ISSN
   0041-9516
University of Dayton Law Review
University of Florida Law Review   ISSN 0041-9583
University of Miami Law Review   ISSN 0041-9818
University of Pennsylvania Law Review   ISSN
   0041-9907
University of Pittsburgh Law Review   ISSN
   0041-9915

Vanderbilt Law Review   ISSN 0042-2533
Virginia Medical

Wake Forest Law Review   ISSN 0043-003X
Wall Street Journal (Eastern Edition)
Washington Post
West Virginia Code
Western Humanities Review   ISSN 0043-3845
Western Journal of Medicine   ISSN 0093-0415
WHO Chronicle   ISSN 0042-9694
Women Lawyers Journal   ISSN 0043-7468
World Medical Journal   ISSN 0049-8122
Worldview

BIOETHICS THESAURUS

BIOETHICS THESAURUS

ABORTED FETUSES
BT   Fetuses
RT   Abortion↓

ABORTION
UF   Abortion laws†
Compulsory abortion†
Mandatory abortion†
NT   Abortion on demand
Illegal abortion
Selective abortion
Therapeutic abortion
RT   Aborted fetuses

Abortion Laws
Use ABORTION, *and*
LEGISLATION

ABORTION ON DEMAND
BT   Abortion

ACTIVE EUTHANASIA
UF   Positive euthanasia
BT   Euthanasia

ADMINISTRATORS
RT   Health personnel↓

ADOLESCENTS
SN   Age designation for humans 13–18
years old
RT   Children↓

ADOPTION

ADULTS
SN   Age designation for humans 19–64
years old

AGE

AGED
SN   Age designation for humans 65 years
old or older
UF   Elderly persons
Homes for the aged†

Agency Review Boards
Use ETHICS COMMITTEES

AGGRESSION
SN   Feelings of hostility or overt hostile

action which harms, or is intended to
harm, a person or object
RT   Violence

AID
UF   Artificial insemination, donor
Heterologous insemination
BT   Artificial insemination
RT   AID children
AIH
Semen donors

AID CHILDREN
SN   Children conceived by means of
artificial insemination using donor
sperm
BT   Children
RT   AID

AID Donors
Use SEMEN DONORS

AIH
UF   Artificial insemination, husband
Homologous insemination
BT   Artificial insemination
RT   AID

ALCOHOL ABUSE
BT   Social problems
RT   Drug abuse
Smoking

Allocation of Resources
Use RESOURCE ALLOCATION

ALLOWING TO DIE
UF   Death with dignity
Indirect euthanasia
Negative euthanasia
Passive euthanasia
Right to die
RT   Euthanasia↓
Prolongation of life
Withholding treatment

ALTERNATIVES

ALTRUISM
RT   Egoism

---

† = See main entry for narrower terms.   ↓ = Consult phrase in alphabetical listing.

SN = Scope Note   UF = Used For   BT = Broader Term   NT = Narrower Term   RT = Related Term

AMNIOCENTESIS
    BT Prenatal diagnosis

ANCIENT HISTORY
    RT Historical aspects

ANESTHESIA

ANIMAL EXPERIMENTATION
    RT Human Experimentation↓

Antenatal Diagnosis
    Use PRENATAL DIAGNOSIS

Antenatal Injuries
    Use PRENATAL INJURIES

ARTIFICIAL GENES
    UF Synthetic genes
    RT Recombinant DNA research

ARTIFICIAL INSEMINATION
    BT Reproductive technologies
    NT AID
          AIH
    RT Frozen semen

Artificial Insemination, Donor
    Use AID

Artificial Insemination, Husband
    Use AIH

ARTIFICIAL ORGANS

ATTITUDES
    NT Attitudes to death
    RT Public opinion
          Religious beliefs

ATTITUDES TO DEATH
    UF Death attitudes
          Denial of death
    BT Attitudes
    RT Death↓

AUDIOVISUAL AIDS

AUTHORITARIANISM
    RT Social dominance

Autonomy
    Use SELF DETERMINATION

AUTOPSIES
    UF Postmortem examinations

BATTERY
    SN Any unlawful beating or touching of a
          person without the person's consent
    BT Malpractice
    RT Negligence

BEGINNING OF LIFE
    SN The time at which human individuality
          begins, as logically distinguished from
          the beginning of biological life
    RT Biological life
          Personhood

BEHAVIOR CONTROL
    NT Operant conditioning↓
          Psychosurgery
    RT Electrical stimulation of the brain
          Electroconvulsive therapy

BEHAVIOR DISORDERS
    NT Hyperkinesis
    RT Brain pathology↓

Behavior Modification
    Use OPERANT CONDITIONING

BEHAVIORAL GENETICS
    UF Psychogenetics
    BT Genetics
    RT Sociobiology
          XYY karyotype

BEHAVIORAL RESEARCH
    RT Biomedical research↓
          Investigators

BENEFICENCE
    UF Kindness

Benefit Harm Calculus
    Use RISKS AND BENEFITS

Biblical Interpretation
    Use SCRIPTURAL INTERPRETATION

BIOETHICAL ISSUES
    SN Generic designation for the content
          of materials which discuss a series
          of bioethical topics

BIOETHICS
    SN A branch of applied ethics which
          studies the value implications of
          practices and developments in the
          life sciences and medicine
    BT Ethics
    NT Medical ethics
          Nursing ethics

Biohazards
    Use HEALTH HAZARDS

BIOLOGICAL CONTAINMENT
    SN In recombinant DNA research,
          the use of enfeebled organisms
          which have a severely limited
          capacity to survive in the absence

↓ = See main entry for narrower term   † = Consult phrase in alphabetical listing

of special laboratory conditions
BT   Containment
RT   Physical containment
Recombinant DNA research

Biological Fathers
Use GENETIC FATHERS

BIOLOGICAL LIFE
UF   Body life
Physical life
RT   Beginning of life
Personhood

Biological Research
Use BIOMEDICAL RESEARCH

BIOLOGICAL WARFARE
BT   War

BIOLOGY
BT   Science
NT   Genetics↓

BIOMEDICAL RESEARCH
UF   Biological research
Medical research
NT   Recombinant DNA research
RT   Behavioral research
Human experimentation↓
Investigators
Science↓

BIOMEDICAL TECHNOLOGIES
BT   Technology
NT   Reproductive technologies↓
RT   Medical devices

Birth Control
Use CONTRACEPTION

Birth Defects
Use CONGENITAL DEFECTS

BIRTH ORDER
SN   The sequence in which children are
born into a family

BIRTH RATE
BT   Statistics

BLACKS
UF   Negroes
BT   Minority groups

BLOOD DONATION
RT   Organ donation

Blood Pressure, High
Use HYPERTENSION

BLOOD TRANSFUSIONS
UF   Transfusions, blood

Body Life
Use BIOLOGICAL LIFE

BODY PARTS AND FLUIDS

BONE MARROW

BRAIN

BRAIN DEATH
UF   Neurological death
BT   Death
RT   Coma

BRAIN PATHOLOGY
BT   Central nervous system diseases
NT   Phenylketonuria
Tay Sachs disease
Temporal lobe epilepsy
RT   Behavior disorders↓

BURN PATIENTS
BT   Patients

Cadaver Transplants
Use ORGAN TRANSPLANTATION, *and*
CADAVERS

CADAVERS
UF   Cadaver transplants†

CANCER
NT   Leukemia

CAPITAL PUNISHMENT
BT   Killing

CAPITALISM
RT   Communism
Democracy
Socialism

CARCINOGENS
BT   Health hazards

Care for the Dying
Use TERMINAL CARE

CARRIERS
SN   Individuals possessing a specified gene
who are capable of transmitting it to
offspring but not of showing its typical
expression
RT   Genetic defects↓

Case Histories
Use CASE REPORTS

CASE REPORTS
UF   Case histories

CASE STUDIES

Catholic Church Teachings
Use ROMAN CATHOLIC ETHICS

SN=Scope Note   UF=Used For   BT=Broader Term   NT=Narrower Term   RT=Related Term

Catholic Hospitals
Use RELIGIOUS HOSPITALS, *and*
ROMAN CATHOLICISM

Catholicism, Roman
Use ROMAN CATHOLICISM

Caucasians
Use WHITES

Celebrities
Use FAMOUS PERSONS

CENTRAL NERVOUS SYSTEM DISEASES
NT Brain pathology↓

CHILD NEGLECT
BT Social problems

CHILDREN
SN Age designation for humans 2–12
years old
NT AID children
Unwanted children
RT Adolescents
Infants
Newborns

Chimeras
Use HYBRIDS

CHRISTIAN ETHICS
BT Religious ethics
NT Eastern Orthodox ethics
Protestant ethics
Roman Catholic ethics
RT Islamic ethics
Jewish ethics

CHROMOSOMAL DISORDERS
BT Genetic defects
NT Down's syndrome
XYY karyotype
RT Karyotyping

CHRONICALLY ILL
BT Patients
RT Critically ill
Terminally ill

Citizen Advocacy
Use PUBLIC ADVOCACY

City Government
Use MUNICIPAL GOVERNMENT

Civil Commitment
Use INVOLUNTARY COMMITMENT

Civil Rights
Use LEGAL RIGHTS

CLERGY
RT Religion↓

Clinical Investigators
Use INVESTIGATORS

Clinical Research
Use HUMAN EXPERIMENTATION

Clinical Trials
Use HUMAN EXPERIMENTATION

CLONES
RT Cloning

CLONING
BT Reproductive technologies
RT Clones

CODES OF ETHICS
UF Ethical codes
RT Medical ethics

COERCION
RT Mandatory programs
Voluntary programs

COMA
RT Brain death

Commitment, Duration
Use DURATION OF COMMITMENT

Commitment, Involuntary
Use INVOLUNTARY COMMITMENT

COMMON GOOD
UF General welfare
Social good

COMMUNICABLE DISEASES
NT Hepatitis
Influenza
Poliomyelitis
Rubella
Venereal diseases↓

COMMUNICATION
SN The exchange or transmission of ideas
between individuals or groups
NT Information dissemination
Interdisciplinary communication
Privileged communication
RT Editorial policies
Journalism
Mass media
Social interaction

COMMUNISM
BT Political systems
RT Capitalism
Democracy
Socialism

↓ = See main entry for narrower term   † = Consult phrase in alphabetical listing

COMMUNITY MEDICINE
SN Health care services which are based in the local community and directed toward the entire population of the community
UF Social medicine
BT Medicine
RT Preventive medicine
    Public health

Community Participation
Use PUBLIC PARTICIPATION

COMMUNITY SERVICES
RT Health care↓
    Health facilities↓

COMPENSATION
SN Payment for injury
RT Remuneration↓

COMPETENCE
SN Mental capacity to make responsible choices
UF Incompetence, mental
    Mental competence
RT Professional competence

COMPREHENSION
RT Recall

Compulsory Abortion
Use ABORTION, *and*
    MANDATORY PROGRAMS

Compulsory Genetic Screening
Use GENETIC SCREENING, *and*
    MANDATORY PROGRAMS

Computerized Data Bases
Use DATA BASES

CONFIDENTIALITY
RT Patient access
    Privacy
    Privileged communication

CONFLICT OF INTEREST

CONGENITAL DEFECTS
SN Defects which are present at birth, regardless of their causation
UF Birth defects
RT Genetic defects↓

CONSCIENCE
RT Religious beliefs

Consent
Use INFORMED CONSENT

CONSENT FORMS
RT Informed consent↓

CONSEQUENCES
RT Intention
    Social impact

Consequentialism
Use TELEOLOGICAL ETHICS

CONSTITUTIONAL AMENDMENTS

CONSTITUTIONAL LAW
BT Law

Consumer Participation
Use PUBLIC PARTICIPATION

Consultation
Use REFERRAL AND CONSULTATION

CONTAINMENT
SN The use of biological or physical means for minimizing or preventing the dissemination of potentially hazardous materials
NT Biological containment
    Physical containment
RT Recombinant DNA research

CONTRACEPTION
UF Birth control
BT Family planning

CONTRACTS
RT Law↓

CONTROL GROUPS
RT Research subjects
    Selection of subjects

Control of One's Body
Use SELF DETERMINATION

COSTS AND BENEFITS
RT Economics
    Risks and benefits
    Socioeconomic factors

COUNSELING
NT Genetic counseling

Criminal Abortion
Use ILLEGAL ABORTION

CRIMINAL LAW
BT Law
RT Law enforcement

CRITICALLY ILL
BT Patients
RT Chronically ill
    Terminally ill

CRYONIC SUSPENSION
SN The preservation of the human body by freezing or supercooling

---

SN = Scope Note   UF = Used For   BT = Broader Term   NT = Narrower Term   RT = Related Term

CULTURAL EVOLUTION
UF Evolution, cultural

CULTURAL PLURALISM
SN The presence of multiple value systems
within a society
UF Pluralism, cultural

CURRICULUM
RT Teaching methods↓

DANGEROUSNESS

DATA BASES
UF Computerized data bases

DEATH
SN The concept of death; or, the process
of dying
UF Dying
NT Brain death
RT Attitudes to death
Determination of death
Terminal care

Death Attitudes
Use ATTITUDES TO DEATH

Death Rate
Use MORTALITY

Death with Dignity
Use ALLOWING TO DIE

Death, Determination
Use DETERMINATION OF DEATH

DEBRIEFING
UF Postexperimental explanation

DECEPTION

DECISION MAKING

Definition of Health
Use HEALTH

DEHUMANIZATION
UF Depersonalization

DEINSTITUTIONALIZED PERSONS
SN Persons who have been released from
institutions
RT Institutionalized persons

DEMOCRACY
BT Political Systems
RT Capitalism
Communism
Socialism

Denial of Death
Use ATTITUDES TO DEATH

DENTISTRY

DEONTOLOGICAL ETHICS
SN Theories of ethics which hold that some
actions are morally obligatory regardless
of their actual or anticipated consequences
BT Normative ethics

Depersonalization
Use DEHUMANIZATION

DETERMINATION OF DEATH
UF Death, determination
RT Death↓

DEVELOPING COUNTRIES

DIABETES
RT Insulin

DIAGNOSIS
NT Prenatal diagnosis↓
Psychiatric diagnosis

DIGNITY
UF Human dignity
Respect for persons

DISADVANTAGED
NT Indigents
RT Minority groups↓

DISCRIMINATION
UF Racism
BT Social problems

DISCLOSURE
UF Reasonable disclosure
Truthtelling
RT Informed consent↓

Disease Rate
Use MORBIDITY

DISSENT
RT Political activity

Distinctively Human
Use PERSONHOOD

DNA Hybridization
Use RECOMBINANT DNA RESEARCH

DNA Recombinants
Use RECOMBINANT DNA RESEARCH

DNA THERAPY
SN Biochemical or mechanical alteration
of genetic material
UF Gene therapy
Genetic engineering
Germ cell alteration
BT Genetic intervention
RT Recombinant DNA research

↓ = See main entry for narrower term   † = Consult phrase in alphabetical listing

Dominance, Social
Use SOCIAL DOMINANCE

DOMINANT GENETIC CONDITIONS
BT   Genetic defects
NT   Huntington's chorea
RT   Recessive genetic conditions↓

DONOR CARDS

DONORS
NT   Organ donors
     Ovum donors
     Semen donors

DOUBLE EFFECT
SN   In ethics, a technical term which refers to
     two types of consequences which may be
     produced by a single action, namely,
     intended consequences and unintended
     side-effects
UF   Principle of double effect

DOWN'S SYNDROME
UF   Mongolism
BT   Chromosomal disorders

Drive Satisfaction
Use MOTIVATION

DRUG ABUSE
BT   Social problems
RT   Alcohol abuse
     Smoking

DRUG INDUSTRY
UF   Pharmaceutical industry
BT   Industry

DRUG SCREENING
SN   Screening to determine whether persons
     have been using drugs
UF   Screening, drug

DRUGS
NT   Psychoactive drugs↓
RT   Placebos

DUCHENNE MUSCULAR DYSTROPHY
BT   Recessive genetic conditions
     Sex linked defects

DUE PROCESS
RT   Equal protection

DURATION OF COMMITMENT
UF   Commitment, duration
RT   Involuntary commitment

Dying
Use DEATH

Dying Patients
Use TERMINALLY ILL

EASTERN ORTHODOX ETHICS
BT   Christian ethics

ECOLOGY
UF   Pollution
RT   Health hazards↓
     Natural resources

ECONOMICS
UF   Financial costs
RT   Costs and benefits
     Resource allocation
     Socioeconomic factors

EDITORIAL POLICIES
UF   Publication policies
RT   Communication↓

EDUCATION
NT   Medical education
     Nursing education
RT   Students
     Teaching methods↓

EEG
UF   Electroencephalogram

Egg Donors
Use OVUM DONORS

EGOISM
SN   A theory of ethics which holds that
     one's only moral obligation is to
     promote one's own good
BT   Teleological ethics
RT   Altruism
     Utilitarianism

Elderly Persons
Use AGED

ELECTRICAL STIMULATION OF THE BRAIN
UF   ESB
RT   Behavior control↓

ELECTROCONVULSIVE THERAPY
UF   Electroshock therapy
     Shock therapy, electric
RT   Behavior control↓

Electroencephalogram
Use EEG

Electromechanical Aids
Use MEDICAL DEVICES

Electroshock Therapy
Use ELECTROCONVULSIVE THERAPY

SN = Scope Note   UF = Used For   BT = Broader Term   NT = Narrower Term   RT = Related Term

EMBRYO TRANSFER
SN Transfer of a human zygote or blastocyst into a uterus following in vitro or in vivo fertilization
BT Reproductive technologies
RT In vitro fertilization
Host mothers

EMERGENCY CARE
BT Health care
RT Resuscitation

Emotional Adjustment
Use SOCIAL ADJUSTMENT

Emotionally Disturbed
Use MENTALLY ILL

Emotional Stress
Use PSYCHOLOGICAL STRESS

EMPLOYMENT

ENGINEERING

Environmental Hazards
Use HEALTH HAZARDS

EPIDEMIOLOGY
RT Medicine↓

Epilepsy, Temporal Lobe
Use TEMPORAL LOBE EPILEPSY

EQUAL PROTECTION
RT Due process
Human equality

Equality
Use HUMAN EQUALITY

Erotic Behavior
Use SEXUALITY

ESB
Use ELECTRICAL STIMULATION OF THE BRAIN

ETHICAL ANALYSIS
BT Methods

Ethical Codes
Use CODES OF ETHICS

ETHICAL RELATIVISM
UF Relativism, ethical

ETHICAL REVIEW
RT Ethics committees
Review committees

Ethical Review Committees
Use ETHICS COMMITTEES

ETHICISTS

ETHICS
NT Bioethics↓
Hedonism
Metaethics
Normative ethics↓
Professional ethics
Religious ethics↓
Situational ethics
RT Morality
Natural law
Philosophy↓
Theology

ETHICS COMMITTEES
UF Agency review boards
Ethical review committees
Institutional review boards
RT Ethical review
Review committees

ETIOLOGY

Eugenic Abortion
Use SELECTIVE ABORTION

EUGENICS
SN The science which studies ways to improve the hereditary characteristics of the human race, especially by means of selective reproduction
BT Genetic intervention
NT Negative eugenics
Positive eugenics
RT Euthenics

EUTHANASIA
BT Killing
NT Active euthanasia
Involuntary euthanasia
Voluntary euthanasia
RT Allowing to die
Prolongation of life
Withholding treatment

EUTHENICS
SN Development of human well being and efficient functioning through improvement of environmental conditions
RT Eugenics↓

EVALUATION
RT Quality control

EVOLUTION
SN Biological evolution
RT Mutation
Natural selection

Evolution, Cultural
Use CULTURAL EVOLUTION

↓ = See main entry for narrower term   † = Consult phrase in alphabetical listing

EYE DISEASES

EXISTENTALISM
    BT Philosophy

Experimental Design
    Use RESEARCH DESIGN

Experimental Subjects
    Use RESEARCH SUBJECTS

EXPERT TESTIMONY

EXTRAORDINARY TREATMENT
    SN Therapeutic measures which cannot be
        obtained without excessive expense,
        pain, or other inconvenience, or
        which, if used, would not offer a
        reasonable hope of benefit
    UF Heroic treatment
    RT Prolongation of life
        Resuscitation

FAMILY

FAMILY MEMBERS
    UF Relatives
    NT Parents↓
        Siblings

FAMILY PLANNING
    NT Contraception
    RT Population control

FAMILY PROBLEMS
    RT Social problems↓

FAMILY RELATIONSHIP
    NT Marital relationship
        Parent child relationship↓

FAMOUS PERSONS
    UF Celebrities

FATHERS
    BT Males
        Parents
    NT Genetic fathers
        Social fathers

Federal Aid
    Use FINANCIAL SUPPORT, *and*
        FEDERAL GOVERNMENT

FEDERAL GOVERNMENT
    UF Federal aid†
    RT Government agencies
        Municipal government
        State government

Fees, Medical
    Use MEDICAL FEES

FEMALES
    UF Women
    NT Mothers↓
        Pregnant women

FERTILITY
    UF Human fertility
        Infertility

FETAL DEVELOPMENT
    UF Gestational age
        Prenatal development

Fetal Experimentation
    Use HUMAN EXPERIMENTATION, *and*
        FETUSES

Fetal Viability
    Use FETUSES, *and*
        VIABILITY

FETUSES
    UF Fetal experimentation†
        Fetal viability†
    NT Aborted fetuses

Financial Costs
    Use ECONOMICS

FINANCIAL SUPPORT
    UF Federal aid†

FLUORIDATION

FOOD
    RT Natural resources
        Nutrition

FORCE FEEDING
    RT Nutrition

Free Will
    Use FREEDOM

FREEDOM
    UF Free will
        Liberty
    NT Self determination

Freedom of Choice
    Use SELF DETERMINATION

FROZEN SEMEN
    RT Artificial insemination↓
        Sperm↓

Fundamental Rights
    Use HUMAN RIGHTS, *or*
        LEGAL RIGHTS

FUTURE GENERATIONS
    UF Obligations to future generations†

SN=Scope Note   UF=Used For   BT=Broader Term   NT=Narrower Term   RT=Related Term

GENE FREQUENCY
SN The percentage of occurrence of a specified gene in the chromosomes of a population

GENE POOL

Gene Therapy
Use DNA THERAPY

GENERALIZATION OF EXPERTISE
SN The rendering of judgments in fields of knowledge which lie beyond the sphere of one's primary subject matter competence
RT Technical expertise

Genetic Abortion
Use SELECTIVE ABORTION

GENETIC COUNSELING
BT Counseling
    Genetic intervention

GENETIC DEFECTS
UF Genetic diseases
NT Chromosomal disorders↓
    Dominant genetic conditions↓
    Recessive genetic conditions↓
    Sex linked defects↓
    Single gene defects↓
RT Carriers
    Congenital defects
    Metabolic diseases↓

Genetic Diseases
Use GENETIC DEFECTS

GENETIC DIVERSITY

Genetic Engineering
Use DNA THERAPY, *or* GENETIC INTERVENTION

GENETIC FATHERS
SN Males who contribute genetic material to offspring
UF Biological fathers
BT Fathers
RT Social fathers

GENETIC IDENTITY
SN Genetic parentage; pedigree
RT Social identity

GENETIC INTERVENTION
SN General term for the modification of inheritable characteristics through various social mechanisms and/or biomedical techniques
UF Genetic engineering
    Genetic manipulation

NT DNA therapy
    Eugenics↓
    Genetic counseling
    Genetic screening
RT Recombinant DNA research

Genetic Manipulation
Use GENETIC INTERVENTION

GENETIC SCREENING
UF Compulsory genetic screening†
    Screening, genetic
BT Genetic intervention
    Mass screening

GENETICS
UF Human genetics
BT Biology
NT Behavioral genetics
RT Mutation

Germ Cell Alteration
Use DNA THERAPY

German Measles
Use RUBELLA

Gestational Age
Use FETAL DEVELOPMENT

Gestational Mothers
Use HOST MOTHERS

GOALS

GOVERNMENT AGENCIES
RT Federal government
    Municipal government
    State government

Government Policy
Use PUBLIC POLICY

GOVERNMENT REGULATION
UF Regulation, government
    State control
BT Social control
RT Judicial action
    Legislation↓

GROUP DISCUSSION
BT Teaching methods

GROUP THERAPY
RT Psychotherapy

Guardians
Use LEGAL GUARDIANS

Halfway Houses
Use RESIDENTIAL FACILITIES

↓ = See main entry for narrower term   † = Consult phrase in alphabetical listing

HANDICAPPED
NT Mentally handicapped↓
   Physically handicapped

Harm
Use INJURIES

HEALTH
UF Definition of health
NT Maternal health
   Mental health

HEALTH CARE
UF Medical care
NT Emergency care
   Patient care↓
RT Community services

HEALTH CARE DELIVERY

HEALTH FACILITIES
NT Hospitals↓
RT Community services
   Residential facilities↓

HEALTH HAZARDS
UF Biohazards
   Environmental hazards
NT Carcinogens
RT Ecology
   Radiation

HEALTH INSURANCE
BT Insurance
RT Life insurance

HEALTH PERSONNEL
NT Medical staff
   Nurses
   Physicians
RT Administrators
   Social workers

HEART DISEASES

Heart Transplantation
Use ORGAN TRANSPLANTATION, *and*
   HEARTS

HEARTS
UF Heart transplantation†

HEDONISM
SN A theory of value which holds that
   pleasure is the good
BT Ethics

Hemodialysis
Use RENAL DIALYSIS

HEMOPHILIA
BT Sex linked defects

HEPATITIS

BT Communicable diseases

Heroic Treatment
Use EXTRAORDINARY TREATMENT

Heterologous Insemination
Use AID

HISTORICAL ASPECTS
RT Ancient history

Homes for the Aged
Use AGED, *and*
   RESIDENTIAL FACILITIES

Homologous Insemination
Use AIH

HOMOSEXUALS
RT Sexuality
   Transsexualism

HOSPICES
RT Hospitals↓
   Terminal care

Hospital Policies
Use INSTITUTIONAL POLICIES, *and*
   HOSPITALS

HOSPITALS
UF Hospital policies†
BT Health facilities
NT Mental institutions
   Private hospitals↓
   Public hospitals
RT Hospices
   Intensive care units

HOST MOTHERS
UF Gestational mothers
   Uterine mothers
BT Mothers
RT Embryo transfer
   Ovum donors

HUMAN CHARACTERISTICS
SN Qualities which distinguish human
   beings from nonhuman species
RT Personhood

HUMAN DEVELOPMENT

Human Dignity
Use DIGNITY

HUMAN EQUALITY
UF Equality
RT Equal protection

HUMAN EXPERIMENTATION
UF Clinical research
   Clinical trials
   Fetal experimentation†

SN = Scope Note   UF = Used For   BT = Broader Term   NT = Narrower Term   RT = Related Term

NT   Nontherapeutic research
     Therapeutic research
RT   Animal experimentation
     Behavioral research
     Biomedical research↓
     Investigators

Human Fertility
Use FERTILITY

Human Genetics
Use GENETICS

Human Reproduction
Use REPRODUCTION

HUMAN RIGHTS
UF   Fundamental rights
     Individual rights
     Right to reproduce†
BT   Rights
RT   Legal rights↓
     Moral obligations↓
     Normative ethics↓

Human Sperm
Use SPERM

Human Values
Use VALUES

HUMANISM
RT   Philosophy↓

HUMANITIES
RT   Social sciences

Humanness
Use PERSONHOOD

HUNTINGTON'S CHOREA
BT   Dominant genetic conditions
     Single gene defects

HYBRIDS
UF   Chimeras

HYPERKINESIS
BT   Behavior disorders

HYPERTENSION
UF   Blood pressure, high

HYPNOSIS

ILLEGAL ABORTION
UF   Criminal abortion
BT   Abortion

Illegitimacy
Use LEGITIMACY

Illness, Self Induced
Use SELF INDUCED ILLNESS

IMMUNIZATION

IMPRISONMENT
RT   Prisoners
     Punishment

IN VITRO FERTILIZATION
UF   Test tube fertilization
BT   Reproductive technologies
RT   Embryo transfer
     Ovum donors
     Products of in vitro fertilization

INCENTIVES
RT   Motivation
     Remuneration↓

INCIDENCE
BT   Statistics

Incompetence, Mental
Use COMPETENCE

INDIGENTS
UF   Poor persons
BT   Disadvantaged
RT   Poverty

Indirect Euthanasia
Use ALLOWING TO DIE

Individual Rights
Use HUMAN RIGHTS

INDUSTRIAL MEDICINE
UF   Occupational medicine
BT   Medicine
RT   Occupational diseases

INDUSTRY
NT   Drug industry

INFANTICIDE
BT   Killing

INFANTS
SN   Age designation for humans 1-23
     months old
RT   Children↓
     Newborns

Infertility
Use FERTILITY

INFLUENZA
BT   Communicable diseases

INFORMAL SOCIAL CONTROL
SN   Restraints on behavior which are
     exerted through social pressure, for

↓ = See main entry for narrower term   † = Consult phrase in alphabetical listing

example, through mores, folkways, conventions, or public sentiment
BT Social control
NT Public opinion

INFORMATION DISSEMINATION
BT Communication

INFORMED CONSENT
UF Consent
Voluntary consent
NT Third party consent↓
RT Consent forms
Disclosure

INJURIES
UF Harm
NT Preconception injuries
Prenatal injuries
RT Torts↓

Insemination, Artificial
Use ARTIFICIAL INSEMINATION

INSTITUTIONAL OBLIGATIONS

INSTITUTIONAL POLICIES
UF Hospital policies†
RT Organizational policies
Public policy

Institutional Review Boards
Use ETHICS COMMITTEES

INSTITUTIONALIZED PERSONS
NT Prisoners
RT Deinstitutionalized persons

INSULIN
RT Diabetes

INSURANCE
NT Health insurance
Life insurance
National health insurance

INTELLIGENCE

INTENSIVE CARE UNITS
RT Hospitals↓

INTENTION
RT Consequences

INTERDISCIPLINARY COMMUNICATION
BT Communication

Interest of the State
Use STATE INTEREST

INTERNATIONAL ASPECTS

INVESTIGATOR SUBJECT RELATIONSHIP
RT Physician nurse relationship

Physician patient relationship
Professional patient relationship

INVESTIGATORS
UF Clinical investigators
BT Research personnel
RT Behavioral research
Biomedical research↓
Human experimentation↓
Physicians

Involuntary Civil Commitment
Use INVOLUNTARY COMMITMENT

INVOLUNTARY COMMITMENT
UF Civil commitment
Commitment, involuntary
Involuntary civil commitment
RT Duration of commitment
Voluntary admission

INVOLUNTARY EUTHANASIA
BT Euthanasia
RT Voluntary euthanasia

Involuntary Programs
Use MANDATORY PROGRAMS

INVOLUNTARY STERILIZATION
BT Sterilization
RT Voluntary sterilization

ISLAMIC ETHICS
BT Religious ethics
RT Christian ethics↓
Jewish ethics

JEHOVAH'S WITNESSES
RT Religion↓

JEWISH ETHICS
BT Religious ethics
RT Christian ethics↓
Islamic ethics

JEWS
BT Minority groups
RT Religion↓

JOURNALISM
RT Communication↓

JUDICIAL ACTION
BT Social control
RT Government regulation
Legislation↓

JUSTICE

JUSTIFIABLE KILLING
BT Killing

SN=Scope Note   UF=Used For   BT=Broader Term   NT=Narrower Term   RT=Related Term

KARYOTYPING
RT Chromosomal disorders↓

Kidney Dialysis
Use RENAL DIALYSIS

KIDNEY DISEASES
UF Renal diseases

Kidney Transplantation
Use ORGAN TRANSPLANTATION, *and*
    KIDNEYS

KIDNEYS
UF Kidney transplantation†
    Renal transplantation†

KILLING
UF Obligation not to kill†
NT Capital punishment
    Euthanasia↓
    Infanticide
    Justifiable killing
    Suicide

Kindness
Use BENEFICENCE

Labeling
Use STIGMATIZATION

LAW
NT Constitutional law
    Criminal law
RT Contracts
    Torts↓

LAW ENFORCEMENT
RT Criminal law

Laws
Use LEGISLATION

LEGAL ASPECTS

LEGAL GUARDIANS
UF Guardians

LEGAL LIABILITY

LEGAL OBLIGATIONS
RT Legal rights↓
    Moral obligations↓

LEGAL RIGHTS
UF Civil rights
    Fundamental rights
BT Rights
NT Property rights
    Right to treatment
RT Human rights
    Legal obligations

LEGISLATION
UF Abortion laws†
    Laws
    Statutes
BT Social control
NT Model legislation
RT Government regulation
    Judicial action

LEGITIMACY
UF Illegitimacy

LEUKEMIA
BT Cancer

Liberty
Use FREEDOM

LIFE EXTENSION
SN The development or use of techniques
    for retarding the process of aging
RT Prolongation of life

LIFE INSURANCE
BT Insurance
RT Health insurance

Life, Quality of
Use QUALITY OF LIFE

Life, Value of
Use VALUE OF LIFE

LITERATURE

LIVING WILLS
SN Written, witnessed declarations in
    which persons request that if they
    become disabled beyond reasonable
    expectation of recovery, they be
    allowed to die rather than be kept
    alive by extraordinary means

LOVE

LSD
BT Psychoactive drugs

MALES
UF Men
NT Fathers↓

MALPRACTICE
UF Medical malpractice
BT Torts
NT Battery
    Negligence

Mandatory Abortion
Use ABORTION, *and*
    MANDATORY PROGRAMS

↓ = See main entry for narrower term   † = Consult phrase in alphabetical listing

MANDATORY PROGRAMS
UF Compulsory abortion†
    Compulsory genetic screening†
    Involuntary programs
    Mandatory abortion†
RT Coercion
    Voluntary programs

MARITAL RELATIONSHIP
BT Family relationship

MARRIED PERSONS
RT Single persons

MASS MEDIA
RT Communication↓

MASS SCREENING
UF Screening, mass
NT Genetic screening

MATERNAL HEALTH
BT Health

MATERNAL LIFE

Medical Care
Use HEALTH CARE

MEDICAL DEVICES
UF Electromechanical aids
RT Biomedical technologies↓

MEDICAL EDUCATION
BT Education
RT Nursing education

MEDICAL ETHICS
BT Bioethics
RT Codes of ethics
    Nursing ethics
    Professional ethics

MEDICAL ETIQUETTE

MEDICAL FEES
UF Fees, medical
BT Remuneration

Medical Indications for Abortion
Use THERAPEUTIC ABORTION

Medical Malpractice
Use MALPRACTICE

MEDICAL RECORDS
BT Records

Medical Research
Use BIOMEDICAL RESEARCH

MEDICAL STAFF
BT Health personnel
RT Nurses

Physicians
Research personnel↓

MEDICINE
NT Community medicine
    Industrial medicine
    Obstetrics and gynecology
    Pediatrics
    Perinatology
    Preventive medicine
    Psychiatry
    Public health
    Radiology
    Surgery↓
RT Epidemiology
    Physicians
    Science↓

Men
Use MALES

Meningomyelocele
Use SPINA BIFIDA

Mental Competencce
Use COMPETENCE

MENTAL HEALTH
UF Mental illness
    Psychiatric indications for abortion†
BT Health

Mental Hospitals
Use MENTAL INSTITUTIONS

Mental Illness
Use MENTAL HEALTH

MENTAL INSTITUTIONS
UF Mental hospitals
BT Hospitals
RT Residential facilities↓

MENTALLY HANDICAPPED
BT Handicapped
NT Mentally ill
    Mentally retarded

MENTALLY ILL
UF Emotionally disturbed
BT Mentally handicapped
RT Schizophrenia

MENTALLY RETARDED
BT Mentally handicapped

METABOLIC DISEASES
NT Phenylketonuria
    Tay Sachs disease
RT Genetic defects↓

METAETHICS
BT Ethics

SN=Scope Note   UF=Used For   BT=Broader Term   NT=Narrower Term   RT=Related Term

METHODS
  NT Ethical analysis
       Teaching methods↓
  RT Research design

MILITARY PERSONNEL

MINORITY GROUPS
  NT Blacks
       Jews
  RT Disadvantged↓

MODEL LEGISLATION
  BT Legislation

Mongolism
  Use DOWN'S SYNDROME

MORAL OBLIGATIONS
  UF Obligation not to kill†
       Obligations
       Obligations to future generations†
       Prima facie obligations
       Responsibilities
  NT Obligations of society
       Obligations to society
  RT Human rights
       Legal obligations
       Normative ethics↓

MORAL POLICY
  SN A way of devising, relating, or
       ranking moral rules
  RT Morality

MORALITY
  RT Ethics↓
       Moral policy

MORBIDITY
  UF Disease rate
  BT Statistics

MORTALITY
  UF Death rate
  BT Statistics

MOTHER CHILD RELATIONSHIP
  BT Parent child relationsip

MOTHER FETUS RELATIONSHIP

MOTHERS
  BT Females
       Parents
  NT Host mothers

MOTIVATION
  UF Drive satisfaction
  RT Incentives

MUNICIPAL GOVERNMENT
  UF City Government

  RT Federal government
       Government agencies
       State government

MUTATION
  RT Evolution
       Genetics↓
       Natural selection

Myelomeningocele
  Use SPINA BIFIDA

NATIONAL HEALTH INSURANCE
  BT Insurance
  RT State medicine

NATIONAL SOCIALISM

NATURAL LAW
  SN A body of law derived from nature
       and considered to be binding upon
       human society in the absence of, or
       in addition to, institutional law
  RT Ethics↓

NATURAL RESOURCES
  RT Ecology
       Food

NATURAL SELECTION
  RT Evolution
       Mutation

NEGATIVE EUGENICS
  BT Eugenics
  RT Positive eugenics

Negative Euthanasia
  Use ALLOWING TO DIE

NEGATIVE REINFORCEMENT
  BT Operant conditioning
  RT Positive reinforcement
       Punishment

NEGLIGENCE
  SN In law or ethics, failure to exercise a
       reasonable degree of care in one's
       actions
  BT Malpractice
  RT Battery

Negroes
  Use BLACKS

Neurological Death
  Use BRAIN DEATH

Neonates
  Use NEWBORNS

NEWBORNS
  SN Age designation for humans during

---

the first month of postnatal life
UF Neonates
RT Children↓
    Infants

NONTHERAPEUTIC RESEARCH
SN Research which is nontherapeutic for
    the subject on whom it is performed
BT Human experimentation
RT Therapeutic research

NORMALITY

NORMATIVE ETHICS
BT Ethics
NT Deontological ethics
    Teleological ethics↓
RT Human rights
    Moral obligations↓

NURSES
BT Health personnel
RT Medical staff
    Physicians

Nursing Care
Use PATIENT CARE

NURSING EDUCATION
BT Education
RT Medical education

NURSING ETHICS
BT Bioethics
RT Medical ethics
    Professional ethics

NURSING HOMES
BT Residential facilities

NUTRITION
RT Food
    Force feeding

Obligation Not to Kill
Use KILLING, *and*
    MORAL OBLIGATIONS

Obligations
Use MORAL OBLIGATIONS

OBLIGATIONS OF SOCIETY
BT Moral obligations
RT Obligations to society

Obligations to Future Generations
Use MORAL OBLIGATIONS, *and*
    FUTURE GENERATIONS

OBLIGATIONS TO SOCIETY
BT Moral obligations
RT Obligations of society

OBSTETRICS AND GYNECOLOGY
BT Medicine

OCCUPATIONAL DISEASES
RT Industrial medicine

Occupational Medicine
Use INDUSTRIAL MEDICINE

Omission of Treatment
Use WITHHOLDING TREATMENT

OPERANT CONDITIONING
SN Modification of behavior through
    the use of positive and/or negative
    reinforcement
UF Behavior modification
BT Behavior control
NT Negative reinforcement
    Positive reinforcement

Organ Banking
Use TISSUE BANKING

ORGAN DONATION
RT Blood donation

ORGAN DONORS
BT Donors

Organ Transplant Recipients
Use TRANSPLANT RECIPIENTS

ORGAN TRANSPLANTATION
UF Cadaver transplants†
    Heart transplantation†
    Kidney transplantation†
    Renal transplantation†
BT Transplantation

ORGANIZATIONAL POLICIES
RT Institutional policies
    Public policy

OVUM DONORS
UF Egg donors
BT Donors
RT Host mothers
    In vitro fertilization
    Semen donors

PAIN
BT Suffering

Papal Teachings
Use ROMAN CATHOLIC ETHICS

PARENT CHILD RELATIONSHIP
BT Family relationship
NT Mother child relationship

PARENTAL CONSENT
BT Third party consent

SN=Scope Note   UF=Used For   BT=Broader Term   NT=Narrower Term   RT=Related Term

PARENTS
    BT Family members
    NT Fathers↓
          Mothers↓

PATERNALISM
    SN In ethics and law, the setting
          of limits on individual autonomy
          in an effort to benefit, or to
          prevent harm to, the person(s)
          whose autonomy is limited

Passive Euthanasia
    Use ALLOWING TO DIE

PATENTS

PATIENT ACCESS
    RT Confidentiality
          Patients' rights

PATIENT ADVOCACY
    RT Public advocacy

PATIENT CARE
    UF Nursing care
    BT Health care
    NT Terminal care

PATIENT PARTICIPATION
    RT Public participation

PATIENTS
    NT Burn patients
          Chronically ill
          Critically ill
          Terminally ill
    RT Research subjects

PATIENTS' RIGHTS
    BT Rights
    RT Patient access

PEDIATRICS
    BT Medicine

PEER REVIEW
    BT Self regulation

PERINATOLOGY
    SN A field of medicine concerned
          with the care of pregnant
          women, fetuses, and newborns
    BT Medicine

PERSONALITY

PERSONHOOD
    UF Distinctively human
          Humanness
          Protectable humanity
    RT Beginning of life

          Biological life
          Human characteristics

Pharmaceutical Industry
    Use DRUG INDUSTRY

PHENYLKETONURIA
    UF PKU
    BT Brain pathology
          Metabolic diseases
          Recessive genetic conditions
          Single gene defects

PHILOSOPHY
    NT Existentialism
          Stoicism
    RT Ethics↓
          Humanism
          Theology

PHYSICAL CONTAINMENT
    SN The use of special laboratory
          designs or equipment and safe
          handling techniques in an effort
          to prevent the dissemination of
          potentially hazardous materials
          within or outside the laboratory
    BT Containment
    RT Biological containment
          Recombinant DNA research

Physical Life
    Use BIOLOGICAL LIFE

PHYSICALLY HANDICAPPED
    BT Handicapped

PHYSICIAN NURSE RELATIONSHIP
    RT Investigator subject relationship
          Physician patient relationship
          Professional patient relationship

PHYSICIAN PATIENT RELATIONSHIP
    RT Investigator subject relationship
          Physician nurse relationship
          Professional patient relationship

PHYSICIAN'S ROLE

PHYSICIANS
    BT Health personnel
    RT Investigators
          Medical staff
          Medicine↓
          Nurses

PKU
    Use PHENYLKETONURIA

PLACEBOS
    RT Drugs↓

↓ = See main entry for narrower term   † = Consult phrase in alphabetical listing

PLACENTAS

Pluralism, cultural
    Use CULTURAL PLURALISM

POLIOMYELITIS
    BT Communicable diseases

POLITICAL ACTIVITY
    RT Dissent

POLITICAL SYSTEMS
    NT Communism
         Democracy
         Socialism

POLITICS

Pollution
    Use ECOLOGY

Poor Persons
    Use INDIGENTS

POPULATION CONTROL
    RT Family planning↓

POPULATION DISTRIBUTION

POPULATION GROWTH

POSITIVE EUGENICS
    BT Eugenics
    RT Negative eugenics

Positive Euthanasia
    Use ACTIVE EUTHANASIA

POSITIVE REINFORCEMENT
    SN The use of rewards in operant
         conditioning
    BT Operant conditioning
    RT Negative reinforcement

Postexperimental Explanation
    Use DEBRIEFING

Postmortem Examinations
    Use AUTOPSIES

POVERTY
    BT Social problems
    RT Indigents

PRECONCEPTION INJURIES
    BT Injuries
    RT Prenatal injuries

PREGNANT WOMEN
    BT Females

PREMATURITY

Prenatal Development
    Use FETAL DEVELOPMENT

PRENATAL DIAGNOSIS
    SN Determination of fetal status
         prior to birth
    UF Antenatal diagnosis
    BT Diagnosis
    NT Amniocentesis

PRENATAL INJURIES
    UF Antenatal injuries
    BT Injuries
    RT Preconception injuries

PREVENTIVE MEDICINE
    BT Medicine
    RT Community medicine
         Public health

Prima Facie Obligations
    Use MORAL OBLIGATIONS

Principle of Double Effect
    Use DOUBLE EFFECT

Principle of Totality
    Use TOTALITY

PRISONERS
    BT Institutionalized persons
    RT Imprisonment

PRIVACY
    RT Confidentiality
         Privileged communication

PRIVATE HOSPITALS
    BT Hospitals
    NT Religious hospitals
    RT Public hospitals

PRIVILEGED COMMUNICATION
    SN In law, a confidential statement
         made to a lawyer, physician, pastor,
         or spouse, which is privileged against
         disclosure in court if the privilege
         is claimed by the client, patient,
         penitent, or spouse
    BT Communication
    RT Confidentiality
         Privacy

Procreation
    Use REPRODUCTION

PRODUCTS OF IN VITRO FERTILIZATION
    RT In vitro fertilization

PROFESSIONAL COMPETENCE
    RT Competence
         Technical expertise

Professional Dominance
    Use SOCIAL DOMINANCE

SN=Scope Note   UF=Used For   BT=Broader Term   NT=Narrower Term   RT=Related Term

PROFESSIONAL ETHICS
BT   Ethics
RT   Medical ethics
     Nursing ethics

PROFESSIONAL ORGANIZATIONS

PROFESSIONAL PATIENT RELATIONSHIP
SN   Interaction between patients and health
     professionals who are not physicians
RT   Investigator subject relationship
     Physician nurse relationship
     Physician patient relationship

Professional Standards
Use STANDARDS

PROGNOSIS

PROGRAM DESCRIPTIONS

PROLONGATION OF LIFE
SN   The use of therapeutic measures
     to prevent or delay the death of
     critically or terminally ill patients
RT   Allowing to die
     Euthanasia↓
     Extraordinary treatment
     Life extension
     Withholding treatment

PROPERTY RIGHTS
BT   Legal rights

Protectable Humanity
Use PERSONHOOD

PROTESTANT ETHICS
BT   Christian ethics
RT   Protestantism

PROTESTANTISM
BT   Religion
RT   Protestant ethics
     Roman Catholicism

Proxy Consent
Use THIRD PARTY CONSENT

PSYCHIATRIC DIAGNOSIS
BT   Diagnosis

Psychiatric Indications for Abortion
Use THERAPEUTIC ABORTION, *and*
     MENTAL HEALTH

Psychiatric Neurosurgery
Use PSYCHOSURGERY

PSYCHIATRY
BT   Medicine
RT   Psychology
     Psychotherapy

PSYCHOACTIVE DRUGS
SN   Pharmacological agents which affect
     the mind or alter states of conciousness
BT   Drugs
NT   LSD

Psychogenetics
Use BEHAVIORAL GENETICS

PSYCHOLOGICAL STRESS
UF   Emotional stress
RT   Social adjustment

PSYCHOLOGY
RT   Psychiatry
     Psychotherapy

PSYCHOSURGERY
UF   Psychiatric neurosurgery
BT   Behavior control
     Surgery

PSYCHOTHERAPY
RT   Group therapy
     Psychiatry
     Psychology

PUBLIC ADVOCACY
SN   Defense of the rights or interests
     of citizens
UF   Citizen advocacy
RT   Patient advocacy

PUBLIC HEALTH
BT   Medicine
RT   Community medicine
     Preventive medicine

PUBLIC HOSPITALS
BT   Hospitals
RT   Private hospitals↓

PUBLIC OPINION
BT   Informal social control
RT   Attitudes↓

PUBLIC PARTICIPATION
UF   Community participation
     Consumer participation
RT   Patient participation

PUBLIC POLICY
UF   Government policy
RT   Institutional policies
     Organizational policies

Publication Policies
Use EDITORIAL POLICIES

PUNISHMENT
RT   Imprisonment
     Negative reinforcement

↓ = See main entry for narrower term   † = Consult phrase in alphabetical listing

QUALITY CONTROL
RT Evaluation

QUALITY OF LIFE
UF Life, quality of
RT Value of life

Racism
Use DISCRIMINATION

RADIATION
RT Health hazards ↓

RADIOLOGY
BT Medicine

RANDOM SELECTION
RT Selection of subjects

Reasonable Disclosure
Use DISCLOSURE

RECALL
RT Comprehension

RECESSIVE GENETIC CONDITIONS
BT Genetic defects
NT Duchenne muscular dystrophy
    Phenylketonuria
    Sickle cell anemia
    Tay Sachs disease
    Thalassemia
RT Dominant genetic conditions ↓

RECOMBINANT DNA RESEARCH
UF DNA hybridization
    DNA recombinants
BT Biomedical research
RT Artificial genes
    Containment ↓
    DNA therapy
    Genetic intervention ↓

RECORDS
NT Medical records

REFERRAL AND CONSULTATION
UF Consultation

Refusal of Treatment
Use TREATMENT REFUSAL

Regulation
Use SOCIAL CONTROL

Regulation, Government
Use GOVERNMENT REGULATION

Relatives
Use FAMILY MEMBERS

Relativism, Ethical
Use ETHICAL RELATIVISM

RELIGION
NT Protestantism
    Roman Catholicism
RT Clergy
    Jehovah's Witnesses
    Jews
    Theology

RELIGIOUS BELIEFS
RT Attitudes ↓
    Conscience

RELIGIOUS ETHICS
BT Ethics
NT Christian ethics ↓
    Islamic ethics
    Jewish ethics

RELIGIOUS HOSPITALS
UF Catholic hospitals †
BT Private hospitals

REMUNERATION
SN Payment for a service
NT Medical fees
RT Compensation
    Incentives

RENAL DIALYSIS
UF Hemodialysis
    Kidney dialysis

Renal Diseases
Use KIDNEY DISEASES

Renal Transplantation
Use ORGAN TRANSPLANTATION, *and*
    KIDNEYS

REPRODUCTION
UF Human reproduction
    Procreation
    Right to reproduce †

REPRODUCTIVE TECHNOLOGIES
BT Biomedical technologies
NT Artificial insemination ↓
    Cloning
    Embryo transfer
    In vitro fertilization
    Sex preselection

RESEARCH DESIGN
UF Experimental design
RT Methods ↓

RESEARCH INSTITUTES
RT Universities

RESEARCH PERSONNEL
SN Professional and nonprofessional personnel

SN = Scope Note   UF = Used For   BT = Broader Term   NT = Narrower Term   RT = Related Term

NT Investigators
RT Medical staff

RESEARCH SUBJECTS
UF Experimental subjects
RT Control groups
   Patients↓
   Volunteers

RESIDENTIAL FACILITIES
SN Long-term care facilities which provide
   supervision and assistance in activities
   of daily living, with medical and
   nursing services when required
UF Halfway houses
   Homes for the aged†
NT Nursing homes
RT Health facilities↓
   Mental institutions

RESOURCE ALLOCATION
UF Allocation of resources
RT Economics
   Scarcity

Respect for Persons
Use DIGNITY

Responsibilities
Use MORAL OBLIGATIONS

RESUSCITATION
RT Emergency care
   Extraordinary treatment

REVIEW COMMITTEES
RT Ethical review
   Ethics committees

Reverence for Life
Use VALUE OF LIFE

Right to Die
Use ALLOWING TO DIE

Right to Life
Use VALUE OF LIFE

Right to Reproduce
Use REPRODUCTION, *and*
   HUMAN RIGHTS

RIGHT TO TREATMENT
BT Legal rights

RIGHTS
NT Human rights
   Legal rights↓
   Patients' rights
   Women's rights

RISKS AND BENEFITS
UF Benefit harm calculus
RT Costs and benefits

ROMAN CATHOLIC ETHICS
UF Catholic Church teachings
   Papal teachings
BT Christian ethics
RT Roman Catholicism

ROMAN CATHOLICISM
UF Catholic hospitals†
   Catholicism, Roman
BT Religion
RT Protestantism
   Roman Catholic ethics

RUBELLA
UF German measles
BT Communicable diseases

Sanctity of Life
Use VALUE OF LIFE

SCARCITY
RT Resource allocation

SCIENCE
NT Biology
RT Biomedical research↓
   Medicine↓
   Technology↓

SCHIZOPHRENIA
RT Mentally ill

Screening, Drug
Use DRUG SCREENING

Screening, Genetic
Use GENETIC SCREENING

Screening, Mass
Use MASS SCREENING

SCRIPTURAL INTERPRETATION
UF Biblical interpretation

Segmentation
Use TWINNING

SELECTION FOR TREATMENT

SELECTION OF SUBJECTS
SN Selection of experimental subjects
   for research purposes
RT Control groups
   Random selection

SELECTIVE ABORTION
SN Abortion of a fetus because it is,
   or may be, defective
UF Eugenic abortion

↓ = See main entry for narrower term   † = Consult phrase in alphabetical listing

Genetic abortion
BT   Abortion

Self Awareness
Use SELF CONCEPT

SELF CONCEPT
UF   Self awareness

SELF DETERMINATION
UF   Autonomy
     Control of one's body
     Freedom of choice
BT   Freedom

SELF INDUCED ILLNESS
UF   Illness, self induced

SELF REGULATION
BT   Social control
NT   Peer review

SEMEN DONORS
UF   AID donors
BT   Donors
RT   AID
     Ovum donors

SEX DETERMINATION
SN   Determination of the sex of a fetus
     in utero subsequent to conception
RT   Sex preselection

SEX LINKED DEFECTS
BT   Genetic defects
NT   Duchenne muscular dystrophy
     Hemophilia

SEX OFFENSES

SEX PRESELECTION
SN   Choosing the sex of a child prior
     to conception
BT   Reproductive technologies
RT   Sex determination

SEX RATIO
SN   The proportion between the number of
     males and the number of females within
     a specified population
BT   Statistics

SEXUALITY
UF   Erotic behavior
RT   Homosexuals
     Transsexualism

Shock Therapy, Electric
Use ELECTROCONVULSIVE THERAPY

SIBLINGS
BT   Family members

SICKLE CELL ANEMIA
BT   Recessive genetic conditions
     Single gene defects
RT   Thalassemia

SINGLE GENE DEFECTS
BT   Genetic defects
NT   Huntington's chorea
     Phenylketonuria
     Sickle cell anemia
     Tay Sachs disease
     Thalassemia

SINGLE PERSONS
RT   Married persons

SITUATIONAL ETHICS
BT   Ethics

SMOKING
BT   Social problems
RT   Alcohol abuse
     Drug abuse

SOCIAL ADJUSTMENT
UF   Emotional adjustment
RT   Psychological stress

SOCIAL CONTROL
UF   Regulation
NT   Government regulation
     Informal social control↓
     Judicial action
     Legislation↓
     Self regulation↓

SOCIAL DETERMINANTS
RT   Socioeconomic factors

SOCIAL DOMINANCE
SN   Superiority of rank or authority in
     relation to other persons or groups
UF   Dominance, social
     Professional dominance
RT   Authoritarianism

SOCIAL FATHERS
SN   Males who function socially as
     fathers to children, regardless of
     their genetic relationship to those
     children
BT   Fathers
RT   Genetic fathers

SOCIAL IDENTITY
SN   Identification with one or more
     social groups
RT   Genetic identity

SOCIAL IMPACT
RT   Consequences

SN = Scope Note   UF = Used For   BT = Broader Term   NT = Narrower Term   RT = Related Term

SOCIAL INTERACTION
RT Communication↓

Social Medicine
Use COMMUNITY MEDICINE

SOCIAL PROBLEMS
NT Alcohol abuse
Child neglect
Discrimination
Drug abuse
Poverty
Smoking
RT Family problems

SOCIAL SCIENCES
RT Humanities

Social Values
Use VALUES

SOCIAL WORKERS
RT Health personnel↓

SOCIAL WORTH
SN Usefulness or importance of a
person or object to society

SOCIALISM
BT Political systems
RT Capitalism
Communism
Democracy

Socialized Medicine
Use STATE MEDICINE

SOCIETY

SOCIOBIOLOGY
SN The systematic study of the
biological basis of social
behavior
RT Behavioral genetics

SOCIOECONOMIC FACTORS
RT Costs and benefits
Economics
Social determinants

SOCIOLOGY OF MEDICINE

SPERM
UF Human sperm
NT X bearing sperm
Y bearing sperm
RT Frozen semen

SPINA BIFIDA
UF Meningomyelocele
Myelomeningocele

SPONSORING AGENCIES

SPOUSAL CONSENT
BT Third party consent

STANDARDS
UF Professional standards

STATE ACTION

State Control
Use GOVERNMENT REGULATION

STATE COURTS

STATE GOVERNMENT
RT Federal government
Government agencies
Municipal government

STATE INTEREST
UF Interest of the state

STATE MEDICINE
UF Socialized medicine
RT National health insurance

STATISTICS
NT Birth rate
Incidence
Morbidity
Mortality
Sex ratio

Statutes
Use LEGISLATION

STERILIZATION
NT Involuntary sterilization
Voluntary sterilization

STIGMATIZATION
UF Labeling

STOICISM
BT Philosophy

STRIKES

STUDENTS
RT Education↓

SUDDEN INFANT DEATH

SUFFERING
SN Includes both mental and physical
suffering
NT Pain

SUICIDE
BT Killing
RT Voluntary euthanasia

SUPREME COURT DECISIONS
SN Decisions by the United States
Supreme Court

↓ = See main entry for narrower term   † = Consult phrase in alphabetical listing

SURGERY
BT Medicine
NT Psychosurgery
Transplantation↓

SURVIVAL
SN Species survival

Synthetic Genes
Use ARTIFICIAL GENES

SYPHILIS
BT Venereal diseases

TAY SACHS DISEASE
BT Brain pathology
Metabolic diseases
Recessive genetic conditions
Single gene defects

TEACHING METHODS
BT Methods
NT Group discussion
RT Curriculum
Education↓

TECHNICAL EXPERTISE
RT Generalization of expertise
Professional competence

TECHNOLOGY
NT Biomedical technologies↓
RT Science↓

TECHNOLOGY ASSESSMENT

TELEOLOGICAL ETHICS
SN Theories of ethics which hold that
the rightness or wrongness of an act
can be determined by assessing the
good and evil consequences which the
act produces
UF Consequentialism
BT Normative ethics
NT Egoism
Utilitarianism

TEMPORAL LOBE EPILEPSY
UF Epilepsy, temporal lobe
BT Brain pathology

TERMINAL CARE
UF Care for the dying
BT Patient care
RT Death↓
Hospices

TERMINALLY ILL
UF Dying patients
BT Patients
RT Chronically ill
Critically ill

Termination of Treatment
Use WITHHOLDING TREATMENT

Test Tube Fertilization
Use IN VITRO FERTILIZATION

THALASSEMIA
BT Recessive genetic conditions
Single gene defects
RT Sickle cell anemia

THEOLOGY
RT Ethics↓
Philosophy↓
Religion↓

THERAPEUTIC ABORTION
UF Medical indications for abortion
Psychiatric indications for abortion†
BT Abortion

THERAPEUTIC RESEARCH
SN Research which is therapeutic for the
subject on whom it is performed
BT Human experimentation
RT Nontherapeutic research

THIRD PARTY CONSENT
UF Proxy consent
BT Informed consent
NT Parental consent
Spousal consent

TISSUE BANKING
UF Organ banking

TORTS
SN In law, private or civil wrongs, other
than breach of contract, for which
the courts will provide a remedy
in the form of an action for damages
NT Malpractice↓
Wrongful death
Wrongful life
RT Injuries↓
Law↓

TORTURE

TOTALITY
SN In Roman Catholic moral theology, a
principle which affirms that a part of
one's body may legitimately be mutilated
for the sake of the body as a whole or
for the sake of one's neighbor, as in the
donation of bodily organs
UF Principle of totality

TOXICITY

Transfusions, Blood
Use BLOOD TRANSFUSIONS

SN = Scope Note   UF = Used For   BT = Broader Term   NT = Narrower Term   RT = Related Term

TRANSPLANT RECIPIENTS
  UF Organ transplant recipients

TRANSPLANTATION
  BT Surgery
  NT Organ transplantation

TRANSSEXUALISM
  RT Homosexuals
      Sexuality

TREATMENT REFUSAL
  UF Refusal of treatment

TRUST
  SN Confidence in or reliance on a
      person or thing

Truthtelling
  Use DISCLOSURE

TWINNING
  SN The division of a zygote into two
      parts, each of which is capable of
      further development; or, the state
      of being a twin
  UF Segmentation

UNIVERSITIES
  RT Research institutes

UNWANTED CHILDREN
  BT Children

Uterine Mothers
  Use HOST MOTHERS

UTILITARIANISM
  BT Teleological ethics
  RT Egoism

VALUE OF LIFE
  UF Life, value of
      Reverence for life
      Right to life
      Sanctity of life
  RT Quality of life

VALUES
  UF Human values
      Social values

VENEPUNCTURE

VENEREAL DISEASES
  BT Communicable diseases
  NT Syphilis

VIABILITY
  UF Fetal viability†

VIOLENCE
  RT Aggression

VOLUNTARY ADMISSION
  RT Involuntary commitment

Voluntary Consent
  Use INFORMED CONSENT

VOLUNTARY EUTHANASIA
  BT Euthanasia
  RT Involuntary euthanasia
      Suicide

VOLUNTARY PROGRAMS
  RT Coercion
      Mandatory programs

VOLUNTARY STERILIZATION
  BT Sterilization
  RT Involuntary sterilization

VOLUNTEERS
  RT Research subjects

WAR
  NT Biological warfare

WEDGE ARGUMENT
  SN Ethical argument which asserts that
      one morally questionable action or
      policy will set a precedent for, or lead
      to, other actions or policies which
      are even more morally questionable

WHITES
  UF Caucasians

WITHHOLDING TREATMENT
  UF Omission of treatment
      Termination of treatment
  RT Allowing to die
      Euthanasia↓
      Prolongation of life

Women
  Use FEMALES

WOMEN'S RIGHTS
  BT Rights

WRONGFUL DEATH
  SN In civil law, a death of a human
      being which is occasioned by the
      negligent or wrongful act of another
  BT Torts

WRONGFUL LIFE
  SN In civil law, a cause of action which
      alleges that a defendant has wrong-
      fully caused a child to be born
  BT Torts

X BEARING SPERM
  SN Sperm cells which carry the X sex
      chromosome and thus produce female

↓ = See main entry for narrower term   † = Consult phrase in alphabetical listing

offspring
BT   Sperm
RT   Y bearing sperm

XYY KARYOTYPE
BT   Chromosomal disorders
RT   Behavioral genetics

Y BEARING SPERM
SN   Sperm cells which carry the Y sex
      chromosome and thus produce male
      offspring
BT   Sperm
RT   X bearing sperm

SUBJECT ENTRY SECTION

SUBJECT ENTRY SECTION

ABORTED FETUSES *See also* FETUSES

ABORTED FETUSES *See under*

LEGAL OBLIGATIONS / PHYSICIANS / ABORTED FETUSES

PATIENT CARE / ABORTED FETUSES

PATIENT CARE / ABORTED FETUSES / SUPREME COURT DECISIONS

ABORTION *See also* SELECTIVE ABORTION

ABORTION

Anderson, Norman. Issues of Life and Death: Abortion, Birth Control, Capital Punishment, Euthanasia. Downers Grove, Ill.: InterVarsity Press, 1977. 130 p. 189 fn. ISBN 0-87784-721-5. BE05787.
AID; *abortion; *artificial insemination; *capital punishment; Christian ethics; *contraception; DNA therapy; *euthanasia; genetic intervention; justifiable killing; organ transplantation; personhood; scriptural interpretation; *sterilization; theology; *value of life

Beauchamp, Tom L.; Walters, LeRoy, eds. Contemporary Issues in Bioethics. Encino, Calif.: Dickenson Publishing Company, 1978. 612 p. References and footnotes included. ISBN 0-8221-0200-5. BE06700.
*abortion; allowing to die; *behavior control; *bioethics; biomedical technologies; brain death; children; codes of ethics; confidentiality; decision making; deontological ethics; *determination of death; *euthanasia; fetuses; genetic intervention; *health; *health care; *human experimentation; *informed consent; involuntary commitment; justice; legal aspects; *medical ethics; natural law; *normative ethics; operant conditioning; patients' rights; personhood; psychosurgery; *resource allocation; rights; selection for treatment; treatment refusal; utilitarianism; Kaimowitz v. Department of Mental Health; Quinlan, Karen

Brown, Harold O. Death before Birth. Nashville: Thomas Nelson, 1977. 168 p. 114 fn. Bibliography: p. 166. BE06673.
*abortion; active euthanasia; allowing to die; *Christian ethics; fathers; fetuses; incidence; international aspects; killing; legal aspects; methods; morbidity; personhood; physicians; politics; pregnant women; scriptural interpretation; self deter-

mination; *Supreme Court decisions; unwanted children; value of life; wedge argument

Callahan, Daniel. Abortion and medical ethics. *In* Barber, Bernard, ed. Medical Ethics and Social Change. Philadelphia: American Academy of Political and Social Science, 1978. p. 116-127. Annals of the American Academy of Political and Social Science, Vol. 437, May 1978. 7 fn. ISBN 0-87761-226-9. BE07004.
*abortion; amniocentesis; fetuses; genetic defects; human experimentation; prenatal diagnosis; rights; *selective abortion; sex determination; Supreme Court decisions; viability

Glover, Jonathan. Causing Death and Saving Lives. New York: Penguin Books, 1977. 328 p. 67 fn. Bibliography: p. 299-324. ISBN 0-1402-2003-8. BE06513.
*abortion; active euthanasia; *allowing to die; capital punishment; consequences; *decision making; double effect; infanticide; involuntary euthanasia; *killing; morality; newborns; paternalism; personhood; prolongation of life; quality of life; self determination; suicide; utilitarianism; value of life; voluntary euthanasia; war; wedge argument; women's rights

National Conference of Catholic Bishops. Committee for Pro-Life Activities. Abortion and Related Topics: A Select Bibliography, 1977. Washington: National Conference of Catholic Bishops, 1977. 30 p. Annotated entries. Bibliography. BE06681.
*abortion; audiovisual aids; death; population control

Potts, Malcolm; Diggory, Peter; Peel, John. A conspectus. *In their* Abortion. New York: Cambridge University Press, 1977. p. 525-550. 21 refs. ISBN 0-521-29150-X. BE06544.
*abortion; attitudes; contraception; fetal development; illegal abortion; physicians; pregnant women; social impact; statistics

Ramsey, Paul. Ethics at the Edges of Life: Medical and Legal Intersections. New Haven: Yale University Press, 1978. 353 p. The Bampton Lectures in America. 302 fn. ISBN 0-300-02137-2. BE06884.
*abortion; adults; *allowing to die; congenital defects; *decision making; *euthanasia; extraordinary treatment; fetuses; law; legal aspects; legislation; medicine; newborns; physician's role; quality of life; *selection for treatment; *Supreme Court decisions; terminal care; *terminally ill;

BE = bioethics accession number   fn. = footnotes   refs. = references

*third party consent; *treatment refusal; *with-holding treatment; California Natural Death Act; Commonwealth v. Edelin; *In re Quinlan; Saikew-icz, Joseph

Simmons, Paul D. Dialogue on abortion. *In* Hollis, Harry N., comp. A Matter of Life and Death: Christian Perspectives. Nashville: Broadman Press, 1977. p. 100-112. 5 fn. ISBN 0-8054-6118-3. BE06533.
   *abortion; fetuses; morality; personhood; Protestantism; religious beliefs; Supreme Court decisions

Thiroux, Jacques P. Abortion. *In his* Ethics: Theory and Practice. Encino, Calif.: Glencoe Press, 1977. p. 145-159. 10 refs. 7 fn. BE05712.
   *abortion; abortion on demand; alternatives; pregnant women; selective abortion; self determination; therapeutic abortion; value of life

Wojcik, Jan. Muted Consent: A Casebook in Modern Medical Ethics. West Lafayette, Ind.: Purdue University, 1978. 164 p. 237 fn. Case study. ISBN 911198-50-4. BE06704.
   *abortion; *behavior control; *case reports; *determination of death; disclosure; drugs; electrical stimulation of the brain; ethics; eugenics; euthanasia; extraordinary treatment; fetuses; *genetic counseling; genetic defects; *genetic intervention; *genetic screening; health care; *human experimentation; *informed consent; involuntary commitment; mentally ill; operant conditioning; pregnant women; psychosurgery; quality of life; reproductive technologies; *resource allocation; rights; selective abortion; value of life; withholding treatment

ABORTION / ADOLESCENTS / LEGAL ASPECTS

Straus, Thomas R. *Planned Parenthood v. Danforth*: resolving the antinomy. *Ohio Northern University Law Review* 4(2): 425-440, 1977. 100 fn. BE06282.
   aborted fetuses; *abortion; *adolescents; fetuses; *legal aspects; *parental consent; patient care; physicians; privacy; *spousal consent; state interest; *Supreme Court decisions; viability; *Planned Parenthood of Central Missouri v. Danforth

ABORTION / ADOLESCENTS / LEGAL RIGHTS

Lenobel, Jeffrey. Constitutional law—abortion—statute requiring spousal and parental consent declared unconstitutional. *Cumberland Law Review* 7(3): 539-550, Winter 1977. 72 fn. BE06286.
   *abortion; *adolescents; *legal aspects; *legal rights; *parental consent; pregnant women; privacy; *spousal consent; state interest; Supreme Court decisions; Planned Parenthood of Central Missouri v. Danforth

Talbert, Jeffrey T. The validity of parental consent statutes after *Planned Parenthood*. *Journal of Urban Law* 54(1): 127-164, Fall 1977. 169 fn. BE05846.
   *abortion; *adolescents; due process; equal protection; government regulation; *legal aspects; *legal rights; parent child relationship; *parental consent; privacy; state government; state interest; Supreme Court decisions; *Planned Parenthood of Central Missouri v. Danforth

Wand, Barbara F. Parental consent abortion statutes: the limits of state power. *Indiana Law Journal* 52(4): 837-850, Summer 1977. 78 fn. BE06313.
   *abortion; *adolescents; *legal aspects; government regulation; *legal rights; *parental consent; privacy; state government; state interest; Planned Parenthood of Central Missouri v. Danforth

ABORTION / ALTERNATIVES

Levy, Richard N. Abortion and its alternatives. *Sh'ma* 8(144): 212-214, 23 Dec 1977. BE06128.
   *abortion; *adoption; *alternatives; financial support; *indigents; legal rights; morality; *unwanted children; value of life

Schulte, Eugene J. Supreme Court abortion decisions—a challenge to use political process. *Hospital Progress* 58(8): 22-23, Aug 1977. 1 fn. BE05765.
   *abortion; *alternatives; discrimination; financial support; indigents; *informal social control; judicial action; *political activity; *public policy; *Roman Catholicism; state government; Supreme Court decisions

ABORTION / ALTERNATIVES / LEGAL ASPECTS

Humber, James M. Abortion, fetal research, and the law. *Social Theory and Practice* 4(2): 127-147, Spring 1977. 28 fn. BE06297.
   aborted fetuses; *abortion; *alternatives; *artificial organs; constitutional amendments; *embryo transfer; *fetuses; government regulation; *human experimentation; informed consent; *legal aspects; legislation; *morality; nontherapeutic research; pain; parental consent; *placentas; rights; selective abortion; state government; therapeutic research; utilitarianism

ABORTION / ATTITUDES

Potts, Malcolm; Diggory, Peter; Peel, John. Introduction. *In their* Abortion. New York: Cambridge University Press, 1977. p. 1-19. 23 refs. ISBN 0-521-29150-X. BE06539.
   *abortion; *attitudes; *case reports; historical aspects; Islamic ethics; Protestant ethics; Roman Catholic ethics

ABORTION / ATTITUDES / GREAT BRITAIN

Anti-abortion myths. *New Humanist* 92(5): 175-178, Jan-Feb 1977. BE06288.
   *abortion; *attitudes; discrimination; fetuses; ille-

BE = bioethics accession number   fn. = footnotes   refs. = references

gal abortion; incidence; *legislation; model legislation; morbidity; mortality; nurses; pregnant women; viability; *Great Britain

ABORTION / ATTITUDES / HAWAII

Steinhoff, Patricia G.; Diamond, Milton. Anti-repeal issues: the right to life. *In their* Abortion Politics: The Hawaii Experience. Honolulu: University Press of Hawaii, 1977. p. 94-117. 68 fn. ISBN 0-8248-0498-8. BE06192.
    *abortion; *attitudes; fetuses; legal rights; *legislation; moral obligations; morality; physicians; political activity; pregnant women; Roman Catholic ethics; state government; *value of life; women's rights; *Hawaii

Steinhoff, Patricia G.; Diamond, Milton. Pro-repeal issues: human control of social problems. *In their* Abortion Politics: The Hawaii Experience. Honolulu: University Press of Hawaii, 1977. p. 71-93. 74 fn. ISBN 0-8248-0498-8. BE06191.
    *abortion; *attitudes; contraception; costs and benefits; economics; illegal abortion; *legislation; legitimacy; physician's role; political activity; poverty; quality of life; religious beliefs; *self determination; social impact; *social problems; state government; unwanted children; women's rights; *Hawaii

ABORTION / ATTITUDES / UNITED STATES

Potts, Malcolm; Diggory, Peter; Peel, John. The American revolution. *In their* Abortion. New York: Cambridge University Press, 1977. p. 332-376. 56 refs. ISBN 0-521-29150-X. BE06541.
    *abortion; *attitudes; incidence; law enforcement; *legal aspects; legislation; medical fees; mortality; physicians; political activity; public opinion; state government; Supreme Court decisions; *United States

ABORTION / AUSTRALIA

Northern Territory (Australia). Abortion. *International Digest of Health Legislation* 28(3): 427-429, 1977. An ordinance relating to offences involving abortion, dated 3 Apr 1974. BE07129.
    *abortion; conscience; fetal development; informed consent; legal obligations; maternal health; maternal life; medical staff; physicians; selective abortion; therapeutic abortion; *Australia; *Northern Territory

ABORTION / BLACKS / SOCIAL IMPACT

Cates, Willard. Legal abortion: are American black women healthier because of it? *Phylon* 38(3): 267-281, Sep 1977. 35 refs. BE06331.
    *abortion; age; attitudes; *blacks; females; fertility;

illegal abortion; incidence; morbidity; *mortality; *social impact

ABORTION / CANADA

Canada. Committee on the Operation of the Abortion Law. Report of the Committee on the Operation of the Abortion Law. Ottawa: Printing and Publishing Supply and Services Canada, 1977. 474 p. BE06672.
    *abortion; *attitudes; *case reports; contraception; economics; *ethics committees; hospitals; illegal abortion; incidence; informed consent; institutional policies; *legislation; married persons; medical fees; mental health; nurses; *physicians; pregnant women; public opinion; sexuality; single persons; statistics; therapeutic abortion; *Canada

Thomas, W.D.S. The Badgley report on the abortion law. *Canadian Medical Association Journal* 116(9): 966, 7 May 1977. 6 fn. BE06487.
    *abortion; economics; ethics committees; hospitals; legislation; therapeutic abortion; *Canada

ABORTION / CASE REPORTS

Potts, Malcolm; Diggory, Peter; Peel, John. Introduction. *In their* Abortion. New York: Cambridge University Press, 1977. p. 1-19. 23 refs. ISBN 0-521-29150-X. BE06539.
    *abortion; *attitudes; *case reports; historical aspects; Islamic ethics; Protestant ethics; Roman Catholic ethics

ABORTION / CHRISTIAN ETHICS

Anderson, Norman. Birth control, sterilisation and abortion. *In his* Issues of Life and Death: Abortion, Birth Control, Capital Punishment, Euthanasia. Downers Grove, Ill.: InterVarsity Press, 1977. p. 58-84. 46 fn. ISBN 0-87784-721-5. BE05790.
    *abortion; *Christian ethics; *contraception; eugenics; legal aspects; methods; *personhood; population control; reproduction; selective abortion; sexuality; *sterilization; therapeutic abortion; voluntary sterilization

Tuck, Christine. Unwanted pregnancy. *In* Vale, J.A., ed. Medicine and the Christian Mind. London: Christian Medical Fellowship Publications, 1975. p. 98-105. 11 refs. ISBN 85111-956-5. BE07072.
    *abortion; *Christian ethics; decision making; killing; maternal health; mental health; physicians

ABORTION / COMMONWEALTH V. EDELIN

Homans, William P. *Commonwealth v. Kenneth Edelin*, a first in criminal prosecution since *Roe v. Wade*. *Criminal Justice Journal* 1(2): 207-232, Spring 1977. 145 fn. BE06296.
    *abortion; expert testimony; fetuses; *law enforcement; legislation; personhood; *physicians; state

---

BE = bioethics accession number   fn. = footnotes   refs. = references

government; *viability; *Commonwealth v. Edelin; Edelin, Kenneth

Ramsey, Paul. The *Edelin* case. *In his* Ethics at the Edges of Life: Medical and Legal Intersections. New Haven: Yale University Press, 1978. p. 94-142. 30 fn. ISBN 0-300-02137-2. BE06887.
*aborted fetuses; *abortion; abortion on demand; criminal law; freedom; judicial action; *killing; law; law enforcement; *legal aspects; legislation; morality; personhood; *physicians; quality of life; reproductive technologies; state courts; *Supreme Court decisions; *viability; *Commonwealth v. Edelin; Edelin, Kenneth; Massachusetts Supreme Judicial Court

ABORTION / CONSCIENCE / INDIA

Minattur, Joseph. Medical termination of pregnancy and conscientious objection. *Journal of the Indian Law Institute* 16(4): 704-709, Oct-Dec 1974. 15 fn. BE07064.
*abortion; *conscience; legislation; *medical staff; religious beliefs; *India; Medical Termination of Pregnancy Act

ABORTION / CONSTITUTIONAL AMENDMENTS

Neuhaus, Richard J.; Garvey, John. Two views on the human life amendment. *U.S. Catholic* 42(4): 28-31, Apr 1977. BE05570.
*abortion; *constitutional amendments; law; *morality; public opinion; religious beliefs; social impact; Supreme Court decisions; value of life

U.S. Congress. House. Committee on the Judiciary. Subcommittee on Civil and Constitutional Rights. Proposed Constitutional Amendments on Abortion. Hearings. Washington: U.S. Government Printing Office, 1976. 449 p. Hearings, 94th Cong., 2d Sess., on proposed constitutional amendments on abortion, 4-5 Feb and 22-26 Mar 1976. Serial No. 46, Part 2, Appendix. BE06558.
*abortion; attitudes; *constitutional amendments; fetuses; legal aspects; legislation; morbidity; mortality; organizational policies; pregnant women; professional organizations; religious ethics; rights; state government; Supreme Court decisions; value of life

U.S. Congress. House. Committee on the Judiciary. Subcommittee on Civil and Constitutional Rights. Proposed Constitutional Amendments on Abortion. Hearings. Washington: U.S. Government Printing Office, 1976. 639 p. Hearings, 94th Cong., 2d Sess., on proposed constitutional amendments on abortion, 4-5 Feb and 22-26 Mar 1976. Serial No. 46, Part 1. Bibliographies included. BE06557.
*abortion; *constitutional amendments; constitutional law; fathers; fetuses; financial support; hospitals; legal aspects; legal rights; legislation; mental health; morality; personhood; pregnant women;

privacy; religious beliefs; spousal consent; state government; Supreme Court decisions; therapeutic abortion; value of life

ABORTION / CRIMINAL LAW / CANADA

Leigh, L.H. Necessity and the case of Dr. Morgentaler. *Criminal Law Review* 1978: 151-158, Mar 1978. 26 fn. BE06817.
*abortion; *criminal law; judicial action; law enforcement; legislation; mental health; *therapeutic abortion; *Canada; *Morgentaler v. The Queen

ABORTION / DISCRIMINATION

Mulhauser, Karen. Federal funds for abortion. *In* Departments of Labor and Health, Education, and Welfare Appropriations for 1978. Hearings. Washington: U.S. Government Printing Office, 1977. p. 127-139. Hearings. U.S. House. Committee on Appropriations. Subcommittee on the Departments of Labor and Health, Education, and Welfare. 95th Cong., 1st Sess. Part 7. BE06559.
*abortion; contraception; *discrimination; economics; federal government; *financial support; illegal abortion; *indigents; legal rights; mortality; organizational policies; professional organizations; *Hyde Amendment; *Medicaid; National Abortion Rights Action League

ABORTION / DISCRIMINATION / MEDICAID

Somers, Anne R. Of abortions, Medicaid and health priorities. *New York Times,* 5 Dec 1977, p. 36. Letter. BE06418.
*abortion; *discrimination; financial support; health care delivery; indigents; national health insurance; *resource allocation; therapeutic abortion; *Medicaid

ABORTION / EQUAL PROTECTION / SUPREME COURT DECISIONS

Carter traps self on abortion issue. *Human Events* 37(27): 500-501, 2 Jul 1977. BE06315.
*abortion; *equal protection; *financial support; indigents; *politics; public hospitals; *Supreme Court decisions; value of life; Medicaid

ABORTION / ETHICS

Smith, David H. The abortion of defective fetuses: some moral considerations. *In* Smith, David H., ed. No Rush to Judgement: Essays on Medical Ethics. Bloomington, Ind.: Indiana University, Poynter Center, 1977. p. 18-43. 50 fn. BE05794.
*abortion; Christian ethics; conscience; contraception; cultural pluralism; decision making; *ethics; fetal development; fetuses; genetic defects; love; *moral obligations; moral policy; normative ethics; personhood; pregnant women; rights; *selective abortion; society; theology; viability

---

BE = bioethics accession number   fn. = footnotes   refs. = references

ABORTION / ETHICS COMMITTEES / CANADA

Krass, M.E. Assessment of the structure and function of the therapeutic abortion committee. *Canadian Medical Association Journal* 116(7): 786+, 9 Apr 1977. 23 refs. BE05562.
*abortion; criminal law; *decision making; *ethics committees; hospitals; international aspects; legal aspects; legislation; mental health; physicians; *therapeutic abortion; *Canada

ABORTION / FATHERS / LEGAL RIGHTS

New Jersey. Superior Court, Chancery Division. Rothenberger v. Doe. 20 Apr 1977. *Atlantic Reporter, 2d Series,* 374: 57-59. BE06268.
*abortion; *fathers; informed consent; *legal rights; pregnant women; privacy; single persons

Swan, George S. Abortion, parental burdens, and the right to choose: is a penumbral or Ninth-Amendment right about to arise from the ashes of a common law liberty? *Case and Comment* 82(3): 30-32, May-Jun 1977. BE06302.
*abortion; *abortion on demand; discrimination; *fathers; financial support; legal obligations; self determination; *legal rights; spousal consent; Supreme Court decisions; unwanted children; women's rights

ABORTION / FETUSES / LEGAL RIGHTS

Chabon, Robert S. The legal status of the unborn child. *Journal of Legal Medicine* 5(5): 22-24, May 1977. 20 refs. BE05607.
*abortion; congenital defects; *fetuses; *legal aspects; *legal rights; personhood; *prenatal injuries; viability; wrongful death; wrongful life

ABORTION / FETUSES / RIGHTS

Newman, Jay. An empirical argument against abortion: discussion article II. *New Scholasticism* 51(3): 384-395, Summer 1977. 3 fn. BE06308.
*abortion; attitudes; *fetuses; killing; *rights; society; value of life

ABORTION / FINANCIAL SUPPORT

Buckley, William F., et al. Ms. Pilpel v. Mr. Hyde. *Human Life Review* 4(1): 89-108, Winter 1978. Reprinted from the transcript of William F. Buckley's "Firing Line" program, originally titled "Abortion: the Hyde Amendment," telecast during the week of 4 Nov 1977. BE06738.
*abortion; attitudes; discrimination; federal government; fetuses; *financial support; indigents; personhood; politics; selective abortion; social impact; state government; Supreme Court decisions; therapeutic abortion; Hyde Amendment; Medicaid

Clymer, Adam. Senate vote forbids using federal funds for most abortions. *New York Times,* 30 Jun 1977, p. A1+. BE06043.

*abortion; *federal government; *financial support; legislation; politics; Supreme Court decisions

Colen, B.D.; MacPherson, Myra. Court clears Hill curb on U.S. abortion aid: rulings muddle issue. *Washington Post,* 30 Jun 1977, p. A1+. BE06044.
*abortion; federal government; *financial support; social impact; state government; Supreme Court decisions; therapeutic abortion; Medicaid

Fenwick, Millicent. Mental retardation and abortion rights. *In* Departments of Labor and Health, Education, and Welfare Appropriations for 1978. Hearings. Washington: U.S. Government Printing Office, 1977. p. 371-373. Hearings. U.S. House. Committee on Appropriations. Subcommittee on the Departments of Labor and Health, Education, and Welfare. 95th Cong., 1st Sess. Part 7. BE06560.
*abortion; discrimination; federal government; economics; *financial support; illegal abortion; *indigents; morbidity; mortality; *Hyde Amendment

Fritchey, Clayton. Medicaid dollars and abortion sense. *Washington Post,* 9 Jul 1977, p. A11. BE06053.
*abortion; costs and benefits; federal government; *financial support; incidence; international aspects

In search of conscience on abortion. *New York Times,* 22 Jun 1977, p. A22. Editorial. BE06036.
*abortion; *financial support; indigents; state government; Supreme Court decisions; Medicaid

Mintz, Morton. Court clears Hill curb on U.S. abortion aid: Medicaid can be denied. *Washington Post,* 30 Jun 1977, p. A1+. BE06045.
*abortion; attitudes; federal government; *financial support; politics; Supreme Court decisions; therapeutic abortion; Medicaid

Mr. Carter's cruel abortion plan. *New York Times,* 13 Jun 1977, p. 28. Editorial. BE06217.
*abortion; discrimination; *federal government; *financial support; indigents; unwanted children

Mulhauser, Karen. Federal funds for abortion. *In* Departments of Labor and Health, Education, and Welfare Appropriations for 1978. Hearings. Washington: U.S. Government Printing Office, 1977. p. 127-139. Hearings. U.S. House. Committee on Appropriations. Subcommittee on the Departments of Labor and Health, Education, and Welfare. 95th Cong., 1st Sess. Part 7. BE06559.
*abortion; contraception; *discrimination; economics; federal government; *financial support; illegal abortion; *indigents; legal rights; mortality; organizational policies; professional organizations; *Hyde Amendment; *Medicaid; National Abortion Rights Action League

The new abortion debate. *Commonweal* 104(15): 451-452, 22 Jul 1977. Editorial. BE05966.
*abortion; alternatives; *federal government;

BE = bioethics accession number   fn. = footnotes   refs. = references

*financial support; indigents; public policy; Roman Catholicism

Packwood, Robert W., et al. Use of federal funds for abortion. [Title provided]. *Congressional Record (Daily Edition)* 123(113): S11030-S11056, 29 Jun 1977. BE06584.
*abortion; costs and benefits; discrimination; *federal government; *financial support; indigents; morality; pregnant women; professional organizations; public opinion; public policy; self determination; therapeutic abortion; unwanted children; value of life; wedge argument; Medicaid

Pritchard, Joel; Steers, Newton. Funding for abortion. *In* Departments of Labor and Health, Education, and Welfare Appropriations for 1978. Hearings. Washington: U.S. Government Printing Office, 1977. p. 612-615. Hearings. U.S. House. Committee on Appropriations. Subcommittee on the Departments of Labor and Health, Education, and Welfare. 95th Cong., 1st Sess. Part 7. BE06562.
*abortion; discrimination; economics; federal government; *financial support; indigents; *Hyde Amendment

Rich, Spencer. Senate measure found to permit most abortions. *Washington Post,* 8 Jul 1977, p. A1+. BE06051.
*abortion; attitudes; federal government; *financial support; legislation; therapeutic abortion

Rich, Spencer. Senate votes curb on use of federal funds for abortions. *Washington Post,* 30 Jun 1977, p. A14. BE06046.
*abortion; attitudes; *federal government; *financial support; politics

Rose, Charlie. Hyde Amendment. *In* Departments of Labor and Health, Education, and Welfare Appropriations for 1978. Hearings. Washington: U.S. Government Printing Office, 1977. p. 601-604. Hearings. U.S. House. Committee on Appropriations. Subcommittee on the Departments of Labor and Health, Education, and Welfare. 95th Cong., 1st Sess. Part 7. 5 fn. BE06561.
*abortion; discrimination; economics; equal protection; *financial support; indigents; religious beliefs; *Hyde Amendment

Russell, Mary. House bans use of U.S. funds in abortion cases. *Washington Post,* 18 Jun 1977, p. A1+. BE06029.
*abortion; *federal government; *financial support; indigents; Hyde Amendment; Medicaid

Senate battle seen over HEW funds, abortion. *Wall Street Journal,* 21 Jun 1977, p. 2. BE06034.
*abortion; economics; federal government; *financial support; legislation; politics

U.S. Department of Health, Education, and Welfare. Federal financial participation in expenditures for abortions funded through various HEW programs.

Federal Register 43(23): 4570-4582, 2 Feb 1978. BE06917.
*abortion; *abortion on demand; disclosure; *federal government; *financial support; government agencies; government regulation; historical aspects; law enforcement; legal aspects; legislation; maternal health; maternal life; medical records; physician's role; politics; therapeutic abortion; Health Care Financing Administration; Office of Human Development Services; Public Health Service; U.S. Congress

U.S. District Court, E.D. Wisconsin. Doe v. Mundy. 2 Sep 1977. *Federal Supplement* 441: 447-452. BE07114.
*abortion; equal protection; *financial support; *institutional policies; *public hospitals; therapeutic abortion

Wicker, Tom. Kitchen-table justice. *New York Times,* 28 Jun 1977, p. 31. BE06042.
*abortion; discrimination; federal government; *financial support; indigents; legislation; social impact; Supreme Court decisions

Will our humanity also be aborted? *New York Times,* 5 Jul 1977, p. 28. Editorial. BE06049.
*abortion; federal government; *financial support; indigents; *social impact; social problems

ABORTION / FINANCIAL SUPPORT / ATTITUDES

Leiman, Sid Z. Subsidized abortion: response to a call. *Sh'ma* 8(143): 197-200, 9 Dec 1977. BE06125.
*abortion; *attitudes; *financial support; indigents; legal rights; morality; public policy; *religion; therapeutic abortion; Medicaid

ABORTION / FINANCIAL SUPPORT / CONSCIENCE

Borowitz, Eugene B. A call for poor to act on conscience. *Sh'ma* 8(143): 200-202, 9 Dec 1977. BE06126.
*abortion; *conscience; *financial support; indigents; legislation; politics; self determination; Medicaid

ABORTION / FINANCIAL SUPPORT / EQUAL PROTECTION

New Jersey. Supreme Court. Planned Parenthood, etc. v. State, etc. 1 Nov 1977. *Atlantic Reporter, 2d Series,* 379: 841-847. 4 fn. BE07118.
*abortion; *equal protection; *financial support; state government; Supreme Court decisions

ABORTION / FINANCIAL SUPPORT / HYDE AMENDMENT

Hyde, Henry J. The heart of the matter. *Human Life Review* 3(3): 90-96, Summer 1977. BE05677.
*abortion; *abortion on demand; *federal govern-

BE = bioethics accession number   fn. = footnotes   refs. = references

ment; *fetuses; *financial support; indigents; killing; politics; *value of life; *Hyde Amendment; *Medicaid

ABORTION / FINANCIAL SUPPORT / LEGAL ASPECTS

Hardy, David T. Privacy and public funding: *Maher v. Roe* as the interaction of *Roe v. Wade* and *Dandridge v. Williams*. *Arizona Law Review* 18(4): 903-938, 1976. 181 fn. BE07098.
*abortion; conscience; discrimination; *due process; economics; *equal protection; *financial support; indigents; *legal aspects; legal rights; resource allocation; state government; state interest

ABORTION / FINANCIAL SUPPORT / MEDIC-AID

Jacobs, Sarah. Abortion—the question is whether the government should pay. *National Journal* 9(19): 713-715, 7 May 1977. BE05620.
*abortion; community services; discrimination; *federal government; *financial support; indigents; politics; social impact; state government; Hyde Amendment; *Medicaid

U.S. Supreme Court. Beal v. Doe. 20 Jun 1977. *United States Reports* 432: 438-463. 22 fn. BE06578.
*abortion; abortion on demand; *equal protection; discrimination; federal government; fetuses; *financial support; indigents; legislation; state government; state interest; *Medicaid

U.S. Supreme Court. Maher v. Roe. 20 Jun 1977. *United States Reports* 432: 464-490. 14 fn. BE06579.
*abortion; abortion on demand; discrimination; *equal protection; *financial support; government regulation; indigents; legislation; pregnant women; self determination; state government; *Medicaid

ABORTION / FINANCIAL SUPPORT / MORAL-ITY

Leiman, Sid Z. Abortion needs concern not subsidy. *Sh'ma* 8(144): 215-216, 23 Dec 1977. BE06130.
*abortion; *financial support; *morality; therapeutic abortion

Siegel, Seymour. Abortion: moral issues beyond legal ones. *Sh'ma* 8(143): 202-205, 9 Dec 1977. BE06127.
*abortion; *financial support; indigents; killing; legal rights; *morality; political activity; self determination

ABORTION / FINANCIAL SUPPORT / PENN-SYLVANIA

Abortion—Pennsylvania Medicaid regulations and procedures denying non-therapeutic abortions to indi-

gent women held inconsistent with Title XIX of the Social Security Act. *Journal of Family Law* 15(3): 587-592, 1976-1977. 12 fn. BE06202.
*abortion; equal protection; *financial support; indigents; state government; *Pennsylvania

ABORTION / FINANCIAL SUPPORT / POLI-TICS

The unborn and the born again. *New Republic* 177(27): 5-6+, 2 Jul 1977. Editorial. BE05737.
*abortion; abortion on demand; beginning of life; *constitutional law; equal protection; *federal government; fetuses; *financial support; government regulation; indigents; killing; *politics; *social impact; state government; *Supreme Court decisions; women's rights; Medicaid

ABORTION / FINANCIAL SUPPORT / SOCIAL IMPACT

Abortion. New York: WNET/13, 1977. 10 p. Transcript of the MacNeil/Lehrer Report, Show No. 3092, Library No. 552, 8 Nov 1977. Available from WNET/13. Box 345, New York 10019. Television script. BE06575.
*abortion; *federal government; *financial support; illegal abortion; *indigents; killing; mortality; pregnant women; self determination; *social impact; therapeutic abortion; Medicaid

Packwood, Robert W. Federal funding of abortions. *Congressional Record (Daily Edition)* 123(64): S5933-S5935, 19 Apr 1977. BE05929.
*abortion; costs and benefits; discrimination; equal protection; *federal government; *financial support; human rights; *indigents; *pregnant women; *social impact

Yankauer, Alfred. Abortions and public policy, II. *American Journal of Public Health* 67(9): 817-818, Sep 1977. 4 refs. Editorial. BE05994.
*abortion; *abortion on demand; economics; *financial support; *indigents; legal aspects; *social impact; Medicaid

ABORTION / FINANCIAL SUPPORT / SU-PREME COURT DECISIONS

The abortion issue. *Washington Post,* 24 Jun 1977, p. A20. Editorial. BE06039.
*abortion; *financial support; indigents; institutional policies; legal rights; public hospitals; *Supreme Court decisions; U.S. Congress

Annas, George J. Let them eat cake. *Hastings Center Report* 7(4): 8-9, Aug 1977. BE05742.
*abortion; *abortion on demand; equal protection; *financial support; *indigents; pregnant women; rights; social impact; *state government; *Supreme Court decisions

Arditti, Rita. Abortion legislation. *Science for the People* 9(5): 19, Sep-Oct 1977. BE05847.
*abortion; *abortion on demand; *financial

BE = bioethics accession number   fn. = footnotes   refs. = references

support; indigents; *state government; *Supreme Court decisions

Callahan, Sidney. The Court and a conflict of principles. *Hastings Center Report* 7(4): 7-8, Aug 1977. BE05741.
*abortion; *abortion on demand; *financial support; *indigents; pregnant women; privacy; rights; *Supreme Court decisions

Fraker, Susan. Abortion: who pays? *Newsweek* 90(1): 12-13, 4 Jul 1977. BE06318.
*abortion; costs and benefits; federal government; *financial support; state government; *Supreme Court decisions; *Medicaid

Goodman, Ellen. A pregnant moment for Justice Powell. *Washington Post,* 23 Jun 1977, p. A25. BE06038.
*abortion; *financial support; indigents; legislation; *Supreme Court decisions

Kraft, Joseph. 'The decisions abort the right to abortion'. *Washington Post,* 26 Jun 1977, p. B7. BE06040.
*abortion; equal protection; *financial support; indigents; legal rights; minority groups; *Supreme Court decisions

Mintz, Morton. High Court backs curb on Medicaid use for abortions. *Washington Post,* 21 Jun 1977, p. A1+. BE06032.
*abortion; federal government; *financial support; government regulation; indigents; state government; *Supreme Court decisions; *Medicaid

Noonan, John T. A half-step forward: the justices retreat on abortion. *Human Life Review* 3(4): 11-18, Fall 1977. BE05995.
*abortion; *abortion on demand; conscience; *financial support; *indigents; maternal health; politics; public policy; *Supreme Court decisions

Oelsner, Lesley. Court rules states may deny Medicaid for some abortions. *New York Times,* 21 Jun 1977, p. 1+. BE06033.
*abortion; abortion on demand; attitudes; *financial support; government regulation; indigents; state government; *Supreme Court decisions; Hyde Amendment; *Medicaid

Segers, Mary C. Abortion and the Supreme Court: some are more equal than others. *Hastings Center Report* 7(4): 5-6, Aug 1977. 1 fn. BE05740.
*abortion; *abortion on demand; discrimination; *financial support; indigents; public opinion; public policy; social impact; *state government; *Supreme Court decisions

Shinn, Roger, et al. Paying for abortion: is the Court wrong? *Christianity and Crisis* 37(14): 202-207, 19 Sep 1977. BE05990.
*abortion; *abortion on demand; discrimination; federal government; *financial support; *indigents;

legal rights; public policy; *Supreme Court decisions

States needn't fund elective abortions for low-income women, Justices rule. *Wall Street Journal,* 21 Jun 1977, p. 2. BE06035.
*abortion; abortion on demand; discrimination; federal government; *financial support; indigents; state government; *Supreme Court decisions; *Medicaid

Supreme Court rules on access to contraception, abortion: states can deny Medicaid benefits, hospital services for elective abortions. *Family Planning/Population Reporter* 6(4): 41+, Aug 1977. 5 fn. BE05768.
*abortion; *abortion on demand; federal government; *financial support; indigents; legal rights; public hospitals; *state government; *Supreme Court decisions; *Medicaid

Will, George F. A shift to the 'appropriate forum'.... *Washington Post,* 23 Jun 1977, p. A25. BE06037.
*abortion; abortion on demand; *financial support; indigents; *Supreme Court decisions

ABORTION / GOVERNMENT REGULATION

Mr. Califano operates on the law. *New York Times,* 27 Dec 1977, p. 34. Editorial. BE06431.
*abortion; federal government; financial support; *government regulation

U.S. Supreme Court. Sendak v. Arnold. 29 Nov 1976. *United States Reports* 429: 968-972. 1 fn. BE07091.
*abortion; *government regulation; standards; state government

ABORTION / GOVERNMENT REGULATION / LEGAL ASPECTS

Sirkis, Judith E. A lesson in judicial abdication: Roe v. Arizona Board of Regents and the right of privacy. *Arizona State Law Journal* 1976(3): 499-524, 1976. 161 fn. BE06632.
*abortion; *abortion on demand; institutional obligations; *government regulation; *institutional policies; *legal aspects; legal rights; legislation; physicians; pregnant women; *privacy; self determination; *public hospitals; sexuality; state courts; *state government; state interest; Supreme Court decisions; universities; *Arizona; *Roe v. Arizona Board of Regents

ABORTION / GOVERNMENT REGULATION / PLANNED PARENTHOOD OF CENTRAL MISSOURI V. DANFORTH

Wodtke, Alexis K. Constitutional law—a state may constitutionally regulate the abortion decision during the first trimester of pregnancy if it can show that the regulation is necessary to protect a compelling state interest.... *Drake Law Review* 26(3): 716-727,

---

BE = bioethics accession number   fn. = footnotes   refs. = references

1976-1977. 89 fn. BE06633.
*abortion; adolescents; disclosure; *government regulation; *informed consent; *parental consent; physicians; *spousal consent; *state government; Supreme Court decisions; Missouri; *Planned Parenthood of Central Missouri v. Danforth

ABORTION / GREAT BRITAIN

Goodhart, C.B., et al. Abortion: the great debate. *Times (London),* 12 Jul 1977, p. 12. Letter. BE06450.
aborted fetuses; *abortion; allowing to die; attitudes; incidence; legislation; patient care; pregnant women; self determination; viability; Abortion Act 1967; Benyon Bill; *Great Britain

ABORTION / HARE, R.M.

Werner, Richard. Hare on abortion. *Analysis* 36(4): 177-181, Jun 1976. 3 fn. BE07101.
*abortion; fetuses; morality; normative ethics; *personhood; *Hare, R.M.

ABORTION / HISTORICAL ASPECTS

Mohr, James C. Abortion in America: The Origins and Evolution of National Policy, 1800-1900. New York: Oxford University Press, 1978. 331 p. References and footnotes: p. 265-328. ISBN 0-19-502249-1. BE06979.
*abortion; *attitudes; clergy; *historical aspects; illegal abortion; incidence; killing; law enforcement; *legislation; *mass media; physician's role; *physicians; public opinion; public policy; *state government; Supreme Court decisions; Roe v. Wade

ABORTION / HYDE AMENDMENT

The Hyde Amendment—again. *Washington Post,* 3 Jun 1977, p. A26. Editorial. BE06026.
*abortion; equal protection; *federal government; *financial support; *indigents; *Hyde Amendment

ABORTION / ICELAND

Iceland. Abortion, sterilization, and sex education. *International Digest of Health Legislation* 28(3): 614-620, 1977. Law of 22 May 1975. BE06464.
*abortion; counseling; legal liability; physicians; sexuality; socioeconomic factors; standards; *sterilization; therapeutic abortion; *Iceland

ABORTION / INFORMAL SOCIAL CONTROL / ROMAN CATHOLICISM

Schulte, Eugene J. Supreme Court abortion decisions—a challenge to use political process. *Hospital Progress* 58(8): 22-23, Aug 1977. 1 fn. BE05765.
*abortion; *alternatives; discrimination; financial support; indigents; *informal social control; judi-

cial action; *political activity; *public policy; *Roman Catholicism; state government; Supreme Court decisions

ABORTION / INSTITUTIONAL POLICIES

U.S. District Court, E.D. Wisconsin. Doe v. Mundy. 2 Sep 1977. *Federal Supplement* 441: 447-452. BE07114.
*abortion; equal protection; *financial support; *institutional policies; *public hospitals; therapeutic abortion

ABORTION / INSTITUTIONAL POLICIES / SUPREME COURT DECISIONS

Kemp, Kathleen A.; Carp, Robert A.; Brady, David W. Abortion and the law: the impact on hospital policy of the *Roe* and *Doe* decisions. *Journal of Health Politics, Policy and Law* 1(3): 319-337, Fall 1976. 23 fn. BE06643.
*abortion; *administrators; attitudes; decision making; *hospitals; *institutional policies; religious hospitals; Roman Catholicism; *social impact; socioeconomic factors; *Supreme Court decisions; Doe v. Bolton; Roe v. Wade

ABORTION / INTERNATIONAL ASPECTS

Skowronski, Marjory. Differing perspectives on abortion. *In her* Abortion and Alternatives. Millbrae, Calif.: Les Femmes Publishing, 1977. p. 9-18. 7 fn. ISBN 0-89087-923-0. BE06190.
*abortion; attitudes; historical aspects; *international aspects; *religious ethics; values

Thomas, John M.; Ryniker, Barbara M.; Kaplan, Milton. Indian abortion law revision and population policy: an overview. *Journal of the Indian Law Institute* 16(4): 513-534, Oct-Dec 1974. 46 fn. BE07060.
*abortion; attitudes; family planning; government regulation; *international aspects; law; *legislation; physicians; *population control; *public policy; religion; socioeconomic factors; therapeutic abortion; *India; *Medical Termination of Pregnancy Act

ABORTION / ISRAEL

Israel. Abortion. *International Digest of Health Legislation* 28(2): 290-292, 1977. The Penal Law Amendment (Termination of Pregnancy) Law, 1977. BE06465.
*abortion; ethics committees; legal liability; physicians; standards; *Israel

ABORTION / LAW ENFORCEMENT / SOCIAL IMPACT

Affleck, Glenn; Thomas, Agnes. The Edelin decision revisited: a survey of the reactions of Connecticut's OB/GYNs. *Connecticut Medicine* 41(10): 637-640,

---

BE = bioethics accession number   fn. = footnotes   refs. = references

Oct 1977. 8 refs. BE06081.
aborted fetuses; * abortion; *attitudes; killing; law enforcement; legal aspects; *obstetrics and gynecology; *physicians; social impact; Edelin, Kenneth

ABORTION / LEGAL ASPECTS

Dworkin, Roger. Legal issues in genetic counseling. *In* Smith, David H., ed. No Rush to Judgement: Essays on Medical Ethics. Bloomington, Ind.: Indiana University, Poynter Center, 1977. p. 44-67. 1 fn. BE05795.
*abortion; competence; constitutional law; disclosure; family members; family relationship; fetuses; *genetic counseling; *genetic screening; informed consent; *legal aspects; legal obligations; legislation; legitimacy; mandatory programs; phenylketonuria; physicians; pregnant women; privacy *AID; rights; sickle cell anemia; state interest; sterilization; *Supreme Court decisions; Roe v. Wade

Gentles, Ian. Do we need a law against abortion? *In* Fairweather, Eugene; Gentles, Ian, eds. The Right to Birth: Some Christian Views on Abortion. Toronto: Anglican Book Centre, 1976. p. 13-24. 21 fn. ISBN 0-919030-14-9. BE06551.
*abortion; Christian ethics; fetuses; *legal aspects; legal rights; legislation; personhood

Hall, James. Abortion: Legal Control. Issue Brief Number IB74019. Washington: Library of Congress, Congressional Research Service, 30 Mar 1977. 11 p. Congressional Research Service Major Issues System. Originally published 5 Feb 1974; updated 30 Mar 1977. Available only through Members of Congress. 13 refs. BE06572
*abortion; constitutional amendments; financial support; historical aspects; *legal aspects; legislation; parental consent; *politics; public hospitals; spousal consent; state government; Supreme Court decisions

Hunter, Ian A. Abortion: reflections on a protracted debate. *Human Life Review* 3(4): 47-67, Fall 1977. 49 fn. BE05998.
*abortion; criminal law; human rights; fetuses; *legal aspects; legislation; maternal health; mental health; *pregnant women; *quality of life; *self determination; *therapeutic abortion; torts; value of life; *Canada

ABORTION / LEGAL ASPECTS / NEW ZEALAND

O'Neill, John S. Fetus-in-Law. Dunedin, N.Z.: Independent Publishing Company, Jun 1976. 148 p. BE07089.
*abortion; criminal law; fetuses; killing; *legal aspects; *legislation; *New Zealand

ABORTION / LEGAL ASPECTS / UNITED STATES

Potts, Malcolm; Diggory, Peter; Peel, John. The American revolution. *In their* Abortion. New York: Cambridge University Press, 1977. p. 332-376. 56 refs. ISBN 0-521-29150-X. BE06541.
*abortion; *attitudes; incidence; law enforcement; *legal aspects; legislation; medical fees; mortality; physicians; political activity; public opinion; state government; Supreme Court decisions; *United States

ABORTION / LEGISLATION / ATTITUDES

Healy, Pat. Abortion: why the doctors are closing ranks against new curbs. *Times (London),* 8 Jul 1977, p. 14. BE06449.
*abortion; *attitudes; *legislation; *physicians; professional organizations; *Great Britain

ABORTION / LEGISLATION / CANADA

Bennett, J.S. Inequities in abortion law found result of attitudes in people and institutions. *Canadian Medical Association Journal* 116(5): 553-554, 5 Mar 1977. BE05915.
*abortion; attitudes; hospitals; institutional policies; *legislation; physicians; statistics; *Canada

ABORTION / LEGISLATION / CONSCIENCE

Ramsey, Paul. Abortion after the law: conscience and its problems. *In his* Ethics at the Edges of Life: Medical and Legal Intersections. New Haven: Yale University Press, 1978. p. 43-93. 48 fn. ISBN 0-300-02137-2. BE06886.
*abortion; *abortion on demand; case reports; *conscience; cultural pluralism; discrimination; employment; federal government; government regulation; fetuses; *institutional policies; judicial action; legal obligations; *legislation; *medical staff; morality; *nurses; *physicians; privacy; *private hospitals; public hospitals; religious ethics; *religious hospitals; rights; Roman Catholicism; social impact; state action; state government; *Supreme Court decisions; therapeutic abortion; value of life; *Curran, Charles E.; Great Britain; New Jersey Supreme Court; *United States

ABORTION / LEGISLATION / EUROPE

Potts, Malcolm; Diggory, Peter; Peel, John. Continental Europe. *In their* Abortion. New York: Cambridge University Press, 1977. p. 377-409. 45 refs. ISBN 0-521-29150-X. BE06542.
*abortion; attitudes; historical aspects; international aspects; *legislation; *Europe

ABORTION / LEGISLATION / GREAT BRITAIN

Garner, Lesley. Wanted: cool heads, not hot air. *Sun-*

BE = bioethics accession number   fn. = footnotes   refs. = references

day Times (London), 20 Feb 1977, p. 42. BE06006.
*abortion; *legislation; political activity; *Great Britain

Potts, Malcolm; Diggory, Peter; Peel, John. Abortion and the law in Britain. *In their* Abortion. New York: Cambridge University Press, 1977. p. 277-297. 30 refs. ISBN 0-521-29150-X. BE06540.
*abortion; historical aspects; illegal abortion; *legislation; political activity; *Great Britain

ABORTION / LEGISLATION / HAWAII

Steinhoff, Patricia G.; Diamond, Milton. Anti-repeal issues: the right to life. *In their* Abortion Politics: The Hawaii Experience. Honolulu: University Press of Hawaii, 1977. p. 94-117. 68 fn. ISBN 0-8248-0498-8. BE06192.
*abortion; *attitudes; fetuses; legal rights; *legislation; moral obligations; morality; physicians; political activity; pregnant women; Roman Catholic ethics; state government; *value of life; women's rights; *Hawaii

Steinhoff, Patricia G.; Diamond, Milton. Pro-repeal issues: human control of social problems. *In their* Abortion Politics: The Hawaii Experience. Honolulu: University Press of Hawaii, 1977. p. 71-93. 74 fn. ISBN 0-8248-0498-8. BE06191.
*abortion; *attitudes; contraception; costs and benefits; economics; illegal abortion; *legislation; legitimacy; physician's role; political activity; poverty; quality of life; religious beliefs; *self determination; social impact; *social problems; state government; unwanted children; women's rights; *Hawaii

ABORTION / LEGISLATION / INDIA

Bose, Asit K. Abortion in India: a legal study. *Journal of the Indian Law Institute* 16(4): 535-548, Oct-Dec 1974. 19 fn. BE07061.
*abortion; *attitudes; *family planning; illegal abortion; *legislation; motivation; social impact; *India; *Medical Termination of Pregnancy Act

Menon, N.R. Madhava. Population policy, law enforcement and the liberalization of abortion: a socio-legal inquiry into the implementation of the abortion law in India. *Journal of the Indian Law Institute* 16(4): 626-648, Oct-Dec 1974. 15 fn. BE07062.
*abortion; attitudes; criminal law; family planning; historical aspects; illegal abortion; *legislation; physicians; public policy; *social impact; socioeconomic factors; *India; *Medical Termination of Pregnancy Act

Thomas, John M.; Ryniker, Barbara M.; Kaplan, Milton. Indian abortion law revision and population policy: an overview. *Journal of the Indian Law Institute* 16(4): 513-534, Oct-Dec 1974. 46 fn. BE07060.
*abortion; attitudes; family planning; government

regulation; *international aspects; law; *legislation; physicians; *population control; *public policy; religion; socioeconomic factors; therapeutic abortion; *India; *Medical Termination of Pregnancy Act

ABORTION / LEGISLATION / INTERNATIONAL ASPECTS

Potts, Malcolm; Diggory, Peter; Peel, John. The experience of other countries. *In their* Abortion. New York: Cambridge University Press, 1977. p. 410-453. 81 refs. ISBN 0-521-29150-X. BE06543.
*abortion; *international aspects; *legislation

ABORTION / LEGISLATION / ISRAEL

Falk, Ze'ev W. The new abortion law of Israel. *Israel Law Review* 13(1): 103-110, Jan 1978. 16 fn. BE06731.
*abortion; informed consent; *legislation; parental consent; physician's role; pregnant women; spousal consent; *Israel

ABORTION / LEGISLATION / ITALY

Gilbert, Sari. Liberal abortion bill approved in Italy. *Washington Post,* 22 Jan 1977, p. A11. BE06211.
*abortion; *legislation; politics; Roman Catholicism; *Italy

ABORTION / MANDATORY PROGRAMS

Swan, George S. State-mandated abortion: has 1984 arrived seven years early? *Law and Liberty* 3(4): 4-6, Spring/Summer 1977. BE06225.
*abortion; financial support; *mandatory programs; *reproduction; self determination; state interest; Supreme Court decisions; *women's rights

ABORTION / MEDICAID

A sad compromise on abortion. *New York Times,* 10 Dec 1977, p. 24. Editorial. BE06423.
*abortion; federal government; financial support; indigents; legislation; *Medicaid

Tolchin, Martin. Compromise is voted by House and Senate in abortion dispute. *New York Times,* 8 Dec 1977, p. A1+. BE06421.
*abortion; *federal government; *financial support; *Medicaid

ABORTION / MEDICAL STAFF / CONSCIENCE

Ramsey, Paul. Abortion after the law: conscience and its problems. *In his* Ethics at the Edges of Life: Medical and Legal Intersections. New Haven: Yale University Press, 1978. p. 43-93. 48 fn. ISBN 0-300-02137-2. BE06886.
*abortion; *abortion on demand; case reports; *conscience; cultural pluralism; discrimination; employment; federal government; government reg-

---

BE = bioethics accession number   fn. = footnotes   refs. = references

ulation; fetuses; *institutional policies; judicial action; legal obligations; *legislation; *medical staff; morality; *nurses; *physicians; privacy; *private hospitals; public hospitals; religious ethics; *religious hospitals; rights; Roman Catholicism; social impact; state action; state government; *Supreme Court decisions; therapeutic abortion; value of life; *Curran, Charles E.; Great Britain; New Jersey Supreme Court; *United States

ABORTION / MENTALLY RETARDED / LEGAL RIGHTS

Schukoske, Jane. Abortion for the severely retarded: a search for authorization. *Mental Disability Law Reporter* 1(6): 485-492, May-Jun 1977. 66 fn. BE05933.
*abortion; competence; economics; fetuses; informed consent; *legal rights; legislation; *mentally retarded; negative eugenics; *pregnant women; privacy; state government; state interest; third party consent

ABORTION / MORAL OBLIGATIONS

McLachlan, Hugh V. Must we accept either the conservative or the liberal view on abortion? *Analysis* 37(4): 197-204, Jun 1977. BE05654.
*abortion; *fetuses; future generations; *moral obligations; pregnant women; personhood; property rights; *rights; self determination; value of life

ABORTION / MORALITY

Canavan, Francis. Simple-minded separationism. *Human Life Review* 3(4): 36-46, Fall 1977. BE05997.
*abortion; fetuses; *law; *morality; personhood; Roman Catholicism; *theology; value of life

Gustafson, Donald. Some reflections on the abortion issue. *Western Humanities Review* 30(3): 181-198, Summer 1976. 1 fn. BE06638.
*abortion; abortion on demand; attitudes; contraception; *fetuses; maternal health; maternal life; moral obligations; *pregnant women; *morality; punishment; *rights; *self determination; *value of life

Lally, Mark. Abortion and deception. *Human Life Review* 3(3): 47-70, Summer 1977. 68 fn. BE05679.
*abortion; discrimination; *fetuses; indigents; legal aspects; *morality; personhood; pregnant women; religious beliefs; rights; self determination; social interaction; unwanted children; *value of life; viability

Novak, David. A call to concern: a response. *Sh'ma* 8(144): 214-215, 23 Dec 1977. BE06129.
*abortion; financial support; Jewish ethics; legal rights; *morality; suffering

Vallance, Elizabeth. The abortion debate. *New Universities Quarterly* 31(3): 358-364, Summer 1977. 2 fn. BE05681.

*abortion; *abortion on demand; *decision making; fetuses; *morality; pregnant women; self determination; *value of life; wedge argument

ABORTION / MORALITY / ATTITUDES

Harrison, Beverly, et al. Continuing the discussion —how to argue about abortion: II. *Christianity and Crisis* 37(21): 311-318, 26 Dec 1977. BE06371.
*abortion; abortion on demand; *attitudes; beginning of life; fetuses; constitutional amendments; *morality; killing; personhood; political activity; public opinion; public policy; Roman Catholicism; socioeconomic factors; therapeutic abortion; value of life

ABORTION / MORGENTALER V. THE QUEEN

Leigh, L.H. Necessity and the case of Dr. Morgentaler. *Criminal Law Review* 1978: 151-158, Mar 1978. 26 fn. BE06817.
*abortion; *criminal law; judicial action; law enforcement; legislation; mental health; *therapeutic abortion; *Canada; *Morgentaler v. The Queen

ABORTION / NEW ZEALAND

Contraception, sterilisation and abortion in New Zealand. *New Zealand Medical Journal* 85(588): 441-445, 25 May 1977. Recommendations from the Report of the Royal Commission of Inquiry into Contraception, Sterilisation and Abortion (Government Printer, Wellington). BE05601.
*abortion; adolescents; alternatives; community services; conscience; *contraception; counseling; education; employment; ethics committees; family planning; health personnel; information dissemination; *legal aspects; mentally handicapped; pregnant women; *public policy; review committees; sexuality; social workers; *sterilization; therapeutic abortion; *New Zealand

New Zealand. Royal Commission of Inquiry. Contraception, Sterilisation, and Abortion in New Zealand: Report of the Royal Commission of Inquiry. Wellington, N.Z.: E.C. Keating, 1977. 455 p. Duncan W. McMullin, Chairman of the Commission. Bibliography: p. 421-445. BE06682.
*abortion; adolescents; adults; *contraception; family planning; fetuses; historical aspects; human experimentation; illegal abortion; law; legal aspects; *legislation; mentally handicapped; methods; morality; pregnant women; public opinion; public policy; *social impact; socioeconomic factors; statistics; *voluntary sterilization; women's rights; *New Zealand

ABORTION / OBSTETRICS AND GYNECOLOGY / ATTITUDES

Affleck, Glenn; Thomas, Agnes. The Edelin decision revisited: a survey of the reactions of Connecticut's OB/GYNs. *Connecticut Medicine* 41(10): 637-640,

---

BE = bioethics accession number   fn. = footnotes   refs. = references

Oct 1977. 8 refs. BE06081.
    aborted fetuses;*abortion; *attitudes; killing; law
    enforcement; legal aspects; *obstetrics and gyne-
    cology; *physicians; social impact; Edelin,
    Kenneth

Nathanson, Constance A.; Becker, Marshall H. The
influence of physicians' attitudes on abortion perfor-
mance, patient management and professional fees.
Family Planning Perspectives 9(4): 158-163, Jul/Aug
1977. 9 refs. 3 fn. BE05731.
    *abortion; *attitudes; bioethical issues; fetal devel-
    opment; legislation; medical fees; *obstetrics and
    gynecology; parental consent; patient care; *physi-
    cians; psychological stress; public policy; religion;
    social adjustment; socioeconomic factors; spousal
    consent; Supreme Court decisions; values; Mary-
    land

ABORTION / PARENTAL CONSENT / LEGAL ASPECTS

Lenobel, Jeffrey. Constitutional law—abortion—statute requiring spousal and parental consent declared
unconstitutional. *Cumberland Law Review* 7(3):
539-550, Winter 1977. 72 fn. BE06286.
    *abortion; *adolescents; *legal aspects; *legal
    rights; *parental consent; pregnant women; pri-
    vacy; *spousal consent; state interest; Supreme
    Court decisions; Planned Parenthood of Central
    Missouri v. Danforth

Talbert, Jeffrey T. The validity of parental consent
statutes after *Planned Parenthood*. *Journal of Urban
Law* 54(1): 127-164, Fall 1977. 169 fn. BE05846.
    *abortion; *adolescents; due process; equal protec-
    tion; government regulation; *legal aspects; *legal
    rights; parent child relationship; *parental con-
    sent; privacy; state government; state interest; Su-
    preme Court decisions; *Planned Parenthood of
    Central Missouri v. Danforth

Wand, Barbara F. Parental consent abortion statutes:
the limits of state power. *Indiana Law Journal*
52(4): 837-850, Summer 1977. 78 fn. BE06313.
    *abortion; *adolescents; *legal aspects; govern-
    ment regulation; *legal rights; *parental consent;
    privacy; state government; state interest; Planned
    Parenthood of Central Missouri v. Danforth

ABORTION / PARENTAL CONSENT / LEGAL RIGHTS

Taylor, Michael A. Abortion statistics and parental
consent: a state-by-state review. *Family Law Reporter* 1(25): 1-10, 6 May 1975. Monograph No. 7.
24 fn. BE07084.
    *abortion; adolescents; incidence; *legal rights;
    legislation; mortality; *parental consent; pregnant
    women; state government; *statistics

ABORTION / PARENTAL CONSENT / PLANNED PARENTHOOD OF CENTRAL MISSOURI V. DANFORTH

Crist-Brown, Blythe. Abortion—possible alternatives
to unconstitutional spousal and parental consent provisions of Missouri's abortion law. *Missouri Law Review* 42(2): 291-297, Spring 1977. 52 fn. BE05949.
    *abortion; adolescents; legislation; *parental con-
    sent; *spousal consent; state government; Supreme
    Court decisions; Missouri; *Planned Parenthood
    of Central Missouri v. Danforth

ABORTION / PARENTAL CONSENT / SUPREME COURT DECISIONS

Newbery, Sue A. The right to abortion: the end of
parental and spousal consent requirements. *Arkansas Law Review* 31(1): 122-126, Spring 1977. 28 fn.
BE05952.
    *abortion; *parental consent; *privacy; *spousal
    consent; *Supreme Court decisions; Planned Par-
    enthood of Central Missouri v. Danforth

Ramsey, Paul. The Supreme Court's bicentennial abortion decision: can the 1973 abortion decisions be
justly hedged? *In his* Ethics at the Edges of Life:
Medical and Legal Intersections. New Haven: Yale
University Press, 1978. p. 3-42. 28 fn. ISBN
0-300-02137-2. BE06885.
    *aborted fetuses; *abortion; adolescents; family re-
    lationship; fathers; fetuses; legal obligations; *leg-
    islation; marital relationship; medical ethics;
    methods; morality; obstetrics and gynecology;
    *parental consent; parents; *patient care; physi-
    cians; pregnant women; privacy; rights; *spousal
    consent; state interest; *Supreme Court decisions;
    *viability; *Missouri; *Planned Parenthood of
    Central Missouri v. Danforth

Straus, Thomas R. *Planned Parenthood v. Danforth*:
resolving the antinomy. *Ohio Northern University
Law Review* 4(2): 425-440, 1977. 100 fn.
BE06282.
    aborted fetuses; *abortion; *adolescents; fetuses;
    *legal aspects; *parental consent; patient care;
    physicians; privacy; *spousal consent; state inter-
    est; *Supreme Court decisions; viability; *Planned
    Parenthood of Central Missouri v. Danforth

ABORTION / PERSONHOOD

Connery, John R. Abortion: a philosophical and historical analysis. *Hospital Progress* 58(4): 49-50, Apr
1977. BE05546.
    *abortion; fetal development; *fetuses; historical
    aspects; *personhood; pregnant women; Roman
    Catholic ethics; self determination

Fairweather, Eugene. The child as neighbour: abortion
as a theological issue. *In* Fairweather, Eugene; Gentles, Ian, eds. The Right to Birth: Some Christian
Views on Abortion. Toronto: Anglican Book Centre, 1976. p. 41-55. 16 fn. ISBN 0-919030-14-9.

BE = bioethics accession number   fn. = footnotes   refs. = references

BE06553.
*abortion; *Christian ethics; *fetuses; historical aspects; *personhood; *value of life

Gillespie, Norman C. Abortion and human rights. *Ethics* 87(3): 237-243, Apr 1977. 11 fn. BE05558.
*abortion; fetal development; *fetuses; *human rights; justice; *morality; *personhood; philosophy; value of life

Newton, Lisa H. Abortion in the law: an essay on absurdity. *Ethics* 87(3): 244-250, Apr 1977. 7 fn. BE05571.
*abortion; constitutional law; *fetuses; government regulation; *personhood; pregnant women; privacy; *rights; state government; Supreme Court decisions

Warren, Mary A. Do potential people have moral rights? *Canadian Journal of Philosophy* 7(2): 275-289, Jun 1977. 10 fn. BE06306.
*abortion; consequences; contraception; ethics; *fetal development; future generations; human development; *human rights; moral obligations; *personhood; population growth; quality of life; reproduction; *self concept; value of life; Golden Rule; *Hare, R.M.

Werner, Richard. Hare on abortion. *Analysis* 36(4): 177-181, Jun 1976. 3 fn. BE07101.
*abortion; fetuses; morality; normative ethics; *personhood; *Hare, R.M.

Wogaman, J. Philip. Abortion as a theological issue. *Washington Post,* 16 Aug 1977, p. A19. BE06393.
*abortion; constitutional amendments; fetal development; fetuses; *personhood; religion

ABORTION / PHYSICIANS / ATTITUDES

Roper, John. Gynaecologists object to abortion on demand. *Times (London),* 13 Jul 1977, p. 5. BE06451.
*abortion; abortion on demand; *attitudes; fetal development; *physicians; Great Britain

ABORTION / PHYSICIANS / NEGLIGENCE

Tennessee. Supreme Court. Olson v. Molzen. 21 Nov 1977. *South Western Reporter, 2d Series,* 558: 429-433. 3 fn. BE06659.
*abortion; contracts; informed consent; *legal liability; *negligence; *physicians

ABORTION / POLITICAL ACTIVITY

Fraker, Susan, et al. Abortion under attack. *Newsweek* 91(23): 36-37+, 5 Jun 1978. BE06973.
*abortion; *attitudes; contraception; financial support; incidence; *political activity; pregnant women; Roman Catholicism; self determination; state government; value of life

Johnston, Laurie. Abortion foes gain support as they intensify campaign. *New York Times,* 23 Oct 1977, sect. 1, p. 1+. BE06407.

*abortion; constitutional amendments; financial support; legal rights; *political activity

ABORTION / POLITICS

Carter traps self on abortion issue. *Human Events* 37(27): 500-501, 2 Jul 1977. BE06315.
*abortion; *equal protection; *financial support; indigents; *politics; public hospitals; *Supreme Court decisions; value of life; Medicaid

ABORTION / POLITICS / ATTITUDES

Ashdown-Sharp, Patricia; Darroch, Sandra. What MPs really think about abortion. *Sunday Times (London),* 26 Jun 1977, p. 6. BE06447.
*abortion; *attitudes; *legislation; *politics; Abortion Act 1967; Benyon Bill; Great Britain

ABORTION / POLITICS / GREAT BRITAIN

Simms, Madeleine. Abortion and democratic politics. *New Humanist* 93(1): 15-17, May-Aug 1977. BE06301.
*abortion; abortion on demand; incidence; informal social control; legislation; *politics; public opinion; Roman Catholicism; socioeconomic factors; women's rights; *Abortion Act 1967; *Great Britain

ABORTION / PREGNANT WOMEN / LEGAL RIGHTS

Skowronski, Marjory. Abortion and the law. *In her* Abortion and Alternatives. Millbrae, Calif.: Les Femmes Publishing, 1977. p. 1-7. 10 fn. ISBN 0-89087-923-0. BE06189.
*abortion; federal government; financial support; government regulation; indigents; *legal rights; *pregnant women; self determination; state government

ABORTION / PRIVACY / SUPREME COURT DECISIONS

Byrn, Robert M. Judicial imperialism. *Hospital Progress* 58(11): 90-97+, Nov 1977. 50 fn. BE06107.
*abortion; fetuses; financial support; hospitals; institutional policies; *judicial action; mental health; parental consent; personhood; *privacy; spousal consent; *Supreme Court decisions; therapeutic abortion; value of life

Byrn, Robert M. Which way for judicial imperialism? *Human Life Review* 3(4): 19-35, Fall 1977. 50 fn. BE05996.
*abortion; financial support; judicial action; legal rights; parental consent; *privacy; public hospitals;

BE = bioethics accession number   fn. = footnotes   refs. = references

spousal consent; state government; state interest; *Supreme Court decisions; therapeutic abortion

ABORTION / PRIVATE HOSPITALS / CONSCIENCE

Ramsey, Paul. Abortion after the law: conscience and its problems. *In his* Ethics at the Edges of Life: Medical and Legal Intersections. New Haven: Yale University Press, 1978. p. 43-93. 48 fn. ISBN 0-300-02137-2. BE06886.
*abortion; *abortion on demand; case reports; *conscience; cultural pluralism; discrimination; employment; federal government; government regulation; fetuses; *institutional policies; judicial action; legal obligations; *legislation; *medical staff; morality; *nurses; *physicians; privacy; *private hospitals; public hospitals; religious ethics; *religious hospitals; rights; Roman Catholicism; social impact; state action; state government; *Supreme Court decisions; therapeutic abortion; value of life; *Curran, Charles E.; Great Britain; New Jersey Supreme Court; *United States

ABORTION / PUBLIC HOSPITALS

U.S. Supreme Court. Poelker v. Doe. 20 Jun 1977. *United States Reports* 432: 519-525. 4 fn. BE06580.
*abortion; financial support; indigents; municipal government; *public hospitals; public policy; *St. Louis; Missouri

ABORTION / PUBLIC HOSPITALS / LEGAL ASPECTS

Sirkis, Judith E. A lesson in judicial abdication: Roe v. Arizona Board of Regents and the right of privacy. *Arizona State Law Journal* 1976(3): 499-524, 1976. 161 fn. BE06632.
*abortion; *abortion on demand; institutional obligations; *government regulation; *institutional policies; *legal aspects; legal rights; legislation; physicians; pregnant women; *privacy; self determination; *public hospitals; sexuality; state courts; *state government; state interest; Supreme Court decisions; universities; *Arizona; *Roe v. Arizona Board of Regents

ABORTION / PUBLIC OPINION

Noonan, John T. The American consensus on abortion. *Human Life Review* 4(1): 60-63, Winter 1978. BE06747.
*abortion; abortion on demand; constitutional amendments; females; males; political activity; *public opinion; Supreme Court decisions

ABORTION / QUALITY OF LIFE

Hunter, Ian A. Abortion: reflections on a protracted debate. *Human Life Review* 3(4): 47-67, Fall 1977. 49 fn. BE05998.

*abortion; criminal law; human rights; fetuses; *legal aspects; legislation; maternal health; mental health; *pregnant women; *quality of life; *self determination; *therapeutic abortion; torts; value of life; *Canada

ABORTION / RELIGION / ATTITUDES

The churches and abortion. *New Humanist* 92(1): 32-35, May-Jun 1976. BE06636.
*abortion; *attitudes; organizational policies; Protestant ethics; *religion; Roman Catholic ethics; Church of England; Church of Scotland; Methodist Church

ABORTION / RELIGIOUS ETHICS

Skowronski, Marjory. Differing perspectives on abortion. *In her* Abortion and Alternatives. Millbrae, Calif.: Les Femmes Publishing, 1977. p. 9-18. 7 fn. ISBN 0-89087-923-0. BE06190.
*abortion; attitudes; historical aspects; *international aspects; *religious ethics; values

ABORTION / RIGHTS

Bethell, Tom. Abortion is not a 'right'. *Washington Post*, 23 Jul 1977, p. A15. BE06059.
*abortion; financial support; indigents; *rights; Supreme Court decisions

Grant, Sheila; Grant, George. Abortion and rights: the value of political freedom. *In* Fairweather, Eugene; Gentles, Ian, eds. The Right to Birth: Some Christian Views on Abortion. Toronto: Anglican Book Centre, 1976. p. 1-12. 13 fn. ISBN 0-919030-14-9. BE06550.
*abortion; fetuses; legal aspects; personhood; quality of life; *rights; value of life; wedge argument

ABORTION / ROMAN CATHOLIC ETHICS / DOUBLE EFFECT

Nicholson, Susan T. Abortion and the Roman Catholic Church. Knoxville, Tenn.: University of Tennessee, Department of Religious Studies, 1978. 109 p. Journal of Religious Ethics, Studies in Religious Ethics: No. 2. 75 fn. Bibliography: p. 99-103. ISBN 0-931886-00-7. BE06980.
*abortion; *allowing to die; *double effect; fetuses; justifiable killing; maternal life; *killing; morality; moral obligations; mother fetus relationship; philosophy; pregnant women; *Roman Catholic ethics; selective abortion; sexuality; *therapeutic abortion; value of life; withholding treatment; Grisez, Germain

ABORTION / ROMAN CATHOLIC ETHICS / HISTORICAL ASPECTS

Connery, John R. *Abortion*: The Development of the Roman Catholic Perspective. Chicago: Loyola University Press, 1977. 336 p. 467 fn. ISBN

BE = bioethics accession number   fn. = footnotes   refs. = references

0-8294-0257-8. BE05878.

*abortion; *ancient history; Christian ethics; double effect; fetuses; *historical aspects; infanticide; *intention; Jewish ethics; *killing; literature; maternal life; personhood; *Roman Catholic ethics; scriptural interpretation; theology; *therapeutic abortion

ABORTION / ROMAN CATHOLICISM / POLITICAL ACTIVITY

Bennett, John; Hoyt, Robert. How to argue about abortion. *Christianity and Crisis* 37(17): 264-266, 14 Nov 1977. BE06105.

*abortion; attitudes; constitutional amendments; financial support; *political activity; Protestantism; *Roman Catholicism

ABORTION / SCRIPTURAL INTERPRETATION

Smyth, Harley. The Bible and the unborn child: reflections on life before birth. *In* Fairweather, Eugene; Gentles, Ian, eds. The Right to Birth: Some Christian Views on Abortion. Toronto: Anglican Book Centre, 1976. p. 25-39. 11 fn. ISBN 0-919030-14-9. BE06552.

*abortion; Christian ethics; fetuses; love; *scriptural interpretation; therapeutic abortion

ABORTION / SOCIAL IMPACT / INDIA

Menon, N.R. Madhava. Population policy, law enforcement and the liberalization of abortion: a socio-legal inquiry into the implementation of the abortion law in India. *Journal of the Indian Law Institute* 16(4): 626-648, Oct-Dec 1974. 15 fn. BE07062.

*abortion; attitudes; criminal law; family planning; historical aspects; illegal abortion; *legislation; physicians; public policy; *social impact; socioeconomic factors; *India; *Medical Termination of Pregnancy Act

ABORTION / SOCIOECONOMIC FACTORS

Lidz, Ruth W. Abortion. *Connecticut Medicine* 42(3): 201-202, March 1978. 5 refs. BE06819.

*abortion; abortion on demand; adolescents; contraception; counseling; education; family planning; females; pregnant women; *socioeconomic factors; unwanted children; women's rights

ABORTION / SPOUSAL CONSENT / LEGAL ASPECTS

Lenobel, Jeffrey. Constitutional law—abortion—statute requiring spousal and parental consent declared unconstitutional. *Cumberland Law Review* 7(3): 539-550, Winter 1977. 72 fn. BE06286.

*abortion; *adolescents; *legal aspects; *legal rights; *parental consent; pregnant women; privacy; *spousal consent; state interest; Supreme

Court decisions; Planned Parenthood of Central Missouri v. Danforth

ABORTION / SPOUSAL CONSENT / LEGAL RIGHTS

Farber, Alan J. The abortion decision: the question of the father's rights. *Family Law Reporter* 2(1): 1-7, 4 Nov 1975. Monograph No. 12. 63 fn. BE07082.

*abortion; *fathers; *legal rights; *spousal consent

ABORTION / SPOUSAL CONSENT / PLANNED PARENTHOOD OF CENTRAL MISSOURI V. DANFORTH

Crist-Brown, Blythe. Abortion—possible alternatives to unconstitutional spousal and parental consent provisions of Missouri's abortion law. *Missouri Law Review* 42(2): 291-297, Spring 1977. 52 fn. BE05949.

*abortion; adolescents; legislation; *parental consent; *spousal consent; state government; Supreme Court decisions; Missouri; *Planned Parenthood of Central Missouri v. Danforth

ABORTION / SPOUSAL CONSENT / SUPREME COURT DECISIONS

Newbery, Sue A. The right to abortion: the end of parental and spousal consent requirements. *Arkansas Law Review* 31(1): 122-126, Spring 1977. 28 fn. BE05952.

*abortion; *parental consent; *privacy; *spousal consent; *Supreme Court decisions; Planned Parenthood of Central Missouri v. Danforth

Ramsey, Paul. The Supreme Court's bicentennial abortion decision: can the 1973 abortion decisions be justly hedged? *In his* Ethics at the Edges of Life: Medical and Legal Intersections. New Haven: Yale University Press, 1978. p. 3-42. 28 fn. ISBN 0-300-02137-2. BE06885.

*aborted fetuses; *abortion; adolescents; family relationship; fathers; fetuses; legal obligations; *legislation; marital relationship; medical ethics; methods; morality; obstetrics and gynecology; *parental consent; parents; *patient care; physicians; pregnant women; privacy; rights; *spousal consent; state interest; *Supreme Court decisions; *viability; *Missouri; *Planned Parenthood of Central Missouri v. Danforth

Straus, Thomas R. *Planned Parenthood v. Danforth*: resolving the antinomy. *Ohio Northern University Law Review* 4(2): 425-440, 1977. 100 fn. BE06282.

aborted fetuses; *abortion; *adolescents; fetuses; *legal aspects; *parental consent; patient care; physicians; privacy; *spousal consent; state interest; *Supreme Court decisions; viability; *Planned Parenthood of Central Missouri v. Danforth

---

BE = bioethics accession number   fn. = footnotes   refs. = references

ABORTION / STATISTICS

Taylor, Michael A. Abortion statistics and parental consent: a state-by-state review. *Family Law Reporter* 1(25): 1-10, 6 May 1975. Monograph No. 7. 24 fn. BE07084.
> *abortion; adolescents; incidence; *legal rights; legislation; mortality; *parental consent; pregnant women; state government; *statistics

ABORTION / SUPREME COURT DECISIONS

Dworkin, Roger. Legal issues in genetic counseling. *In* Smith, David H., ed. No Rush to Judgement: Essays on Medical Ethics. Bloomington, Ind.: Indiana University, Poynter Center, 1977. p. 44-67. 1 fn. BE05795.
> *abortion; competence; constitutional law; disclosure; family members; family relationship; fetuses; *genetic counseling; *genetic screening; informed consent; *legal aspects; legal obligations; legislation; legitimacy; mandatory programs; phenylketonuria; physicians; pregnant women; privacy; *AID; rights; sickle cell anemia; state interest; sterilization; *Supreme Court decisions; Roe v. Wade

ABORTION / SUPREME COURT DECISIONS / PUBLIC OPINION

Blake, Judith. The Supreme Court's abortion decisions and public opinion in the United States. *Population and Development Review* 3(1-2): 45-62, Mar-Jun 1977. 18 fn. BE05912.
> *abortion; constitutional amendments; federal government; fetuses; financial support; personhood; *public opinion; spousal consent; *Supreme Court decisions

ABORTION / SUPREME COURT DECISIONS / SOCIAL IMPACT

The unborn and the born again. *New Republic* 177(27): 5-6+, 2 Jul 1977. Editorial. BE05737.
> *abortion; abortion on demand; beginning of life; *constitutional law; equal protection; *federal government; fetuses; *financial support; government regulation; indigents; killing; *politics; *social impact; state government; *Supreme Court decisions; women's rights; Medicaid

ABORTION / VALUE OF LIFE

Bandow, Doug. Should society condone abortion as 'convenient'? *Human Events* 37(49): 14, 3 Dec 1977. BE06505.
> *abortion; *abortion on demand; fetuses; killing; unwanted children; *value of life; women's rights

Evans, M. Stanton. The death brigade marches on. *Human Events* 37(32): 609, 6 Aug 1977. BE05755.
> *abortion; *value of life; wedge argument

Gustafson, Donald. Some reflections on the abortion issue. *Western Humanities Review* 30(3): 181-198,

Summer 1976. 1 fn. BE06638.
> *abortion; abortion on demand; attitudes; contraception; *fetuses; maternal health; maternal life; moral obligations; *pregnant women; *morality; punishment; *rights; self determination; *value of life

Henle, Robert J. The 'demystification' of life. *Commonweal* 104(15): 457-460, 22 Jul 1977. BE05959.
> *abortion; fetuses; legal aspects; morality; religious beliefs; society; *value of life

Hyde, Henry J. The heart of the matter. *Human Life Review* 3(3): 90-96, Summer 1977. BE05677.
> *abortion; *abortion on demand; *federal government; *fetuses; *financial support; indigents; killing; politics; *value of life; *Hyde Amendment; *Medicaid

Hyde, Henry J. The humanity of the unborn. *Washington Post*, 25 Jul 1977, p. A21. BE06060.
> *abortion; fetuses; killing; personhood; *value of life

Lally, Mark. Abortion and deception. *Human Life Review* 3(3): 47-70, Summer 1977. 68 fn. BE05679.
> *abortion; discrimination; *fetuses; indigents; legal aspects; *morality; personhood; pregnant women; religious beliefs; rights; self determination; social interaction; unwanted children; *value of life; viability

ABORTION / VALUE OF LIFE / CHRISTIAN ETHICS

Fairweather, Eugene. The child as neighbour: abortion as a theological issue. *In* Fairweather, Eugene; Gentles, Ian, eds. The Right to Birth: Some Christian Views on Abortion. Toronto: Anglican Book Centre, 1976. p. 41-55. 16 fn. ISBN 0-919030-14-9. BE06553.
> *abortion; *Christian ethics; *fetuses; historical aspects; *personhood; *value of life

Fairweather, Eugene; Gentles, Ian, eds. The Right to Birth: Some Christian Views on Abortion. Toronto: Anglican Book Centre, 1976. 76 p. 82 fn. ISBN 0-919030-14-9. BE06549.
> *abortion; *Christian ethics; fetuses; fetal development; legal aspects; personhood; rights; scriptural interpretation; therapeutic abortion; *value of life; wedge argument; women's rights

ABORTION / VALUE OF LIFE / JEWISH ETHICS

Los Angeles Catholic-Jewish Respect Life Committee. Respect for life: Jewish and Roman Catholic reflections on abortion and related issues. *Catholic Mind* 76(1320): 54-64, Feb 1978. BE06783.
> *abortion; fetal development; fetuses; goals; *Jewish ethics; personhood; *Roman Catholic ethics; Supreme Court decisions; *value of life

---

BE = bioethics accession number fn. = footnotes refs. = references

ABORTION / VALUE OF LIFE / PROPERTY RIGHTS

O'Driscoll, L.H. Abortion, property rights, and the right to life. *Personalist* 58(2): 99-114, Apr 1977. 26 fn. BE05573.
　　*abortion; fetuses; justice; *moral obligations; pregnant women; *property rights; risks and benefits; self concept; self determination; *value of life

ABORTION / VALUE OF LIFE / ROMAN CATHOLIC ETHICS

Los Angeles Catholic-Jewish Respect Life Committee. Respect for life: Jewish and Roman Catholic reflections on abortion and related issues. *Catholic Mind* 76(1320): 54-64, Feb 1978. BE06783.
　　*abortion; fetal development; fetuses; goals; *Jewish ethics; personhood; *Roman Catholic ethics; Supreme Court decisions; *value of life

ABORTION / VALUE OF LIFE / WEDGE ARGUMENT

Koop, C. Everett. The sanctity of life. *Journal of the Medical Society of New Jersey* 75(1): 62-67, Jan 1978. 1 fn. BE06740.
　　*abortion; Christian ethics; congenital defects; newborns; quality of life; religious beliefs; *value of life; *wedge argument

ABORTION / VIABILITY / LEGAL ASPECTS

Ramsey, Paul. The *Edelin* case. *In his* Ethics at the Edges of Life: Medical and Legal Intersections. New Haven: Yale University Press, 1978. p. 94-142. 30 fn. ISBN 0-300-02137-2. BE06887.
　　*aborted fetuses; *abortion; abortion on demand; criminal law; freedom; judicial action; *killing; law; law enforcement; *legal aspects; legislation; morality; personhood; *physicians; quality of life; reproductive technologies; state courts; *Supreme Court decisions; *viability; *Commonwealth v. Edelin; Edelin, Kenneth; Massachusetts Supreme Judicial Court

U.S. District Court, D. South Carolina, Columbia Division. Floyd v. Anders. 4 Nov 1977. *Federal Supplement* 440: 535-540. 5 fn. BE07120.
　　*aborted fetuses; *abortion; constitutional law; fetal development; fetuses; government regulation; *killing; *law enforcement; *legal aspects; legal liability; legal rights; legislation; personhood; *physicians; pregnant women; state government; Supreme Court decisions; *viability; Roe v. Wade; *South Carolina

ABORTION / WEDGE ARGUMENT

Adamek, Raymond J. It has happened here. *Human Life Review* 3(4): 74-85, Fall 1977. 43 fn. BE05999.
　　*abortion; coercion; euthanasia; Jews; killing; *na-

tional socialism; population control; public opinion; *wedge argument

ABORTION / WEST GERMANY

Getler, Michael. Bonn abortion law: precise, tormenting. *Washington Post,* 2 Jul 1977, p. A11. BE06047.
　　*abortion; fetuses; *legislation; personhood; politics; public opinion; *West Germany

ABORTION / YUGOSLAVIA

Socialist Republic of Slovenia (Yugoslavia). Abortion, sterilization, contraception, and artificial insemination. *International Digest of Health Legislation* 28(4): 1112-1115, 1977. Law of 20 Apr 1977. BE07131.
　　*abortion; abortion on demand; age; *artificial insemination; competence; confidentiality; contraception; counseling; decision making; fertility; fetal development; rights; reproduction; semen donors; therapeutic abortion; third party consent; *voluntary sterilization; *Socialist Republic of Slovenia; *Yugoslavia

ACTIVE EUTHANASIA

Cahill, Lisa S. Comment on "euthanasia". *Linacre Quarterly* 44(4): 299-300, Nov 1977. Letter. BE06108.
　　*active euthanasia; Christian ethics; extraordinary treatment; totality; value of life

Lasagna, Louis. Effecting euthanasia should not be secretive. *National Observer* 28 Apr 1973, p. 13. BE06600.
　　*active euthanasia; allowing to die; withholding treatment

Montague, Phillip. The morality of active and passive euthanasia. *Ethics in Science and Medicine* 5(1): 39-45, 1978. 8 fn. BE06908.
　　*active euthanasia; *allowing to die; *killing; moral obligations; philosophy

Sullivan, Thomas D. Active and passive euthanasia: an impertinent distinction? *Human Life Review* 3(3): 40-46, Summer 1977. 5 fn. BE05680.
　　*active euthanasia; *allowing to die; extraordinary treatment; intention; killing; terminally ill; withholding treatment; Rachels, James

Trammell, Richard L. The presumption against taking life. *Journal of Medicine and Philosophy* 3(1): 53-67, Mar 1978. 13 refs. 7 fn. BE06829.
　　*active euthanasia; *allowing to die; *decision making; extraordinary treatment; intention; killing; morality; quality of life; value of life

ACTIVE EUTHANASIA / AGED / ATTITUDES

Kraus, A.S., et al. Potential interest of the elderly in active euthanasia. *Canadian Family Physician* 23: 123+, Mar 1977. 11 refs. BE06581.

BE = bioethics accession number　　　fn. = footnotes　　　refs. = references

*active euthanasia; *aged; *attitudes; public opinion; terminally ill

ACTIVE EUTHANASIA / ETHICS

Maguire, Daniel C. Death and the moral domain. *St. Luke's Journal of Theology* 20(3): 197-216, Jun 1977. 58 fn. BE06305.
*active euthanasia; *allowing to die; *Christian ethics; death; ethicists; *ethics; *euthanasia; justifiable killing; *killing; legislation; moral obligations; pain; philosophy; suffering; suicide; value of life; *Church of England

ACTIVE EUTHANASIA / MORALITY

Beauchamp, Tom L. A reply to Rachels on active and passive euthanasia. *In* Beauchamp, Tom L.; Perlin, Seymour, eds. Ethical Issues in Death and Dying. Englewood Cliffs, N.J.: Prentice-Hall, 1978. p. 246-258. 10 fn. ISBN 0-13-290114-5. BE06881.
*active euthanasia; *allowing to die; justifiable killing; *morality; utilitarianism; wedge argument

Thiroux, Jacques P. Euthanasia and allowing someone to die. *In his* Ethics: Theory and Practice. Encino, Calif.: Glencoe Press, 1977. p. 121-144. 19 refs. 8 fn. BE05711.
*active euthanasia; *allowing to die; congenital defects; decision making; hospices; living wills; *morality; newborns; obligations of society; pain; *quality of life; self determination; suffering; terminal care; terminally ill; value of life; voluntary euthanasia; wedge argument

ACTIVE EUTHANASIA / NEWBORNS / SPINA BIFIDA

Reid, Robert. Spina bifida: the fate of the untreated. *Hastings Center Report* 7(4): 16-19, Aug 1977. BE05745.
*active euthanasia; allowing to die; decision making; *newborns; parents; physicians; quality of life; *selection for treatment; selective abortion; *spina bifida; values; wedge argument; withholding treatment

ACTIVE EUTHANASIA / VALUE OF LIFE

Mahoney, Edward J. The morality of terminating life vs. allowing to die. *Louvain Studies* 6(3): 256-272, Spring 1977. 23 fn. BE06851.
*active euthanasia; *allowing to die; *coma; *congenital defects; decision making; *newborns; terminal care; *terminally ill; *value of life; values; withholding treatment

ADMINISTRATORS *See under*

HEALTH CARE DELIVERY / ADMINISTRATORS / ETHICS

HUMAN EXPERIMENTATION / ADMINISTRATORS / HOSPITALS

ADMINISTRATORS *See under (cont'd.)*

MORAL OBLIGATIONS / ADMINISTRATORS / NURSING HOMES

ADOLESCENTS / PSYCHIATRY / CONTRACTS

Smith, Susannah; Miller, Derek. Contractual agreements with adolescents? *American Journal of Psychiatry* 134(10): 1160-1161, Oct 1977. 4 refs. Letter. BE06499.
*adolescents; *contracts; legal aspects; mentally ill; moral obligations; physician patient relationship; *psychiatry

ADOLESCENTS *See under*

ABORTION / ADOLESCENTS / LEGAL ASPECTS

ABORTION / ADOLESCENTS / LEGAL RIGHTS

BEHAVIORAL RESEARCH / ADOLESCENTS / SEXUALITY

CONFIDENTIALITY / MEDICAL RECORDS / ADOLESCENTS

CONTRACEPTION / ADOLESCENTS

CONTRACEPTION / ADOLESCENTS / LEGAL ASPECTS

CONTRACEPTION / ADOLESCENTS / LEGAL RIGHTS

HEALTH CARE / ADOLESCENTS / LEGAL ASPECTS

HEALTH CARE / ADOLESCENTS / LEGAL RIGHTS

HEALTH CARE / ADOLESCENTS / OKLAHOMA

INFORMED CONSENT / ADOLESCENTS / CANADA

INFORMED CONSENT / ADOLESCENTS / LEGAL ASPECTS

INFORMED CONSENT / ADOLESCENTS / LEGAL RIGHTS

INVOLUNTARY COMMITMENT / ADOLESCENTS / DUE PROCESS

INVOLUNTARY COMMITMENT / ADOLESCENTS / LEGAL ASPECTS

INVOLUNTARY STERILIZATION / ADOLESCENTS / MENTALLY RETARDED

PARENTAL CONSENT / ADOLESCENTS / LEGAL ASPECTS

PSYCHOACTIVE DRUGS / ADOLESCENTS / CALIFORNIA

PSYCHOTHERAPY / ADOLESCENTS

TERMINAL CARE / ADOLESCENTS

VOLUNTARY ADMISSION / ADOLESCENTS / MENTALLY ILL

AGED / ETHICS / DIGNITY

Kanoti, George A. Needed: a geriatric ethic. *Hospital Progress* 58(9): 104+, Sep 1977. BE05780.
*aged; *attitudes; *dignity; communication;

*ethics; health personnel; nursing homes; social interaction; *society; socioeconomic factors

AGED / SOCIETY / ATTITUDES

Kanoti, George A. Needed: a geriatric ethic. *Hospital Progress* 58(9): 104+, Sep 1977. BE05780.
*aged; *attitudes; *dignity; communication; *ethics; health personnel; nursing homes; social interaction; *society; socioeconomic factors

AGED   See under

ACTIVE EUTHANASIA / AGED / ATTITUDES
ALLOWING TO DIE / AGED
OPERANT CONDITIONING / AGED / MENTALLY ILL
RESOURCE ALLOCATION / AGED / NURSING HOMES

AGGRESSION   See under

BEHAVIORAL RESEARCH / XYY KARYOTYPE / AGGRESSION

AHA PATIENT BILL OF RIGHTS

Hunter, Thomas H., et al. What Rights Do Patients Have? [Videorecording]. Charlottesville, Va.: University of Virginia Medical Center, Health Sciences Library, 1973. 1 cassette; approx. 60 min.; sound; black and white; 3/4 in. Tape of the Medical Center Hour, 3 Dec 1973. Panel discussion moderated by Thomas H. Hunter. BE07043.
attitudes; communication; decision making; dehumanization; health care; paternalism; hospitals; patient participation; patients; *patients' rights; physician patient relationship; physicians; professional organizations; self determination; *AHA Patient Bill of Rights

AID

Anderson, Norman. Genetic engineering and artificial insemination. *In his* Issues of Life and Death: Abortion, Birth Control, Capital Punishment, Euthanasia. Downers Grove, Ill.: InterVarsity Press, 1977. p. 34-57. 43 fn. ISBN 0-87784-721-5. BE05789.
*AID; AID children; AIH; adoption; artificial genes; Christian ethics; cloning; *DNA therapy; eugenics; family relationship; genetic defects; in vitro fertilization; legal aspects; spousal consent; theology

Templeton, Allan. AID—what are the problems? *Midwife, Health Visitor and Community Nurse* 13(7): 208+, Jul 1977. 3 refs. BE06321.
*AID; AID children; attitudes; evaluation; legitimacy; married persons; physicians; semen donors

AID CHILDREN / LEGITIMACY

Life after death. *Economist* 264(6985): 23-24, 16 Jul 1977. BE06319.
*AID children; *AIH; death; fathers; *frozen semen; *legitimacy

AID / ANGLICAN CHURCH OF CANADA

Creighton, Phyllis. Artificial Insemination by Donor: A Study of Ethics, Medicine, and Law in Our Technological Society. Toronto: Anglican Book Centre, 1977. 84 p. 143 fn. ISBN 0-919030-21-1. BE06512.
*AID; AID children; Christian ethics; confidentiality; embryo transfer; frozen semen; legal liability; legitimacy; marital relationship; physicians; psychological stress; selection of subjects; semen donors; single persons; values; *Anglican Church of Canada

AID / GENETIC FATHERS / LEGAL RIGHTS

New Jersey. Juvenile and Domestic Relations Court, Cumberland County. C.M. v. C.C. 19 Jul 1977. *Atlantic Reporter, 2d Series,* 377: 821-825. BE06267.
*AID; AID children; *genetic fathers; *legal rights; legitimacy; *semen donors; single persons

AID / LEGAL ASPECTS

Dworkin, Roger. Legal issues in genetic counseling. *In* Smith, David H., ed. No Rush to Judgement: Essays on Medical Ethics. Bloomington, Ind.: Indiana University, Poynter Center, 1977. p. 44-67. 1 fn. BE05795.
*abortion; competence; constitutional law; disclosure; family members; family relationship; fetuses; *genetic counseling; *genetic screening; informed consent; *legal aspects; legal obligations; legislation; legitimacy; mandatory programs; phenylketonuria; physicians; pregnant women; privacy *AID; rights; sickle cell anemia; state interest; sterilization; *Supreme Court decisions; Roe v. Wade

AIH / FROZEN SEMEN

Cusine, D.J. Artificial insemination with the husband's semen after the husband's death. *Journal of Medical Ethics* 3(4): 163-165, Dec 1977. 10 refs. BE06251.
*AIH; children; *death; *fathers; *frozen semen; legal aspects; legitimacy; sperm; tissue banking; Great Britain

Life after death. *Economist* 264(6985): 23-24, 16 Jul 1977. BE06319.
*AID children; *AIH; death; fathers; *frozen semen; *legitimacy

ALCOHOL ABUSE   See under

CONFIDENTIALITY / MEDICAL RECORDS / ALCOHOL ABUSE
HUMAN EXPERIMENTATION / ALCOHOL ABUSE
HUMAN EXPERIMENTATION / ALCOHOL

BE = bioethics accession number   fn. = footnotes   refs. = references

ALCOHOL ABUSE *See under (cont'd.)*
ABUSE / RISKS AND BENEFITS

HUMAN EXPERIMENTATION / INFORMED CONSENT / ALCOHOL ABUSE

HUMAN EXPERIMENTATION / SELECTION OF SUBJECTS / ALCOHOL ABUSE

ALLOWING TO DIE   *See also* EUTHANASIA, PROLONGATION OF LIFE

ALLOWING TO DIE

Cassell, Eric J.   What is the function of medicine? *In* McMullin, Ernan, ed.   Death and Decision.   Boulder, Colo.: Westview Press, 1978.   p. 35-44.   American Association for the Advancement of Science Selected Symposium 18.   4 refs.   ISBN 0-89158-152-9.   BE06708.
*allowing to die; chronically ill; critically ill; goals; medicine; physician patient relationship; *self determination; terminally ill; *treatment refusal

Glover, Jonathan.   Causing Death and Saving Lives. New York: Penguin Books, 1977.   328 p.   67 fn. Bibliography: p. 299-324.   ISBN 0-1402-2003-8. BE06513.
*abortion; active euthanasia; *allowing to die; capital punishment; consequences; *decision making; double effect; infanticide; involuntary euthanasia; *killing; morality; newborns; paternalism; personhood; prolongation of life; quality of life; self determination; suicide; utilitarianism; value of life; voluntary euthansia; war; wedge argument; women's rights

Hausman, David B.   On abandoning life support: an alternative proposal. *Man and Medicine: Journal of Values and Ethics in Health Care* 2(3): 169-188, Spring 1977.   Comments on essay on p. 178-188.   13 fn.   BE05920.
*allowing to die; brain death; coma; decision making; *determination of death; *justifiable killing; organ transplantation; resource allocation; scarcity; utilitarianism; value of life; *withholding treatment; Harvard Committee on Brain Death

May, William F.   The right to die and the obligation to care: allowing to die, killing for mercy, and suicide. *In* Smith, David H., ed.   No Rush to Judgement: Essays on Medical Ethics.   Bloomington, Ind.: Indiana University, Poynter Center, 1977.   p. 68-92.   13 fn.   BE05796.
*active euthanasia; *allowing to die; Christian ethics; critically ill; *death; decision making; extraordinary treatment; hospitals; informed consent; institutional policies; moral obligations; obligations of society; physicians; prolongation of life; quality of life; resuscitation; suffering; suicide; *terminal care; terminally ill; theology; value of life; withholding treatment

May, William F.   The right to die and the obligation to care: allowing to die, killing for mercy, and suicide.

In McMullin, Ernan, ed.   Death and Decision. Boulder, Colo.: Westview Press, 1978.   p. 111-130. American Association for the Advancement of Science Selected Symposium 18.   13 fn.   ISBN 0-89158-152-9.   BE06710.
*allowing to die; euthanasia; hospitals; institutional policies; *rights; prolongation of life; self determination; *suicide; terminal care; terminally ill; theology; withholding treatment

McMullin, Ernan, ed.   Death and Decision.   Boulder, Colo.: Westview Press, 1978.   154 p.   American Association for the Advancement of Science Selected Symposium 18.   17 refs.   144 fn.   ISBN 0-89158-152-9.   BE06705.
*allowing to die; brain death; determination of death; *euthanasia; human rights; legal aspects; self determination; social adjustment; terminally ill; treatment refusal; value of life

Montague, Phillip.   The morality of active and passive euthanasia. *Ethics in Science and Medicine* 5(1): 39-45, 1978.   8 fn.   BE06908.
*active euthanasia; *allowing to die; *killing; moral obligations; philosophy

Ramsey, Paul.   Ethics at the Edges of Life: Medical and Legal Intersections.   New Haven: Yale University Press, 1978.   353 p.   The Bampton Lectures in America.   302 fn.   ISBN 0-300-02137-2.   BE06884.
*abortion; adults; *allowing to die; congenital defects; *decision making; *euthanasia; extraordinary treatment; fetuses; law; legal aspects; legislation; medicine; newborns; physician's role; quality of life; *selection for treatment; *Supreme Court decisions; terminal care; *terminally ill; *third party consent; *treatment refusal; *withholding treatment; California Natural Death Act; Commonwealth v. Edelin; *In re Quinlan; Saikewicz, Joseph

Ramsey, Paul.   "Euthanasia" and dying well enough. *In his* Ethics at the Edges of Life: Medical and Legal Intersections.   New Haven: Yale University Press, 1978.   p. 145-188.   45 fn.   ISBN 0-300-02137-2. BE06888.
active euthanasia; *adults; *allowing to die; chronically ill; coma; competence; congenital defects; death; *decision making; diagnosis; drugs; *euthanasia; *extraordinary treatment; human rights; legal aspects; medical ethics; moral obligations; newborns; pain; physicians; prognosis; quality of life; religious ethics; *selection for treatment; social interaction; spina bifida; *standards; terminal care; *terminally ill; theology; *third party consent; *treatment refusal; value of life; withholding treatment; McCormick, Richard A.; Veatch, Robert M.

Reed, Nicholas.   Recent thinking about voluntary euthanasia. *New Humanist* 92(5): 173-174, Jan-Feb 1977.   BE06290.
active euthanasia; *allowing to die; family members; extraordinary treatment; legislation; living

---

BE = bioethics accession number   fn. = footnotes   refs. = references

wills; physicians; terminally ill; treatment refusal; *voluntary euthanasia; withholding treatment

Rosner, Fred. The use and abuse of heroic measures to prolong dying. *Journal of Religion and Health* 17(1): 8-18, Jan 1978. 45 fn. BE06750.
*allowing to die; brain death; determination of death; *extraordinary treatment; Jewish ethics; legal aspects; patients; physician's role; prolongation of life; Roman Catholic ethics; self determination; terminally ill; value of life

Sullivan, Thomas D. Active and passive euthanasia: an impertinent distinction? *Human Life Review* 3(3): 40-46, Summer 1977. 5 fn. BE05680.
*active euthanasia; *allowing to die; extraordinary treatment; intention; killing; terminally ill; withholding treatment; Rachels, James

Trammell, Richard L. The presumption against taking life. *Journal of Medicine and Philosophy* 3(1): 53-67, Mar 1978. 13 refs. 7 fn. BE06829.
*active euthanasia; *allowing to die; *decision making; extraordinary treatment; intention; killing; morality; quality of life; value of life

Wakin, Edward. Is the right-to-die wrong? *U.S. Catholic* 43(3): 6-12, Mar 1978. BE06831.
active euthanasia; *allowing to die; case reports; coma; determination of death; extraordinary treatment; living wills; *prolongation of life; religious ethics; terminally ill; withholding treatment

Weir, Robert F., ed. Ethical Issues in Death and Dying. New York: Columbia University Press, 1977. 405 p. References and footnotes included. ISBN 0-231-04307-4. BE06696.
*allowing to die; brain death; cancer; congenital defects; *determination of death; diagnosis; *disclosure; *euthanasia; legal aspects; newborns; physicians; *suicide; *terminally ill; treatment refusal

ALLOWING TO DIE / AGED

Masterman, John. How the right to die could improve the quality of life. *Times (London),* 20 Oct 1976, p. 10. BE06615.
*aged; *allowing to die; chronically ill; suicide

Raymond, Edward A. A plea to let the old die peaceably. *New York Times,* 25 Sep 1977, sect. 4, p. 15. BE06399.
*aged; *allowing to die; costs and benefits; *prolongation of life; terminally ill

ALLOWING TO DIE / ANGLICAN CHURCH OF CANADA

Trumbull, Robert. Anglican report in Canada leans toward euthanasia. *New York Times,* 28 Jul 1977, p. A12. BE06061.
*allowing to die; *mentally retarded; newborns; *organizational policies; *terminally ill; *withholding treatment; *Anglican Church of Canada

ALLOWING TO DIE / BRAIN DEATH

Fowler, A.W. Switching off. *British Medical Journal* 2(6096): 1223, 5 Nov 1977. 1 fn. Letter. BE06112.
*allowing to die; *brain death; *decision making; family members; physicians; withholding treatment

ALLOWING TO DIE / BRAIN DEATH / LAW ENFORCEMENT

Massachusetts. Supreme Judicial Court, Suffolk. Commonwealth v. Golston. 26 Aug 1977. *North Eastern Reporter, 2d Series,* 366: 744-752. 12 fn. BE06263.
*allowing to die; *brain death; killing; *law enforcement; withholding treatment

ALLOWING TO DIE / BRAIN DEATH / LEGAL LIABILITY

Question of death goes to courts to protect hospitals. *Hospitals* 51(3): 20, 1 Feb 1977. BE06291.
*allowing to die; *brain death; hospitals; *legal liability; physicians; withholding treatment

ALLOWING TO DIE / BRAIN PATHOLOGY / LEGAL ASPECTS

Mueller, Daniel M. Involuntary passive euthanasia of brain-stem-damaged patients: the need for legislation—an analysis and a proposal. *San Diego Law Review* 14(5): 1277-1298, Jul 1977. 100 fn. BE05965.
*allowing to die; brain pathology; brain death; determination of death; killing; law enforcement; *legal aspects; legal liability; living wills; model legislation; patients; *physicians; *withholding treatment

ALLOWING TO DIE / BURN PATIENTS

Pratt, David S., et al. Autonomy for severely burned patients. *New England Journal of Medicine* 297(21): 1182-1183, 24 Nov 1977. Letter. BE06097.
*allowing to die; *burn patients; costs and benefits; critically ill; *extraordinary treatment; prognosis; *self determination; *terminally ill; withholding treatment

ALLOWING TO DIE / CALIFORNIA NATURAL DEATH ACT

Bleich, J. David. Legal immorality not personal freedom. *Sh'ma* 7(132): 101-102, 15 Apr 1977. BE05540.
*allowing to die; Jewish ethics; legislation; living wills; self determination; state government; *value of life; *California Natural Death Act

Horowitz, Elliot. The California Natural Death Act and us. *Sh'ma* 7(132): 93-94, 15 Apr 1977.

BE = bioethics accession number   fn. = footnotes   refs. = references

BE05535.
*allowing to die; legislation; living wills; physician's role; self determination; state government; *terminally ill; withholding treatment; *California Natural Death Act

Lebacqz, Karen. On 'natural death'. *Hastings Center Report* 7(2): 14, Apr 1977. 1 fn. BE05563.
*allowing to die; legislation; living wills; state government; terminally ill; treatment refusal; withholding treatment; *California Natural Death Act

Ramsey, Paul. The California Natural Death Act. *In his* Ethics at the Edges of Life: Medical and Legal Intersections. New Haven: Yale University Press, 1978. p. 318-332. 22 fn. ISBN 0-300-02137-2. BE06893.
*allowing to die; extraordinary treatment; involuntary euthanasia; legal aspects; legal liability; *legislation; living wills; physician's role; physicians; prolongation of life; Roman Catholic ethics; socioeconomic factors; state government; terminally ill; third party consent; withholding treatment; *California Natural Death Act; United States

Winslade, William J. Thoughts on technology and death: an appraisal of California's Natural Death Act. *DePaul Law Review* 26(4): 717-742, Summer 1977. 71 fn. BE06491.
*allowing to die; attitudes to death; competence; evaluation; legislation; living wills; physician's role; state government; terminally ill; treatment refusal; withholding treatment; *California Natural Death Act

ALLOWING TO DIE / COMA

Schwager, Robert L. Life, death, and the irreversibly comatose. *In* Beauchamp, Tom L.; Perlin, Seymour, eds. Ethical Issues in Death and Dying. Englewood Cliffs, N.J.: Prentice-Hall, 1978. p. 38-50. 32 fn. ISBN 0-13-290114-5. BE06880.
*allowing to die; *brain death; *coma; determination of death; *extraordinary treatment; patients; *withholding treatment

ALLOWING TO DIE / COMA / LEGAL ASPECTS

McKenney, Edward J. Death and dying in Tennessee. *Memphis State University Law Review* 7(4): 503-554, Summer 1977. 343 fn. BE06307.
*allowing to die; *brain death; *coma; *determination of death; euthanasia; *extraordinary treatment; family members; judicial action; *legal aspects; legal liability; *legislation; *living wills; organ transplantation; physician patient relationship; physician's role; privacy; state government; rights; state interest; terminally ill; third party consent; treatment refusal; *withholding treatment; *Dockery v. Dockery; In re Quinlan; *Tennessee; *Tennessee Natural Death Act

ALLOWING TO DIE / CONGENITAL DEFECTS / CONSTITUTIONAL LAW

Law, Sylvia. Constitutional right to privacy. *In* Swinyard, Chester A., ed. Decision Making and the Defective Newborn. Proceedings of a Conference on Spina Bifida and Ethics. Springfield, Ill.: Charles C. Thomas, 1978. p. 384-395. 7 fn. ISBN 0-398-03662-4. BE06958.
abortion; *allowing to die; *congenital defects; *constitutional law; contraception; decision making; equal protection; legal rights; legislation; morality; *newborns; parent child relationship; parents; personhood; physicians; *privacy; religious beliefs; *Supreme Court decisions; *withholding treatment

ALLOWING TO DIE / DECISION MAKING

Hellegers, André E., et al. What would 'reasonable' people do? *New York Times,* 18 May 1977, p. A24. Letter. BE06654.
*allowing to die; *decision making; social interaction; standards; Quinlan, Karen

Hunter, Thomas H.; Smith, William. Attitudes Toward Death and Dying. [Videorecording]. Charlottesville, Va.: University of Virginia Medical Center, Health Sciences Library, 1974. 1 cassette; 30 min.; sound; color; 3/4 in. Tape of the program "People, Places, Things," 27 Feb 1974, produced by Jefferson Cable Corp., Channel 10, Charlottesville. BE07045.
*allowing to die; attitudes to death; autopsies; brain death; *decision making; disclosure; economics; family members; living wills; *moral obligations; organ donation; *physician's role; *physicians; prolongation of life; *terminally ill

McCormick, Richard A.; Hellegers, André E. The specter of Joseph Saikewicz: mental incompetence and the law. *America* 138(12): 257-260, 1 Apr 1978. BE06849.
aged; *allowing to die; *competence; *decision making; drugs; family members; *judicial action; leukemia; *mentally retarded; physicians; prolongation of life; terminally ill; *withholding treatment; *Saikewicz, Joseph

Who shall make the ultimate decision? *New York Times,* 2 May 1977, p. 32. Editorial. BE06653.
*allowing to die; *decision making; physicians; *withholding treatment; Quinlan, Karen

ALLOWING TO DIE / HOSPITALS / CALIFORNIA NATURAL DEATH ACT

Friedman, Emily. California hospitals design Natural Death Act procedures. *Hospitals* 51(22): 62-65, 16 Nov 1977. BE06354.
*allowing to die; *hospitals; *institutional policies; living wills; physician's role; records; renal dialysis; *California Natural Death Act

---

BE = bioethics accession number   fn. = footnotes   refs. = references

ALLOWING TO DIE / HUMANISM

Csank, James F. The right to a natural death. *Human Life Review* 4(1): 44-54, Winter 1978. 19 fn. BE06725.
  abortion; *allowing to die; *coma; decision making; extraordinary treatment; *humanism; *judicial action; law; legal guardians; physician's role; privacy; state interest; terminally ill; third party consent; withholding treatment; *In re Quinlan; New Jersey Supreme Court

ALLOWING TO DIE / IN RE QUINLAN

Becker, Douglas; Fleming, Robert; Overstreet, Rebecca. The legal aspects of the right to die: before and after the *Quinlan* decision. *Kentucky Law Journal* 65(4): 823-879, 1977. 281 fn. BE06629.
  *allowing to die; competence; constitutional law; criminal law; euthanasia; informed consent; *judicial action; legal guardians; legal liability; legal rights; legislation; living wills; model legislation; organ donation; physicians; *privacy; religious beliefs; self determination; state interest; *terminally ill; third party consent; *treatment refusal; *withholding treatment; California Natural Death Act; *In re Quinlan

Coburn, Daniel R. In re Quinlan: a practical overview. *Arkansas Law Review* 31(1): 59-74, Spring 1977. 79 fn. BE05948.
  *allowing to die; brain death; decision making; fathers; *judicial action; legal guardians; legal rights; living wills; public opinion; *terminally ill; *withholding treatment; *In re Quinlan

Csank, James F. The right to a natural death. *Human Life Review* 4(1): 44-54, Winter 1978. 19 fn. BE06725.
  abortion; *allowing to die; *coma; decision making; extraordinary treatment; *humanism; *judicial action; law; legal guardians; physician's role; privacy; state interest; terminally ill; third party consent; withholding treatment; *In re Quinlan; New Jersey Supreme Court

Gold, Jay A. The Quinlan case: a review of two books. [Title provided]. *American Journal of Law and Medicine* 3(1): 89-94, Spring 1977. BE05919.
  *allowing to die; brain death; *determination of death; personhood; state courts; terminally ill; *withholding treatment; *In re Quinlan; New Jersey Supreme Court; *Quinlan, Karen

Horan, Dennis J. The Quinlan case. *Linacre Quarterly* 44(2): 168-176, May 1977. 18 fn. BE05618.
  *allowing to die; *decision making; informed consent; legal liability; physicians; privacy; resuscitation; terminally ill; treatment refusal; *withholding treatment; *In re Quinlan; *New Jersey Supreme Court

ALLOWING TO DIE / INSTITUTIONALIZED PERSONS / MENTALLY RETARDED

Massachusetts. Supreme Judicial Court. Jones v. Saikewicz. 9 Jul 1976. Unpublished court decision. 3 p. Docket No. SJC-711, Appeals Court No. 76-369, Hampshire Probate No. 45596. BE06209.
  *allowing to die; *institutionalized persons; *mentally retarded; *withholding treatment; *Saikewicz, Joseph

Massachusetts. Supreme Judicial Court, Hampshire. Superintendent of Belchertown v. Saikewicz. 28 Nov 1977. *North Eastern Reporter, 2d Series,* 370: 417-435. 20 fn. BE06210.
  *allowing to die; competence; drugs; *institutionalized persons; *legal rights; leukemia; *mentally retarded; privacy; prolongation of life; state interest; terminally ill; treatment refusal; *withholding treatment; *Saikewicz, Joseph

ALLOWING TO DIE / JUDICIAL ACTION

Annas, George J. The incompetent's right to die: the case of Joseph Saikewicz. *Hastings Center Report* 8(1): 21-23, Feb 1978. 1 fn. BE06769.
  *allowing to die; competence; ethics committees; *judicial action; *legal rights; leukemia; mentally retarded; *self determination; state interest; terminally ill; *treatment refusal; *withholding treatment; *Saikewicz, Joseph

Becker, Douglas; Fleming, Robert; Overstreet, Rebecca. The legal aspects of the right to die: before and after the *Quinlan* decision. *Kentucky Law Journal* 65(4): 823-879, 1977. 281 fn. BE06629.
  *allowing to die; competence; constitutional law; criminal law; euthanasia; informed consent; *judicial action; legal guardians; legal liability; legal rights; legislation; living wills; model legislation; organ donation; physicians; *privacy; religious beliefs; self determination; state interest; *terminally ill; third party consent; *treatment refusal; *withholding treatment; California Natural Death Act; *In re Quinlan

Rothenberg, Leslie S. Demands for life and requests for death: the judicial dilemma. *In* McMullin, Ernan, ed. Death and Decision. Boulder, Colo.: Westview Press, 1978. p. 131-154. American Association for the Advancement of Science Selected Symposium 18. 66 fn. ISBN 0-89158-152-9. BE06711.
  *allowing to die; blood transfusions; decision making; Jehovah's Witnesses; *judicial action; legal aspects; surgery; *treatment refusal; withholding treatment

ALLOWING TO DIE / JUDICIAL ACTION / MASSACHUSETTS SUPREME JUDICIAL COURT

Curran, William J. The Saikewicz decision. *New England Journal of Medicine* 298(9): 499-500, 2 Mar 1978. 5 refs. BE06965.

BE = bioethics accession number   fn. = footnotes   refs. = references

*allowing to die; *decision making; ethics committees; family members; *judicial action; mentally handicapped; physicians; *state courts; *terminally ill; withholding treatment; *Massachusetts Supreme Judicial Court; *Saikewicz, Joseph

Relman, Arnold S. The Saikewicz decision: judges as physicians. *New England Journal of Medicine* 298(9): 508-509, 2 Mar 1978. 6 refs. Editorial. BE06966.
*allowing to die; *decision making; family members; *judicial action; mentally handicapped; physicians; *state courts; *terminally ill; withholding treatment; *Massachusetts Supreme Judicial Court; *Saikewicz, Joseph

ALLOWING TO DIE / KIDNEY DISEASES

Some kidney patients over 45 'not treated'. *Times (London)*, 19 May 1977, p. 4. BE06441.
*allowing to die; *kidney diseases; renal dialysis; *resource allocation; scarcity; *withholding treatment

ALLOWING TO DIE / KILLING

Maguire, Daniel C. Death and the moral domain. *St. Luke's Journal of Theology* 20(3): 197-216, Jun 1977. 58 fn. BE06305.
*active euthanasia; *allowing to die; *Christian ethics; death; ethicists; *ethics; *euthanasia; justifiable killing; *killing; legislation; moral obligations; pain; philosophy; suffering; suicide; value of life; *Church of England

ALLOWING TO DIE / LEGAL ASPECTS

Becker, Douglas; Fleming, Robert; Overstreet, Rebecca. The legal aspects of the right to die: before and after the *Quinlan* decision. *Kentucky Law Journal* 65(4): 823-879, 1977. 281 fn. BE06629.
*allowing to die; competence; constitutional law; criminal law; euthanasia; informed consent; *judicial action; legal guardians; legal liability; legal rights; legislation; living wills; model legislation; organ donation; physicians; *privacy; religious beliefs; self determination; state interest; *terminally ill; third party consent; *treatment refusal; *withholding treatment; California Natural Death Act; *In re Quinlan

Huber, Robert. Legal considerations of cardiopulmonary resuscitation. *In* Safar, P., ed. Advances in Cardiopulmonary Resuscitation. New York: Springer-Verlag, 1977. p. 246-249. 2 refs. ISBN 0-387-90234-1. BE06574.
*allowing to die; *legal aspects; informed consent; *legal liability; medical staff; physicians; *resuscitation; surgery; withholding treatment

Kennedy, Ian M. Switching off life support machines: the legal implications. *Criminal Law Review* 1977: 443-452, Aug 1977. 33 fn. BE06493.
*allowing to die; brain death; chronically ill; coma; criminal law; informed consent; *legal aspects; re-

suscitation; terminally ill; treatment refusal; withholding treatment

Sharpe, Gilbert. Listening for the death-bells. *Canadian Nurse* 74(1): 20-23, Jan 1978. BE06912.
*allowing to die; decision making; family members; hospitals; *legal aspects; legal liability; nurses; physicians; resuscitation; terminally ill; treatment refusal; withholding treatment

ALLOWING TO DIE / LEGAL GUARDIANS

Sklar, Zachary; Coburn, Daniel. R. Life against death: an interview with Karen Ann Quinlan's guardian. *Juris Doctor* 7(11): 27-31, Dec 1977. BE06369.
*allowing to die; fathers; *decision making; law; *legal guardians; mass media; physicians; terminally ill; *Quinlan, Karen

ALLOWING TO DIE / LEGISLATION

Beresford, H. Richard. The Quinlan decision: problems and legislative alternatives. *Annals of Neurology* 2(1): 74-81, Jul 1977. 26 fn. BE05819.
*allowing to die; brain pathology; coma; decision making; determination of death; ethics committees; family members; judicial action; killing; legislation; living wills; physicians; terminally ill; withholding treatment; *In re Quinlan

Goldberg, Joel H. The extraordinary confusion over "the right to die". *Medical Economics* 54(1): 121-122, 10 Jan 1977. BE05903.
*allowing to die; euthanasia; extraordinary treatment; *legislation; living wills; *state government; terminally ill

Raible, Jane A. The right to refuse treatment and natural death legislation. *Medicolegal News* 5(4): 6-8+, Fall 1977. 9 fn. BE06245.
*adults; *allowing to die; aged; attitudes; brain death; children; competence; decision making; legal aspects; legal rights; *legislation; *living wills; medical records; physicians; public opinion; pregnant women; religion; state government; terminal care; *terminally ill; treatment refusal; withholding treatment

Right-to-die legislation. *Lancet* 2(8033): 347, 13 Aug 1977. BE05764.
*allowing to die; *legislation; living wills; *state government

Spinella, Nicholas A. Update on opposition to death-with-dignity legislation. *Hospital Progress* 58(7): 70-72+, Jul 1977. 1 fn. BE05736.
*allowing to die; decision making; determination of death; extraordinary treatment; hospitals; *legal aspects; legal liability; *legislation; living wills; moral obligations; physician's role; physicians; political activity; state government; terminally ill; withholding treatment

Tobin, Charles. A statement on legislation concerning death and dying. *Catholic Mind* 75(1312): 7-10, Apr 1977. BE05584.

BE = bioethics accession number   fn. = footnotes   refs. = references

*allowing to die; *decision making; extraordinary treatment; determination of death; *legislation; living wills; physician's role; professional organizations; prolongation of life; public policy; terminally ill; wedge argument; New York State Catholic Conference

Veatch, Robert M. Death and dying: the legislative options. *Hastings Center Report* 7(5): 5-8, Oct 1977. BE06074.
    active euthanasia; *allowing to die; competence; *decision making; family members; legal guardians; *legislation; *living wills; physicians; model legislation; state government; *terminally ill; *treatment refusal; withholding treatment

ALLOWING TO DIE / LEGISLATION / SOCIAL IMPACT

Friedman, Emily. 'Natural death' laws cause hospitals few problems. *Hospitals* 52(10): 124+, 16 May 1978. 7 refs. BE06969.
    *allowing to die; hospitals; *legislation; *living wills; medical records; physicians; *social impact; *state government

ALLOWING TO DIE / LEGISLATION / SOUTH CAROLINA

Ackerman, Terrence F. "Death with dignity" legislation in South Carolina: an appraisal. *Journal of the South Carolina Medical Association* 73(8): 364-366, Aug 1977. BE05832.
    *allowing to die; *competence; decision making; family members; *legislation; living wills; physicians; prolongation of life; *terminally ill; treatment refusal; *withholding treatment; *South Carolina

ALLOWING TO DIE / LEGISLATION / TENNESSEE

McKenney, Edward J. Death and dying in Tennessee. *Memphis State University Law Review* 7(4): 503-554, Summer 1977. 343 fn. BE06307.
    *allowing to die; *brain death; *coma; *determination of death; euthanasia; *extraordinary treatment; family members; judicial action; *legal aspects; legal liability; *legislation; *living wills; organ transplantation; physician patient relationship; physician's role; privacy; state government; rights; state interest; terminally ill; third party consent; treatment refusal; *withholding treatment; *Dockery v. Dockery; In re Quinlan; *Tennessee; *Tennessee Natural Death Act

ALLOWING TO DIE / LIVING WILLS

Colen, B.D. Quinlan alive one year after leaving respirator. *Washington Post,* 1 Apr 1977, p. C6. BE06008.
    *allowing to die; legislation; *living wills; state government; withholding treatment; Quinlan, Karen

Death laws too new to predict legal consequences for physicians. *American Medical News* 20(25): 11-12, 20 Jun 1977. BE05632.
    *allowing to die; *legal liability; legislation; *living wills; *physicians; state government; terminally ill

Hollowell, Edward E. The right to die: how legislation is defining the right. *Journal of Practical Nursing* 27(10): 20-21+, Oct 1977. BE06347.
    *allowing to die; *legislation; *living wills; *state government; *terminally ill

ALLOWING TO DIE / LIVING WILLS / ARKANSAS

Arkansas. Death with dignity. *Arkansas Statutes Annotated,* Chapter 38, Sects. 82-3801 to 82-3804. Approved 30 Mar 1977. 2 p. BE06655.
    *allowing to die; legal liability; *living wills; third party consent; treatment refusal; *Arkansas

ALLOWING TO DIE / LIVING WILLS / LEGAL ASPECTS

Akers, Stephen R. The living will: already a practical alternative. *Texas Law Review* 55(4): 665-717, Mar 1977. 306 fn. BE05911.
    *allowing to die; attitudes; competence; extraordinary treatment; family members; hospitals; *legal aspects; legal liability; *living wills; model legislation; physicians; public opinion; social impact; standards; terminally ill; treatment refusal; withholding treatment

ALLOWING TO DIE / LIVING WILLS / LEGISLATION

Alsofrom, Judy. States tackle flood of 'right-to-die' bills. *American Medical News* 20(25): 11-12, 20 Jun 1977. BE05631.
    *allowing to die; *legislation; *living wills; physician's role; *state government; terminally ill; withholding treatment

Evans, Franklin J. The right to die—a basic constitutional right. *Journal of Legal Medicine* 5(8): 17-20, Aug 1977. 26 fn. BE05754.
    active euthanasia; *allowing to die; attitudes; *legislation; *living wills; physicians; wedge argument

Zucker, Karin W. Legislatures provide for death with dignity. *Journal of Legal Medicine* 5(8): 21-24, Aug 1977. 7 fn. BE05770.
    *allowing to die; *legislation; *living wills; *state government; terminally ill; withholding treatment

ALLOWING TO DIE / LIVING WILLS / NEVADA

Nevada. Withholding or withdrawal of life-sustaining procedures. *Nevada Revised Statutes,* Sects.

BE = bioethics accession number   fn. = footnotes   refs. = references

449.540-449.690, 1977. 5 p. BE06658.
*allowing to die; legal liability; *living wills; physicians; terminally ill; *withholding treatment; *Nevada

ALLOWING TO DIE / LIVING WILLS / NORTH CAROLINA

North Carolina. Right to natural death; brain death. *General Statutes of North Carolina,* Article 23, Sect. 90-320 to 90-322. Effective 1 Jul 1977. 5 p. BE06218.
*allowing to die; *brain death; extraordinary treatment; *living wills; *terminally ill; *withholding treatment; *North Carolina

ALLOWING TO DIE / LIVING WILLS / TEXAS

Texas. Natural death act. *Texas Revised Civil Statutes Annotated (Vernon),* Article 4590h. Effective 29 Aug 1977. 5 p. BE06219.
*allowing to die; health personnel; legal liability; *living wills; physicians; *terminally ill; *withholding treatment; *Texas

ALLOWING TO DIE / MENTALLY RETARDED / COMPETENCE

Ramsey, Paul. The strange case of Joseph Saikewicz. *In his* Ethics at the Edges of Life: Medical and Legal Intersections. New Haven: Yale University Press, 1978. p. 300-317. 9 fn. Case study. ISBN 0-300-02137-2. BE06892.
aged; *allowing to die; *competence; *decision making; drugs; *judicial action; legal guardians; leukemia; *mentally retarded; physicians; prognosis; prolongation of life; *quality of life; standards; *terminally ill; *third party consent; treatment refusal; *withholding treatment; *In re Quinlan; *Massachusetts Supreme Judicial Court; *Saikewicz, Joseph

ALLOWING TO DIE / MENTALLY RETARDED / JUDICIAL ACTION

McCormick, Richard A.; Hellegers, André E. The specter of Joseph Saikewicz: mental incompetence and the law. *America* 138(12): 257-260, 1 Apr 1978. BE06849.
aged; *allowing to die; *competence; *decision making; drugs; family members; *judicial action; leukemia; *mentally retarded; physicians; prolongation of life; terminally ill; *withholding treatment; *Saikewicz, Joseph

ALLOWING TO DIE / MENTALLY RETARDED / LEGAL ASPECTS

Ayd, Frank J. Treatment for the terminally ill incompetent: who decides—courts or physicians? *Medical-Moral Newsletter* 15(4): 13-16, Apr 1978. BE06836.
aged; *allowing to die; competence; *decision making; drugs; judicial action; *legal aspects; leukemia;
*mentally retarded; *terminally ill; treatment refusal; withholding treatment; *Saikewicz, Joseph

Brant, Jonathan. The right to die in peace: substituted consent and the mentally incompetent. *Suffolk University Law Review* 11(4): 959-973, Spring 1977. 105 fn. BE06234.
aged; *allowing to die; children; *competence; *decision making; judicial action; *legal aspects; legal guardians; legal rights; leukemia; *mentally retarded; organ donation; parental consent; privacy; quality of life; state interest; suffering; terminally ill; *third party consent; treatment refusal; *withholding treatment; *Jones v. Saikewicz

Schultz, Stephen; Swartz, William; Appelbaum, Judith C. Deciding right-to-die cases involving incompetent patients: Jones v. Saikewicz. *Suffolk University Law Review* 11(4): 936-958, Spring 1977. 95 fn. BE06233.
*allowing to die; competence; constitutional law; *decision making; institutionalized persons; *legal aspects; legal guardians; legal rights; leukemia; *mentally retarded; physician's role; privacy; quality of life; right to treatment; state courts; *third party consent; value of life; *withholding treatment; *Jones v. Saikewicz

ALLOWING TO DIE / MORAL OBLIGATIONS

Clouser, K. Danner. Allowing or causing: another look. *Annals of Internal Medicine* 87(5): 622-624, Nov 1977. BE06109.
active euthanasia; *allowing to die; extraordinary treatment; living wills; *moral obligations; physicians; *terminally ill; *withholding treatment

Russell, Bruce. On the relative strictness of negative and positive duties. *American Philosophical Quarterly* 14(2): 87-97, Apr 1977. 28 fn. BE06300.
*allowing to die; alternatives; double effect; *killing; *moral obligations; philosophy

ALLOWING TO DIE / MORALITY

Beauchamp, Tom L. A reply to Rachels on active and passive euthanasia. *In* Beauchamp, Tom L.; Perlin, Seymour, eds. Ethical Issues in Death and Dying. Englewood Cliffs, N.J.: Prentice-Hall, 1978. p. 246-258. 10 fn. ISBN 0-13-290114-5. BE06881.
*active euthanasia; *allowing to die; justifiable killing; *morality; utilitarianism; wedge argument

Thiroux, Jacques P. Euthanasia and allowing someone to die. *In his* Ethics: Theory and Practice. Encino, Calif.: Glencoe Press, 1977. p. 121-144. 19 refs. 8 fn. BE05711.
*active euthanasia; *allowing to die; congenital defects; decision making; hospices; living wills; *morality; newborns; obligations of society; pain; *quality of life; self determination; suffering; terminal care; terminally ill; value of life; voluntary euthanasia; wedge argument

---

BE = bioethics accession number   fn. = footnotes   refs. = references

ALLOWING TO DIE / NEW JERSEY SUPREME COURT / IN RE QUINLAN

Ramsey, Paul. In the matter of *Quinlan*. *In his* Ethics at the Edges of Life: Medical and Legal Intersections. New Haven: Yale University Press, 1978. p. 268-299. 30 fn. Case study. ISBN 0-300-02137-2. BE06891.
*allowing to die; coma; decision making; ethics committees; *extraordinary treatment; family members; intention; *involuntary euthanasia; killing; *judicial action; law; legal aspects; legal guardians; legal liability; physician's role; physicians; privacy; prognosis; prolongation of life; *quality of life; religion; *religious beliefs; Roman Catholic ethics; state courts; state interest; terminally ill; *third party consent; treatment refusal; withholding treatment; *In re Quinlan; *New Jersey Supreme Court

ALLOWING TO DIE / NEWBORNS

Anderson, Norman. The prolongation of life, transplant surgery, euthanasia and suicide. *In his* Issues of Life and Death: Abortion, Birth Control, Capital Punishment, Euthanasia. Downers Grove, Ill.: InterVarsity Press, 1977. p. 85-107. 19 fn. ISBN 0-87784-721-5. BE05791.
abortion; active euthanasia; adults; *allowing to die; brain death; congenital defects; decision making; determination of death; *euthanasia; fetuses; hearts; infanticide; *kidneys; legal aspects; morality; *newborns; organ donors; *organ transplantation; physicians; prenatal diagnosis; prolongation of life; suffering; suicide; *terminally ill; voluntary euthanasia

Fost, Norman. Proxy consent for seriously ill newborns. *In* Smith, David H., ed. No Rush to Judgement: Essays on Medical Ethics. Bloomington, Ind.: Indiana University, Poynter Center, 1977. p. 1-17. 23 refs. BE05793.
*allowing to die; congenital defects; costs and benefits; *decision making; ethics; ethics committees; informed consent; involuntary euthanasia; living wills; mentally retarded; *newborns; normative ethics; parental consent; parents; physicians; resource allocation; *third party consent; voluntary euthanasia

ALLOWING TO DIE / NEWBORNS / CHRISTIAN ETHICS

Tripp, John. Ethical problems in paediatrics. *In* Vale, J.A., ed. Medicine and the Christian Mind. London: Christian Medical Fellowship Publications, 1975. p. 106-113. 11 refs. ISBN 85111-956-5. BE07073.
*allowing to die; *Christian ethics; *children; *congenital defects; family problems; *human experimentation; *newborns; physicians; withholding treatment

ALLOWING TO DIE / NEWBORNS / CONGENITAL DEFECTS

Avery, Gordon B. The morality of drastic intervention. *In* Avery, Gordon B., ed. Neonatology: Pathophysiology and Management of the Newborn. Philadelphia: J.B. Lippincott, 1975. p. 11-14. 10 refs. ISBN 0-397-50331-8. BE07076.
*allowing to die; *decision making; *congenital defects; family members; *newborns; physicians; psychological stress

Etzioni, Amitai. The next Quinlan case. *Human Behavior* 5(4): 10-11, Apr 1976. BE06203.
*allowing to die; *congenital defects; decision making; *newborns; public participation; terminally ill

Hemphill, Michael; Freeman, John M. Ethical aspects of care of the newborn with serious neurological disease. *Clinics in Perinatology* 4(1): 201-209, Mar 1977. 27 refs. BE05943.
*allowing to die; *congenital defects; *decision making; ethics committees; mentally retarded; *newborns; parents; patient advocacy; physically handicapped; physicians; quality of life; *withholding treatment

Milunsky, Aubrey. Should the defective newborn be allowed to die? *In his* Know Your Genes. Boston: Houghton Mifflin, 1977. p. 289-297. ISBN 0-395-25374-8. BE05887.
*allowing to die; *congenital defects; *decision making; ethics committees; genetic defects; legal aspects; *newborns; parents; physicians; state government; *withholding treatment

Pence, Gregory E. Between cold logic and naive compassion—on allowing defective babies to die. *Bioethics Digest* 2(7): 1-7, Nov 1977. 23 fn. BE06248.
*allowing to die; *congenital defects; decision making; ethics; family members; *newborns; parents; physicians; *quality of life; self concept; socioeconomic factors; suffering; withholding treatment

Ramsey, Paul. The benign neglect of defective infants. *In his* Ethics at the Edges of Life: Medical and Legal Intersections. New Haven: Yale University Press, 1978. p. 189-227. 54 fn. ISBN 0-300-02137-2. BE06889.
active euthanasia; *allowing to die; Christian ethics; *congenital defects; ethics; extraordinary treatment; genetic defects; human characteristics; human experimentation; involuntary euthanasia; justice; killing; moral obligations; medical ethics; *newborns; pain; parents; *patient care; personhood; physicians; prognosis; quality of life; selection for treatment; self concept; social interaction; socioeconomic factors; spina bifida; *terminally ill; theology; third party consent; *value of life; Jonas, Hans; McCormick, Richard A.

Ramsey, Paul. An ingathering of other reasons for neonatal infanticide. *In his* Ethics at the Edges of Life:

BE = bioethics accession number   fn. = footnotes   refs. = references

Medical and Legal Intersections. New Haven: Yale University Press, 1978. p. 228-267. 36 fn. ISBN 0-300-02137-2. BE06890.

abortion; *allowing to die; *congenital defects; *costs and benefits; ethical analysis; fetal development; fetuses; health care; human equality; *intensive care units; killing; medical ethics; *newborns; normality; personhood; prematurity; prognosis; public policy; *quality of life; *resource allocation; resuscitation; *selection for treatment; socioeconomic factors; Supreme Court decisions; *value of life; withholding treatment; *Roe v. Wade; *Sanoma Conference

Shaw, Anthony. The ethics of proxy consent. *In* Swinyard, Chester A., ed. Decision Making and the Defective Newborn. Proceedings of a Conference on Spina Bifida and Ethics. Springfield, Ill.: Charles C. Thomas, 1978. p. 589-597. 11 refs. ISBN 0-398-03662-4. BE06964.

adolescents; *allowing to die; blood transfusions; children; *congenital defects; *decision making; family relationship; informed consent; legal aspects; *newborns; parental consent; *patient care; *physician's role; quality of life; society; *third party consent; treatment refusal; withholding treatment

Shaw, Anthony; Randolph, Judson G.; Manard, Barbara. Ethical issues in pediatric surgery: a national survey of pediatricians and pediatric surgeons. *Pediatrics* 60(4-Part 2): 588-599, Oct 1977. BE06587.

abortion; *allowing to die; *attitudes; *congenital defects; *decision making; Down's syndrome; extraordinary treatment; legal aspects; *newborns; parental consent; parents; *pediatrics; *physicians; prolongation of life; spina bifida; *surgery; *treatment refusal; *withholding treatment

Todres, I. David, et al. Pediatricians' attitudes affecting decision-making in defective newborns. *Pediatrics* 60(2): 197-201, Aug 1977. 13 refs. BE06328.

age; *allowing to die; *attitudes; case reports; *congenital defects; decision making; Down's syndrome; *newborns; parental consent; *patient care; *pediatrics; religion; Roman Catholicism; spina bifida; surgery; treatment refusal; withholding treatment

Tripp, John. Ethical problems in paediatrics. *In* Vale, J.A., ed. Medicine and the Christian Mind. London: Christian Medical Fellowship Publications, 1975. p. 106-113. 11 refs. ISBN 85111-956-5. BE07073.

*allowing to die; *Christian ethics; *children; *congenital defects; family problems; *human experimentation; *newborns; physicians; withholding treatment

ALLOWING TO DIE / NEWBORNS / DOWN'S SYNDROME

Diamond, Eugene F. The deformed child's right to life. *In* Horan, Dennis J.; Mall, David, eds. Death, Dy-

ing, and Euthanasia. Washington: University Publications of America, 1977. p. 127-138. 10 refs. ISBN 0-89093-139-9. BE05687.

*allowing to die; case reports; *congenital defects; *Down's syndrome; mentally retarded; moral obligations; *newborns; physician's role; quality of life; rights; selective abortion; value of life; *withholding treatment

ALLOWING TO DIE / NEWBORNS / GENETIC DEFECTS

Pauli, Richard M.; Cassell, Eric J. Nurturing a defective newborn. *Hastings Center Report* 8(1): 13-14, Feb 1978. Case study. BE06791.

*allowing to die; communication; counseling; *decision making; extraordinary treatment; *genetic defects; *newborns; *parent child relationship; parental consent; parents; physician's role; prolongation of life; resuscitation; suffering

Yeaworth, Rosalee C. The agonizing decisions in mental retardation. *American Journal of Nursing* 77(5): 864-867, May 1977. 13 refs. BE05593.

*allowing to die; coercion; *amniocentesis; decision making; economics; *genetic defects; *newborns; mentally retarded; nurses; *resource allocation; *selective abortion; sterilization; value of life

ALLOWING TO DIE / NEWBORNS / SPINA BIFIDA

Cooperman, Earl M. Meningomyelocele: to treat or not to treat. *Canadian Medical Association Journal* 116(12): 1339-1340, 18 Jun 1977. 12 refs. BE05645.

*allowing to die; amniocentesis; decision making; family members; legal aspects; *newborns; psychological stress; resource allocation; selection for treatment; *spina bifida; *withholding treatment

Searle, J.F.; Brewer, Colin. Life with spina bifida. *British Medical Journal* 2(6103): 1670, 24-31 Dec 1977. Letter. BE06510.

*allowing to die; *newborns; *spina bifida; withholding treatment

Spina bifida babies are 'killed unlawfully'—SPUC. *Nursing Times* 74(3): 86, 19 Jan 1978. BE06758.

*allowing to die; *newborns; *spina bifida; withholding treatment

Swinyard, Chester A., ed. Decision Making and the Defective Newborn. Proceedings of a Conference on Spina Bifida and Ethics. Springfield, Ill.: Charles C. Thomas, 1978. 649 p. References and footnotes included. ISBN 0-398-03662-4. BE06951.

*allowing to die; attitudes; congenital defects; *decision making; ethics; extraordinary treatment; informed consent; legal aspects; medical ethics; *newborns; parental consent; *patient care; physician's role; *quality of life; *selection for treat-

---

BE = bioethics accession number   fn. = footnotes   refs. = references

ment; socioeconomic factors; *spina bifida; withholding treatment

Zachary, R.B. Life with spina bifida. *British Medical Journal* 2(6100): 1460-1462, 3 Dec 1977. 1 ref. BE06199.

> *allowing to die; morbidity; *newborns; quality of life; selection for treatment; *spina bifida; wedge argument; *withholding treatment

ALLOWING TO DIE / PHYSICIAN'S ROLE

Bates, Richard C. It's *our* right to pull the plug. *Medical Economics* 54(10): 162-166, 16 May 1977. BE05603.

> *allowing to die; clergy; *decision making; ethics committees; family members; living wills; *physician's role; physicians; terminally ill; withholding treatment

Brody, Howard, et al. The complexities of dying. *New England Journal of Medicine* 296(21): 1237-1240, 26 May 1977. 10 refs. Letter. BE05606.

> *allowing to die; biomedical technologies; costs and benefits; *decision making; *physician's role; prolongation of life; quality of life; resource allocation; scarcity; terminal care; terminally ill; value of life; *values; *withholding treatment

Hunter, Thomas H.; Smith, William. Attitudes Toward Death and Dying. [Videorecording]. Charlottesville, Va.: University of Virginia Medical Center, Health Sciences Library, 1974. 1 cassette; 30 min.; sound; color; 3/4 in. Tape of the program "People, Places, Things," 27 Feb 1974, produced by Jefferson Cable Corp., Channel 10, Charlottesville. BE07045.

> *allowing to die; attitudes to death; autopsies; brain death; *decision making; disclosure; economics; family members; living wills; *moral obligations; organ donation; *physician's role; *physicians; prolongation of life; *terminally ill

Van den Berg, Jan H. Medical Power and Medical Ethics. New York: W.W. Norton, 1978. 91 p. 2 fn. ISBN 0-393-06428-X. BE06984.

> active euthanasia; aged; *allowing to die; *biomedical technologies; congenital defects; disclosure; historical aspects; medical ethics; newborns; patients; *physician's role; *prolongation of life; self determination; surgery; terminally ill; withholding treatment

ALLOWING TO DIE / PHYSICIANS / ATTITUDES

Crane, Diana. Consensus and controversy in medical practice: the dilemma of the critically ill patient. *In* Barber, Bernard, ed. Medical Ethics and Social Change. Philadelphia: American Academy of Political and Social Science, 1978. p. 99-110. Annals of the American Academy of Political and Social Science, Vol. 437, May 1978. 13 fn. ISBN 0-87761-226-9. BE07002.

> *allowing to die; *attitudes; chronically ill; brain death; decision making; legal aspects; mentally retarded; *physicians; terminally ill; *withholding treatment

Shaw, Anthony; Randolph, Judson G.; Manard, Barbara. Ethical issues in pediatric surgery: a national survey of pediatricians and pediatric surgeons. *Pediatrics* 60(4-Part 2): 588-599, Oct 1977. BE06587.

> abortion; *allowing to die; *attitudes; *congenital defects; *decision making; Down's syndrome; extraordinary treatment; legal aspects; *newborns; parental consent; parents; *pediatrics; *physicians; prolongation of life; spina bifida; *surgery; *treatment refusal; *withholding treatment

ALLOWING TO DIE / PHYSICIANS / LEGAL ASPECTS

DeMere, McCarthy. My position is to save lives, not to terminate them. *Journal of Legal Medicine* 5(8): 35-37, Aug 1977. BE05753.

> *allowing to die; coma; *decision making; *legal aspects; patients; *physicians; *withholding treatment

ALLOWING TO DIE / PREGNANT WOMEN / COMA

Colen, B.D. Pregnant woman in coma dies after heart attack. *Washington Post,* 7 Dec 1977, p. C12. BE06420.

> *allowing to die; *coma; *pregnant women; prolongation of life

ALLOWING TO DIE / QUALITY OF LIFE

Drinkwater, C.K.; Thorne, Susan; Wilson, Michael. Strive officiously to keep alive? *Journal of Medical Ethics* 3(4): 189-193, Dec 1977. 1 ref. Case study. BE06253.

> *allowing to die; central nervous system diseases; decision making; family members; legal aspects; legislation; mentally handicapped; physician's role; physicians; prolongation of life; *quality of life; review committees; self determination; social control; terminally ill; values; withholding treatment; General Medical Council; Great Britain

Ramsey, Paul. In the matter of *Quinlan*. *In his* Ethics at the Edges of Life: Medical and Legal Intersections. New Haven: Yale University Press, 1978. p. 268-299. 30 fn. Case study. ISBN 0-300-02137-2. BE06891.

> *allowing to die; coma; decision making; ethics committees; *extraordinary treatment; family members; intention; *involuntary euthanasia; killing; *judicial action; law; legal aspects; legal guardians; legal liability; physician's role; physicians; privacy; prognosis; prolongation of life; *quality of life; religion; *religious beliefs; Roman Catholic ethics; state courts; state interest; terminally ill; *third party consent; treatment refusal; withholding treatment; *In re Quinlan; *New Jersey Supreme Court

ALLOWING TO DIE / QUALITY OF LIFE / JUDICIAL ACTION

Ramsey, Paul. The strange case of Joseph Saikewicz.

BE = bioethics accession number   fn. = footnotes   refs. = references

In his Ethics at the Edges of Life: Medical and Legal Intersections. New Haven: Yale University Press, 1978. p. 300-317. 9 fn. Case study. ISBN 0-300-02137-2. BE06892.

aged; *allowing to die; *competence; *decision making; drugs; *judicial action; legal guardians; leukemia; *mentally retarded; physicians; prognosis; prolongation of life; *quality of life; standards; *terminally ill; *third party consent; treatment refusal; *withholding treatment; *In re Quinlan; *Massachusetts Supreme Judicial Court; *Saikewicz, Joseph

ALLOWING TO DIE / SELF DETERMINATION

Cassell, Eric J. The function of medicine. *Hastings Center Report* 7(6): 16-19, Dec 1977. BE06134.
*allowing to die; *chronically ill; *critically ill; Jehovah's Witnesses; motivation; physician's role; *self determination; suicide; *treatment refusal

Schelling, Thomas C. Strategic relationships in dying. *In* McMullin, Ernan, ed. Death and Decision. Boulder, Colo.: Westview Press, 1978. p. 63-73. American Association for the Advancement of Science Selected Symposium 18. ISBN 0-89158-152-9. BE06709.
*allowing to die; costs and benefits; decision making; moral obligations; rights; *self determination; terminally ill; value of life

ALLOWING TO DIE / SPINA BIFIDA / DECISION MAKING

Duff, Raymond S. A physician's role in the decision-making process: a physician's experience. *In* Swinyard, Chester A., ed. Decision Making and the Defective Newborn. Proceedings of a Conference on Spina Bifida and Ethics. Springfield, Ill.: Charles C. Thomas, 1978. p. 194-219. 18 refs. Case study. ISBN 0-398-03662-4. BE06954.
*allowing to die; alternatives; biomedical technologies; *decision making; disclosure; ethics; extraordinary treatment; family members; family problems; *handicapped; hospitals; informed consent; institutional policies; killing; legal aspects; medical staff; medicine; *newborns; parents; *physician's role; patient care; prognosis; physicians; psychological stress; public opinion; quality of life; risks and benefits; social impact; *spina bifida; terminally ill; withholding treatment

ALLOWING TO DIE / SPINA BIFIDA / LEGAL LIABILITY

Robertson, John A. Legal issues in nontreatment of defective newborns. *In* Swinyard, Chester A., ed. Decision Making and the Defective Newborn. Proceedings of a Conference on Spina Bifida and Ethics. Springfield, Ill.: Charles C. Thomas, 1978. p. 359-383. 8 refs. 9 fn. ISBN 0-398-03662-4. BE06957.

*allowing to die; *alternatives; child neglect; contracts; *congenital defects; criminal law; decision making; costs and benefits; financial support; killing; hospitals; law enforcement; *legal liability; *legal obligations; legal rights; medical staff; moral obligations; nurses; *newborns; obligations of society; *parental consent; parents; physician patient relationship; *physicians; resource allocation; *spina bifida; *treatment refusal; value of life; *withholding treatment

ALLOWING TO DIE / SWITZERLAND

Swiss medical academy issues guidelines for doctors to discontinue treatment for dying patients. *New York Times,* 21 Apr 1977, p. A15. BE06016.
*allowing to die; physician's role; terminally ill; withholding treatment; *Switzerland

ALLOWING TO DIE / TERMINALLY ILL

Cawley, Michele A. Euthanasia: should it be a choice? *American Journal of Nursing* 77(5): 859-861, May 1977. 4 refs. BE05592.
*allowing to die; case reports; *decision making; family members; physician's role; self determination; terminal care; *terminally ill; treatment refusal

Gonda, Thomas A. Coping with dying and death. *Geriatrics* 32(9): 71-73, Sep 1977. 7 refs. BE05985.
aged; *allowing to die; attitudes to death; legislation; physician's role; psychological stress; state government; *terminally ill; California Natural Death Act

Padgett, Jack F. Is there a right to die? *Bioethics Digest* 2(4): 1-4, Aug 1977. 34 fn. BE05763.
*allowing to die; legal rights; patients' rights; physician's role; *self determination; *terminally ill; *treatment refusal

Shannon, Thomas A. Caring for the dying patient: what guidance from the guidelines? *Hastings Center Report* 7(3): 28-30, Jun 1977. BE05666.
*allowing to die; competence; *decision making; economics; ethics committees; *hospitals; family members; *institutional policies; patients' rights; physicians; prognosis; resuscitation; self determination; standards; *terminally ill; treatment refusal; *withholding treatment

Tendler, Moshe D. Torah ethics prohibit natural death. *Sh'ma* 7(132): 97-99, 15 Apr 1977. BE05538.
*allowing to die; coma; Jewish ethics; living wills; *morality; prolongation of life; *terminally ill; *withholding treatment

ALLOWING TO DIE / TERMINALLY ILL / JEWISH ETHICS

Cohn, Hillel. Natural death—humane, just and Jewish. *Sh'ma* 7(132): 99-101, 15 Apr 1977. BE05539.

BE = bioethics accession number   fn. = footnotes   refs. = references

*allowing to die; decision making; *Jewish ethics; *living wills; prolongation of life; *terminally ill; withholding treatment

Siegel, Seymour. Jewish law permits natural death. *Sh'ma* 7(132): 96-97, 15 Apr 1977. BE05537.
*allowing to die; *Jewish ethics; legislation; *living wills; state government; *terminally ill; withholding treatment; California Natural Death Act

ALLOWING TO DIE / TERMINALLY ILL / JUDICIAL ACTION

Coburn, Daniel R. In re Quinlan: a practical overview. *Arkansas Law Review* 31(1): 59-74, Spring 1977. 79 fn. BE05948.
*allowing to die; brain death; decision making; fathers; *judicial action; legal guardians; legal rights; living wills; public opinion; *terminally ill; *withholding treatment; *In re Quinlan

ALLOWING TO DIE / TERMINALLY ILL / LEGAL ASPECTS

Harrison, C.P. Medicine, terminal illness and the law. *Canadian Medical Association Journal* 117(5): 514+, 3 Sep 1977. 7 refs. BE06333.
*allowing to die; intention; *legal aspects; physician's role; *terminally ill; withholding treatment

Sackett, Walter W.; Evans, Franklin J. The right to die. *Journal of Legal Medicine* 5(11): 17+, Nov 1977. Letter. BE06585.
*allowing to die; decision making; *legal aspects; legislation; physicians; *terminally ill

ALLOWING TO DIE / TERMINALLY ILL / LIVING WILLS

Otten, Alan L. Death rights. *Wall Street Journal (Eastern Edition)*, 8 Dec 1977, p. 18. BE06459.
*allowing to die; decision making; family members; legislation; *living wills; *terminally ill

ALLOWING TO DIE / TERMINALLY ILL / SWITZERLAND

Swiss Academy of Medical Sciences. Swiss guidelines on care of the dying. *Hastings Center Report* 7(3): 30-31, Jun 1977. BE05671.
active euthanasia; *allowing to die; decision making; legal aspects; physicians; terminal care; *terminally ill; *withholding treatment; *Switzerland

Wright, Irving S. Guidelines concerning euthanasia: dying patient. *New York State Journal of Medicine* 78(1): 61-63, Jan 1978. Translation of a report prepared by the Swiss Academy of the Medical Sciences. 4 fn. BE06764.
active euthanasia; *allowing to die; physician's role; self determination; *terminally ill; treatment refusal; *Switzerland

ALLOWING TO DIE / THIRD PARTY CONSENT

Fost, Norman. Proxy consent for seriously ill newborns. *In* Smith, David H., ed. No Rush to Judgement: Essays on Medical Ethics. Bloomington, Ind.: Indiana University, Poynter Center, 1977. p. 1-17. 23 refs. BE05793.
*allowing to die; congenital defects; costs and benefits; *decision making; ethics; ethics committees; informed consent; involuntary euthanasia; living wills; mentally retarded; *newborns; normative ethics; parental consent; parents; physicians; resource allocation; *third party consent; voluntary euthanasia

Ramsey, Paul. In the matter of *Quinlan*. *In his* Ethics at the Edges of Life: Medical and Legal Intersections. New Haven: Yale University Press, 1978. p. 268-299. 30 fn. Case study. ISBN 0-300-02137-2. BE06891.
*allowing to die; coma; decision making; ethics committees; *extraordinary treatment; family members; intention; *involuntary euthanasia; killing; *judicial action; law; legal aspects; legal guardians; legal liability; physician's role; physicians; privacy; prognosis; prolongation of life; *quality of life; religion; *religious beliefs; Roman Catholic ethics; state courts; state interest; terminally ill; *third party consent; treatment refusal; withholding treatment; *In re Quinlan; *New Jersey Supreme Court

ALLOWING TO DIE / THIRD PARTY CONSENT / JUDICIAL ACTION

Kindregan, Charles P. The court as forum for life and death decisions: reflections on procedures for substituted consent. *Suffolk University Law Review* 11(4): 919-935, Spring 1977. 78 fn. BE06232.
*allowing to die; competence; institutionalized persons; *judicial action; legal aspects; leukemia; mentally retarded; organ donation; parental consent; patient care; *third party consent; *withholding treatment; *Saikewicz, Joseph

ALLOWING TO DIE / VALUE OF LIFE

Mahoney, Edward J. The morality of terminating life vs. allowing to die. *Louvain Studies* 6(3): 256-272, Spring 1977. 23 fn. BE06851.
*active euthanasia; *allowing to die; *coma; *congenital defects; decision making; *newborns; terminal care; *terminally ill; *value of life; values; withholding treatment

ALTERNATIVES *See under*

ABORTION / ALTERNATIVES

ABORTION / ALTERNATIVES / LEGAL ASPECTS

INFORMED CONSENT / SURGERY / ALTERNATIVES

BE = bioethics accession number   fn. = footnotes   refs. = references

ALTERNATIVES *See under (cont'd.)*

TREATMENT REFUSAL / SPINA BIFIDA / ALTERNATIVES

ALTRUISM   *See under*

BLOOD DONATION / ALTRUISM

SOCIOBIOLOGY / ALTRUISM

AMERICAN FEDERATION FOR CLINICAL RESEARCH   *See under*

RECOMBINANT DNA RESEARCH / AMERICAN FEDERATION FOR CLINICAL RESEARCH

AMERICAN HEART ASSOCIATION   *See under*

HEALTH CARE / AMERICAN HEART ASSOCIATION

PREVENTIVE MEDICINE / AMERICAN HEART ASSOCIATION

AMERICAN INDIANS   *See under*

INVOLUNTARY STERILIZATION / AMERICAN INDIANS

AMERICAN MEDICAL ASSOCIATION   *See under*

CODES OF ETHICS / AMERICAN MEDICAL ASSOCIATION

PROFESSIONAL ETHICS / AMERICAN MEDICAL ASSOCIATION

RECOMBINANT DNA RESEARCH / AMERICAN MEDICAL ASSOCIATION

AMNIOCENTESIS   *See also* PRENATAL DIAGNOSIS

AMNIOCENTESIS / ATTITUDES

Finley, Sara C., et al. Participants' reaction to amniocentesis and prenatal genetic studies. *Journal of the American Medical Association* 238(22): 2377-2379, 28 Nov 1977. 7 refs. BE06111.

*amniocentesis; *attitudes; genetic counseling; genetic defects; married persons; prenatal diagnosis

AMNIOCENTESIS / RISKS AND BENEFITS

Milunsky, Aubrey. Prenatal genetic diagnosis: risks and needs. *In* Lubs, Herbert A.; de la Cruz, Felix, eds. Genetic Counseling: A Monograph of the Nat'l. Inst. of Child Health and Human Development. New York: Raven Press, 1977. p. 477-494. 10 refs. ISBN 0-89004-150-4. BE05800.

abortion; *amniocentesis; costs and benefits; fetuses; genetic defects; health facilities; informed consent; mortality; obstetrics and gynecology; *risks and benefits; standards; statistics

Wyatt, Philip R. Who's for amniocentesis? *Lancet* 1(8025): 1315, 18 Jun 1977. 3 refs. Letter.

BE05675.

age; *amniocentesis; chromosomal disorders; genetic defects; pregnant women; *risks and benefits

AMNIOCENTESIS / SELECTIVE ABORTION / CHRISTIAN ETHICS

Vaux, Kenneth L. Biomedical issues and reproduction. *In* Hollis, Harry N., comp. A Matter of Life and Death: Christian Perspectives. Nashville: Broadman Press, 1977. p. 68-78. ISBN 0-8054-6118-3. BE06531.

*amniocentesis; *Christian ethics; genetic defects; health; *selective abortion; values

AMNIOCENTESIS / SELECTIVE ABORTION / GENETIC DEFECTS

Yeaworth, Rosalee C. The agonizing decisions in mental retardation. *American Journal of Nursing* 77(5): 864-867, May 1977. 13 refs. BE05593.

*allowing to die; coercion; *amniocentesis; decision making; economics; *genetic defects; *newborns; mentally retarded; nurses; *resource allocation; *selective abortion; sterilization; value of life

ANESTHESIA   *See under*

DISCLOSURE / ANESTHESIA / RISKS AND BENEFITS

INFORMED CONSENT / ANESTHESIA

ANGLICAN CHURCH OF CANADA   *See under*

AID / ANGLICAN CHURCH OF CANADA

ALLOWING TO DIE / ANGLICAN CHURCH OF CANADA

BIOETHICAL ISSUES / ANGLICAN CHURCH OF CANADA

ANIMAL EXPERIMENTATION / EMBRYO TRANSFER

Sullivan, Walter. Woman gives birth to baby conceived outside the body. *New York Times,* 26 Jul 1978, p. A1+. BE07018.

*animal experimentation; *embryo transfer; *in vitro fertilization

ARISTOTLE   *See under*

MEDICINE / ETHICS / ARISTOTLE

ARKANSAS   *See under*

ALLOWING TO DIE / LIVING WILLS / ARKANSAS

ARTIFICIAL INSEMINATION

Anderson, Norman. Issues of Life and Death: Abortion, Birth Control, Capital Punishment, Euthanasia. Downers Grove, Ill.: InterVarsity Press, 1977.

BE = bioethics accession number   fn. = footnotes   refs. = references

130 p. 189 fn. ISBN 0-87784-721-5. BE05787.
AID; *abortion; *artificial insemination; *capital punishment; Christian ethics; *contraception; DNA therapy; *euthanasia; genetic intervention; justifiable killing; organ transplantation; personhood; scriptural interpretation; *sterilization; theology; *value of life

ARTIFICIAL INSEMINATION / ROMAN CATHOLIC ETHICS

Hellegers, André E.; McCormick, Richard A. Unanswered questions on test tube life. *America* 139(4): 74-78, 12-19 Aug 1978. BE07030.
AID; AIH; abortion; *artificial insemination; biomedical technologies; embryo transfer; *in vitro fertilization; marital relationship; *morality; newborns; products of in vitro fertilization; reproduction; resource allocation; risks and benefits; *Roman Catholic ethics; social impact

ARTIFICIAL INSEMINATION / YUGOSLAVIA

Socialist Republic of Slovenia (Yugoslavia). Abortion, sterilization, contraception, and artificial insemination. *International Digest of Health Legislation* 28(4): 1112-1115, 1977. Law of 20 Apr 1977. BE07131.
*abortion; abortion on demand; age; *artificial insemination; competence; confidentiality; contraception; counseling; decision making; fertility; fetal development; rights; reproduction; semen donors; therapeutic abortion; third party consent; *voluntary sterilization; *Socialist Republic of Slovenia; *Yugoslavia

ARTIFICIAL INSEMINATION, DONOR See
AID

ATTITUDES TO DEATH

Miller, Albert J.; Acri, Michael J. Medical profession and nursing experiences. *In their* Death: A Bibliographical Guide. Metuchen, N.J.: Scarecrow Press, 1977. p. 95-174. Bibliography. ISBN 0-8108-1025-5. BE06186.
*attitudes to death; cancer; communication; *death; determination of death; disclosure; family members; nurses; physicians; psychological stress; *terminal care; terminally ill

Woodward, Kenneth L., et al. Living with dying. *Newsweek* 91(18): 52-56+, 1 May 1978. BE06972.
*attitudes to death; family members; historical aspects; *hospices; psychological stress; *terminal care; terminally ill

ATTITUDES TO DEATH / NURSES / NURSING EDUCATION

Hopping, Betty L. Nursing students' attitudes toward death. *Nursing Research* 26(6): 443-447, Nov-Dec 1977. 19 refs. BE06501.

*attitudes to death; curriculum; death; *nurses; *nursing education; *students; terminal care

ATTITUDES   *See under*
ABORTION / ATTITUDES
ABORTION / ATTITUDES / GREAT BRITAIN
ABORTION / ATTITUDES / HAWAII
ABORTION / ATTITUDES / UNITED STATES
ABORTION / FINANCIAL SUPPORT / ATTITUDES
ABORTION / LEGISLATION / ATTITUDES
ABORTION / MORALITY / ATTITUDES
ABORTION / OBSTETRICS AND GYNECOLOGY / ATTITUDES
ABORTION / PHYSICIANS / ATTITUDES
ABORTION / POLITICS / ATTITUDES
ABORTION / RELIGION / ATTITUDES
ACTIVE EUTHANASIA / AGED / ATTITUDES
AGED / SOCIETY / ATTITUDES
ALLOWING TO DIE / PHYSICIANS / ATTITUDES
AMNIOCENTESIS / ATTITUDES
CONFIDENTIALITY / MEDICAL RECORDS / ATTITUDES
EUTHANASIA / STUDENTS / ATTITUDES
FAMILY PLANNING / PHYSICIANS / ATTITUDES
GENETIC COUNSELING / ATTITUDES
GENETIC SCREENING / OBSTETRICS AND GYNECOLOGY / ATTITUDES
NEWBORNS / SPINA BIFIDA / ATTITUDES
PATIENT CARE / PEDIATRICS / ATTITUDES
PATIENT CARE / PLACEBOS / ATTITUDES
POPULATION CONTROL / PHYSICIANS / ATTITUDES
RECOMBINANT DNA RESEARCH / ATTITUDES
RECOMBINANT DNA RESEARCH / GOVERNMENT REGULATION / ATTITUDES
RECOMBINANT DNA RESEARCH / INVESTIGATORS / ATTITUDES
RECOMBINANT DNA RESEARCH / SOCIAL CONTROL / ATTITUDES
REPRODUCTION / GENETIC DEFECTS / ATTITUDES
TERMINAL CARE / PHYSICIANS / ATTITUDES
TREATMENT REFUSAL / PHYSICIANS / ATTITUDES

AUSTRALIA   *See under*
ABORTION / AUSTRALIA
MEDICAL ETHICS / MEDICAL EDUCATION / AUSTRALIA
ORGAN TRANSPLANTATION / AUSTRALIA
ORGAN TRANSPLANTATION / LEGAL ASPECTS / AUSTRALIA
TRANSPLANTATION / AUSTRALIA
VOLUNTARY EUTHANASIA / LEGAL ASPECTS / AUSTRALIA

BE = bioethics accession number   fn. = footnotes   refs. = references

AUTHORITARIANISM *See under*

PHYSICIAN PATIENT RELATIONSHIP / AUTHORITARIANISM

AUTOPSIES / LEGAL RIGHTS / MARYLAND

Curran, William J. Religious objection to a medicolegal autopsy: a case and a statute. *New England Journal of Medicine* 297(5): 260-261, 4 Aug 1977. 3 refs. BE05752.
   *autopsies; *coercion; *Jews; *legal rights; property rights; *religious beliefs; state government; *Maryland

BEGINNING OF LIFE

Pastrana, Gabriel. Personhood and the beginning of human life. *Thomist* 41(2): 247-294, Apr 1977. 75 fn. BE05574.
   abortion; *beginning of life; brain; fetal development; fetuses; *personhood; moral policy; philosophy; self concept; social interaction; twinning; viability

BEHAVIOR CONTROL *See also* ELECTRICAL STIMULATION OF THE BRAIN, OPERANT CONDITIONING, PSYCHOSURGERY

BEHAVIOR CONTROL

Beauchamp, Tom L.; Walters, LeRoy, eds. Contemporary Issues in Bioethics. Encino, Calif.: Dickenson Publishing Company, 1978. 612 p. References and footnotes included. ISBN 0-8221-0200-5. BE06700.
   *abortion; allowing to die; *behavior control; *bioethics; biomedical technologies; brain death; children; codes of ethics; confidentiality; decision making; deontological ethics; *determination of death; *euthanasia; fetuses; genetic intervention; *health; *health care; *human experimentation; *informed consent; involuntary commitment; justice; legal aspects; *medical ethics; natural law; *normative ethics; operant conditioning; patients' rights; personhood; psychosurgery; *resource allocation; rights; selection for treatment; treatment refusal; utilitarianism; Kaimowitz v. Department of Mental Health; Quinlan, Karen

Chavkin, Samuel. The Mind Stealers: Psychosurgery and Mind Control. Boston: Houghton Mifflin, 1978. 228 p. 400 fn. ISBN 0-395-26381-6. BE06977.
   *adolescents; aggression; *behavior control; behavioral genetics; brain pathology; children; *electrical stimulation of the brain; evaluation; human experimentation; imprisonment; law enforcement; institutionalized persons; legal aspects; legal rights; *operant conditioning; *prisoners; prognosis; psychoactive drugs; *psychosurgery; socioeconomic factors; *violence; Project START

Clark, Henry, ed. The Ethics of Experience and Behav-

ior Control. Los Angeles: University of Southern California, Center for the Humanities, 1976. 65 p. Report of the colloquium sponsored by the USC Center for the Humanities, 12-14 Feb 1976. BE06548.
   *behavior control; humanism; legal aspects; operant conditioning; politics; psychoactive drugs; self determination; social control; social impact; technology assessment

Erickson, Richard C. Walden III: toward an ethics of changing behavior. *Journal of Religion and Health* 16(1): 7-14, Jan 1977. 10 fn. BE06481.
   *behavior control; human rights; negative reinforcement; positive reinforcement; psychotherapy; self determination; social interaction; society

McNamara, J. Regis. Socioethical considerations in behavior therapy research and practice. *Behavior Modification* 2(1): 3-24, Jan 1978. 61 refs. BE06743.
   *behavior control; children; contracts; costs and benefits; evaluation; goals; informed consent; investigator subject relationship; legal rights; methods; operant conditioning; privacy; risks and benefits; selection of subjects; stigmatization; values

Packard, Vance. The People Shapers. Boston: Little, Brown, 1977. 398 p. 249 refs. ISBN 0-316-68750-2. BE06518.
   AID; aggression; *behavior control; behavior disorders; cloning; brain; electrical stimulation of the brain; embryo transfer; eugenics; employment; frozen semen; *genetic intervention; host mothers; genetic screening; human characteristics; hypnosis; in vitro fertilization; intelligence; life extension; legal aspects; mass media; operant conditioning; organ transplantation; personality; privacy; psychoactive drugs; psychological stress; psychosurgery; selective abortion; sex determination; sex preselection; *social control; *social impact

Valenstein, Elliot S. Brain control: scientific, ethical and political considerations. *In* Ellison, Craig W., ed. Modifying Man: Implications and Ethics. Washington: University Press of America, 1978. p. 143-168. ISBN 0-8191-0302-0. BE06670.
   aggression; *behavior control; brain; brain pathology; *electrical stimulation of the brain; ethics committees; evaluation; government regulation; human experimentation; mentally ill; political activity; *psychosurgery; social determinants; *violence

Wojcik, Jan. Muted Consent: A Casebook in Modern Medical Ethics. West Lafayette, Ind.: Purdue University, 1978. 164 p. 237 fn. Case study. ISBN 911198-50-4. BE06704.
   *abortion; *behavior control; *case reports; *determination of death; disclosure; drugs; electrical stimulation of the brain; ethics; eugenics; euthanasia; extraordinary treatment; fetuses; *genetic counseling; genetic defects; *genetic intervention; *genetic screening; health care; *human experimentation; *informed consent; involuntary com-

---

mitment; mentally ill; operant conditioning; pregnant women; psychosurgery; quality of life; reproductive technologies; *resource allocation; rights; selective abortion; value of life; withholding treatment

BEHAVIOR CONTROL / BLACKS

Lombard, Rudy. The rise of drugs in behavior modification programs. *In* U.S. National Commission for Protection of Human Subjects.... Research Involving Those Institutionalized as Mentally Infirm: Appendix. Washington: U.S. Government Printing Office, 1978. p. 4.1-4.4. DHEW Publication No. (OS) 78-0007. Paper prepared for the National Minority Conference on Human Experimentation, 6-8 Jan 1976. BE06899.
   *behavior control; *blacks; drugs; political activity; psychosurgery

BEHAVIOR CONTROL / CENTRAL INTELLIGENCE AGENCY

Control C.I.A., not behavior. *New York Times,* 5 Aug 1977, p. A20. Editorial. BE06382.
   *behavior control; federal government; government agencies; government regulation; psychoactive drugs; *Central Intelligence Agency

Crewdson, John M. Abuses in testing of drugs by C.I.A. to be panel focus. *New York Times,* 20 Sep 1977, p. 1+. BE06385.
   *behavior control; federal government; government agencies; *human experimentation; program descriptions; *psychoactive drugs; *Central Intelligence Agency

Horrock, Nicholas M. Drugs tested by C.I.A. on mental patients. *New York Times,* 3 Aug 1977, p. A1+. BE06379.
   *behavior control; federal government; government agencies; *human experimentation; LSD; mentally ill; *psychoactive drugs; *Central Intelligence Agency

Horrock, Nicholas M. Eighty institutions used in C.I.A. mind studies. *New York Times,* 4 Aug 1977, p. A17. BE06380.
   *behavior control; federal government; financial support; government agencies; universities; *Central Intelligence Agency

Horrock, Nicholas M. Private institutions used in C.I.A. effort to control behavior. *New York Times,* 2 Aug 1977, p. 1+. BE06378.
   *behavior control; federal government; financial support; government agencies; human experimentation; LSD; physician's role; program descriptions; universities; *Central Intelligence Agency

Mindboggling. *Washington Post,* 19 Aug 1977, p. A26. Editorial. BE06384.
   *behavior control; federal government; government agencies; human experimentation; *Central Intelligence Agency

Reid, T.R. Range of mind-control efforts revealed in CIA documents. *Washington Post,* 23 Sep 1977, p. A17. BE06387.
   behavior control; federal government; government agencies; *Central Intelligence Agency

Rensberger, Royce. Ethical questions in mind-control experiments. *New York Times,* 4 Aug 1977, p. A17. BE06381.
   *behavior control; human experimentation; informed consent; risks and benefits; *Central Intelligence Agency

Thomas, Jo. C.I.A. sought to spray drug on partygoers. *New York Times,* 21 Sep 1977, p. A11. BE06386.
   *behavior control; federal government; government agencies; *human experimentation; *LSD; *Central Intelligence Agency

Treaster, Joseph B. Researchers say that students were among 200 who took LSD in tests financed by C.I.A. in early 50's. *New York Times,* 9 Aug 1977, p. 21. BE06383.
   *behavior control; federal government; government agencies; *human experimentation; *LSD; students; *Central Intelligence Agency

BEHAVIOR CONTROL / CHILDREN / HYPERKINESIS

Appleman, Michael A. The Legal Issues Involved in the Use of Stimulants on Hyperactive School Children. Ann Arbor: University Microfilms International, 1974. 158 p. Dissertation, Ph.D. in Philosophy, Graduate School of the University of Minnesota, Jul 1974. Order No. 75-2081. 243 fn. Bibliography: p. 151-158. BE06660.
   *behavior control; *children; codes of ethics; coercion; diagnosis; disclosure; education; *hyperkinesis; informed consent; *legal aspects; legal rights; *legislation; *psychoactive drugs; risks and benefits; *state government; students

BEHAVIOR CONTROL / COERCION

Goldiamond, Israel. Protection of human subjects and patients. *In* Krapfl, Jon E.; Vargas, Ernest A., eds. Behaviorism and Ethics. Kalamazoo: Behaviordelia, 1977. p. 129-195. Commentary by Pamela Meadowcroft on p. 188-195. 41 refs. ISBN 0-914-47425-1. BE06546.
   *behavior control; *coercion; consequences; ethical review; *human experimentation; goals; intention; informed consent; investigator subject relationship; *operant conditioning; remuneration; risks and benefits; self determination

---

BE = bioethics accession number   fn. = footnotes   refs. = references

BEHAVIOR CONTROL / COERCION / LEGAL ASPECTS

Maley, Roger; Hayes, Steven C. Coercion and control: ethical and legal issues. *In* Krapfl, Jon E.; Vargas, Ernest A., eds. Behaviorism and Ethics. Kalamazoo: Behaviordelia, 1977. p. 265-288. Commentary by Roy A. Moxley on p. 285-288. 25 refs. ISBN 0-914-47425-1. BE06547.
> *behavior control; *coercion; education; involuntary commitment; judicial action; *legal aspects; treatment refusal

BEHAVIOR CONTROL / FINANCIAL SUPPORT / LAW ENFORCEMENT ASSISTANCE ADMINISTRATION

Oelsner, Lesley. U.S. bars crime fund use on behavior modification. *New York Times,* 15 Feb 1974, p. 66. BE06604.
> *behavior control; *federal government; *financial support; government regulation; negative reinforcement; *prisoners; *Law Enforcement Assistance Administration

BEHAVIOR CONTROL / GOVERNMENT REGULATION

DiGiacomo, Robert E. Behavior modification: toward the understanding and reform of federal policy. *Corrective and Social Psychiatry and Journal of Behavior Technology Methods and Therapy* 23(4): 101-110, 1977. 37 refs. BE06279.
> *behavior control; codes of ethics; coercion; due process; federal government; financial support; *government regulation; informed consent; *legal rights; *prisoners; privacy; punishment

BEHAVIOR CONTROL / HÄRING, BERNARD

Soane, Brendan. The literature of medical ethics: Bernard Häring. *Journal of Medical Ethics* 3(2): 85-92, Jun 1977. 37 fn. BE05669.
> AIH; *behavior control; biology; Christian ethics; contraception; genetic intervention; goals; *health; health care; medical ethics; medicine; *normality; operant conditioning; philosophy; psychology; Roman Catholic ethics; *self determination; *teleological ethics; *totality; values; voluntary sterilization; *Häring, Bernard

BEHAVIOR CONTROL / HUMAN RIGHTS

Oates, Wayne E. Bioethical issues in behavior control. *In* Hollis, Harry N., comp. A Matter of Life and Death: Christian Perspectives. Nashville: Broadman Press, 1977. p. 32-49. 17 fn. ISBN 0-8054-6118-3. BE06529.
> *behavior control; Christian ethics; electroconvulsive therapy; electrical stimulation of the brain; goals; group therapy; *human rights; negative rein-

forcement; psychoactive drugs; psychosurgery; psychotherapy; values

BEHAVIOR CONTROL / INSTITUTIONALIZED PERSONS / LEGAL RIGHTS

Wood, W. Scott. Behavior modification and civil rights. *In* Krapfl, Jon E.; Vargas, Ernest A., eds. Behaviorism and Ethics. Kalamazoo: Behaviordelia, 1977. p. 101-118. Commentary by Margret M. Baltes on p. 114-118. 35 refs. ISBN 0-914-47425-1. BE06545.
> *behavior control; *due process; goals; *institutionalized persons; *legal rights; mentally handicapped; self regulation; social control; values

BEHAVIOR CONTROL / PRISONERS / LEGAL RIGHTS

Delgado, Richard. Organically induced behavioral change in correctional institutions: release decisions and the "new man" phenomenon. *Southern California Law Review* 50(2): 215-270, Jan 1977. 340 fn. BE05902.
> aggression; *behavior control; decision making; *duration of commitment; *electrical stimulation of the brain; imprisonment; *legal rights; *prisoners; *psychoactive drugs; *psychosurgery; punishment; violence; Eighth Amendment

DiGiacomo, Robert E. Behavior modification: toward the understanding and reform of federal policy. *Corrective and Social Psychiatry and Journal of Behavior Technology Methods and Therapy* 23(4): 101-110, 1977. 37 refs. BE06279.
> *behavior control; codes of ethics; coercion; due process; federal government; financial support; *government regulation; informed consent; *legal rights; *prisoners; privacy; punishment

BEHAVIOR CONTROL / PRISONERS / PROJECT START

Oelsner, Lesley. U.S. ends project on jail inmates. *New York Times,* 7 Feb 1974, p. 12. BE06603.
> *behavior control; federal government; legal rights; *prisoners; *Project START

BEHAVIOR CONTROL / PSYCHIATRY / PHYSICIAN'S ROLE

Whitlock, F.A. The ethics of psychosurgery. *In* Smith, J. Sydney; Kiloh, L.G., eds. Psychosurgery and Society. New York: Pergamon Press, 1977. p. 129-135. A Symposium Organized by the Neuropsychiatric Institute, Sydney, 26-27 Sep 1974. 17 refs. ISBN 0-08-021836-9. BE06169.
> *behavior control; brain pathology; informed consent; law enforcement; mental health; mentally ill; morality; *physician's role; prisoners; *psychiatry; *psychosurgery; self determination; violence

---

BE = bioethics accession number   fn. = footnotes   refs. = references

BEHAVIOR CONTROL / PSYCHOACTIVE DRUGS / LEGAL ASPECTS

Appleman, Michael A. The Legal Issues Involved in the Use of Stimulants on Hyperactive School Children. Ann Arbor: University Microfilms International, 1974. 158 p. Dissertation, Ph.D. in Philosophy, Graduate School of the University of Minnesota, Jul 1974. Order No. 75-2081. 243 fn. Bibliography: p. 151-158. BE06660.
*behavior control; *children; codes of ethics; coercion; diagnosis; disclosure; education; *hyperkinesis; informed consent; *legal aspects; legal rights; *legislation; *psychoactive drugs; risks and benefits; *state government; students

BEHAVIOR CONTROL / PSYCHOTHERAPY / RELIGIOUS BELIEFS

McLemore, Clinton W.; Court, John H. Religion and psychotherapy—ethics, civil liberties, and clinical savvy: a critique. *Journal of Consulting and Clinical Psychology* 45(6): 1172-1175, Dec 1977. 11 refs. BE06373.
*behavior control; case reports; mentally ill; *psychotherapy; *religious beliefs; values

BEHAVIOR CONTROL / RIGHTS

Kobler, Arthur L. Civil liberties. *In* Wolman, Benjamin B., ed. International Encyclopedia of Psychiatry, Psychology, Psychoanalysis, and Neurology. Volume 3. New York: Van Nostrand Reinhold, 1977. p. 151-154. 4 refs. ISBN 0-918228-01-8. BE06564.
*behavior control; *confidentiality; *human experimentation; human rights; legal rights; medical records; operant conditioning; *rights; self determination; social control

BEHAVIOR CONTROL / SELF DETERMINATION

London, Perry. Behavior control, values and the future. *In* Ellison, Craig W., ed. Modifying Man: Implications and Ethics. Washington: University Press of America, 1978. p. 189-208. 8 fn. ISBN 0-8191-0302-0. BE06671.
*behavior control; behavior disorders; coercion; common good; drugs; freedom; homosexuals; human rights; mentally ill; *operant conditioning; politics; psychotherapy; *self determination; *values

BEHAVIOR MODIFICATION *See* OPERANT CONDITIONING

BEHAVIORAL GENETICS

Cooke, Robert. Improving on Nature: The Brave New World of Genetic Engineering. New York: Quadrangle/The New York Times Book Company, 1977. 248 p. ISBN 0-8129-0667-5. BE05699.
amniocentesis; *behavioral genetics; artificial genes; biological containment; biology; cancer; cloning; DNA therapy; embryo transfer; evolution; federal government; *food; genetic defects; *genetic intervention; genetic screening; genetics; government regulation; health hazards; in vitro fertilization; mutation; physical containment; prenatal diagnosis; *recombinant DNA research; risks and benefits; social impact

BEHAVIORAL GENETICS / BLACKS

Littlewood, Roland; Lipsedge, Maurice; Adiseshiah, Mohankumar. Science, race, and intelligence. *British Medical Journal* 2(6097): 1286, 12 Nov 1977. 12 refs. Letter. BE06116.
*behavioral genetics; behavioral research; *blacks; historical aspects; human characteristics; *intelligence; psychiatry

BEHAVIORAL GENETICS / DISCRIMINATION

Science for the People. Ann Arbor Editorial Collective. Biology as a Social Weapon. Minneapolis: Burgess Publishing Company, 1977. 154 p. 339 refs. ISBN 0-8087-4534-4. BE06157.
aggression; *behavioral genetics; *discrimination; ecology; eugenics; evaluation; females; historical aspects; human characteristics; human equality; *intelligence; males; minority groups; politics; population control; sexuality; social determinants; *sociobiology; violence; *XYY karyotype

BEHAVIORAL GENETICS / INFORMAL SOCIAL CONTROL

Hunter, Thomas H.; Davis, Bernard D.; Fletcher, Joseph. Is Behavioral Genetics Taboo? The Neo-Lysenkoism. [Videorecording]. Charlottesville, Va.: University of Virginia Medical Center, Health Sciences Library, 1976. 1 cassette; 60 min.; sound; black and white; 3/4 in. Tape of the Medical Center Hour, 13 Oct 1976. Panel discussion moderated by Thomas H. Hunter. BE07050.
*behavioral genetics; ethics; eugenics; *evolution; freedom; genetic diversity; genetic intervention; genetic screening; *genetics; historical aspects; *informal social control; intelligence; *politics; political activity; science; sociobiology; socioeconomic factors; XYY karyotype; Lysenkoism; *Science for the People

BEHAVIORAL GENETICS / INTELLIGENCE / MORALITY

Davis, Bernard D. The moralistic fallacy. *Nature* 272(5652): 390, 30 Mar 1978. BE06806.
attitudes; *behavioral genetics; discrimination; evolution; freedom; genetics; *intelligence; *morality; science; social impact

---

BE = bioethics accession number   fn. = footnotes   refs. = references

BEHAVIORAL GENETICS / POLITICS

Hunter, Thomas H.; Davis, Bernard D.; Fletcher, Joseph. Is Behavioral Genetics Taboo? The Neo-Lysenkoism. [Videorecording]. Charlottesville, Va.: University of Virginia Medical Center, Health Sciences Library, 1976. 1 cassette; 60 min.; sound; black and white; 3/4 in. Tape of the Medical Center Hour, 13 Oct 1976. Panel discussion moderated by Thomas H. Hunter. BE07050.
*behavioral genetics; ethics; eugenics; *evolution; freedom; genetic diversity; genetic intervention; genetic screening; *genetics; historical aspects; *informal social control; intelligence; *politics; political activity; science; sociobiology; socioeconomic factors; XYY karyotype; Lysenkoism; *Science for the People

BEHAVIORAL GENETICS / VALUES

McShea, Robert J. Biology and ethics. *Ethics* 88(2): 139-149, Jan 1978. 7 fn. BE06744.
*behavioral genetics; biology; metaethics; moral obligations; philosophy; *values

BEHAVIORAL GENETICS *See under*

SOCIOBIOLOGY / BEHAVIORAL GENETICS

BEHAVIORAL RESEARCH

Wax, Murray L. Field workers and research subjects: who needs protection? *Hastings Center Report* 7(4): 29-32, Aug 1977. 13 fn. BE05769.
*behavioral research; deception; informed consent; investigator subject relationship; methods; risks and benefits; social impact; social sciences

BEHAVIORAL RESEARCH / ADOLESCENTS / SEXUALITY

Nilson, Donald R.; Steinfels, Margaret O. Parental consent and a teenage sex survey. *Hastings Center Report* 7(3): 13-15, Jun 1977. Case study. BE05658.
*adolescents; *behavioral research; confidentiality; informed consent; *parental consent; research design; *sexuality; values

BEHAVIORAL RESEARCH / BLACKS

Jackson, Jacquelyne J. Informed consent: ethical issues in behavioral research. *In* U.S. National Commission for Protection of Human Subjects.... Research Involving Those Institutionalized as Mentally Infirm: Appendix. Washington: U.S. Government Printing Office, 1978. p. 5.1-5.18. DHEW Publication No. (OS) 78-0007. Paper prepared for the National Minority Conference on Human Experimentation, 6-8 Jan 1976. 3 refs. BE06900.
*behavioral research; *blacks; federal government; government regulation; *human experimentation; *informed consent

BEHAVIORAL RESEARCH / CENTRAL INTELLIGENCE AGENCY

Roark, Anne C. The CIA on the campus. *Chronicle of Higher Education* 15(4): 7, 26 Sep 1977. BE05784.
*behavior control; biomedical research; *behavioral research; drugs; financial support; government agencies; human experimentation; politics; *universities; *Central Intelligence Agency; Project MKULTRA

BEHAVIORAL RESEARCH / CONFIDENTIALITY

Knerr, Charles R. Confidentiality of Social Science Research Sources and Data: Analysis of a Public Policy Problem. Ann Arbor: University Microfilms International, 1976. 312 p. Dissertation, Ph.D. in Public Administration, Graduate School of Syracuse University, May 1977. Order No. 77-30,735. 381 fn. Bibliography: p. 287-312. BE06661.
attitudes; *behavioral research; *case studies; *disclosure; *confidentiality; investigators; *legal aspects; legislation; methods; privacy; privileged communication; research subjects; statistics

BEHAVIORAL RESEARCH / DECEPTION

Bok, Sissela. Deceptive social science research. *In* her Lying: Moral Choice in Public and Private Life. New York: Pantheon Books, 1978. p. 182-202. 19 fn. ISBN 0-394-41370-0. BE06701.
alternatives; attitudes; *behavioral research; codes of ethics; debriefing; *deception; ethics committees; human experimentation; informed consent; investigators; moral obligations; psychology; research subjects; risks and benefits; social impact; standards; volunteers

Milgram, Stanley. Subject reaction: the neglected factor in the ethics of experimentation. *Hastings Center Report* 7(5): 19-23, Oct 1977. 8 fn. BE06078.
*attitudes; *behavioral research; *deception; evaluation; informed consent; psychological stress; psychology; *research subjects; social impact

Shipley, Thorne. Misinformed consent: an enigma in modern social science research. *Ethics in Science and Medicine* 4(3-4): 93-106, 1977. 21 fn. BE06066.
*behavioral research; communication; debriefing; *deception; *informed consent; investigator subject relationship; *methods; psychology; *research design; selection of subjects; social sciences; trust; volunteers

BEHAVIORAL RESEARCH / DECEPTION / COSTS AND BENEFITS

Eisner, Margaret S. Ethical problems in social psychological experimentation in the laboratory. *Canadian*

BE = bioethics accession number   fn. = footnotes   refs. = references

Psychological Review 18(3): 233-241, Jul 1977. 67 refs. BE05957.
> *behavioral research; *costs and benefits; *deception; decision making; investigators; *methods; psychology; research design; research subjects

BEHAVIORAL RESEARCH / DECEPTION / METHODS

West, Stephen G.; Gunn, Steven P. Some issues of ethics and social psychology. *American Psychologist* 33(1): 30-38, Jan 1978. 44 refs. 1 fn. BE06762.
> *behavioral research; *deception; decision making; investigators; *methods; motivation; *psychology; research subjects; risks and benefits; social sciences

BEHAVIORAL RESEARCH / DEONTOLOGICAL ETHICS

Schlenker, Barry R.; Forsyth, Donelson R. On the ethics of psychological research. *Journal of Experimental Social Psychology* 13(4): 369-396, Jul 1977. 34 refs. 5 fn. BE05823.
> *behavioral research; consequences; costs and benefits; deception; *deontological ethics; *ethical relativism; injuries; evaluation; moral policy; morality; psychology; research subjects; *teleological ethics

BEHAVIORAL RESEARCH / DRUG ABUSE / RIGHTS

Meyer, Roger E. Subjects' rights, freedom of inquiry, and the future of research in the addictions. *American Journal of Psychiatry* 134(8): 899-903, Aug 1977. 11 refs. BE06239.
> alcohol abuse; *behavioral research; case reports; *drug abuse; human experimentation; informed consent; investigators; medical ethics; newborns; politics; pregnant women; prisoners; *public advocacy; *public participation; research subjects; review committees; rights; social impact; society

BEHAVIORAL RESEARCH / ETHICAL RELATIVISM

Schlenker, Barry R.; Forsyth, Donelson R. On the ethics of psychological research. *Journal of Experimental Social Psychology* 13(4): 369-396, Jul 1977. 34 refs. 5 fn. BE05823.
> *behavioral research; consequences; costs and benefits; deception; *deontological ethics; *ethical relativism; injuries; evaluation; moral policy; morality; psychology; research subjects; *teleological ethics

BEHAVIORAL RESEARCH / ETHICS

Warwick, Donald P. Social sciences and ethics. *Hastings Center Report* 7(6): Special Suppl. 8-10, Dec 1977. BE06155.
> *behavioral research; deception; *education; *ethics; science; *social sciences; universities; values

BEHAVIORAL RESEARCH / ETHICS / PUBLIC OPINION

Apple, William. Ethical problems for subject or experimenter? *American Psychologist* 32(8): 683-684, Aug 1977. 9 refs. 1 fn. Letter. BE05829.
> *behavioral research; *ethics; evaluation; *public opinion

Sieber, Joan E. What is meant by ethics? *American Psychologist* 32(8): 684-685, Aug 1977. 1 ref. Letter. BE05830.
> *behavioral research; *ethics; methods; *public opinion

Wilson, David W.; Donnerstein, Edward. The public as an ethical consultant: a reply to Apple and to Sieber. *American Psychologist* 32(8): 685-686, Aug 1977. 3 refs. Letter. BE05831.
> *behavioral research; *ethics; investigators; *public opinion; risks and benefits

BEHAVIORAL RESEARCH / GOVERNMENT REGULATION

Chalkley, Donald T. Federal constraints: earned or unearned? *American Journal of Psychiatry* 134(8): 911-913, Aug 1977. 12 refs. 1 fn. BE06242.
> *behavioral research; confidentiality; constitutional law; ethical review; ethics committees; *federal government; *government regulation; historical aspects; human experimentation; informed consent; judicial action; legislation; privacy; public policy; social sciences; Department of Health, Education, and Welfare; Merriken v. Cressman

BEHAVIORAL RESEARCH / GOVERNMENT REGULATION / NORWAY

Oyen, Orjar. Social research and the protection of privacy: a review of the Norwegian development. *Acta Sociologica* 19(3): 249-262, 1976. 25 refs. 36 fn. BE06631.
> *behavioral research; *confidentiality; data bases; financial support; government agencies; *government regulation; legislation; privacy; social impact; *social sciences; *Norway

BEHAVIORAL RESEARCH / GOVERNMENT REGULATION / SOCIAL IMPACT

Robins, Lee N. Problems in follow-up studies. *American Journal of Psychiatry* 134(8): 904-907, Aug 1977. 3 refs. BE06240.
> adults; behavior disorders; *behavioral research; children; confidentiality; epidemiology; family; *government regulation; *informed consent; mentally ill; methods; records; research design; research subjects; *social impact

---

BE = bioethics accession number   fn. = footnotes   refs. = references

BEHAVIORAL RESEARCH / INFORMED CONSENT

Shipley, Thorne. Misinformed consent: an enigma in modern social science research. *Ethics in Science and Medicine* 4(3-4): 93-106, 1977. 21 fn. BE06066.
  *behavioral research; communication; debriefing; *deception; *informed consent; investigator subject relationship; *methods; psychology; *research design; selection of subjects; social sciences; trust; volunteers

BEHAVIORAL RESEARCH / INFORMED CONSENT / EVALUATION

Singer, Eleanor. Informed consent: consequences for response rate and response quality in social surveys. *American Sociological Review* 43(2): 144-162, Apr 1978. 57 refs. 21 fn. BE06946.
  *behavioral research; confidentiality; disclosure; *evaluation; *informed consent; research design

BEHAVIORAL RESEARCH / INFORMED CONSENT / RISKS AND BENEFITS

Lasser, Eric S.; Silverman, Lloyd H. Ethical considerations in pathology-intensifying research. *American Psychologist* 32(7): 577-578, Jul 1977. 10 refs. 2 fn. Letter. BE05728.
  *behavioral research; *disclosure; *informed consent; psychological stress; psychology; research subjects; risks and benefits

BEHAVIORAL RESEARCH / INFORMED CONSENT / SOCIAL IMPACT

Robins, Lee N. Problems in follow-up studies. *American Journal of Psychiatry* 134(8): 904-907, Aug 1977. 3 refs. BE06240.
  adults; behavior disorders; *behavioral research; children; confidentiality; epidemiology; family; *government regulation; *informed consent; mentally ill; methods; records; research design; research subjects; *social impact

BEHAVIORAL RESEARCH / INSTITUTIONALIZED PERSONS / MENTALLY HANDICAPPED

Protecting behavioral research subjects. *Science News* 113(9): 134, 4 Mar 1978. BE06925.
  *behavioral research; federal government; government regulation; *institutionalized persons; *mentally handicapped; National Commission for Protection of Human Subjects

Tannenbaum, Arnold S.; Cooke, Robert A. University of Michigan. Survey Research Center. Report on the mentally infirm. *In* U.S. National Commission for Protection of Human Subjects.... Research Involving Those Institutionalized as Mentally Infirm: Appendix. Washington: U.S. Government Printing Office, 1978. p. 1.1-1.117. DHEW Publication No.

(OS) 78-0007. 6 fn. BE06896.
  attitudes; *behavioral research; consent forms; ethical review; ethics committees; *human experimentation; informed consent; *institutionalized persons; investigators; *mentally handicapped; nontherapeutic research; research subjects; risks and benefits; selection of subjects; *statistics; therapeutic research; third party consent

BEHAVIORAL RESEARCH / PARENTAL CONSENT / SEXUALITY

Nilson, Donald R.; Steinfels, Margaret O. Parental consent and a teenage sex survey. *Hastings Center Report* 7(3): 13-15, Jun 1977. Case study. BE05658.
  *adolescents; *behavioral research; confidentiality; informed consent; *parental consent; research design; *sexuality; values

BEHAVIORAL RESEARCH / PRESIDENT'S COMMISSION FOR PROTECTION OF HUMAN SUBJECTS

U.S. Congress. Senate. President's Commission for the Protection of Human Subjects of Biomedical and Behavioral Research Act of 1978. S. 2579, 95th Cong., 2d Sess., 23 Feb 1978. 17 p. By Edward Kennedy, et al. Referred to the Committee on Human Resources. BE06718.
  *behavioral research; confidentiality; determination of death; ethics committees; federal government; genetic counseling; government agencies; *government regulation; *human experimentation; informed consent; privacy; *program descriptions; records; *President's Commission for Protection of Human Subjects

BEHAVIORAL RESEARCH / PRISONERS

California. Biomedical and behavioral research. *California Penal Code (Deering)*, Title 2.1, Sects. 3500-3524. Approved 1 Oct 1977. 6 p. BE06656.
  *behavioral research; drugs; ethics committees; *human experimentation; informed consent; legal liability; *prisoners; *California

BEHAVIORAL RESEARCH / PRIVACY

Kelman, Herbert C. Privacy and research with human beings. *Journal of Social Issues* 33(3): 169-195, 1977. 20 refs. BE07031.
  *behavioral research; *confidentiality; *deception; investigators; moral obligations; motivation; *privacy; psychological stress; research subjects; self determination

BEHAVIORAL RESEARCH / PRIVACY / DIGNITY

Koocher, Gerald P. Bathroom behavior and human dignity. *Journal of Personality and Social Psychology* 35(2): 120-121, Feb 1977. 3 refs. BE05908.

BE = bioethics accession number   fn. = footnotes   refs. = references

*behavioral research; costs and benefits; *dignity; editorial policies; *privacy; psychology; research subjects

BEHAVIORAL RESEARCH / PROFESSIONAL ETHICS / EVALUATION

McNamara, J. Regis; Woods, Kathryn M. Ethical considerations in psychological research: a comparative review. *Behavior Therapy* 8(4): 703-708, Sep 1977. 15 refs. BE06582.
*behavioral research; debriefing; deception; editorial policies; *evaluation; informed consent; *journalism; *professional ethics; psychology; standards

BEHAVIORAL RESEARCH / PUBLIC PARTICIPATION / DRUG ABUSE

Meyer, Roger E. Subjects' rights, freedom of inquiry, and the future of research in the addictions. *American Journal of Psychiatry* 134(8): 899-903, Aug 1977. 11 refs. BE06239.
alcohol abuse; *behavioral research; case reports; *drug abuse; human experimentation; informed consent; investigators; medical ethics; newborns; politics; pregnant women; prisoners; *public advocacy; *public participation; research subjects; review committees; rights; social impact; society

BEHAVIORAL RESEARCH / TELEOLOGICAL ETHICS

Schlenker, Barry R.; Forsyth, Donelson R. On the ethics of psychological research. *Journal of Experimental Social Psychology* 13(4): 369-396, Jul 1977. 34 refs. 5 fn. BE05823.
*behavioral research; consequences; costs and benefits; deception; *deontological ethics; *ethical relativism; injuries; evaluation; moral policy; morality; psychology; research subjects; *teleological ethics

BEHAVIORAL RESEARCH / TERMINALLY ILL

Simpson, Michael A.; Mount, Balfour M. Living with the dying. *Canadian Medical Association Journal* 117(1): 14-15, 9 Jul 1977. 2 refs. Letter. BE05970.
*behavioral research; *deception; *evaluation; hospitals; medical staff; psychological stress; *terminal care; *terminally ill; volunteers

BEHAVIORAL RESEARCH / XYY KARYOTYPE / AGGRESSION

Witkin, Herman A.; Goodenough, Donald R.; Hirschhorn, Kurt. XYY men: are they criminally aggressive? *Sciences* 17(6): 10-13, Oct 1977. BE05851.
*aggression; behavioral genetics; *behavioral research; criminal law; intelligence; methods; socioeconomic factors; *XYY karyotype

BEHAVIORAL RESEARCH   *See under*
CONFIDENTIALITY / BEHAVIORAL RESEARCH
CONFIDENTIALITY / BEHAVIORAL RESEARCH / NORWAY

BIOETHICAL ISSUES

Barber, Bernard, ed. Medical Ethics and Social Change. Philadelphia: American Academy of Political and Social Science, 1978. 201 p. Annals of the American Academy of Political and Social Science, Vol. 437, May 1978. 303 fn. ISBN 0-87761-226-9. BE06994.
abortion; allowing to die; *bioethical issues; critically ill; decision making; disclosure; drugs; epidemiology; health care delivery; historical aspects; human experimentation; informed consent; justice; legal aspects; *medical ethics; medical education; physicians; preventive medicine; public policy; resource allocation; selective abortion; self induced illness; social impact; withholding treatment

Duncan, A.S.; Dunstan, G.R.; Welbourn, R.B. Dictionary of Medical Ethics. London: Darton, Longman and Todd, 1977. 336 p. ISBN 0-232-51302-3. BE06677.
*bioethical issues; medical ethics

Fulton, Gere B. Bioethics and health education: some issues of the biological revolution. *Journal of School Health* 47(4): 205-211, Apr 1977. 21 refs. 2 fn. BE05556.
abortion; amniocentesis; *bioethical issues; brain death; cadavers; determination of death; embryo transfer; fetuses; genetic defects; human experimentation; in vitro fertilization; personhood; selective abortion

Harrell, George T. Some moral aspects of health care. *Maryland State Medical Journal* 26(12): 66-69, Dec 1977. BE06509.
abortion; *bioethical issues; death; drugs; genetic intervention; health; health care; medicine; population control; public opinion; religion; science; terminal care

Heyer, Robert, ed. Medical/Moral Problems. New York: Paulist Press, 1976. 64 p. ISBN 0-8091-2058-5. BE06516.
allowing to die; *bioethical issues; decision making; family members; genetic counseling; health care; human experimentation; informed consent; medical ethics; mentally retarded; newborns; organ transplantation; physician's role; psychotherapy; religion; rights; Roman Catholic ethics; terminally ill; value of life

Marcolongo, Francis J. Moral Choices in Contemporary Society: A Study Guide for Courses by Newspaper. Del Mar, Calif.: Publisher's Inc., 1977. 70 p. Study guide to accompany a series of sixteen weekly newspaper articles and an anthology; intended to integrate the themes of the newspaper and anthology articles. Bibliographies included. ISBN 0-89163-024-4. BE06536.

BE = bioethics accession number   fn. = footnotes   refs. = references

abortion; aged; *bioethical issues; biomedical research; discrimination; education; employment; ethics; family; industry; law; law enforcement; *morality; politics; science; sexuality

Robbins, Dennis A., comp. Ethical Dimensions of Biomedicine and Related Areas. Unpublished document. 1978. 335 p. Univ. of North Carolina, Symposium on Ethical Issues in Biomedicine and Related Areas, Wilmington, N.C., 12-14 Jan 1978. References and footnotes included. BE06982.
> allowing to die; *bioethical issues; decision making; disclosure; ethics; family members; fetuses; genetic counseling; genetic intervention; hospices; human experimentation; informed consent; nurses; patient care; personhood; physicians; recombinant DNA research; rights; risks and benefits; selective abortion; terminal care; terminally ill

Sobel, Lester A., ed. Medical Science and the Law: The Life and Death Controversy. New York: Facts on File, 1977. 185 p. ISBN 0-87196-286-1. BE06002.
> abortion; allowing to die; behavior control; *bioethical issues; carcinogens; contraception; drugs; genetic intervention; government regulation; health hazards; human experimentation; legislation; politics; psychosurgery; public policy; religious beliefs; sterilization; Supreme Court decisions; surgery; value of life

Spicker, Stuart F.; Engelhardt, H. Tristram, eds. Philosophical Medical Ethics: Its Nature and Significance. Proceedings. Boston: D. Reidel, 1977. 252 p. Proceedings of the Third Trans-Disciplinary Symposium on Philosophy and Medicine, Farmington, Connecticut, 11-13 Dec 1975. References and footnotes included. ISBN 90-277-0772-3. BE05858.
> abortion; *bioethical issues; children; cultural pluralism; euthanasia; evolution; goals; human experimentation; informed consent; *medical ethics; medicine; moral obligations; morality; parental consent; personhood; *philosophy; physician patient relationship; physician's role; prisoners; rights; self determination; suicide; utilitarianism; value of life; voluntary euthanasia; wedge argument

Taylor, Nancy K., comp. Bibliography of Society, Ethics and the Life Sciences: Supplement for 1977-78. Hastings-on-Hudson, N.Y.: Institute of Society, Ethics and the Life Sciences, 1977. 26 p. Full bibliography compiled by Sharmon Sollitto and Robert M. Veatch. Bibliography. BE06688.
> behavior control; *bioethical issues; bioethics; death; euthanasia; genetic intervention; health care delivery; human experimentation; *medical ethics; population control

BIOETHICAL ISSUES / ANGLICAN CHURCH OF CANADA

Hames, Jerry. Canadian Anglicans debate life-and-death issues. *Christian Century* 94(28): 772-773, 14

Sep 1977. BE05778.
> abortion; allowing to die; *bioethical issues; brain death; clergy; congenital defects; journalism; newborns; *organizational policies; politics; Protestantism; value of life; *Anglican Church of Canada

BIOETHICAL ISSUES / CHRISTIAN ETHICS

Hollis, Harry N. Biomedical ethics: an overview. *In* Hollis, Harry N., comp. A Matter of Life and Death: Christian Perspectives. Nashville: Broadman Press, 1977. p. 10-20. 10 fn. ISBN 0-8054-6118-3. BE06528.
> *bioethical issues; *Christian ethics; theology

Hollis, Harry N., comp. A Matter of Life and Death: Christian Perspectives. Nashville: Broadman Press, 1977. 143 p. 82 fn. ISBN 0-8054-6118-3. BE06527.
> abortion; behavior control; *bioethical issues; *Christian ethics; clergy; genetic defects; genetic intervention; human experimentation; organ transplantation; psychotherapy; reproductive technologies; selective abortion

BIOETHICAL ISSUES / DECISION MAKING

Shallenberger, Clyde R. Facing the brave new world. *Brethren Life and Thought* 21(3): 165-171, Summer 1976. BE06639.
> allowing to die; *bioethical issues; *decision making; ethics committees; genetic counseling; in vitro fertilization; involuntary sterilization; organ transplantation

Wampler, J. Paul. The physician and his patient: ethical decisions. *Brethren Life and Thought* 21(3): 185-190, Summer 1976. BE06640.
> abortion; allowing to die; *bioethical issues; *decision making; living wills; prolongation of life; terminally ill; treatment refusal; withholding treatment

BIOETHICAL ISSUES / INTENSIVE CARE UNITS

Powers, Samuel R. Ethical and legal considerations in the intensive care unit. *In* American College of Surgeons. Committee on Pre- and Postoperative Care, ed. Manual of Surgical Care. Philadelphia: W.B. Saunders, 1977. p. 191-199. 7 refs. ISBN 0-7216-1180-X. BE06524.
> allowing to die; *bioethical issues; bone marrow; decision making; economics; extraordinary treatment; family members; human experimentation; informed consent; *intensive care units; legal aspects; physician's role; prolongation of life; review committees; selection for treatment; suffering; terminal care; terminally ill; transplantation; withholding treatment

BIOETHICAL ISSUES / LAW / CANADA

Canada. Law Reform Commission. Protection of Life

Project: Project Description. Unpublished document. May 1977. 20 p. Available from the Law Reform Commission, 130 Albert Street, Ottawa, Ontario K1A 0L6. BE05697.
> behavior control; *bioethical issues; criminal law; determination of death; euthanasia; human experimentation; *law; methods; value of life; *Canada

BIOETHICAL ISSUES / MEDICAL RECORDS / LEGAL ASPECTS

Hayt, Emanuel. Medicolegal Aspects of Hospital Records. Second Edition. Berwyn, Ill.: Physicians' Record Company, 1977. 519 p. Footnotes included. ISBN 0-917036-13-1. BE06678.
> abortion; adults; artificial insemination; autopsies; blood transfusions; children; *confidentiality; consent forms; contraception; data bases; disclosure; famous persons; federal government; *hospitals; *informed consent; *legal aspects; legal obligations; legal rights; legislation; mass media; *medical records; medical staff; organ donation; organ transplantation; patients; physicians; privacy; *privileged communication; radiology; review committees; state government; sterilization

BIOETHICAL ISSUES / NURSES

Smith, Olive. Dilemmas in hospital nursing practice. *Nursing Mirror* 145(3): 22-24, 21 Jul 1977. BE05817.
> abortion; allowing to die; attitudes to death; *bioethical issues; cancer; confidentiality; conscience; contraception; disclosure; drugs; *nurses; nursing ethics; prolongation of life; suffering; terminally ill

BIOETHICAL ISSUES / PHYSICIANS

Harrell, George T. Moral choice in the daily practice of medicine. *Pennsylvania Medicine* 80(5): 45-50, May 1977. BE05930.
> abortion; aged; *bioethical issues; contraception; *decision making; determination of death; genetic defects; genetic intervention; genetic screening; health; human experimentation; medicine; physician's role; *physicians; prenatal diagnosis; self induced illness; sterilization

BIOETHICAL ISSUES / PUBLIC POLICY / PUBLIC OPINION

New Jersey Institute of Technology. Center for Technology Assessment; Policy Research Incorporated. A Comprehensive Study of the Ethical, Legal, and Social Implications of Advances in Biomedical and Behavioral Research and Technology. Springfield, Va.: National Technical Information Service, 1977. 297 p. Study conducted for the National Commission for the Protection of Human Subjects of Biomedical and Behavioral Research, Final Report, Jul 1977. PB-270 185. BE05889.
> behavior control; *bioethical issues; *biomedical technologies; data bases; economics; genetic screening; government regulation; health care; human experimentation; human rights; life extension; methods; privacy; *public opinion; public participation; *public policy; reproductive technologies; research design; resource allocation; social impact; values

BIOETHICS *See also* MEDICAL ETHICS, PROFESSIONAL ETHICS

BIOETHICS

Beauchamp, Tom L.; Walters, LeRoy, eds. Contemporary Issues in Bioethics. Encino, Calif.: Dickenson Publishing Company, 1978. 612 p. References and footnotes included. ISBN 0-8221-0200-5. BE06700.
> *abortion; allowing to die; *behavior control; *bioethics; biomedical technologies; brain death; children; codes of ethics; confidentiality; decision making; deontological ethics; *determination of death; *euthanasia; fetuses; genetic intervention; *health; *health care; *human experimentation; *informed consent; involuntary commitment; justice; legal aspects; *medical ethics; natural law; *normative ethics; operant conditioning; patients' rights; personhood; psychosurgery; *resource allocation; rights; selection for treatment; treatment refusal; utilitarianism; Kaimowitz v. Department of Mental Health; Quinlan, Karen

Zegel, Vikki A.; Parratt, Donna, eds. U.S. Library of Congress. Congressional Research Service. CRS Bioethics Workshop for Congress, 16 Jun 1977. Washington: U.S. Library of Congress, Congressional Research Service, 16 Dec 1977. 68 p. Report No. 78-1 SP. Available only through Members of Congress. 23 fn. Bibliography p. 64-68. BE06198.
> *bioethics; biomedical research; federal government; government regulation; health care; human experimentation; justice; medicine; public policy; recombinant DNA research; resource allocation; science

BIOETHICS / EDUCATION

Colen, B.D. Doctors, nurses study medical problem ethics. *Washington Post*, 20 Jun 1977, p. C2. BE06030.
> *bioethics; *education; medical ethics; nurses; physicians; program descriptions; *Kennedy Institute of Ethics

BIOETHICS / INSTITUTE OF SOCIETY, ETHICS AND THE LIFE SCIENCES

Kernan, Michael. Man, medicine, morality and the power of rights: the dilemma of bioethics. *Washington Post*, 3 Apr 1977, p. L1+. BE06010.
> bioethical issues; *bioethics; program descriptions;

BE = bioethics accession number   fn. = footnotes   refs. = references

research institutes; rights; *Institute of Society, Ethics and the Life Sciences

BIOLOGICAL LIFE　*See under*

VALUE OF LIFE / BIOLOGICAL LIFE

BIOLOGICAL WARFARE / DEPARTMENT OF THE ARMY

U.S. Congress. Senate. Committee on Human Resources. Subcommittee on Health and Scientific Research. Biological Testing Involving Human Subjects by the Department of Defense, 1977. Hearings. Washington: U.S. Government Printing Office, 1977. 297 p. Hearings...95th Cong., 1st Sess., on Examination of Serious Deficiencies in the Defense Department's Efforts to Protect the Human Subjects of Drug Research, 8 Mar and 23 May 1977. BE06521.
　　　*biological warfare; federal government; government agencies; historical aspects; human experimentation; program descriptions; *Department of the Army

BIOLOGY / SOCIAL IMPACT

Rosenfeld, Albert. When man becomes as God: the biological prospect. *Saturday Review* 5(6): 15-20, 10 Dec 1977. BE06376.
　　　behavioral genetics; *biology; evolution; genetics; recombinant DNA research; risks and benefits; *science; social determinants; *social impact

BIOMEDICAL RESEARCH / COSTS AND BENEFITS

Coleman, Vernon. Paper Doctors: A Critical Assessment of Medical Research. London: Maurice Temple Smith, 1977. 170 p. Bibliography: p. 162-164. ISBN 0-85117-1095. BE06675.
　　　*biomedical research; cancer; *costs and benefits; *dehumanization; drugs; ecology; economics; financial support; health hazards; genetic intervention; *human experimentation; incentives; investigators; journalism; mentally ill; morbidity; organ transplantation; *patient care; physicians; professional organizations; prolongation of life; psychosurgery; volunteers

Rescher, Nicholas. Ethical issues regarding the delivery of health-care services. *Connecticut Medicine* 41(8): 501-506, Aug 1977. 6 fn. BE05973.
　　　abortion; *biomedical research; *costs and benefits; economics; future generations; *health care delivery; human equality; moral obligations; *obligations of society; resource allocation; self induced illness; state medicine

BIOMEDICAL RESEARCH / DECISION MAKING

Hunter, Thomas H.; Wenzel, Richard; Fletcher, Jo-

seph. How Does One Determine Acceptable Risks? [Videorecording]. Charlottesville, Va.: University of Virginia Medical Center, Health Sciences Library, 1976. 1 cassette; 60 min.; sound; black and white; 3/4 in. Tape of the Medical Center Hour, 1 Dec 1976. Panel discussion moderated by Thomas H. Hunter. BE07051.
　　　alternatives; *biomedical research; consequences; *decision making; health care; hospitals; *patient care; *physicians; quality of life; recombinant DNA research; *risks and benefits; statistics; *surgery; teleological ethics; values

BIOMEDICAL RESEARCH / FINANCIAL SUPPORT

Coste, Chris. The brave new world of biomedical research. *New Physician* 26(7): 18-22, Jul 1977. BE06317.
　　　*biomedical research; conflict of interest; *federal government; *financial support; government regulation; politics; public policy; science; *social control

BIOMEDICAL RESEARCH / GOVERNMENT REGULATION / LEGAL ASPECTS

Ladimer, Irving. Root and branch: legal aspects of biomedical studies in man and other animals. *In* The Future of Animals, Cells, Models, and Systems in Research, Development, Education, and Testing. Washington: National Academy of Sciences, 1977. p. 296-317. Proceedings of a Symposium, National Research Council—National Academy of Sciences, 22-23 Oct 1975. 18 refs. ISBN 0-309-02603-2. BE05883.
　　　*biomedical research; federal government; *government regulation; human experimentation; investigators; *legal aspects; legal liability; *legislation; methods; property rights; self regulation; state government

BIOMEDICAL RESEARCH / JOURNALISM

Culliton, Barbara J. Science, society and the press. *New England Journal of Medicine* 296(25): 1450-1453, 23 Jun 1977. 2 fn. BE05646.
　　　*biomedical research; federal government; financial support; government regulation; *interdisciplinary communication; investigators; *journalism; medicine; professional organizations; public participation; risks and benefits; *science

BIOMEDICAL RESEARCH / PUBLIC PARTICIPATION

Powledge, Tabitha M.; Dach, Leslie, eds. Biomedical Research and the Public: Proceedings of a Conference Co-sponsored by the Institute of Society, Ethics and the Life Sciences and Case Western Reserve University, 1-3 Apr 1976. Washington: U.S. Government Printing Office, 1977. 154 p. Edited transcript pre-

---

BE = bioethics accession number　　　　fn. = footnotes　　　　refs. = references

pared for the Subcommittee on Health and Scientific Research of the Committee on Human Resources, U.S. Senate. 37 fn. BE06004.
  *biomedical research; decision making; economics; federal government; freedom; government agencies; government regulation; *human experimentation; investigators; politics; *public participation; public policy; resource allocation; science; self regulation; *social control

BIOMEDICAL RESEARCH / SOCIAL CONTROL

Coste, Chris. The brave new world of biomedical research. *New Physician* 26(7): 18-22, Jul 1977. BE06317.
  *biomedical research; conflict of interest; *federal government; *financial support; government regulation; politics; public policy; science; *social control

Powledge, Tabitha M.; Dach, Leslie, eds. Biomedical Research and the Public: Proceedings of a Conference Co-sponsored by the Institute of Society, Ethics and the Life Sciences and Case Western Reserve University, 1-3 Apr 1976. Washington: U.S. Government Printing Office, 1977. 154 p. Edited transcript prepared for the Subcommittee on Health and Scientific Research of the Committee on Human Resources, U.S. Senate. 37 fn. BE06004.
  *biomedical research; decision making; economics; federal government; freedom; government agencies; government regulation; *human experimentation; investigators; politics; *public participation; public policy; resource allocation; science; self regulation; *social control

Thorup, Oscar A., et al. Sociobiology: Are There Areas of Forbidden Knowledge? [Videorecording]. Charlottesville, Va.: University of Virginia Medical Center, Health Sciences Library, 1978. 1 cassette; 60 min.; sound; black and white; 3/4 in. Tape of the Medical Center Hour, 1 Mar 1978. Panel discussion moderated by Oscar A. Thorup. BE07056.
  behavioral genetics; *biomedical research; discrimination; ethics committees; government regulation; informal social control; intelligence; legal aspects; risks and benefits; *social control; *sociobiology; Science for the People; Wilson, Edward O.

BIOMEDICAL RESEARCH / SOCIAL CONTROL / HEALTH HAZARDS

Sutton, R.N.P. Science versus safety: who should judge the balance? *Nature* 270(5633): 90, 10 Nov 1977. BE06124.
  *biomedical research; *health hazards; industry; physical containment; *social control

BIOMEDICAL RESEARCH / SOCIAL CONTROL / RISKS AND BENEFITS

Stokes, Joseph, et al. Science with a halo or hubris? *New England Journal of Medicine* 296(24): 1417-1418, 16 Jun 1977. 2 refs. Letter. BE05670.
  *biomedical research; cloning; recombinant DNA research; *risks and benefits; science; *social control

BIOMEDICAL RESEARCH / SOCIAL CONTROL / SOCIAL IMPACT

Baltimore, David. Limiting science: a biologist's perspective. *Daedalus* 107(2): 37-45, Spring 1978. 4 fn. BE06935.
  biology; *biomedical research; DNA therapy; recombinant DNA research; risks and benefits; *science; *social control; *social impact

Eisenberg, Leon. The social imperatives of medical research. *Science* 198(4322): 1105-1110, 16 Dec 1977. 43 fn. BE06137.
  *biomedical research; control groups; developing countries; diagnosis; historical aspects; *human experimentation; methods; morbidity; mortality; patient care; physicians; risks and benefits; *social control; *social impact; surgery; thalassemia

BIOMEDICAL RESEARCH   See under
TORTURE / BIOMEDICAL RESEARCH

BIOMEDICAL TECHNOLOGIES / CHRISTIAN ETHICS

Ellison, Craig W., ed. Modifying Man: Implications and Ethics. Washington: University Press of America, 1978. 294 p. 203 fn. ISBN 0-8191-0302-0. BE06666.
  behavior control; *biomedical technologies; *Christian ethics; decision making; dignity; electrical stimulation of the brain; genetic intervention; justice; Protestant ethics; psychosurgery; public policy; reproductive technologies; scriptural interpretation; values

BIOMEDICAL TECHNOLOGIES / COSTS AND BENEFITS

Hiatt, Howard H. Too much medical technology? *Wall Street Journal (Eastern Edition)*, 24 Jun 1976, p. 16. BE06612.
  *biomedical technologies; *costs and benefits; decision making; diagnosis; health care; physician's role; resource allocation

BIOMEDICAL TECHNOLOGIES / DECISION MAKING

Callahan, Daniel. Control technology, values and the future. *In* Ellison, Craig W., ed. Modifying Man: Implications and Ethics. Washington: University Press of America, 1978. p. 39-49. ISBN 0-8191-0302-0. BE06668.
  *biomedical technologies; *decision making; population control; prolongation of life; reproductive technologies; science; social control; values

---

BE = bioethics accession number   fn. = footnotes   refs. = references

Ellison, Craig W. The ethics of human engineering. *In* Ellison, Craig W., ed. Modifying Man: Implications and Ethics. Washington: University Press of America, 1978. p. 3-35. 40 fn. ISBN 0-8191-0302-0. BE06667.
behavior control; *biomedical technologies; Christian ethics; *decision making; *ethical analysis; *human characteristics; human experimentation; reproductive technologies; values

BIOMEDICAL TECHNOLOGIES / FUTURE GENERATIONS / CHRISTIAN ETHICS

Hollis, Harry N. Biomedical ethics and the future: the response of the church. *In* Hollis, Harry N., comp. A Matter of Life and Death: Christian Perspectives. Nashville: Broadman Press, 1977. p. 130-143. 26 fn. ISBN 0-8054-6118-3. BE06535.
biomedical research; *biomedical technologies; *Christian ethics; *future generations; religion

BIOMEDICAL TECHNOLOGIES / INDUSTRY / SOCIAL IMPACT

Callahan, Daniel. Moral dilemma for business. *New York Times,* 13 Nov 1977, sect. 3, p. 1+. BE06410.
*biomedical technologies; costs and benefits; decision making; government regulation; *industry; recombinant DNA research; resource allocation; *social impact

BIOMEDICAL TECHNOLOGIES / SOCIAL IMPACT

Bodmer, W.F. Social concern and biological advances. *Journal of the Royal Society of Arts* 125(5248): 180-194, Mar 1977. Discussion follows essay on p. 192-194. 5 refs. BE06293.
aged; *biomedical technologies; communication; education; *genetic diversity; genetics; human equality; *intelligence; mass media; *population growth; *recombinant DNA research; science; social determinants; *social impact

Reid, Robert. The moral dilemmas of the biological revolution. *Times (London),* 28 Jun 1977, p. 14. BE06041.
amniocentesis; *biomedical technologies; genetic defects; *prenatal diagnosis; *selective abortion; sex determination; *social impact

Smith, Harmon L. Threats to the individual. *Social Science and Medicine* 11(8-9): 449-451, May 1977. BE05932.
abortion; *biomedical technologies; *decision making; health care; organ transplantation; reproductive technologies; resource allocation; social control; *social impact; sterilization

BIOMEDICAL TECHNOLOGIES *See under*

PROLONGATION OF LIFE / BIOMEDICAL TECHNOLOGIES / PHYSICIAN'S ROLE

BIRTH CONTROL *See* CONTRACEPTION

BLACKS *See under*

ABORTION / BLACKS / SOCIAL IMPACT

BEHAVIOR CONTROL / BLACKS

BEHAVIORAL GENETICS / BLACKS

BEHAVIORAL RESEARCH / BLACKS

POPULATION CONTROL / BLACKS / VALUES

VOLUNTARY STERILIZATION / BLACKS

BLOOD DONATION / ALTRUISM

Culyer, A.J. Blood and altruism: an economic review. *In* Johnson, David B., ed. Blood Policy: Issues and Alternatives. Washington: American Enterprise Institute for Public Policy Research, 1977. p. 39-58. Conference on Blood Policy, Washington, D.C., 1976. 48 fn. ISBN 0-8447-2105-0. BE06676.
*altruism; *blood donation; community services; *economics; hepatitis; hospitals; industry; legal liability; risks and benefits; values; Titmuss, Richard M.

Plant, Raymond. Gifts, exchanges and the political economy of health care. Part I: Should blood be bought and sold? *Journal of Medical Ethics* 3(4): 166-173, Dec 1977. 40 fn. BE06257.
*altruism; *blood donation; disclosure; donors; economics; health care; motivation; public policy; *remuneration; self determination; social impact; social interaction; values; voluntary programs; *Gift Relationship*; *Titmuss, Richard

BLOOD DONATION / ECONOMICS

Culyer, A.J. Blood and altruism: an economic review. *In* Johnson, David B., ed. Blood Policy: Issues and Alternatives. Washington: American Enterprise Institute for Public Policy Research, 1977. p. 39-58. Conference on Blood Policy, Washington, D.C., 1976. 48 fn. ISBN 0-8447-2105-0. BE06676.
*altruism; *blood donation; community services; *economics; hepatitis; hospitals; industry; legal liability; risks and benefits; values; Titmuss, Richard M.

BLOOD DONATION / REMUNERATION

Plant, Raymond. Gifts, exchanges and the political economy of health care. Part I: Should blood be bought and sold? *Journal of Medical Ethics* 3(4): 166-173, Dec 1977. 40 fn. BE06257.
*altruism; *blood donation; disclosure; donors; economics; health care; motivation; public policy; *remuneration; self determination; social impact; social interaction; values; voluntary programs; *Gift Relationship*; *Titmuss, Richard

---

BE = bioethics accession number   fn. = footnotes   refs. = references

BLOOD TRANSFUSIONS *See under*

TREATMENT REFUSAL / BLOOD TRANSFU-
SIONS / RELIGIOUS BELIEFS

BODY PARTS AND FLUIDS *See under*

CLONING / BODY PARTS AND FLUIDS

BRAIN DEATH *See also* COMA

BRAIN DEATH

Black, Peter M. Definitions of brain death. *In* Beauchamp, Tom L.; Perlin, Seymour, eds. Ethical Issues
in Death and Dying. Englewood Cliffs, N.J.: Prentice-Hall, 1978. p. 5-10. 9 refs. ISBN 0-13-
290114-5. BE06879.
*brain death; determination of death; standards

Too many tests cloud brain-death diagnosis. *Medical
World News* 18(26): 24-25, 26 Dec 1977. BE06377.
allowing to die; *brain death; determination of
death; EEG; prognosis

Veith, Frank J., et al. Brain death. I. A status report
of medical and ethical considerations. *Journal of the
American Medical Association* 238(15): 1651-1655, 10
Oct 1977. 42 refs. BE06095.
*brain death; costs and benefits; determination of
death; EEG; Jewish ethics; legislation; Protestant
ethics; Roman Catholic ethics; Harvard Committee on Brain Death

Woolsey, Robert M. Death of the brain. *Missouri
Medicine* 74(9): 540-541+, Sep 1977. 10 refs.
BE05993.
*brain death; coma; legal aspects; withholding
treatment

BRAIN DEATH / LEGAL ASPECTS

Veith, Frank J., et al. Brain death. II. A status report
of legal considerations. *Journal of the American
Medical Association* 238(16): 1744-1748, 17 Oct
1977. 21 fn. BE06096.
*brain death; determination of death; killing; law
enforcement; *legal aspects; legislation; standards;
state government

BRAIN DEATH / NORTH CAROLINA

North Carolina. Right to natural death; brain death.
General Statutes of North Carolina, Article 23, Sect.
90-320 to 90-322. Effective 1 Jul 1977. 5 p.
BE06218.
*allowing to die; *brain death; extraordinary treat-
ment; *living wills; *terminally ill; *withholding
treatment; *North Carolina

BRAIN DEATH / STANDARDS

Powner, David J.; Snyder, James V.; Grenvik, Ake.
Brain death certification: a review. *Critical Care
Medicine* 5(5): 230-233, Sep-Oct 1977. 79 refs.

BE06336.
*brain death; determination of death; EEG; legal
aspects; *standards

Samuel, V.N. Brain death. *Diseases of the Nervous
System* 38(9): 691-693, Sep 1977. 13 refs.
BE06337.
allowing to die; *brain death; costs and benefits;
EEG; legal aspects; *standards

BRAIN DEATH *See under*

ALLOWING TO DIE / BRAIN DEATH
ALLOWING TO DIE / BRAIN DEATH / LAW EN-
FORCEMENT
ALLOWING TO DIE / BRAIN DEATH / LEGAL LI-
ABILITY
DETERMINATION OF DEATH / BRAIN DEATH
KILLING / BRAIN DEATH / LAW ENFORCE-
MENT
ORGAN TRANSPLANTATION / BRAIN DEATH
PROLONGATION OF LIFE / PREGNANT
WOMEN / BRAIN DEATH

BRAIN PATHOLOGY *See under*

ALLOWING TO DIE / BRAIN PATHOLOGY / LE-
GAL ASPECTS
LEGAL LIABILITY / BRAIN PATHOLOGY / CRIM-
INAL LAW
PATIENT CARE / CHRONICALLY ILL / BRAIN
PATHOLOGY

BRAIN *See under*

ORGAN TRANSPLANTATION / PERSONHOOD /
BRAIN

BROWN, LOUISE *See under*

IN VITRO FERTILIZATION / BROWN, LOUISE

BURN PATIENTS *See under*

ALLOWING TO DIE / BURN PATIENTS
PATIENT CARE / BURN PATIENTS
SELF DETERMINATION / BURN PATIENTS

CADAVERS *See under*

HUMAN EXPERIMENTATION / CADAVERS / IN-
JURIES
ORGAN DONATION / CADAVERS
ORGAN TRANSPLANTATION / CADAVERS
ORGAN TRANSPLANTATION / CADAVERS / HU-
MAN TISSUE ACT 1961
ORGAN TRANSPLANTATION / CADAVERS /
PORTUGAL

CALIFORNIA NATURAL DEATH ACT *See under*

ALLOWING TO DIE / CALIFORNIA NATURAL
DEATH ACT

BE = bioethics accession number   fn. = footnotes   refs. = references

CALIFORNIA NATURAL DEATH ACT *See under (cont'd.)*

ALLOWING TO DIE / HOSPITALS / CALIFORNIA NATURAL DEATH ACT

CALIFORNIA   *See under*

ELECTROCONVULSIVE THERAPY / GOVERNMENT REGULATION / CALIFORNIA

ELECTROCONVULSIVE THERAPY / LEGISLATION / CALIFORNIA

GOVERNMENT REGULATION / PSYCHOACTIVE DRUGS / CALIFORNIA

HUMAN EXPERIMENTATION / CALIFORNIA

INFORMED CONSENT / PRISONERS / CALIFORNIA

PRIVILEGED COMMUNICATION / LEGISLATION / CALIFORNIA

PSYCHOACTIVE DRUGS / ADOLESCENTS / CALIFORNIA

PSYCHOACTIVE DRUGS / PRISONERS / CALIFORNIA

RECOMBINANT DNA RESEARCH / GOVERNMENT REGULATION / CALIFORNIA

RECOMBINANT DNA RESEARCH / REVIEW COMMITTEES / CALIFORNIA

TREATMENT REFUSAL / INSTITUTIONALIZED PERSONS / CALIFORNIA

CAMBRIDGE   *See under*

RECOMBINANT DNA RESEARCH / GOVERNMENT REGULATION / CAMBRIDGE

RECOMBINANT DNA RESEARCH / PUBLIC ADVOCACY / CAMBRIDGE

RECOMBINANT DNA RESEARCH / PUBLIC PARTICIPATION / CAMBRIDGE

RECOMBINANT DNA RESEARCH / SOCIAL CONTROL / CAMBRIDGE

CANADA   *See under*

ABORTION / CANADA

ABORTION / CRIMINAL LAW / CANADA

ABORTION / ETHICS COMMITTEES / CANADA

ABORTION / LEGISLATION / CANADA

BIOETHICAL ISSUES / LAW / CANADA

INFORMED CONSENT / ADOLESCENTS / CANADA

INFORMED CONSENT / LEGAL LIABILITY / CANADA

PATIENT CARE / PARENTAL CONSENT / CANADA

CANCER   *See under*

COMMUNICATION / TERMINALLY ILL / CANCER

DISCLOSURE / CANCER / COMMUNICATION

DISCLOSURE / CANCER / DIAGNOSIS

DISCLOSURE / CANCER / PHYSICIAN'S ROLE

CANCER *See under (cont'd.)*

DISCLOSURE / CANCER / PROGNOSIS

DISCLOSURE / TERMINALLY ILL / CANCER

HUMAN EXPERIMENTATION / CANCER / COMMON GOOD

MASS SCREENING / CANCER

PATIENT CARE / CANCER / ECONOMICS

PATIENT CARE / CHILDREN / CANCER

PATIENT CARE / JEHOVAH'S WITNESSES / CANCER

PATIENT CARE / LAETRILE / CANCER

PATIENT CARE / PSYCHOACTIVE DRUGS / CANCER

SELF DETERMINATION / LAETRILE / CANCER

CAPITAL PUNISHMENT

Anderson, Norman. Issues of Life and Death: Abortion, Birth Control, Capital Punishment, Euthanasia. Downers Grove, Ill.: InterVarsity Press, 1977. 130 p. 189 fn. ISBN 0-87784-721-5. BE05787.
AID; *abortion; *artificial insemination; *capital punishment; Christian ethics; *contraception; DNA therapy; *euthanasia; genetic intervention; justifiable killing; organ transplantation; personhood; scriptural interpretation; *sterilization; theology; *value of life

CAPITAL PUNISHMENT / DRUGS

Berlyn, Simon. Execution by the needle. *New Scientist* 75(1069): 676-677, 15 Sep 1977. BE05772.
*capital punishment; *drugs; physician's role; state government; *venepuncture

CARCINOGENS / MASS MEDIA / SOCIAL IMPACT

Weisburger, John H. Social and ethical implications of claims for cancer hazards. *Medical and Pediatric Oncology* 3(2): 137-140, 1977. BE06478.
biomedical research; cancer; *carcinogens; food; *mass media; nutrition; smoking; *social impact

CARCINOGENS   *See under*

RECOMBINANT DNA RESEARCH / CARCINOGENS

CARE FOR THE DYING   *See* TERMINAL CARE

CASE REPORTS   *See under*

ABORTION / CASE REPORTS

DEATH / CASE REPORTS

INVOLUNTARY COMMITMENT / CASE REPORTS / USSR

BE = bioethics accession number   fn. = footnotes   refs. = references

CATHOLIC HOSPITAL ASSOCIATION *See under*

WITHHOLDING TREATMENT / CATHOLIC HOSPITAL ASSOCIATION

CATHOLICISM, ROMAN *See* ROMAN CATHOLICISM

CENTRAL INTELLIGENCE AGENCY *See under*

BEHAVIOR CONTROL / CENTRAL INTELLIGENCE AGENCY

BEHAVIORAL RESEARCH / CENTRAL INTELLIGENCE AGENCY

HUMAN EXPERIMENTATION / CENTRAL INTELLIGENCE AGENCY

HUMAN EXPERIMENTATION / DRUGS / CENTRAL INTELLIGENCE AGENCY

CHILD NEGLECT / DIAGNOSIS / LEGAL ASPECTS

Curran, William J. Failure to diagnose battered-child syndrome. *New England Journal of Medicine* 296(14): 795-796, 7 Apr 1977. 4 refs. BE05549.
    *child neglect; *diagnosis; disclosure; injuries; *legal aspects; *legal liability; parents; *physicians

CHILD NEGLECT / PHYSICIAN'S ROLE

Tangen, Ottar. Medical ethics and child abuse. *Scandanavian Journal of Social Medicine* 5(2): 85-90, 1977. 22 refs. BE06283.
    *case reports; *child neglect; diagnosis; fathers; mothers; parent child relationship; *physician's role

CHILDREN *See also* ADOLESCENTS, AID CHILDREN, INFANTS, NEWBORNS

CHILDREN / LEGAL RIGHTS

Mnookin, Robert H. Children's rights: legal and ethical dilemmas. *Pharos* 41(2): 2-7, Apr 1978. 12 fn. BE06941.
    age; allowing to die; *children; competence; conflict of interest; decision making; *health care; involuntary commitment; *legal rights; newborns; *parent child relationship; parental consent

CHILDREN / PATIENTS' RIGHTS

Suran, Bernard G.; Lavigne, John V. Rights of children in pediatric settings: a survey of attitudes. *Pediatrics* 60(5): 715-720, Nov 1977. 12 refs. BE06852.
    *attitudes; *children; health personnel; legislation; nurses; parents; *patient care; *patients' rights; physicians

CHILDREN *See under*
BEHAVIOR CONTROL / CHILDREN / HYPERKIN-

CHILDREN *See under (cont'd.)*
ESIS

HEALTH CARE / CHILDREN / LEGAL ASPECTS

HUMAN EXPERIMENTATION / CHILDREN

HUMAN EXPERIMENTATION / CHILDREN / CHRISTIAN ETHICS

HUMAN EXPERIMENTATION / CHILDREN / COMMUNICATION

HUMAN EXPERIMENTATION / CHILDREN / DRUGS

HUMAN EXPERIMENTATION / CHILDREN / GREAT BRITAIN

HUMAN EXPERIMENTATION / CHILDREN / LEGAL ASPECTS

HUMAN EXPERIMENTATION / CHILDREN / LEGAL RIGHTS

HUMAN EXPERIMENTATION / CHILDREN / MENTALLY HANDICAPPED

HUMAN EXPERIMENTATION / CHILDREN / NATIONAL COMMISSION FOR PROTECTION OF HUMAN SUBJECTS

HUMAN EXPERIMENTATION / CHILDREN / RISKS AND BENEFITS

HUMAN EXPERIMENTATION / CHILDREN / STANDARDS

HUMAN EXPERIMENTATION / CHILDREN / VALUES

IMMUNIZATION / CHILDREN / COMPENSATION

INFORMED CONSENT / CHILDREN

INFORMED CONSENT / CHILDREN / GREAT BRITAIN

INVOLUNTARY COMMITMENT / CHILDREN / LEGAL RIGHTS

INVOLUNTARY COMMITMENT / CHILDREN / MENTALLY HANDICAPPED

ORGAN DONATION / CHILDREN

ORGAN DONATION / CHILDREN / LEGAL ASPECTS

PARENTAL CONSENT / CHILDREN

PATIENT CARE / CHILDREN / CANCER

PATIENT CARE / CHILDREN / GENETIC DEFECTS

PATIENT CARE / CHILDREN / LEUKEMIA

PROLONGATION OF LIFE / CHILDREN / GENETIC DEFECTS

TERMINAL CARE / CHILDREN / COMMUNICATION

TERMINAL CARE / CHILDREN / PHYSICIAN'S ROLE

TRANSPLANTATION / CHILDREN / RISKS AND BENEFITS

CHRISTIAN ETHICS *See also* JEWISH ETHICS, ROMAN CATHOLIC ETHICS

BE = bioethics accession number   fn. = footnotes   refs. = references

CHRISTIAN ETHICS *See under*
   ABORTION / CHRISTIAN ETHICS
   ABORTION / VALUE OF LIFE / CHRISTIAN ETHICS
   ALLOWING TO DIE / NEWBORNS / CHRISTIAN ETHICS
   AMNIOCENTESIS / SELECTIVE ABORTION / CHRISTIAN ETHICS
   BIOETHICAL ISSUES / CHRISTIAN ETHICS
   BIOMEDICAL TECHNOLOGIES / CHRISTIAN ETHICS
   BIOMEDICAL TECHNOLOGIES / FUTURE GENERATIONS / CHRISTIAN ETHICS
   CONTRACEPTION / CHRISTIAN ETHICS
   EUTHANASIA / CHRISTIAN ETHICS
   EUTHANASIA / VALUE OF LIFE / CHRISTIAN ETHICS
   HUMAN EXPERIMENTATION / CHILDREN / CHRISTIAN ETHICS
   HUMAN EXPERIMENTATION / CHRISTIAN ETHICS
   HUMAN EXPERIMENTATION / FETUSES / CHRISTIAN ETHICS
   MEDICAL ETHICS / CHRISTIAN ETHICS
   MEDICINE / CHRISTIAN ETHICS
   PROLONGATION OF LIFE / CHRISTIAN ETHICS
   STERILIZATION / CHRISTIAN ETHICS

CHRONICALLY ILL *See under*
   PATIENT CARE / CHRONICALLY ILL / BRAIN PATHOLOGY
   RESUSCITATION / NURSES / CHRONICALLY ILL

CHURCH OF ENGLAND *See under*
   TERMINAL CARE / CHURCH OF ENGLAND

CIVIL COMMITMENT *See* INVOLUNTARY COMMITMENT

CIVIL RIGHTS *See* LEGAL RIGHTS

CLERGY *See under*
   DISCLOSURE / CLERGY / MEDICAL RECORDS

CLINICAL RESEARCH *See* HUMAN EXPERIMENTATION

CLONCE V. RICHARDSON *See under*
   OPERANT CONDITIONING / CLONCE V. RICHARDSON

CLONING

The cloning of a man: debate begins. *Science News* 113(11): 164, 18 Mar 1978. BE06810.
   animal experimentation; biomedical research;

clones; *cloning; disclosure; political activity; Rorvik, David

Cloning. New York: WNET/13, 1978. 9 p. Transcript of the MacNeil/Lehrer Report, Show No. 3200, Library No. 660, 7 Apr 1978. Available from WNET/13, Box 345, New York, N.Y., 10019. Television script. BE07006.
   animal experimentation; *cloning; federal government; genetic diversity; genetic intervention; government regulation; methods; social impact

Culliton, Barbara J. The clone ranger. *Columbia Journalism Review* 17(2): 58-62, Jul/Aug 1978. BE07009.
   *cloning; journalism; literature; methods; *In His Image: The Cloning of a Man*; Rorvik, David

Hellegers, André E. Book on cloning is a hoax. *Washington Post*, 4 Jun 1978, p. B3. BE07007.
   *cloning; literature; methods; *In His Image: The Cloning of a Man*; Rorvik, David

Johnson, Herschel. Cloning: can science make carbon copies of you? *Ebony* 33(9): 95-96+, Jul 1978. BE07008.
   *cloning; federal government; methods; personality; risks and benefits; *In His Image: The Cloning of a Man*; Rorvik, David

Rorvik, David M. In His Image: The Cloning of a Man. Philadelphia: J.B. Lippincott, 1978. 239 p. 61 fn. Bibliography: p. 223-234. ISBN 0-397-01255-1. BE06703.
   *case reports; *cloning; methods

CLONING / BODY PARTS AND FLUIDS

Gould, Donald. Human spare parts. *New Statesman* 93(2392): 82-83, 21 Jan 1977. BE05904.
   biomedical technologies; *body parts and fluids; *cloning

CLONING / RORVIK, DAVID

Culliton, Barbara J. Scientists dispute book's claim that human clone has been born. *Science* 199(4335): 1314-1316, 24 Mar 1978. 1 fn. BE06812.
   *attitudes; *clones; *cloning; disclosure; editorial policies; government agencies; human experimentation; *investigators; journalism; political activity; *In His Image: The Cloning of a Man*; J.B. Lippincott; *Rorvik, David

CODES OF ETHICS *See also* MEDICAL ETHICS, PROFESSIONAL ETHICS

CODES OF ETHICS

Berlant, Jeffrey L. Medical ethics and professional monopoly. *In* Barber, Bernard, ed. Medical Ethics and Social Change. Philadelphia: American Academy of Political and Social Science, 1978. p. 49-61. Annals of the American Academy of Political and

---

BE = bioethics accession number   fn. = footnotes   refs. = references

Social Science, Vol. 437, May 1978. 23 fn. ISBN 0-87761-226-9. BE06999.
*codes of ethics; *historical aspects; *medical ethics; organizational policies; physicians; professional ethics; professional organizations

Burns, Chester R., ed. Legacies in Ethics and Medicine. New York: Science History Publications, 1977. 326 p. References and footnotes included. ISBN 0-88202-166-4. BE05683.
*codes of ethics; historical aspects; *medical ethics; moral obligations; physicians

CODES OF ETHICS / AMERICAN MEDICAL ASSOCIATION

Nortell, Bruce. AMA Judicial Council activities. *Journal of the American Medical Association* 239(14): 1396-1397, 3 Apr 1978. BE06944.
*codes of ethics; medical ethics; professional organizations; *American Medical Association

Physician comment sought on changes in ethics code. *Connecticut Medicine* 42(2): 139, Feb 1978. BE06914.
*codes of ethics; *medical ethics; physicians; professional organizations; *American Medical Association; Principles of Medical Ethics

CODES OF ETHICS / HAWAII DECLARATION

Blomquist, Clarence D. From the Oath of Hippocrates to the Declaration of Hawaii. *Ethics in Science and Medicine* 4(3-4): 139-149, 1977. Introductory essay to a draft of an international code of ethics for psychiatrists prepared for the World Psyciatric Assoc. and the CIBA Foundation meeting...London, June 1976. 39 fn. BE06069.
*codes of ethics; historical aspects; law; *medical ethics; medical etiquette; paternalism; physician patient relationship; physician's role; professional ethics; religion; utilitarianism; *Hawaii Declaration; *Hippocratic Oath; World Psychiatric Association

CODES OF ETHICS / INDUSTRIAL MEDICINE

Occupational MD's issue ethics code. *International Journal of Occupational Health and Safety* 45(6): 10, Nov-Dec 1976. BE06628.
*codes of ethics; conflict of interest; *industrial medicine; medical ethics; standards; American Occupational Medical Association

CODES OF ETHICS / NURSING ETHICS

American Nurses' Association. Perspectives on the Code for Nurses. Kansas City, Mo.: American Nurses' Association, 1978. 60 p. ANA Publication Code: G-132 3M 3/78. Papers presented at the American Nurses' Association Convention, Atlantic City, N.J., Jun 1976. 13 refs. 56 fn. BE06981.
administrators; *codes of ethics; historical aspects; moral obligations; nurses; *nursing ethics; *American Nurses Association Code for Nurses

Jarvis, Peter. Some comments on the Rcn code of professional conduct. *Nursing Mirror* 145(21): 27-28, 24 Nov 1977. BE06358.
*codes of ethics; *nursing ethics; patient care; strikes; Royal College of Nursing

Royal College of Nursing (Rcn) code of professional conduct: a discussion document. *Journal of Medical Ethics* 3(3): 115-123, Sep 1977. Commentary on the discussion, p. 119-123. 10 refs. BE05980.
*codes of ethics; health personnel; informed consent; morality; nurses; *nursing ethics; patient care; patients; physician nurse relationship; physicians; self determination; social interaction; strikes; withholding treatment; Great Britain; *Royal College of Nursing

CODES OF ETHICS / PSYCHIATRY

Masserman, Jules H. American Psychiatric Association (APA): ethical standards. *In* Wolman, Benjamin B., ed. International Encyclopedia of Psychiatry, Psychology, Psychoanalysis, and Neurology. Volume 1. New York: Van Nostrand Reinhold, 1977. p. 456-461. ISBN 0-918228-01-8. BE06563.
*codes of ethics; confidentiality; medical etiquette; medical fees; obligations to society; physician patient relationship; *psychiatry; standards

CODES OF ETHICS / PSYCHOLOGY

Trachtman, Gilbert M. Ethical standards for school psychologists. *In* Wolman, Benjamin B., ed. International Encyclopedia of Psychiatry, Psychology, Psychoanalysis, and Neurology. Volume 4. New York: Van Nostrand Reinhold, 1977. p. 383-385. 11 refs. ISBN 0-918228-01-8. BE06567.
*codes of ethics; *education; *psychology; *American Psychological Association

CODES OF ETHICS / SCIENCE

Cournand, André. The code of the scientist and its relationship to ethics. *Science* 198(4318): 699-705, 18 Nov 1977. 23 fn. BE06110.
*codes of ethics; deception; ethics; moral obligations; *professional ethics; *science

CODES OF ETHICS *See under*

HUMAN EXPERIMENTATION / PRISONERS / CODES OF ETHICS

TORTURE / CODES OF ETHICS

COERCION *See also* MANDATORY PROGRAMS

BE = bioethics accession number   fn. = footnotes   refs. = references

COERCION / INSTITUTIONALIZED PERSONS / MENTALLY ILL

Lundeen, Paul A. Pennsylvania's new Mental Health Procedures Act: due process and the right to treatment for the mentally ill. *Dickinson Law Review* 81(3): 627-647, Spring 1977. 141 fn. BE05924.
adolescents; *coercion; competence; confidentiality; dangerousness; *due process; *institutionalized persons; involuntary commitment; legal rights; legislation; medical records; *mentally ill; *patient care; prisoners; *right to treatment; state government; voluntary admission; *Mental Health Procedures Act; *Pennsylvania

COERCION / INSTITUTIONALIZED PERSONS / PENNSYLVANIA

Lundeen, Paul A. Pennsylvania's new Mental Health Procedures Act: due process and the right to treatment for the mentally ill. *Dickinson Law Review* 81(3): 627-647, Spring 1977. 141 fn. BE05924.
adolescents; *coercion; competence; confidentiality; dangerousness; *due process; *institutionalized persons; involuntary commitment; legal rights; legislation; medical records; *mentally ill; *patient care; prisoners; *right to treatment; state government; voluntary admission; *Mental Health Procedures Act; *Pennsylvania

COERCION   See under
BEHAVIOR CONTROL / COERCION
BEHAVIOR CONTROL / COERCION / LEGAL ASPECTS
ELECTROCONVULSIVE THERAPY / COERCION
HUMAN EXPERIMENTATION / COERCION
PATIENT CARE / COERCION
STERILIZATION / COERCION
TREATMENT REFUSAL / COERCION / PSYCHOACTIVE DRUGS

COMA   See also BRAIN DEATH

COMA   See under
ALLOWING TO DIE / COMA
ALLOWING TO DIE / COMA / LEGAL ASPECTS
ALLOWING TO DIE / PREGNANT WOMEN / COMA
EUTHANASIA / COMA / COSTS AND BENEFITS
PROLONGATION OF LIFE / PREGNANT WOMEN / COMA

COMMITMENT   See INVOLUNTARY COMMITMENT

COMMON GOOD   See under
HUMAN EXPERIMENTATION / CANCER / COMMON GOOD
POPULATION CONTROL / PUBLIC POLICY /

COMMON GOOD   See under (cont'd.)
COMMON GOOD
PRIVILEGED COMMUNICATION / PSYCHIATRY / COMMON GOOD

COMMONWEALTH V. EDELIN   See under
ABORTION / COMMONWEALTH V. EDELIN

COMMUNICATION / TERMINALLY ILL / CANCER

Brewin, Thurstan B. The cancer patient: communication and morale. *British Medical Journal* 2(6103): 1623-1627, 24-31 Dec 1977. 15 refs. BE06506.
attitudes; *cancer; attitudes to death; *communication; chronically ill; diagnosis; *disclosure; medical staff; nurses; *physicians; *prognosis; psychological stress; *terminally ill

COMMUNICATION   See under
DISCLOSURE / CANCER / COMMUNICATION
HUMAN EXPERIMENTATION / CHILDREN / COMMUNICATION
INFORMED CONSENT / COMMUNICATION / CONSENT FORMS
TERMINAL CARE / CHILDREN / COMMUNICATION
TERMINAL CARE / COMMUNICATION

COMMUNITY SERVICES   See under
HEALTH CARE DELIVERY / MENTALLY ILL / COMMUNITY SERVICES
HEALTH CARE DELIVERY / NURSES / COMMUNITY SERVICES

COMPENSATION   See under
HUMAN EXPERIMENTATION / COMPENSATION
HUMAN EXPERIMENTATION / COMPENSATION / LEGAL ASPECTS
HUMAN EXPERIMENTATION / RESEARCH SUBJECTS / COMPENSATION
IMMUNIZATION / CHILDREN / COMPENSATION
IMMUNIZATION / COMPENSATION
PRENATAL INJURIES / COMPENSATION / LEGAL RIGHTS
WRONGFUL LIFE / COMPENSATION / LEGAL ASPECTS
WRONGFUL LIFE / PARENTS / COMPENSATION

COMPETENCE   See under
ALLOWING TO DIE / MENTALLY RETARDED / COMPETENCE
INFORMED CONSENT / COMPETENCE
TREATMENT REFUSAL / COMPETENCE
TREATMENT REFUSAL / CRITICALLY ILL / COMPETENCE

BE = bioethics accession number   fn. = footnotes   refs. = references

COMPREHENSION *See under*

HUMAN EXPERIMENTATION / INFORMED CONSENT / COMPREHENSION

TREATMENT REFUSAL / COMPREHENSION

COMPULSORY PROGRAMS *See* MANDATORY PROGRAMS

CONFIDENTIALITY *See also* INFORMED CONSENT, PATIENT ACCESS, PRIVACY, PRIVILEGED COMMUNICATION

CONFIDENTIALITY / BEHAVIORAL RESEARCH

Fields, Cheryl M. A growing problem for researchers: protecting privacy. *Chronicle of Higher Education* 14(10): 1+, 2 May 1977. BE05610.
  *behavioral research; *confidentiality; disclosure; investigators; legal aspects; privacy; research subjects; social sciences

CONFIDENTIALITY / BEHAVIORAL RESEARCH / NORWAY

Oyen, Orjar. Social research and the protection of privacy: a review of the Norwegian development. *Acta Sociologica* 19(3): 249-262, 1976. 25 refs. 36 fn. BE06631.
  *behavioral research; *confidentiality; data bases; financial support; government agencies; *government regulation; legislation; privacy; social impact; *social sciences; *Norway

CONFIDENTIALITY / DANGEROUSNESS

Hunter, Thomas H., et al. Medical confidentiality——who protects the victims? *Pharos* 40(2): 24-29, Apr 1977. Edited by Mary B. Wagner from a transcript of a conference. 6 refs. BE05561.
  *confidentiality; *dangerousness; *disclosure; legal liability; legal obligations; *obligations to society; physician patient relationship; *physicians; privacy; psychiatry; psychotherapy; trust

Marks, Cathy L. Torts—psychiatry and the law—duty to warn potential victim of a homicidal patient. *New York Law School Law Review* 22(4): 1011-1022, 1977. 51 fn. BE06072.
  *confidentiality; *dangerousness; *disclosure; legal liability; physician patient relationship; prognosis; psychiatry; torts

Olsen, Theodore A. Imposing a duty to warn on psychiatrists—a judicial threat to the psychiatric profession. *University of Colorado Law Review* 48(2): 283-310, Winter 1977. 133 fn. BE05942.
  *confidentiality; *dangerousness; *disclosure; involuntary commitment; *legal aspects; *legal obligations; mentally ill; physician patient relationship; privacy; prognosis; *psychiatry; psy-

chotherapy; stigmatization; violence; Tarasoff v. Regents

Roth, Loren H.; Meisel, Alan. Dangerousness, confidentiality, and the duty to warn. *American Journal of Psychiatry* 134(5): 508-511, May 1977. 29 refs. 1 fn. BE05627.
  case reports; *confidentiality; *dangerousness; *disclosure; informed consent; involuntary commitment; legal liability; physician patient relationship; prognosis; *psychiatry; violence; Tarasoff v. Regents

CONFIDENTIALITY / DANGEROUSNESS / LEGAL ASPECTS

Lee, John V. Psychiatry—torts—a psychotherapist who knows or should know his patient intends violence to another incurs a duty to warn. *Cumberland Law Review* 7(3): 551-559, Winter 1977. 51 fn. BE06285.
  *confidentiality; *dangerousness; disclosure; *legal aspects; legal liability; legal obligations; psychotherapy; Tarasoff v. Regents

CONFIDENTIALITY / DANGEROUSNESS / PSYCHOTHERAPY

American Academy of Psychiatry and the Law. The Tarasoff Decision. [Sound recording]. San Francisco: Convention Cassettes, 22 Oct 1977. 2 cassettes; 60 min. Tapes No. 5 and 6. Taped during a convention of the American Academy of Psychiatry and the Law, San Francisco, 22-24 Oct 1977. BE07038.
  *confidentiality; *dangerousness; decision making; *disclosure; expert testimony; involuntary commitment; judicial action; law; *legal aspects; legal liability; obligations to society; professional patient relationship; prognosis; *psychotherapy; *Tarasoff v. Regents

CONFIDENTIALITY / FAMOUS PERSONS / HEALTH

Hunter, Thomas H., et al. The Health of Public Figures: What Should Be Disclosed? [Videorecording]. Charlottesville, Va.: University of Virginia Medical Center, Health Sciences Library, 1974. 1 cassette; 60 min.; sound; black and white; 3/4 in. Tape of the Medical Center Hour, 7 Jan 1974. Panel discussion moderated by Thomas H. Hunter. BE07044.
  competence; *confidentiality; conflict of interest; *disclosure; *famous persons; *health; legislation; obligations to society; physician patient relationship; physicians; *politics; privacy; rights; trust

CONFIDENTIALITY / FAMOUS PERSONS / MORTALITY

Humphrey case renews controversy. *American Medical News* 20(46): 13, 21 Nov 1977. BE06357.

BE = bioethics accession number   fn. = footnotes   refs. = references

cancer; *confidentiality; *famous persons; journalism; *mortality; prognosis

CONFIDENTIALITY / HOMOSEXUALS

Callahan, Daniel; Cassell, Eric J.; Veatch, Robert M. The homosexual husband and physician confidentiality. *Hastings Center Report* 7(2): 15-17, Apr 1977. Case study. BE05545.

*confidentiality; *disclosure; *homosexuals; marital relationship; moral obligations; *physician patient relationship; physician's role; single persons

CONFIDENTIALITY / INSTITUTIONALIZED PERSONS

New Hampshire. Supreme Court. State v. Kupchun. 27 May 1977. *Atlantic Reporter, 2d Series,* 373: 1325-1328. BE06266.

*confidentiality; *institutionalized persons; legal aspects; *medical records; privileged communication

CONFIDENTIALITY / MEDICAL RECORDS

Aldrich, Robert F. Health Records and Confidentiality: An Annotated Bibliography with Abstracts. Washington: National Commission on Confidentiality of Health Records, Oct 1977. 76 p. Bibliography. BE05874.

alcohol abuse; behavioral research; *confidentiality; criminal law; dangerousness; data bases; disclosure; drug abuse; employment; family members; group therapy; human experimentation; insurance; legal aspects; *medical records; military personnel; privileged communication; psychiatry; public health; students

Bernstein, Arthur H. How accessible are medical records? *Hospitals* 51(21): 154+, 1 Nov 1977. BE06106.

*confidentiality; *disclosure; *legal rights; *medical records; *patients; physician patient relationship; privileged communication

Bloom, Mark. Are you spilling patients' secrets? *Medical World News* 18(18): 55-56+, 5 Sep 1977. BE05773.

*confidentiality; *data bases; *disclosure; health insurance; hospitals; industry; legal aspects; *medical records; physician's role; privacy

The erosion of confidentiality: a threat that won't go away. *American Medical News* 20(21): Impact 3-4, 23 May 1977. BE06227.

*confidentiality; data bases; health insurance; legal aspects; *medical records; privacy; PSRO's

Hofmann, Paul B. Data release procedures ensure clergy's access, patients' privacy. *Hospitals* 51(8): 71-73, 16 Apr 1977. BE05804.

*clergy; *confidentiality; counseling; *disclosure;

hospitals; informed consent; *medical records; patients' rights; program descriptions

Holden, Constance. Health records and privacy: what would Hippocrates say? *Science* 198(4315): 382, 28 Oct 1977. BE06088.

*confidentiality; employment; epidemiology; legal aspects; *medical records; privacy

New Hampshire. Supreme Court. State v. Kupchun. 27 May 1977. *Atlantic Reporter, 2d Series,* 373: 1325-1328. BE06266.

*confidentiality; *institutionalized persons; legal aspects; *medical records; privileged communication

Smejda, Hellena. Patient record privacy: can we guarantee it? *Hospital Financial Management* 31(4): 12-13+, Apr 1977. 15 fn. BE06485.

*confidentiality; *data bases; disclosure; federal government; *government regulation; hospitals; legislation; *medical records; privacy; public policy; social impact; *Privacy Act 1974

Thorup, Oscar A., et al. Privacy and the Computer: Everything You Know about Yourself but Hoped They'd Never Find Out. [Videorecording]. Charlottesville, Va.: University of Virginia Medical Center, Health Sciences Library, 1978. 1 cassette; 60 min.; sound; black and white; 3/4 in. Tape of the Medical Center Hour, 15 Feb 1978. Panel discussion moderated by Oscar A. Thorup. BE07055.

*confidentiality; *data bases; disclosure; health personnel; informed consent; insurance; legal aspects; *medical records; moral obligations; patient access; *privacy

U.S. Congress. House. A bill to amend...the Social Security Act to provide for the confidentiality of personal medical information created or maintained by medical-care institutions providing services under...Medicare....H.R. 8283, 95th Cong. 1st Sess., 13 Jul 1977. 25 p. By Edward I. Koch. Referred to the Committees on Interstate and Foreign Commerce and Ways and Means. BE06468.

*confidentiality; disclosure; hospitals; legal aspects; *medical records; residential facilities

U.S. Congress. Senate. Committee on Finance. Subcommittee on Health. Confidentiality of Medical Records. Hearing. Washington: U.S. Government Printing Office, 1977. 66 p. Hearing, 95th Cong., 1st Sess., 15 Sep 1977. BE06690.

biomedical research; *confidentiality; disclosure; epidemiology; federal government; government regulation; informed consent; *medical records

U.S. Privacy Protection Study Commission. Record keeping in the medical-care relationship. *In its* Personal Privacy in an Information Society. The Report of the Privacy Protection Study Commission. Washington: U.S. Government Printing Office, Jul 1977. p. 277-317. 74 fn. BE05713.

*confidentiality; data bases; disclosure; federal government; government regulation; insurance;

BE = bioethics accession number   fn. = footnotes   refs. = references

legislation; *medical records; patients; physicians; *privacy; *public policy; state government

Vuori, Hannu. Privacy, confidentiality and automated health information systems. *Journal of Medical Ethics* 3(4): 174-178, Dec 1977. 12 refs. BE06259.
*confidentiality; *data bases; health care; *medical records; *privacy; social impact; technology

Westin, Alan F. Patients' rights: computers and health records. *Hospital Progress* 58(4): 56-61+, Apr 1977. 2 fn. BE05590.
*confidentiality; *data bases; *disclosure; informed consent; international aspects; *medical records; patients' rights; privacy; program descriptions; rights

CONFIDENTIALITY / MEDICAL RECORDS / ADOLESCENTS

Hofmann, Adele D.; Becker, R.D.; Gabriel, H. Paul. Consent and confidentiality. *In their* The Hospitalized Adolescent: A Guide to Managing the Ill and Injured Youth. New York: Free Press, 1976. p. 211-232. 1 fn. ISBN 0-02-914790-5. BE07088.
*adolescents; *confidentiality; emergency care; *health care; hospitals; informed consent; institutional policies; *legal aspects; legal rights; *medical records; *parental consent

CONFIDENTIALITY / MEDICAL RECORDS / ALCOHOL ABUSE

Blume, Sheila B. Ethics of record-keeping. *Alcoholism: Clinical and Experimental Research* 1(4): 301-303, Oct 1977. 8 refs. BE06344.
*alcohol abuse; *confidentiality; data bases; disclosure; *medical records

CONFIDENTIALITY / MEDICAL RECORDS / ATTITUDES

Rosen, Catherine E. The compliant client. *Man and Medicine: The Journal of Values and Ethics in Health Care* 3(1): 1-16, 1978. 24 fn. BE06909.
*attitudes; community services; *confidentiality; data bases; *disclosure; informed consent; *medical records; *mental health; *patients; state government

CONFIDENTIALITY / MEDICAL RECORDS / DRUG ABUSE

New York City. Family Court, Queens County. Matter of Doe Children. 7 Mar 1978. *New York Supplement, 2d Series,* 402: 958-961. BE06714.
alcohol abuse; *child neglect; *confidentiality; disclosure; *drug abuse; legal rights; *medical records; parents; patients; physician patient relationship; privileged communication

New York. Court of Appeals. People v. Newman. 31 May 1973. *New York Supplement, 2d Series,* 345: 502-512. 8 fn. BE06602.

*confidentiality; disclosure; *drug abuse; law enforcement; legislation; *medical records; *privileged communication

Weimer, William; Thomas, Marthalee. Toward a hospital policy on drug abuse patients. *Hospitals* 51(1): 81-82+, 1 Jan 1977. BE05905.
*alcohol abuse; *confidentiality; *disclosure; *drug abuse; *hospitals; informed consent; *institutional policies; law enforcement; legal aspects; *medical records

CONFIDENTIALITY / MEDICAL RECORDS / DRUGS

Brown, Dennis C. Privacy: drug use reporting requirements unconstitutional. *University of Dayton Law Review* 2(1): 127-136, Winter 1977. 59 fn. BE06284.
*confidentiality; *constitutional law; data bases; *disclosure; *drugs; due process; *medical records; physician patient relationship; *privacy; state government; Supreme Court decisions; *Roe v. Ingraham

CONFIDENTIALITY / MEDICAL RECORDS / EPIDEMIOLOGY

Curran, William J. The privacy protection report and epidemiological research. *American Journal of Public Health* 68(2): 173, Feb 1978. BE06773.
biomedical research; *confidentiality; disclosure; *epidemiology; *medical records; patient access; privacy

Stebbings, James H., et al. Privacy protection in epidemiology: a technological quibble. *American Journal of Epidemiology* 106(5): 433-435, Nov 1977. 6 refs. Letter. BE06504.
*confidentiality; *data bases; *epidemiology; government regulation; *medical records; privacy; research subjects

CONFIDENTIALITY / MEDICAL RECORDS / FAMOUS PERSONS

Fabro, Fred. Physicians and the press. *Connecticut Medicine* 42(1): 63-64, Jan 1978. Editorial. BE06730.
cancer; *confidentiality; diagnosis; disclosure; family members; *famous persons; informed consent; journalism; medical ethics; *medical records; *physician's role; physicians; *politics; privacy; prognosis

CONFIDENTIALITY / MEDICAL RECORDS / GENETIC DEFECTS

Riskin, Leonard L.; Reilly, Philip. Remedies for improper disclosure of genetic data. *Rutgers-Camden Law Journal* 8(3): 480-506, Spring 1977. 164 fn. BE06483.
behavioral genetics; *confidentiality; data bases;

BE = bioethics accession number   fn. = footnotes   refs. = references

*disclosure; federal government; *genetic defects; genetic screening; *legal aspects; government agencies; legal liability; legislation; *medical records; physicians; privacy; state government

CONFIDENTIALITY / MEDICAL RECORDS / GREAT BRITAIN

Dua-Agyemang, J.E. Disclosure of medical notes. *Medicine, Science, and the Law* 17(3): 187-189, Jul 1977. 1 ref. BE05956.
*confidentiality; *disclosure; hospitals; law; *legal aspects; legislation; *medical records; physicians; *Great Britain

CONFIDENTIALITY / MEDICAL RECORDS / HOSPITALS

Hayt, Emanuel. Medicolegal Aspects of Hospital Records. Second Edition. Berwyn, Ill.: Physicians' Record Company, 1977. 519 p. Footnotes included. ISBN 0-917036-13-1. BE06678.
abortion; adults; artificial insemination; autopsies; blood transfusions; children; *confidentiality; consent forms; contraception; data bases; disclosure; famous persons; federal government; *hospitals; *informed consent; *legal aspects; legal obligations; legal rights; legislation; mass media; *medical records; medical staff; organ donation; organ transplantation; patients; physicians; privacy; *privileged communication; radiology; review committees; state government; sterilization

Weimer, William; Thomas, Marthalee. Toward a hospital policy on drug abuse patients. *Hospitals* 51(1): 81-82+, 1 Jan 1977. BE05905.
*alcohol abuse; *confidentiality; *disclosure; *drug abuse; *hospitals; informed consent; *institutional policies; law enforcement; legal aspects; *medical records

CONFIDENTIALITY / MEDICAL RECORDS / ILLINOIS

Shlensky, Ronald. Informed consent and confidentiality: proposed new approaches in Illinois. *American Journal of Psychiatry* 134(12): 1416-1418, Dec 1977. BE06150.
*confidentiality; disclosure; *informed consent; legislation; *medical records; *mental health; psychotherapy; *Illinois

CONFIDENTIALITY / MEDICAL RECORDS / INDUSTRIAL MEDICINE

Fishbein, Gershon. Job hazards and privacy. *Washington Post*, 8 Aug 1977, p. A23. BE06391.
cancer; *confidentiality; disclosure; federal government; government agencies; *industrial medicine; informed consent; *medical records; privacy; National Institute for Occupational Safety and Health

Klutas, Edna M. Confidentiality of medical information. *Occupational Health Nursing* 25(4): 14-17, Apr

1977. 6 fn. BE06299.
*confidentiality; disclosure; *industrial medicine; legal obligations; *medical records; moral obligations; *nurses; privacy; privileged communication

Levine, Carol. Sharing secrets: health records and health hazards. *Hastings Center Report* 7(6): 13-15, Dec 1977. BE06142.
*confidentiality; data bases; disclosure; employment; *health hazards; health insurance; *industrial medicine; *medical records; privacy

Should company MDs release records? *Medical World News* 18(22): 44+, 14 Nov 1977. BE06122.
*confidentiality; employment; epidemiology; *industrial medicine; *medical records; physician patient relationship; privacy

CONFIDENTIALITY / MEDICAL RECORDS / LEGAL ASPECTS

Blum, John D.; Gertman, Paul M.; Rabinow, Jean. Confidentiality and peer review. *In their* PSROs and the Law. Germantown, Md.: Aspen Systems Corporation, 1977. p. 131-162. 103 fn. ISBN 0-912862-39-4. BE06182.
*confidentiality; *legal aspects; legislation; *medical records; peer review; privacy; privileged communication; public advocacy; state government; Supreme Court decisions; Freedom of Information Act; Privacy Act 1974; *PSRO's

Butcher, Charles. The erosion of professional confidence. *World Medical Journal* 24(5): 71-73, Sep-Oct 1977. BE06330.
*confidentiality; drug abuse; *legal aspects; *medical records; physician patient relationship; social workers

Creighton, Helen. The right of privacy: cases and research problem. *Supervisor Nurse* 8(11): 62-64, Nov 1977. 25 fn. BE06592.
behavioral research; *confidentiality; *legal aspects; legislation; *medical records; *privacy; social sciences; state government

Dua-Agyemang, J.E. Disclosure of medical notes. *Medicine, Science, and the Law* 17(3): 187-189, Jul 1977. 1 ref. BE05956.
*confidentiality; *disclosure; hospitals; law; *legal aspects; legislation; *medical records; physicians; *Great Britain

MacDonald, Michael G.; Meyer, Kathryn C. Confidentiality between patient and doctor. *Mount Sinai Journal of Medicine* 44(4): 568-569, Jul-Aug 1977. 1 fn. BE05963.
*confidentiality; *disclosure; drugs; *legal aspects; *medical records; privacy; state government; New York

Riskin, Leonard L.; Reilly, Philip. Remedies for improper disclosure of genetic data. *Rutgers-Camden Law Journal* 8(3): 480-506, Spring 1977. 164 fn. BE06483.

BE = bioethics accession number   fn. = footnotes   refs. = references

behavioral genetics; *confidentiality; data bases; *disclosure; federal government; *genetic defects; genetic screening; *legal aspects; government agencies; legal liability; legislation; *medical records; physicians; privacy; state government

CONFIDENTIALITY / MEDICAL RECORDS / MENTAL HEALTH

Illman, John. Vanishing medical secrets. *Sunday Times (London),* 21 Aug 1977, p. 11. BE06454.
 *confidentiality; data bases; hospitals; *medical records; *mental health; psychiatry

New York. Court of Appeals. Volkman v. Miller. 29 Mar 1977. *New York Supplement, 2d Series,* 394: 631-632. BE07108.
 *confidentiality; disclosure; *medical records; *mental health; state government; New York

Rosen, Catherine E. The compliant client. *Man and Medicine: The Journal of Values and Ethics in Health Care* 3(1): 1-16, 1978. 24 fn. BE06909.
 *attitudes; community services; *confidentiality; data bases; *disclosure; informed consent; *medical records; *mental health; *patients; state government

Rosen, Catherine E. To sign or not to sign: a study of client responses to release-of-information forms. *Hospital and Community Psychiatry* 29(3): 161+, Mar 1978. BE06928.
 community services; *confidentiality; disclosure; informed consent; *medical records; *mental health; mentally ill; patient participation

CONFIDENTIALITY / MEDICAL RECORDS / NEW YORK

New York. Legislature. Senate-Assembly. An act to amend the executive law, in relation to confidentiality of medical information. S.4855-A/A.7081-A, 1977-1978 Regular Sess., 24 Mar 1977. 7 p. By Sens. Pisani and Nicolosi, et al. Referred to...Senate Committee on Finance and...Assembly Committee on Government Operations. BE05916.
 *confidentiality; *disclosure; informed consent; *medical records; peer review; physician patient relationship; *New York

CONFIDENTIALITY / MEDICAL RECORDS / PRIVACY PROTECTION STUDY COMMISSION

Legal teeth for medical privacy? *Medical World News* 18(14): 14-15, 11 Jul 1977. BE05729.
 *confidentiality; disclosure; government regulation; hospitals; *medical records; patient access; physicians; privacy; state government; *Privacy Protection Study Commission

CONFIDENTIALITY / MEDICAL RECORDS / PSYCHIATRY

Confidential records under threat: psychiatrists urge DHSS to take forceful action. *Nursing Times* 73(34): 1298, 25 Aug 1977. BE05837.
 *confidentiality; data bases; hospitals; institutional policies; *medical records; *psychiatry

Massachusetts. Supreme Judicial Court, Suffolk. Doe v. Commissioner of Mental Health. 10 May 1977. *North Eastern Reporter, 2d Series,* 362: 920-923. 5 fn. BE06264.
 *confidentiality; *disclosure; *fathers; informed consent; *medical records; *psychiatry

McKerrow, L.W. Policies for releasing psychiatric records. *Dimensions in Health Service* 55(2): 38+, Feb 1978. BE06786.
 competence; *confidentiality; disclosure; informed consent; legislation; *medical records; *psychiatry

Walshe-Brennan, K.S. Confidentiality in the N.H.S. *Journal of the Irish Medical Association* 71(3): 65-66, 28 Feb 1978. 2 refs. BE06802.
 *confidentiality; government agencies; *medical records; property rights; *psychiatry; *Great Britain; National Health Service

CONFIDENTIALITY / MEDICAL RECORDS / SOCIAL IMPACT

Shapiro, Sam. Privacy, research, and the health of the public. *American Journal of Public Health* 66(11): 1050-1051, Nov 1976. 2 refs. Editorial. BE07100.
 biomedical research; *confidentiality; epidemiology; *medical records; privacy; *social impact; Privacy Act 1974

CONFIDENTIALITY / OCCUPATIONAL DISEASES

Bronson, Gail. Workers' right to know. *Wall Street Journal (Eastern Edition),* 1 Jul 1977, p. 4. BE06448.
 *confidentiality; *disclosure; economics; employment; health hazards; morbidity; mortality; *occupational diseases

CONFIDENTIALITY / PHYSICIAN PATIENT RELATIONSHIP

Hunter, Thomas H., et al. The Physician as Double Agent. [Videorecording]. Charlottesville, Va.: University of Virginia Medical Center, Health Sciences Library, 1977. 1 cassette; 60 min.; sound; color; 3/4 in. Tape of the Medical Center Hour, 5 Jan 1977. Panel discussion moderated by Thomas H. Hunter. BE07052.
 case reports; *confidentiality; conflict of interest; dangerousness; decision making; disclosure; employment; informed consent; legal aspects; legal obligations; medical education; mentally ill; moral obligations; *obligations to society; *physician pa-

tient relationship; *physicians; psychiatric diagnosis; psychiatry; students; teaching methods; universities

CONFIDENTIALITY / PHYSICIAN PATIENT RELATIONSHIP / INSURANCE

Harnes, Jack R. Confidentiality. *New York State Journal of Medicine* 77(8): 1326-1327, Jul 1977. 2 fn. BE05820.
  *confidentiality; *insurance; legal aspects; *medical records; *physician patient relationship

CONFIDENTIALITY / PRIVACY / CONSTITUTIONAL LAW

Brown, Dennis C. Privacy: drug use reporting requirements unconstitutional. *University of Dayton Law Review* 2(1): 127-136, Winter 1977. 59 fn. BE06284.
  *confidentiality; *constitutional law; data bases; *disclosure; *drugs; due process; *medical records; physician patient relationship; *privacy; state government; Supreme Court decisions; *Roe v. Ingraham

CONFIDENTIALITY / PRIVACY / LEGAL ASPECTS

Kelty, Miriam F. Ethical issues and requirements for sex research with humans: confidentiality. *In* Masters, William H.; Johnson, Virginia E.; Kolodny, Robert C., eds. Ethical Issues in Sex Therapy and Research. Boston: Little, Brown, 1977. p. 84-118. Includes discussion by Richard Green and others on p. 106-118. 36 refs. ISBN 0-316-549-835. BE06162.
  *confidentiality; data bases; disclosure; education; federal government; government agencies; government regulation; *human experimentation; journalism; *legal aspects; legal rights; *privacy; *sexuality; XYY karyotype

CONFIDENTIALITY / PSRO'S / LEGAL ASPECTS

Blum, John D.; Gertman, Paul M.; Rabinow, Jean. Confidentiality and peer review. *In their* PSROs and the Law. Germantown, Md.: Aspen Systems Corporation, 1977. p. 131-162. 103 fn. ISBN 0-912862-39-4. BE06182.
  legislation; *confidentiality; *legal aspects; *medical records; peer review; privacy; privileged communication; public advocacy; state government; Supreme Court decisions; Freedom of Information Act; Privacy Act 1974; *PSRO's

CONFIDENTIALITY / PSYCHIATRY

California. Court of Appeal, First District, Division 2. Bellah v. Greenson. 5 Oct 1977. *California Reporter* 141: 92-95. 2 fn. BE07117.

  *confidentiality; dangerousness; disclosure; *legal liability; negligence; parents; *psychiatry; *suicide; torts

CONFIDENTIALITY / PSYCHIATRY / LEGAL ASPECTS

Weisstub, David N. Confidentiality and the mental health professional. *Canadian Psychiatric Association Journal* 22(6): 319-323, Oct 1977. 1 fn. BE06352.
  *confidentiality; disclosure; health personnel; *legal aspects; mental health; privacy; privileged communication; *psychiatry

CONFIDENTIALITY / PSYCHIATRY / LITERATURE

New York. Supreme Court, New York County. Doe v. Roe. 21 Nov 1977. *New York Supplement, 2d Series,* 400: 668-680. 9 fn. BE07122.
  compensation; *confidentiality; constitutional law; contracts; *disclosure; insurance; legal obligations; legal rights; legislation; *literature; mentally ill; *patients; physician patient relationship; physicians; *privacy; *psychiatry; psychotherapy; public policy; state government; torts

CONFIDENTIALITY / PSYCHOTHERAPY / LEGAL ASPECTS

American Academy of Psychiatry and the Law. The Tarasoff Decision. [Sound recording]. San Francisco: Convention Cassettes, 22 Oct 1977. 2 cassettes; 60 min. Tapes No. 5 and 6. Taped during a convention of the American Academy of Psychiatry and the Law, San Francisco, 22-24 Oct 1977. BE07038.
  *confidentiality; *dangerousness; decision making; *disclosure; expert testimony; involuntary commitment; judicial action; law; *legal aspects; legal liability; obligations to society; professional patient relationship; prognosis; *psychotherapy; *Tarasoff v. Regents

CONFIDENTIALITY / PSYCHOTHERAPY / LEGAL LIABILITY

Eger, Charles L. Psychotherapists' liability for extrajudicial breaches of confidentiality. *Arizona Law Review* 18(4): 1061-1094, 1976. 171 fn. BE07096.
  compensation; *confidentiality; contracts; dangerousness; *disclosure; *legal liability; legal obligations; malpractice; privacy; professional patient relationship; psychological stress; *psychotherapy; torts; trust

CONFIDENTIALITY / RIGHTS

Kobler, Arthur L. Civil liberties. *In* Wolman, Benjamin B., ed. International Encyclopedia of Psychiatry, Psychology, Psychoanalysis, and Neurology. Volume 3. New York: Van Nostrand Reinhold, 1977. p. 151-154. 4 refs. ISBN 0-918228-01-8.

BE = bioethics accession number   fn. = footnotes   refs. = references

BE06564.
  *behavior control; *confidentiality; *human experimentation; human rights; legal rights; medical records; operant conditioning; *rights; self determination; social control

CONFIDENTIALITY *See under*
  BEHAVIORAL RESEARCH / CONFIDENTIALITY
  HUMAN EXPERIMENTATION / CONFIDENTIALITY / SEXUALITY

CONFLICT OF INTEREST *See under*
  MORAL OBLIGATIONS / PHYSICIAN PATIENT RELATIONSHIP / CONFLICT OF INTEREST
  PSYCHIATRY / CONFLICT OF INTEREST / VALUES
  RECOMBINANT DNA RESEARCH / GOVERNMENT AGENCIES / CONFLICT OF INTEREST

CONGENITAL DEFECTS *See also* GENETIC DEFECTS

CONGENITAL DEFECTS *See under*
  ALLOWING TO DIE / CONGENITAL DEFECTS / CONSTITUTIONAL LAW
  ALLOWING TO DIE / NEWBORNS / CONGENITAL DEFECTS
  NEWBORNS / CONGENITAL DEFECTS / PHYSICIAN'S ROLE
  PATIENT CARE / NEWBORNS / CONGENITAL DEFECTS
  PHYSICIANS / CONGENITAL DEFECTS / LEGAL LIABILITY
  TREATMENT REFUSAL / NEWBORNS / CONGENITAL DEFECTS
  WITHHOLDING TREATMENT / NEWBORNS / CONGENITAL DEFECTS

CONSCIENCE *See under*
  ABORTION / CONSCIENCE / INDIA
  ABORTION / FINANCIAL SUPPORT / CONSCIENCE
  ABORTION / LEGISLATION / CONSCIENCE
  ABORTION / MEDICAL STAFF / CONSCIENCE
  ABORTION / PRIVATE HOSPITALS / CONSCIENCE

CONSENT FORMS *See under*
  INFORMED CONSENT / COMMUNICATION / CONSENT FORMS
  INFORMED CONSENT / CONSENT FORMS
  NEGATIVE REINFORCEMENT / MENTALLY RETARDED / CONSENT FORMS
  PSYCHOTHERAPY / INFORMED CONSENT / CONSENT FORMS

CONSTITUTIONAL AMENDMENTS *See under*
  ABORTION / CONSTITUTIONAL AMENDMENTS

CONSTITUTIONAL LAW *See under*
  ALLOWING TO DIE / CONGENITAL DEFECTS / CONSTITUTIONAL LAW
  CONFIDENTIALITY / PRIVACY / CONSTITUTIONAL LAW
  SCIENCE / CONSTITUTIONAL LAW / FREEDOM

CONTAINMENT *See under*
  RECOMBINANT DNA RESEARCH / CONTAINMENT
  RECOMBINANT DNA RESEARCH / CONTAINMENT / HEALTH HAZARDS
  RECOMBINANT DNA RESEARCH / CONTAINMENT / STANDARDS
  RECOMBINANT DNA RESEARCH / GOVERNMENT REGULATION / CONTAINMENT
  RECOMBINANT DNA RESEARCH / PUBLIC POLICY / CONTAINMENT

CONTRACEPTION *See also* FAMILY PLANNING

CONTRACEPTION / ADOLESCENTS

Goodman, Ellen. Pregnancy prevention: a right to be informed. *Washington Post,* 21 Jun 1977, p. A19. BE06031.
  *adolescents; *contraception; *legal rights; pregnant women; statistics; Supreme Court decisions

CONTRACEPTION / ADOLESCENTS / LEGAL ASPECTS

U.S. District Court, W.D. Michigan, S.D. Doe v. Irwin. 23 Nov 1977. *Federal Supplement* 441: 1247-1261. 20 fn. BE07123.
  abortion; *adolescents; age; competence; constitutional law; *contraception; decision making; education; family; human development; health facilities; informal social control; law; *legal aspects; *legal rights; motivation; *parents; *parental consent; public policy; religious beliefs; sexuality; state action; state interest; Supreme Court decisions

CONTRACEPTION / ADOLESCENTS / LEGAL RIGHTS

U.S. Court of Appeals, Eighth Circuit. M.S. v. Wermers. 29 Jun 1977. *Federal Reporter, 2d Series,* 557: 170-179. 11 fn. BE06273.
  *adolescents; *contraception; disclosure; *legal guardians; *legal rights; parental consent; *parents; privacy

CONTRACEPTION / CHRISTIAN ETHICS

Anderson, Norman. Birth control, sterilisation and abortion. *In his* Issues of Life and Death: Abortion,

---

BE = bioethics accession number   fn. = footnotes   refs. = references

Birth Control, Capital Punishment, Euthanasia. Downers Grove, Ill.: InterVarsity Press, 1977. p. 58-84. 46 fn. ISBN 0-87784-721-5. BE05790.
*abortion; *Christian ethics; *contraception; eugenics; legal aspects; methods; *personhood; population control; reproduction; selective abortion; sexuality; *sterilization; therapeutic abortion; voluntary sterilization

CONTRACEPTION / GOVERNMENT REGULATION

Morality—and the law. *Washington Post,* 13 Jun 1977, p. A22. Editorial. BE06028.
adolescents; *contraception; *government regulation; morality; state government; Supreme Court decisions; New York

CONTRACEPTION / NEW ZEALAND

Contraception, sterilisation and abortion in New Zealand. *New Zealand Medical Journal* 85(588): 441-445, 25 May 1977. Recommendations from the Report of the Royal Commission of Inquiry into Contraception, Sterilisation and Abortion (Government Printer, Wellington). BE05601.
*abortion; adolescents; alternatives; community services; conscience; *contraception; counseling; education; employment; ethics committees; family planning; health personnel; information dissemination; *legal aspects; mentally handicapped; pregnant women; *public policy; review committees; sexuality; social workers; *sterilization; therapeutic abortion; *New Zealand

New Zealand. Royal Commission of Inquiry. Contraception, Sterilisation, and Abortion in New Zealand: Report of the Royal Commission of Inquiry. Wellington, N.Z.: E.C. Keating, 1977. 455 p. Duncan W. McMullin, Chairman of the Commission. Bibliography: p. 421-445. BE06682.
*abortion; adolescents; adults; *contraception; family planning; fetuses; historical aspects; human experimentation; illegal abortion; law; legal aspects; *legislation; mentally handicapped; methods; morality; pregnant women; public opinion; public policy; *social impact; socioeconomic factors; statistics; *voluntary sterilization; women's rights; *New Zealand

CONTRACEPTION / PARENTAL CONSENT / LEGAL RIGHTS

U.S. District Court, W.D. Michigan, S.D. Doe v. Irwin. 23 Nov 1977. *Federal Supplement* 441: 1247-1261. 20 fn. BE07123.
abortion; *adolescents; age; competence; constitutional law; *contraception; decision making; education; family; human development; health facilities; informal social control; law; *legal aspects; *legal rights; motivation; *parents; *parental consent; public policy; religious beliefs;

sexuality; state action; state interest; Supreme Court decisions

CONTRACTS *See under*
ADOLESCENTS / PSYCHIATRY / CONTRACTS
OPERANT CONDITIONING / PATIENT CARE / CONTRACTS

CONTROL GROUPS *See under*
HUMAN EXPERIMENTATION / CONTROL GROUPS / HISTORICAL ASPECTS

COSTS AND BENEFITS *See under*
BEHAVIORAL RESEARCH / DECEPTION / COSTS AND BENEFITS
BIOMEDICAL RESEARCH / COSTS AND BENEFITS
BIOMEDICAL TECHNOLOGIES / COSTS AND BENEFITS
EUTHANASIA / COMA / COSTS AND BENEFITS
GENETIC SCREENING / SPINA BIFIDA / COSTS AND BENEFITS
HEALTH CARE DELIVERY / MASS SCREENING / COSTS AND BENEFITS
HEALTH CARE / COSTS AND BENEFITS
HEALTH CARE / RESOURCE ALLOCATION / COSTS AND BENEFITS
MASS SCREENING / SPINA BIFIDA / COSTS AND BENEFITS
NATIONAL HEALTH INSURANCE / COSTS AND BENEFITS
PATIENT CARE / COSTS AND BENEFITS
PREVENTIVE MEDICINE / COSTS AND BENEFITS
RENAL DIALYSIS / COSTS AND BENEFITS
RESOURCE ALLOCATION / HYPERTENSION / COSTS AND BENEFITS
RESOURCE ALLOCATION / MEDICINE / COSTS AND BENEFITS
RESUSCITATION / INTENSIVE CARE UNITS / COSTS AND BENEFITS
SOCIAL CONTROL / PREVENTIVE MEDICINE / COSTS AND BENEFITS

CRIMINAL LAW *See under*
ABORTION / CRIMINAL LAW / CANADA
KILLING / FETUSES / CRIMINAL LAW
LEGAL LIABILITY / BRAIN PATHOLOGY / CRIMINAL LAW

CRITICALLY ILL *See under*
TREATMENT REFUSAL / CRITICALLY ILL
TREATMENT REFUSAL / CRITICALLY ILL / COMPETENCE
TREATMENT REFUSAL / CRITICALLY ILL / LEGAL RIGHTS

BE = bioethics accession number   fn. = footnotes   refs. = references

CRYONIC SUSPENSION

Miller, Albert J.; Acri, Michael J. Science. *In their* Death: A Bibliographical Guide. Metuchen, N.J.: Scarecrow Press, 1977. p. 205-210. Bibliography. ISBN 0-8108-1025-5. BE06187.
*cryonic suspension; *death; determination of death; *life extension; mortality; *science

CURRICULUM *See under*
MEDICAL ETHICS / NURSING EDUCATION / CURRICULUM

DANGEROUSNESS *See under*
CONFIDENTIALITY / DANGEROUSNESS
CONFIDENTIALITY / DANGEROUSNESS / LEGAL ASPECTS
CONFIDENTIALITY / DANGEROUSNESS / PSYCHOTHERAPY
INVOLUNTARY COMMITMENT / DANGEROUSNESS / HAWAII
INVOLUNTARY COMMITMENT / DANGEROUSNESS / LEGAL ASPECTS
PRIVILEGED COMMUNICATION / PSYCHOTHERAPY / DANGEROUSNESS

DEATH *See also* ATTITUDES TO DEATH, BRAIN DEATH, DETERMINATION OF DEATH, EUTHANASIA, SUICIDE, TERMINAL CARE

DEATH

Brown, Stuart M., et al. 'Natural death': clarifying the definition. *Hastings Center Report* 7(6): 39-40, Dec 1977. Letter. BE06132.
age; attitudes to death; *death; psychological stress; quality of life; value of life

Callahan, Daniel. On defining a 'natural death'. *Hastings Center Report* 7(3): 32-37, Jun 1977. 19 fn. BE05640.
aged; attitudes to death; *death; goals; life extension; moral obligations; obligations of society; public policy; rights; value of life

Dinsmore, Janet L., ed. Death and Dying: An Examination of Legislative and Policy Issues. Washington: Georgetown University, Health Policy Center, 1977. 68 p. Papers from a Conference held in Washington, D.C., 30 Jun 1976, co-sponsored by the Health Policy Center, Georgetown University, and the American Association for the Advancement of Science. BE06523.
*allowing to die; cancer; *death; *determination of death; living wills; mortality; organ transplantation

Fulton, Robert, comp. Death, Grief and Bereavement: A Bibliography, 1845-1975. New York: Arno Press, 1977. 253 p. A volume in the Arno Press collection, The Literature of Death and Dying. Bibliography. ISBN 0-405-09570-8. BE05700.

abortion; attitudes to death; *death; euthanasia; psychological stress; terminal care; terminally ill

Horan, Dennis J.; Mall, David, eds. Death, Dying, and Euthanasia. Washington: University Publications of America, 1977. 821 p. Bibliographies included. ISBN 0-89093-139-9. BE05685.
active euthanasia; allowing to die; brain death; congenital defects; *death; decision making; determination of death; Down's syndrome; *euthanasia; extraordinary treatment; killing; legal aspects; legislation; national socialism; newborns; spina bifida; suicide; terminal care; terminally ill; treatment refusal; withholding treatment

Miller, Albert J.; Acri, Michael J. Medical profession and nursing experiences. *In their* Death: A Bibliographical Guide. Metuchen, N.J.: Scarecrow Press, 1977. p. 95-174. Bibliography. ISBN 0-8108-1025-5. BE06186.
*attitudes to death; cancer; communication; *death; determination of death; disclosure; family members; nurses; physicians; psychological stress; *terminal care; terminally ill

Pattison, E. Mansell, ed. The Experience of Dying. Englewood Cliffs, N.J.: Prentice-Hall, 1977. 335 p. Refs. incl. Bibliography: 327-335. ISBN 0-13-294611-4. BE06683.
*adolescents; *adults; *aged; *attitudes to death; burn patients; cancer; *case reports; *children; chronically ill; congenital defects; *death; family; hemophilia; family members; infants; intensive care units; kidneys; leukemia; medical staff; organ transplantation; physically handicapped; psychological stress; psychology; religious beliefs; renal dialysis; social interaction; terminal care; terminally ill

Sell, Irene L., comp. Dying and Death; An Annotated Bibliography. New York: Tiresias Press, 1977. 144 p. Bibliography. ISBN 0-913292-36-2. BE06685.
audiovisual aids; *death; nurses; *terminal care

U.S. National Library of Medicine. Euthanasia and the Right to Die. January 1975 through August 1977. Bethesda, Md.: National Library of Medicine, 1977. 10 p. Literature Search No. 77-13, prepared by Leonard J. Bahlman and Geraldine D. Nowak. 207 citations. Bibliography. BE06694.
allowing to die; attitudes to death; *death; *euthanasia; health personnel; human rights; legal aspects; terminal care

DEATH WITH DIGNITY *See* ALLOWING TO DIE

DEATH / CASE REPORTS

Pattison, E. Mansell, ed. The Experience of Dying. Englewood Cliffs, N.J.: Prentice-Hall, 1977. 335 p. Refs. incl. Bibliography: 327-335. ISBN 0-13-294611-4. BE06683.
*adolescents; *adults; *aged; *attitudes to death; burn patients; cancer; *case reports; *children;

BE = bioethics accession number   fn. = footnotes   refs. = references

chronically ill; congenital defects; *death; family; hemophilia; family members; infants; intensive care units; kidneys; leukemia; medical staff; organ transplantation; physically handicapped; psychological stress; psychology; religious beliefs; renal dialysis; social interaction; terminal care; terminally ill

DEATH / EDUCATION

Kastenbaum, Robert J. Dying—innovations in care. *In his* Death, Society, and Human Experience. St. Louis: C.V. Mosby, 1977. p. 217-240. 14 refs. ISBN 0-8016-2617-X. BE06185.
*communication; *death; *education; family members; goals; *hospices; nursing education; pain; patients' rights; standards; suffering; *terminal care; *terminally ill

Ulin, Richard O. Death and Dying Education. Washington: National Education Association, 1977. 72 p. 26 fn. Bibliography: p. 61-72. ISBN 0-8106-1815-X. BE06689.
adolescents; aged; children; *curriculum; *death; *education; literature

DEATH / MEDICAL EDUCATION

Dickinson, George E. Death education in U.S. medical schools. *Journal of Medical Education* 51(2): 134-136, Feb 1976. 2 refs. BE07095.
curriculum; *death; *medical education

DEATH / PHYSICIAN'S ROLE

Stephens, G. Gayle; Baker, Sherwood. The ethical dimensions of behavior. *In* Conn, Robert E.; Rakel, Robert E.; Johnson, Thomas W., eds. Family Practice. Philadelphia: W.B. Saunders, 1973. p. 219-230. 12 refs. ISBN 0-7216-2665-3. BE07058.
counseling; *death; decision making; determination of death; ethics; *philosophy; *physician's role; *physicians; *psychological stress

DEATH / SCIENCE

Miller, Albert J.; Acri, Michael J. Science. *In their* Death: A Bibliographical Guide. Metuchen, N.J.: Scarecrow Press, 1977. p. 205-210. Bibliography. ISBN 0-8108-1025-5. BE06187.
*cryonic suspension; *death; determination of death; *life extension; mortality; *science

DECEPTION / PHYSICIANS / TERMINALLY ILL

Bok, Sissela. Lies to the sick and dying. *In her* Lying: Moral Choice in Public and Private Life. New York: Pantheon Books, 1978. p. 220-241. 32 fn. ISBN 0-394-41370-0. BE06702.
attitudes; attitudes to death; codes of ethics; *death; *deception; diagnosis; disclosure; medical

ethics; paternalism; *physicians; prognosis; risks and benefits; terminal care; *terminally ill

DECEPTION *See under*

BEHAVIORAL RESEARCH / DECEPTION

BEHAVIORAL RESEARCH / DECEPTION / COSTS AND BENEFITS

BEHAVIORAL RESEARCH / DECEPTION / METHODS

PATIENT CARE / DECEPTION

DECISION MAKING *See under*

ALLOWING TO DIE / DECISION MAKING

ALLOWING TO DIE / SPINA BIFIDA / DECISION MAKING

BIOETHICAL ISSUES / DECISION MAKING

BIOMEDICAL RESEARCH / DECISION MAKING

BIOMEDICAL TECHNOLOGIES / DECISION MAKING

ETHICAL ANALYSIS / DECISION MAKING

HEALTH CARE / DECISION MAKING

KILLING / DECISION MAKING

MEDICINE / DECISION MAKING

NEWBORNS / SPINA BIFIDA / DECISION MAKING

PATIENT CARE / DECISION MAKING

PATIENT CARE / SPINA BIFIDA / DECISION MAKING

PEDIATRICS / DECISION MAKING

PSYCHOACTIVE DRUGS / DECISION MAKING

PSYCHOSURGERY / DECISION MAKING

TERMINAL CARE / DECISION MAKING

DEINSTITUTIONALIZED PERSONS / MENTALLY ILL / SOCIAL IMPACT

Klerman, Gerald L. Better but not well: social and ethical issues in the deinstitutionalization of the mentally ill. *Schizophrenia Bulletin* 3(4): 617-631, 1977. 38 refs. BE06280.
biomedical technologies; *community services; dehumanization; *deinstitutionalized persons; *health care delivery; historical aspects; mental health; institutionalized persons; mental institutions; *mentally ill; *psychoactive drugs; psychotherapy; quality of life; schizophrenia; *social impact; state government

DEINSTITUTIONALIZED PERSONS *See under*

LEGAL OBLIGATIONS / PHYSICIANS / DEINSTITUTIONALIZED PERSONS

PSYCHIATRY / DEINSTITUTIONALIZED PERSONS / LEGAL LIABILITY

DELIVERY OF HEALTH CARE *See* HEALTH CARE DELIVERY

BE = bioethics accession number   fn. = footnotes   refs. = references

DENMARK *See under*

HUMAN EXPERIMENTATION / ETHICS COMMIT-
TEES / DENMARK

RECOMBINANT DNA RESEARCH / GOVERN-
MENT REGULATION / DENMARK

DEONTOLOGICAL ETHICS *See under*

BEHAVIORAL RESEARCH / DEONTOLOGICAL
ETHICS

DEPARTMENT OF THE ARMY *See under*

BIOLOGICAL WARFARE / DEPARTMENT OF THE
ARMY

HUMAN EXPERIMENTATION / DEPARTMENT
OF THE ARMY / LSD

DEPARTMENT OF THE NAVY *See under*

HUMAN EXPERIMENTATION / DEPARTMENT
OF THE NAVY

DETERMINATION OF DEATH *See also* BRAIN
DEATH

DETERMINATION OF DEATH

Beauchamp, Tom L.; Perlin, Seymour, eds. Ethical Issues in Death and Dying. Englewood Cliffs, N.J.:
Prentice-Hall, 1978. 368 p. References and footnotes included. ISBN 0-13-290114-5. BE06878.
active euthanasia; allowing to die; brain death;
coma; decision making; *determination of death;
disclosure; *euthanasia; informed consent; killing;
legal aspects; legislation; living wills; moral obligations; *morality; newborns; *patients' rights; *suicide; terminal care; *terminally ill; theology;
treatment refusal; *value of life; values; withholding treatment; Quinlan, Karen

Beauchamp, Tom L.; Walters, LeRoy, eds. Contemporary Issues in Bioethics. Encino, Calif.: Dickenson
Publishing Company, 1978. 612 p. References and
footnotes   included.   ISBN   0-8221-0200-5.
BE06700.
*abortion; allowing to die; *behavior control; *bio-
ethics; biomedical technologies; brain death; children; codes of ethics; confidentiality; decision
making; deontological ethics; *determination of
death; *euthanasia; fetuses; genetic intervention;
*health; *health care; *human experimentation;
*informed consent; involuntary commitment; jus-
tice; legal aspects; *medical ethics; natural law;
*normative ethics; operant conditioning; patients'
rights; personhood; psychosurgery; *resource allocation; rights; selection for treatment; treatment
refusal; utilitarianism; Kaimowitz v. Department
of Mental Health; Quinlan, Karen

Congdon, Howard K. Toward a definition of death. *In*
his The Pursuit of Death. Nashville: Abingdon,
1977. p. 29-39. ISBN 0-687-34915-X. BE05698.
brain death; death; *determination of death

Dunstan, G.R.; Thomson, W.A.R. Definitions of
death. *In* Camps, Francis E.; Robinson, Ann E.;
Lucas, Bernard G., eds. Gradwohl's Legal Medicine. Third Edition. Bristol, England: John Wright,
1976. p. 50-56. Distributed by Yearbook Medical
Publications, Chicago, Illinois. 24 refs. ISBN
0-8151-1430-3. BE06201.
brain death; *determination of death; legal aspects;
organ transplantation; value of life

Hausman, David B. On abandoning life support: an
alternative proposal. *Man and Medicine: Journal of
Values and Ethics in Health Care* 2(3): 169-188,
Spring 1977. Comments on essay on p. 178-188. 13
fn. BE05920.
*allowing to die; brain death; coma; decision mak-
ing; *determination of death; *justifiable killing;
organ transplantation; resource allocation; scarcity; utilitarianism; value of life; *withholding
treatment; Harvard Committee on Brain Death

Horan, Dennis J. Introduction. *In* Horan, Dennis J.;
Mall, David, eds. Death, Dying, and Euthanasia.
Washington: University Publications of America,
1977. p. xi-xxii. ISBN 0-89093-139-9. BE05686.
brain death; congenital defects; *determination of
death; *euthanasia; national socialism; newborns;
resuscitation; suicide; terminal care; withholding
treatment

**Mukai, Christine; Josephs, Hilary K.; Nakasone,
Linda.** Towards a Definition of Death. Honolulu:
Hawaii Legislative Reference Bureau, 1977. 181 p.
Report No. 1, 1977. 293 fn. BE06680.
*brain death; coma; criminal law; *determination
of death; EEG; economics; extraordinary treatment; killing; *legal aspects; *legislation; medicine;
organ transplantation; politics; religious ethics;
Roman Catholic ethics; standards; *state government; withholding treatment; Cameron, Alice;
*Hawaii; Quinlan, Karen; *United States

Weir, Robert F., ed. Ethical Issues in Death and Dying. New York: Columbia University Press, 1977.
405 p. References and footnotes included. ISBN
0-231-04307-4. BE06696.
*allowing to die; brain death; cancer; congenital
defects; *determination of death; diagnosis; *disclosure; *euthanasia; legal aspects; newborns; physicians; *suicide; *terminally ill; treatment refusal

DETERMINATION OF DEATH / BRAIN DEATH

Engelhardt, H. Tristram. Definitions of death: where to
draw the lines and why. *In* McMullin, Ernan, ed.
Death and Decision. Boulder, Colo.: Westview
Press, 1978. p. 15-34. American Association for the
Advancement of Science Selected Symposium 18. 40
fn. ISBN 0-89158-152-9. BE06707.
*brain death; *determination of death; historical
aspects; legal aspects; personhood; state government

Isaacs, Leonard. Death, where is thy distinguishing?

BE = bioethics accession number   fn. = footnotes   refs. = references

Hastings Center Report 8(1): 5-8, Feb 1978. 1 fn. BE06779.

*brain death; *determination of death; legal aspects; legislation; state government; Harvard Committee on Brain Death

Lamb, David. Diagnosing death. *Philosophy and Public Affairs* 7(2): 144-153, Winter 1978. 15 fn. BE06742.

allowing to die; *brain death; *determination of death; organ transplantation

DETERMINATION OF DEATH / HAWAII

Mukai, Christine; Josephs, Hilary K.; Nakasone, Linda. Towards a Definition of Death. Honolulu: Hawaii Legislative Reference Bureau, 1977. 181 p. Report No. 1, 1977. 293 fn. BE06680.

*brain death; coma; criminal law; *determination of death; EEG; economics; extraordinary treatment; killing; *legal aspects; *legislation; medicine; organ transplantation; politics; religious ethics; Roman Catholic ethics; standards; *state government; withholding treatment; Cameron, Alice; *Hawaii; Quinlan, Karen; *United States

DETERMINATION OF DEATH / JEWISH ETHICS

Bleich, J. David. Survey of recent halakhic periodical literature: time of death legislation. *Tradition: A Journal of Orthodox Thought* 16(4): 130-139, Summer 1977. BE06489.

allowing to die; brain death; *determination of death; *Jewish ethics; standards

DETERMINATION OF DEATH / LEGAL ASPECTS

McKenney, Edward J. Death and dying in Tennessee. *Memphis State University Law Review* 7(4): 503-554, Summer 1977. 343 fn. BE06307.

*allowing to die; *brain death; *coma; *determination of death; euthanasia; *extraordinary treatment; family members; judicial action; *legal aspects; legal liability; *legislation; *living wills; organ transplantation; physician patient relationship; physician's role; privacy; state government; rights; state interest; terminally ill; third party consent; treatment refusal; *withholding treatment; *Dockery v. Dockery; In re Quinlan; *Tennessee; *Tennessee Natural Death Act

Rothman, Daniel A.; Rothman, Nancy L. Death. *In their* The Professional Nurse and the Law. Boston: Little, Brown, 1977. p. 125-141. 32 fn. ISBN 0-316-75768-3. BE06520.

active euthanasia; allowing to die; brain death; *determination of death; *euthanasia; *legal aspects; legal rights; organ transplantation; religious beliefs; *treatment refusal; withholding treatment

DETERMINATION OF DEATH / LEGISLATION / UNITED STATES

Mukai, Christine; Josephs, Hilary K.; Nakasone, Linda. Towards a Definition of Death. Honolulu: Hawaii Legislative Reference Bureau, 1977. 181 p. Report No. 1, 1977. 293 fn. BE06680.

*brain death; coma; criminal law; *determination of death; EEG; economics; extraordinary treatment; killing; *legal aspects; *legislation; medicine; organ transplantation; politics; religious ethics; Roman Catholic ethics; standards; *state government; withholding treatment; Cameron, Alice; *Hawaii; Quinlan, Karen; *United States

DETERMINATION OF DEATH / PERSONHOOD / SELF CONCEPT

Schiffer, R.B. The concept of death: tradition and alternative. *Journal of Medicine and Philosophy* 3(1): 24-37, Mar 1978. 18 refs. 4 fn. BE06826.

brain; brain death; death; *determination of death; *human characteristics; *personhood; *self concept; Harvard Committee on Brain Death

DETERMINATION OF DEATH / WEST VIRGINIA

West Virginia. Uniform Anatomical Gift Act. Definitions. *West Virginia Code,* Section 16-19-1, 1976. 2 p. BE06606.

body parts and fluids; *determination of death; organ donation; organ donors; *West Virginia

DEVELOPING COUNTRIES *See under*

DHEW GUIDELINES *See under*

DIAGNOSIS *See under*

DIGNITY *See under*

BE = bioethics accession number   fn. = footnotes   refs. = references

DISCLOSURE / ANESTHESIA / RISKS AND BENEFITS

Lankton, James W.; Batchelder, Barron M.; Ominsky, Alan J. Emotional responses to detailed risk disclosure for anesthesia: a prospective, randomized study. *Anesthesiology* 46(4): 294-296, April 1977. 8 fn. BE06484.
*anesthesia; attitudes; *disclosure; evaluation; informed consent; patients; recall; *risks and benefits

DISCLOSURE / CANCER / COMMUNICATION

McIntosh, Jim. Communication and Awareness in a Cancer Ward. New York: Prodist, 1977. 210 p. 39 fn. Bibliography: p. 205-208. ISBN 0-88202-109-5. BE05702.
attitudes; *cancer; case reports; *communication; *deception; diagnosis; *disclosure; family members; nurses; *patients; physicians; prognosis; psychological stress; social adjustment; social interaction

DISCLOSURE / CANCER / DIAGNOSIS

Stone, Brenda. Telling the patient he has cancer. *American Medical News* 20(37): 1+, 19 Sep 1977. BE05786.
adults; attitudes to death; *cancer; children; communication; *diagnosis; *disclosure; physicians; prognosis; psychological stress; terminal care

DISCLOSURE / CANCER / PHYSICIAN'S ROLE

Hanganu, Ecaterina; Popa, G. Cancer and truth. *Journal of Medical Ethics* 3(2): 74-75, Jun 1977. 10 refs. BE05649.
*cancer; dignity; *disclosure; *patients; *physician's role; psychological stress; self concept; terminally ill

DISCLOSURE / CANCER / PROGNOSIS

Brewin, Thurstan B. The cancer patient: communication and morale. *British Medical Journal* 2(6103): 1623-1627, 24-31 Dec 1977. 15 refs. BE06506.
attitudes; *cancer; attitudes to death; *communication; chronically ill; diagnosis; *disclosure; medical staff; nurses; *physicians; *prognosis; psychological stress; *terminally ill

DISCLOSURE / CLERGY / MEDICAL RECORDS

Hofmann, Paul B. Data release procedures ensure clergy's access, patients' privacy. *Hospitals* 51(8): 71-73, 16 Apr 1977. BE05804.
*clergy; *confidentiality; counseling; *disclosure; hospitals; informed consent; *medical records; patients' rights; program descriptions

DISCLOSURE / LEGAL ASPECTS

Katz, Jay. Informed consent—a fairy tale? Law's vi-

sion. *University of Pittsburgh Law Review* 39(2): 137-174, Winter 1977. 113 fn. BE06526.
battery; decision making; *disclosure; *informed consent; *judicial action; *law; *legal aspects; legal liability; legal obligations; malpractice; negligence; patient participation; patients; physicians; *risks and benefits; *self determination; standards

DISCLOSURE / PHYSICIAN PATIENT RELATIONSHIP

Siegel, Seymour. Some reflections on telling the truth. *Linacre Quarterly* 44(3): 229-239, Aug 1977. 22 fn. BE05840.
dignity; *disclosure; injuries; *philosophy; *physician patient relationship; physician's role; psychological stress; *scriptural interpretation; terminally ill; values

DISCLOSURE / TERMINALLY ILL

Heberden, P. Should the doctor tell? *South African Medical Journal* 52(21): 829, 12 Nov 1977. BE06356.
communication; *disclosure; *physicians; psychological stress; *terminally ill

Weir, Robert F., ed. Ethical Issues in Death and Dying. New York: Columbia University Press, 1977. 405 p. References and footnotes included. ISBN 0-231-04307-4. BE06696.
*allowing to die; brain death; cancer; congenital defects; *determination of death; diagnosis; *disclosure; *euthanasia; legal aspects; newborns; physicians; *suicide; *terminally ill; treatment refusal

DISCLOSURE / TERMINALLY ILL / CANCER

Wilkes, Eric. Quality of life: effects of the knowledge of diagnosis in terminal illness. *Nursing Times* 73(39): 1506-1507, 29 Sep 1977. 4 refs. BE06340.
*cancer; diagnosis; *disclosure; evaluation; nurses; *quality of life; *terminally ill

DISCLOSURE *See under*

IMMUNIZATION / DISCLOSURE / RISKS AND BENEFITS
PATIENT CARE / DISCLOSURE
RECOMBINANT DNA RESEARCH / DISCLOSURE

DISCRIMINATION *See under*

ABORTION / DISCRIMINATION
ABORTION / DISCRIMINATION / MEDICAID
BEHAVIORAL GENETICS / DISCRIMINATION
EUGENICS / DISCRIMINATION / HISTORICAL ASPECTS

DISSENT / PSYCHIATRY / USSR

Reich, Walter. Soviet psychiatry on trial. *Commentary* 65(1): 40-48, Jan 1978. 1 fn. BE06749.

BE = bioethics accession number   fn. = footnotes   refs. = references

case reports; *dissent; involuntary commitment; *organizational policies; politics; *professional organizations; *psychiatric diagnosis; *psychiatry; schizophrenia; *USSR; *World Psychiatric Association

DISSENT *See under*

INVOLUNTARY COMMITMENT / DISSENT / USSR

PSYCHIATRY / DISSENT / USSR

DNA HYBRIDIZATION *See* RECOMBINANT DNA RESEARCH

DNA RECOMBINANTS *See* RECOMBINANT DNA RESEARCH

DOUBLE EFFECT *See under*

ABORTION / ROMAN CATHOLIC ETHICS / DOUBLE EFFECT

DOWN'S SYNDROME *See under*

ALLOWING TO DIE / NEWBORNS / DOWN'S SYNDROME

GENETIC COUNSELING / DOWN'S SYNDROME

DRUG ABUSE *See under*

BEHAVIORAL RESEARCH / DRUG ABUSE / RIGHTS

BEHAVIORAL RESEARCH / PUBLIC PARTICIPATION / DRUG ABUSE

CONFIDENTIALITY / MEDICAL RECORDS / DRUG ABUSE

DRUG INDUSTRY *See under*

HUMAN EXPERIMENTATION / DRUG INDUSTRY / VOLUNTEERS

DRUGS *See also* PSYCHOACTIVE DRUGS

DRUGS *See under*

CAPITAL PUNISHMENT / DRUGS

CONFIDENTIALITY / MEDICAL RECORDS / DRUGS

HUMAN EXPERIMENTATION / CHILDREN / DRUGS

HUMAN EXPERIMENTATION / DRUGS

HUMAN EXPERIMENTATION / DRUGS / CENTRAL INTELLIGENCE AGENCY

PATIENT CARE / DRUGS / RISKS AND BENEFITS

DUE PROCESS *See under*

INVOLUNTARY COMMITMENT / ADOLESCENTS / DUE PROCESS

EAST GERMANY *See under*

ORGAN TRANSPLANTATION / EAST GERMANY

ECOLOGY *See under*

POPULATION CONTROL / ECOLOGY

RECOMBINANT DNA RESEARCH / ECOLOGY / SOCIAL IMPACT

ECONOMICS *See under*

BLOOD DONATION / ECONOMICS

HEALTH CARE DELIVERY / ECONOMICS

HEALTH CARE / PUBLIC POLICY / ECONOMICS

HEALTH CARE / RESOURCE ALLOCATION / ECONOMICS

HEALTH CARE / VALUE OF LIFE / ECONOMICS

PATIENT CARE / CANCER / ECONOMICS

RENAL DIALYSIS / ECONOMICS

EDELIN, KENNETH

Robertson, John A. After *Edelin*: little guidance. *Hastings Center Report* 7(3): 15-17+, Jun 1977. 1 fn. BE05663.
 *aborted fetuses; abortion; judicial action; killing; legal liability; *legal obligations; *patient care; *physicians; viability; *Edelin, Kenneth

EDUCATION *See under*

BIOETHICS / EDUCATION

DEATH / EDUCATION

ETHICS / SOCIAL SCIENCES / EDUCATION

MEDICAL ETHICS / EDUCATION

MEDICAL ETHICS / HEALTH PERSONNEL / EDUCATION

MEDICAL ETHICS / MEDICAL STAFF / EDUCATION

MEDICAL ETHICS / PHYSICIANS / EDUCATION

ELECTRICAL STIMULATION OF THE BRAIN

Valenstein, Elliot S. Brain control: scientific, ethical and political considerations. *In* Ellison, Craig W., ed. Modifying Man: Implications and Ethics. Washington: University Press of America, 1978. p. 143-168. ISBN 0-8191-0302-0. BE06670.
 aggression; *behavior control; brain; brain pathology; *electrical stimulation of the brain; ethics committees; evaluation; government regulation; human experimentation; mentally ill; political activity; *psychosurgery; social determinants; *violence

ELECTROCONVULSIVE THERAPY

Anderson, J.F., et al. ECT and the media. *British Medical Journal* 2(6094): 1081-1082, 22 Oct 1977. Letter. BE06082.

BE = bioethics accession number   fn. = footnotes   refs. = references

*electroconvulsive therapy; evaluation; mass media; mentally ill; psychiatry

Blachly, P.H. New developments in electroconvulsive therapy. *Diseases of the Nervous System* 37(6): 356-358, Jun 1976. 16 refs. BE07092.
*electroconvulsive therapy; informed consent; legislation; mentally ill; privacy; state government; California

The Royal College of Psychiatrists' memorandum on the use of electroconvulsive therapy. *British Journal of Psychiatry* 131: 261-272, Sep 1977. 40 refs. BE05987.
*electroconvulsive therapy; *evaluation; informed consent; mentally ill; methods; morbidity; mortality; schizophrenia

ELECTROCONVULSIVE THERAPY / COERCION

Hodgkinson, Neville. Guidelines welcomed, but doubts linger over compulsion. *Times (London),* 5 Aug 1977, p. 2. BE06453.
*coercion; *electroconvulsive therapy; informed consent; institutionalized persons; mentally ill

ELECTROCONVULSIVE THERAPY / GOVERNMENT REGULATION / CALIFORNIA

Aden, Gary C. Effects of restrictive legislation of electroconvulsive therapy at a freestanding psychiatric hospital. *Diseases of the Nervous System* 38(4): 230-233, Apr 1977. 2 refs. BE05541.
case reports; *electroconvulsive therapy; ethics committees; *government regulation; incidence; institutionalized persons; *legislation; mentally ill; psychiatry; *social impact; *state government; *California

ELECTROCONVULSIVE THERAPY / INFORMED CONSENT

Salzman, Carl. ECT and ethical psychiatry. *American Journal of Psychiatry* 134(9): 1006-1009, Sep 1977. 13 refs. BE06244.
case reports; coercion; competence; *electroconvulsive therapy; *informed consent; institutionalized persons; involuntary commitment; legal guardians; legal rights; *mentally ill; paternalism; *patients' rights; right to treatment; risks and benefits; self determination; *treatment refusal

ELECTROCONVULSIVE THERAPY / INSTITUTIONALIZED PERSONS / MENTALLY ILL

New Jersey. Juvenile and Domestic Relations Court, Essex County. Matter of W.S. 16 May 1977. *Atlantic Reporter, 2d Series,* 377: 969-974. 1 fn. BE07109.
competence; *electroconvulsive therapy; emergency care; informed consent; *institutionalized persons; *mentally ill; *third party consent

ELECTROCONVULSIVE THERAPY / LEGISLATION / CALIFORNIA

Moore, R.A. The electroconvulsive therapy fight in California. *Journal of Forensic Sciences* 22(4): 845-850, Oct 1977. 6 refs. BE06350.
*electroconvulsive therapy; informed consent; *legal rights; *legislation; mentally ill; patients; psychiatry; state government; *California

ELECTROCONVULSIVE THERAPY / PATIENTS' RIGHTS

Salzman, Carl. ECT and ethical psychiatry. *American Journal of Psychiatry* 134(9): 1006-1009, Sep 1977. 13 refs. BE06244.
case reports; coercion; competence; *electroconvulsive therapy; *informed consent; institutionalized persons; involuntary commitment; legal guardians; legal rights; *mentally ill; paternalism; *patients' rights; right to treatment; risks and benefits; self determination; *treatment refusal

ELECTROCONVULSIVE THERAPY / RISKS AND BENEFITS

Illman, John. ECT: therapy or trauma? *Nursing Times* 73(32): 1226-1227, 11 Aug 1977. 1 fn. BE05842.
attitudes; *electroconvulsive therapy; *evaluation; psychiatry; *risks and benefits

ELECTROCONVULSIVE THERAPY / ROYAL COLLEGE OF PSYCHIATRISTS

'Unwilling' ECT patients covered in guidelines from psychiatrists. *Nursing Times* 73(32): 1220, 11 Aug 1977. BE05841.
*electroconvulsive therapy; government regulation; *informed consent; *organizational policies; *professional organizations; Great Britain; *Royal College of Psychiatrists

ELECTROCONVULSIVE THERAPY / THIRD PARTY CONSENT

New Jersey. Juvenile and Domestic Relations Court, Essex County. Matter of W.S. 16 May 1977. *Atlantic Reporter, 2d Series,* 377: 969-974. 1 fn. BE07109.
competence; *electroconvulsive therapy; emergency care; informed consent; *institutionalized persons; *mentally ill; *third party consent

ELECTROCONVULSIVE THERAPY / TREATMENT REFUSAL

Hodgkinson, Neville. Call for prevention of forced electric shock treatment. *Times (London),* 1 Aug 1977, p. 2. BE06452.
coercion; *electroconvulsive therapy; institutional-

BE = bioethics accession number   fn. = footnotes   refs. = references

ized persons; legal aspects; mentally ill; *treatment refusal

ELECTROCONVULSIVE THERAPY / TREATMENT REFUSAL / PATIENTS' RIGHTS

Salzman, Carl. ECT and ethical psychiatry. *American Journal of Psychiatry* 134(9): 1006-1009, Sep 1977. 13 refs. BE06244.
case reports; coercion; competence; *electroconvulsive therapy; *informed consent; institutionalized persons; involuntary commitment; legal guardians; legal rights; *mentally ill; paternalism; *patients' rights; right to treatment; risks and benefits; self determination; *treatment refusal

EMBRYO TRANSFER / VALUE OF LIFE

Will, George F. Irreverent test tubes. *Washington Post,* 30 Jul 1978, p. B7. BE07025.
biomedical technologies; *embryo transfer; in vitro fertilization; *value of life

EMBRYO TRANSFER   *See under*

ANIMAL EXPERIMENTATION / EMBRYO TRANSFER

IN VITRO FERTILIZATION / EMBRYO TRANSFER

EMERGENCY CARE / LEGAL OBLIGATIONS

Annas, George J. Beyond the Good Samaritan: should doctors be required to provide essential services? *Hastings Center Report* 8(2): 16-17, Apr 1978. 3 fn. BE06835.
*emergency care; *legal obligations; moral obligations; *patient care; physician patient relationship; *physicians

EMERGENCY CARE   *See under*

PROLONGATION OF LIFE / EMERGENCY CARE

ENGLAND   *See under*

INFANTICIDE / HISTORICAL ASPECTS / ENGLAND

EPIDEMIOLOGY   *See under*

CONFIDENTIALITY / MEDICAL RECORDS / EPIDEMIOLOGY

HUMAN EXPERIMENTATION / EPIDEMIOLOGY

HUMAN EXPERIMENTATION / EPIDEMIOLOGY / LEGAL LIABILITY

EQUAL PROTECTION   *See under*

ABORTION / EQUAL PROTECTION / SUPREME COURT DECISIONS

ABORTION / FINANCIAL SUPPORT / EQUAL PROTECTION

ETHICAL ANALYSIS / DECISION MAKING

Clarke, Dolores D.; Clarke, Desmond M. Definitions and ethical decisions. *Journal of Medical Ethics* 3(4): 186-188, Dec 1977. 5 refs. BE06250.
abortion; brain death; *decision making; determination of death; *ethical analysis; ethics; fetuses; medical ethics; methods; personhood

ETHICAL RELATIVISM   *See under*

BEHAVIORAL RESEARCH / ETHICAL RELATIVISM

ETHICAL REVIEW   *See under*

HUMAN EXPERIMENTATION / ETHICAL REVIEW

PSYCHOSURGERY / ETHICAL REVIEW

ETHICS   *See also* BIOETHICS, CHRISTIAN ETHICS, DEONTOLOGICAL ETHICS, JEWISH ETHICS, MEDICAL ETHICS, MORALITY, NURSING ETHICS, PROFESSIONAL ETHICS, ROMAN CATHOLIC ETHICS, TELEOLOGICAL ETHICS, UTILITARIANISM

ETHICS COMMITTEES   *See under*

ABORTION / ETHICS COMMITTEES / CANADA

HUMAN EXPERIMENTATION / ETHICS COMMITTEES

HUMAN EXPERIMENTATION / ETHICS COMMITTEES / DENMARK

HUMAN EXPERIMENTATION / ETHICS COMMITTEES / HOSPITALS

INFORMED CONSENT / ETHICS COMMITTEES

PATIENT CARE / ETHICS COMMITTEES

PATIENT CARE / ETHICS COMMITTEES / INSTITUTIONALIZED PERSONS

PSYCHOSURGERY / ETHICS COMMITTEES / NEW SOUTH WALES

ETHICS / NURSES

Boyd, Kenneth. The nature of ethics. *Nursing Mirror* 145(3): 14-16, 21 Jul 1977. 7 refs. BE05813.
bioethical issues; decision making; *ethics; *nurses; nursing ethics; philosophy; physicians; politics; social interaction

ETHICS / NURSING EDUCATION

Langham, Paul. Open forum: on teaching ethics to nurses. *Nursing Forum* 16(3-4): 220-227, 1977. 3 refs. 1 fn. BE06475.
decision making; *ethics; *nursing education; nursing ethics; teaching methods

ETHICS / SOCIAL SCIENCES / EDUCATION

Warwick, Donald P. Social sciences and ethics. *Hastings Center Report* 7(6): Special Suppl. 8-10, Dec 1977. BE06155.

BE = bioethics accession number   fn. = footnotes   refs. = references

*behavioral research; deception; *education; *ethics; science; *social sciences; universities; values

ETHICS *See under*
   ABORTION / ETHICS
   ACTIVE EUTHANASIA / ETHICS
   AGED / ETHICS / DIGNITY
   BEHAVIORAL RESEARCH / ETHICS
   BEHAVIORAL RESEARCH / ETHICS / PUBLIC
     OPINION
   HEALTH CARE DELIVERY / ADMINISTRA-
     TORS / ETHICS
   MEDICAL EDUCATION / ETHICS
   MEDICINE / ETHICS
   MEDICINE / ETHICS / ARISTOTLE
   SELECTIVE ABORTION / ETHICS

EUGENICS   *See also* NEGATIVE EUGENICS

EUGENICS

Howard, Ted; Rifkin, Jeremy. Who Should Play God? The Artificial Creation of Life and What It Means for the Future of the Human Race. New York: Dell Publishing Company, 1977. 272 p. 526 fn. ISBN 0-440-19504-7. BE05882.
   AID; behavioral genetics; biomedical technologies; cloning; DNA therapy; drugs; embryo transfer; *eugenics; genetic screening; *genetic intervention; government regulation; health hazards; historical aspects; in vitro fertilization; industry; involuntary sterilization; population control; recombinant DNA research; *reproductive technologies; risks and benefits; selective abortion; social control; *social impact

Medawar, P.B.; Medawar, J.S. Eugenics. *In their* The Life Science: Current Ideas of Biology. London: Wildwood House, 1977. p. 56-65. 2 fn. ISBN 0-7045-0243-7. BE05884.
   biomedical technologies; chromosomal disorders; dominant genetic conditions; *eugenics; gene pool; negative eugenics; recessive genetic conditions

Murphy, Edmond A.; Chase, Gary A.; Rodriguez, Alejandro. Genetic intervention: some social, psychological, and philosophical aspects. *In* Cohen, Bernice H.; Lilienfeld, Abraham M.; Huang, P.C., eds. Genetic Issues in Public Health and Medicine. Springfield, Ill.: Charles C. Thomas, 1978. p. 358-398. 47 refs. ISBN 0-398-03659-4. BE06985.
   amniocentesis; *eugenics; family planning; gene pool; *genetic counseling; genetic defects; human rights; involuntary sterilization; parents; reproduction; selective abortion

EUGENICS / DISCRIMINATION / HISTORICAL ASPECTS

Chase, Allan. The Legacy of Malthus: The Social Costs of the New Scientific Racism. New York: Alfred A.

Knopf, 1977. 704 p. 651 fn. ISBN 0-394-48045-7. BE06674.
   behavioral genetics; *discrimination; ecology; *eugenics; evaluation; *historical aspects; human development; indigents; intelligence; involuntary sterilization; mentally retarded; minority groups; natural law; politics; population control; preventive medicine; public health; science; social impact; socioeconomic factors; *Malthus, Thomas

EUGENICS / HISTORICAL ASPECTS / GERMANY

Graham, Loren R. Science and values: the eugenics movement in Germany and Russia in the 1920s. *American Historical Review* 82(5): 1133-1164, Dec 1977. 60 fn. BE06508.
   attitudes; biology; communism; discrimination; *eugenics; genetics; *historical aspects; national socialism; natural selection; *political systems; politics; professional organizations; *science; social determinants; social impact; socialism; technology; *values; *Germany; Lemarckism; Mendelism; Twentieth Century; *USSR

EUGENICS / HISTORICAL ASPECTS / GREAT BRITAIN

Searle, G.R. Eugenics and Politics in Britain, 1900-1914. Leyden, The Netherlands: Noordhoff International Publishing, 1976. 147 p. 485 fn. ISBN 90-286-0236-4. BE06664.
   attitudes; behavioral genetics; discrimination; *eugenics; *historical aspects; intelligence; *negative eugenics; legal aspects; politics; *positive eugenics; social determinants; socioeconomic factors; *Great Britain

EUGENICS / HISTORICAL ASPECTS / USSR

Graham, Loren R. Science and values: the eugenics movement in Germany and Russia in the 1920s. *American Historical Review* 82(5): 1133-1164, Dec 1977. 60 fn. BE06508.
   attitudes; biology; communism; discrimination; *eugenics; genetics; *historical aspects; national socialism; natural selection; *political systems; politics; professional organizations; *science; social determinants; social impact; socialism; technology; *values; *Germany; Lemarckism; Mendelism; Twentieth Century; *USSR

EUGENICS / POLITICAL SYSTEMS / VALUES

Graham, Loren R. Political ideology and genetic theory: Russia and Germany in the 1920's. *Hastings Center Report* 7(5): 30-39, Oct 1977. 33 fn. BE06080.
   disadvantaged; *eugenics; genetics; *historical aspects; *political systems; politics; *science; social impact; socioeconomic factors; *values; *Germany; *USSR

BE = bioethics accession number   fn. = footnotes   refs. = references

Graham, Loren R. Science and values: the eugenics movement in Germany and Russia in the 1920s. *American Historical Review* 82(5): 1133-1164, Dec 1977. 60 fn. BE06508.
> attitudes; biology; communism; discrimination; *eugenics; genetics; *historical aspects; national socialism; natural selection; *political systems; politics; professional organizations; *science; social determinants; social impact; socialism; technology; *values; *Germany; Lemarckism; Mendelism; Twentieth Century; *USSR

EUROPE *See under*
> ABORTION / LEGISLATION / EUROPE
> RECOMBINANT DNA RESEARCH / EUROPE
> RECOMBINANT DNA RESEARCH / GOVERNMENT REGULATION / EUROPE

EUROPEAN SCIENCE FOUNDATION *See under*
> RECOMBINANT DNA RESEARCH / EUROPEAN SCIENCE FOUNDATION

EUTHANASIA *See also* ACTIVE EUTHANASIA, ALLOWING TO DIE, INFANTICIDE, INVOLUNTARY EUTHANASIA, LIVING WILLS, PROLONGATION OF LIFE, SUICIDE, TERMINAL CARE, VOLUNTARY EUTHANASIA

EUTHANASIA

Anderson, Norman. Issues of Life and Death: Abortion, Birth Control, Capital Punishment, Euthanasia. Downers Grove, Ill.: InterVarsity Press, 1977. 130 p. 189 fn. ISBN 0-87784-721-5. BE05787.
> AID; *abortion; *artificial insemination; *capital punishment; Christian ethics; *contraception; DNA therapy; *euthanasia; genetic intervention; justifiable killing; organ transplantation; personhood; scriptural interpretation; *sterilization; theology; *value of life

Beauchamp, Tom L.; Perlin, Seymour, eds. Ethical Issues in Death and Dying. Englewood Cliffs, N.J.: Prentice-Hall, 1978. 368 p. References and footnotes included. ISBN 0-13-290114-5. BE06878.
> active euthanasia; allowing to die; brain death; coma; decision making; *determination of death; disclosure; *euthanasia; informed consent; killing; legal aspects; legislation; living wills; moral obligations; *morality; newborns; *patients' rights; *suicide; terminal care; *terminally ill; theology; treatment refusal; *value of life; values; withholding treatment; Quinlan, Karen

Beauchamp, Tom L.; Walters, LeRoy, eds. Contemporary Issues in Bioethics. Encino, Calif.: Dickenson Publishing Company, 1978. 612 p. References and footnotes included. ISBN 0-8221-0200-5. BE06700.
> *abortion; allowing to die; *behavior control; *bioethics; biomedical technologies; brain death; children; codes of ethics; confidentiality; decision making; deontological ethics; *determination of death; *euthanasia; fetuses; genetic intervention; *health; *health care; *human experimentation; *informed consent; involuntary commitment; justice; legal aspects; *medical ethics; natural law; *normative ethics; operant conditioning; patients' rights; personhood; psychosurgery; *resource allocation; rights; selection for treatment; treatment refusal; utilitarianism; Kaimowitz v. Department of Mental Health; Quinlan, Karen

Greeley, John. Euthanasia: The Debate. Cincinnati: Pamphlet Publications, 1977. 32 p. BE05701.
> *allowing to die; case reports; decision making; *determination of death; *euthanasia; legal aspects; living wills; organ transplantation; physicians; quality of life; terminally ill

Horan, Dennis J. Introduction. *In* Horan, Dennis J.; Mall, David, eds. Death, Dying, and Euthanasia. Washington: University Publications of America, 1977. p. xi-xxii. ISBN 0-89093-139-9. BE05686.
> brain death; congenital defects; *determination of death; *euthanasia; national socialism; newborns; resuscitation; suicide; terminal care; withholding treatment

Horan, Dennis J.; Mall, David, eds. Death, Dying, and Euthanasia. Washington: University Publications of America, 1977. 821 p. Bibliographies included. ISBN 0-89093-139-9. BE05685.
> active euthanasia; allowing to die; brain death; congenital defects; *death; decision making; determination of death; Down's syndrome; *euthanasia; extraordinary treatment; killing; legal aspects; legislation; national socialism; newborns; spina bifida; suicide; terminal care; terminally ill; treatment refusal; withholding treatment

Luxton, R.W. Against euthanasia: the fallacies behind the pressure for 'mercy killing'. *Sunday Times (London)*, 5 Jun 1977, p. 21. BE06446.
> allowing to die; *euthanasia; hospices; legal aspects; terminally ill

McIntyre, Russell L. Euthanasia: a soft paradigm for medical ethics. *Linacre Quarterly* 45(1): 41-54, Feb 1978. 44 fn. BE06785.
> Christian ethics; deontological ethics; decision making; *euthanasia; moral obligations; natural law; pain; prolongation of life; utilitarianism; value of life; wedge argument

McMullin, Ernan, ed. Death and Decision. Boulder, Colo.: Westview Press, 1978. 154 p. American Association for the Advancement of Science Selected Symposium 18. 17 refs. 144 fn. ISBN 0-89158-152-9. BE06705.
> *allowing to die; brain death; determination of death; *euthanasia; human rights; legal aspects; self determination; social adjustment; terminally ill; treatment refusal; value of life

Parker, M. George. You are a child of the universe: you

---

have a right to be here. *Manitoba Law Journal* 7(3): 151-182, 1977. 55 fn. Bibliography: p. 180-182. BE06281.

active euthanasia; allowing to die; determination of death; *euthanasia; legal aspects; law enforcement; legislation; living wills; pain; physician patient relationship; physicians; suffering; value of life; voluntary euthanasia; wedge argument; withholding treatment

Purtilo, Ruth B. Bioethics and euthanasia. *American Journal of Occupational Therapy* 31(6): 394-396, Jul 1977. 6 refs. BE06229.

active euthanasia; allowing to die; *euthanasia; terminally ill

Ramsey, Paul. Ethics at the Edges of Life: Medical and Legal Intersections. New Haven: Yale University Press, 1978. 353 p. The Bampton Lectures in America. 302 fn. ISBN 0-300-02137-2. BE06884.

*abortion; adults; *allowing to die; congenital defects; *decision making; *euthanasia; extraordinary treatment; fetuses; law; legal aspects; legislation; medicine; newborns; physician's role; quality of life; *selection for treatment; *Supreme Court decisions; terminal care; *terminally ill; *third party consent; *treatment refusal; *withholding treatment; California Natural Death Act; Commonwealth v. Edelin; *In re Quinlan; Saikewicz, Joseph

Ramsey, Paul. "Euthanasia" and dying well enough. *In his* Ethics at the Edges of Life: Medical and Legal Intersections. New Haven: Yale University Press, 1978. p. 145-188. 45 fn. ISBN 0-300-02137-2. BE06888.

active euthanasia; *adults; *allowing to die; chronically ill; coma; competence; congenital defects; death; *decision making; diagnosis; drugs; *euthanasia; *extraordinary treatment; human rights; legal aspects; medical ethics; moral obligations; newborns; pain; physicians; prognosis; quality of life; religious ethics; *selection for treatment; social interaction; spina bifida; *standards; terminal care; *terminally ill; theology; *third party consent; *treatment refusal; value of life; withholding treatment; McCormick, Richard A.; Veatch, Robert M.

Smith, David H. Fatal choices: recent discussions of dying. *Hastings Center Report* 7(2): 8-10, Apr 1977. 16 fn. BE05582.

active euthanasia; allowing to die; *decision making; determination of death; *euthanasia; patients; physician's role; self determination; treatment refusal; *Veatch, Robert M.

U.S. National Library of Medicine. Euthanasia and the Right to Die. January 1975 through August 1977. Bethesda, Md.: National Library of Medicine, 1977. 10 p. Literature Search No. 77-13, prepared by Leonard J. Bahlman and Geraldine D. Nowak. 207 citations. Bibliography. BE06694.

allowing to die; attitudes to death; *death; *eutha-

nasia; health personnel; human rights; legal aspects; terminal care

Weir, Robert F., ed. Ethical Issues in Death and Dying. New York: Columbia University Press, 1977. 405 p. References and footnotes included. ISBN 0-231-04307-4. BE06696.

*allowing to die; brain death; cancer; congenital defects; *determination of death; diagnosis; *disclosure; *euthanasia; legal aspects; newborns; physicians; *suicide; *terminally ill; treatment refusal

EUTHANASIA / CHRISTIAN ETHICS

Maguire, Daniel C. Death and the moral domain. *St. Luke's Journal of Theology* 20(3): 197-216, Jun 1977. 58 fn. BE06305.

*active euthanasia; *allowing to die; *Christian ethics; death; ethicists; *ethics; *euthanasia; justifiable killing; *killing; legislation; moral obligations; pain; philosophy; suffering; suicide; value of life; *Church of England

EUTHANASIA / COMA / COSTS AND BENEFITS

Suckiel, Ellen K. Death and benefit in the permanently unconscious patient; a justification of euthanasia. *Journal of Medicine and Philosophy* 3(1): 38-58, Mar 1978. 16 refs. 11 fn. BE06828.

active euthanasia; allowing to die; *coma; consequences; *costs and benefits; determination of death; *euthanasia; injuries; killing; *involuntary euthanasia; moral obligations; morality; normative ethics; patients; physicians; prolongation of life; *value of life; Harvard Committee on Brain Death

EUTHANASIA / LEGAL ASPECTS

Rothman, Daniel A.; Rothman, Nancy L. Death. *In their* The Professional Nurse and the Law. Boston: Little, Brown, 1977. p. 125-141. 32 fn. ISBN 0-316-75768-3. BE06520.

active euthanasia; allowing to die; brain death; *determination of death; *euthanasia; *legal aspects; legal rights; organ transplantation; religious beliefs; *treatment refusal; withholding treatment

EUTHANASIA / LEGAL ASPECTS / GREAT BRITAIN

Martin, Andrew J. Law on mercy killing. *Times (London),* 9 Oct 1976, p. 13. Letter. BE06613.

*euthanasia; *legal aspects; legislation; *Great Britain

EUTHANASIA / MORAL OBLIGATIONS / RIGHTS

Williams, Peter C. Rights and the alleged right of innocents to be killed. *Ethics* 87(4): 383-394, Jul 1977. 29 fn. BE05739.

active euthanasia; allowing to die; consequences; ethics; *euthanasia; family members; *human

---

BE = bioethics accession number   fn. = footnotes   refs. = references

rights; *killing; law; legal rights; *moral obligations; physicians; *rights; suffering; terminally ill

EUTHANASIA / PHYSICIANS

Shumiatcher, Morris C. Medical heroics and the good death. *Canadian Medical Association Journal* 117(5): 520+, 3 Sep 1977. BE06338.
    abortion; allowing to die; decision making; ethics committees; *euthanasia; legal aspects; *physicians; value of life

EUTHANASIA / PROTESTANT ETHICS / RAMSEY, PAUL

Cahill, Lisa G. Euthanasia: A Catholic and a Protestant Perspective. Chicago: University of Chicago, 1976. 375 p. Dissertation, Ph.D., Divinity School, University of Chicago, Mar 1976. 515 fn. Bibliography: p. 365-375. BE07086.
    active euthanasia; allowing to die; brain death; Christian ethics; death; double effect; ethics; *euthanasia; extraordinary treatment; historical aspects; killing; love; morality; natural law; personhood; prolongation of life; *Protestant ethics; *Roman Catholic ethics; suicide; theology; totality; *Nolan, Kieran; *Ramsey, Paul

EUTHANASIA / RIGHTS

Troyer, John. Euthanasia, the right to life, and moral structures: a reply to Professor Kohl. *In* Spicker, Stuart F.; Engelhardt, H. Tristram, eds. Philosophical Medical Ethics: Its Nature and Significance. Proceedings. Boston: D. Reidel, 1977. p. 85-95. Proceedings of the Third Trans-Disciplinary Symposium on Philosophy and Medicine, Farmington, Connecticut, 11-13 Dec 1975. 5 refs. 6 fn. ISBN 90-277-0772-3. BE05865.
    *euthanasia; killing; legal rights; *rights; suicide; *value of life

EUTHANASIA / ROMAN CATHOLIC ETHICS / NOLAN, KIERAN

Cahill, Lisa G. Euthanasia: A Catholic and a Protestant Perspective. Chicago: University of Chicago, 1976. 375 p. Dissertation, Ph.D., Divinity School, University of Chicago, Mar 1976. 515 fn. Bibliography: p. 365-375. BE07086.
    active euthanasia; allowing to die; brain death; Christian ethics; death; double effect; ethics; *euthanasia; extraordinary treatment; historical aspects; killing; love; morality; natural law; personhood; prolongation of life; *Protestant ethics; *Roman Catholic ethics; suicide; theology; totality; *Nolan, Kieran; *Ramsey, Paul

EUTHANASIA / STUDENTS / ATTITUDES

Adams, Gerald R.; Bueche, Nancy; Schvaneveldt, Jay D. Contemporary views of euthanasia: a regional assessment. *Social Biology* 25(1): 62-68, Spring 1978. 9 refs. BE06918.
    active euthanasia; allowing to die; *attitudes; *euthanasia; public opinion; *socioeconomic factors; *students; universities; *United States

EUTHANASIA / TERMINALLY ILL

Anderson, Norman. The prolongation of life, transplant surgery, euthanasia and suicide. *In his* Issues of Life and Death: Abortion, Birth Control, Capital Punishment, Euthanasia. Downers Grove, Ill.: InterVarsity Press, 1977. p. 85-107. 19 fn. ISBN 0-87784-721-5. BE05791.
    abortion; active euthanasia; adults; *allowing to die; brain death; congenital defects; decision making; determination of death; *euthanasia; fetuses; hearts; infanticide; *kidneys; legal aspects; morality; *newborns; organ donors; *organ transplantation; physicians; prenatal diagnosis; prolongation of life; suffering; suicide; *terminally ill; voluntary euthanasia

Leonard, Graham; Martin, Andrew J. The doctor and the dying patient. *Times (London)*, 31 Dec 1976, p. 15. Graham Leonard is Bishop of Truron. BE06624.
    *euthanasia; legal aspects; legislation; physicians; rights; suffering; terminal care; *terminally ill; Church of England

EUTHANASIA / VALUE OF LIFE

Troyer, John. Euthanasia, the right to life, and moral structures: a reply to Professor Kohl. *In* Spicker, Stuart F.; Engelhardt, H. Tristram, eds. Philosophical Medical Ethics: Its Nature and Significance. Proceedings. Boston: D. Reidel, 1977. p. 85-95. Proceedings of the Third Trans-Disciplinary Symposium on Philosophy and Medicine, Farmington, Connecticut, 11-13 Dec 1975. 5 refs. 6 fn. ISBN 90-277-0772-3. BE05865.
    *euthanasia; killing; legal rights; *rights; suicide; *value of life

EUTHANASIA / VALUE OF LIFE / CHRISTIAN ETHICS

Whytehead, Lawrence; Chidwick, Paul F., eds. Dying: Considerations Concerning the Passage from Life to Death. An Interim Report by the Task Force on Human Life, Winnipeg, Manitoba. Winnipeg, Manitoba: Anglican Church of Canada, Jun 1977. 25 p. 21 fn. Bibliography: p. 23-25. BE06197.
    active euthanasia; allowing to die; *Christian ethics; congenital defects; decision making; *euthanasia; legal aspects; newborns; personhood; quality of life; social interaction; terminal care; terminally ill; *value of life; withholding treatment

EVALUATION *See under*
BEHAVIORAL RESEARCH / INFORMED

BE = bioethics accession number   fn. = footnotes   refs. = references

EVALUATION *See under (cont'd.)*
CONSENT / EVALUATION

BEHAVIORAL RESEARCH / PROFESSIONAL ETHICS / EVALUATION

LAETRILE / EVALUATION

MEDICAL ETHICS / STUDENTS / EVALUATION

PATIENT CARE / PLACEBOS / EVALUATION

PSYCHOSURGERY / EVALUATION

PSYCHOSURGERY / EVALUATION / NEW SOUTH WALES

PSYCHOSURGERY / NATIONAL COMMISSION FOR PROTECTION OF HUMAN SUBJECTS / EVALUATION

QUALITY OF LIFE / EVALUATION

SELECTION FOR TREATMENT / SPINA BIFIDA / EVALUATION

EXPERIMENTATION WITH HUMAN SUBJECTS *See* HUMAN EXPERIMENTATION

EXPERT TESTIMONY *See under*

INFORMED CONSENT / EXPERT TESTIMONY / MALPRACTICE

EXTRAORDINARY TREATMENT / JEWISH ETHICS

Bleich, J. David. Theological considerations in the care of defective newborns. *In* Swinyard, Chester A., ed. Decision Making and the Defective Newborn. Proceedings of a Conference on Spina Bifida and Ethics. Springfield, Ill.: Charles C. Thomas, 1978. p. 512-561. 106 fn. ISBN 0-398-03662-4. BE06963.
adults; contracts; *extraordinary treatment; financial support; informed consent; *Jewish ethics; judicial action; legal liability; legal obligations; *moral obligations; *newborns; pain; *patient care; physician patient relationship; *physicians; *prolongation of life; Roman Catholic ethics; *scriptural interpretation; *spina bifida; terminally ill; theology; treatment refusal; *value of life; withholding treatment

EXTRAORDINARY TREATMENT / NEWBORNS / SPINA BIFIDA

Lorber, John. Ethical concepts in the treatment of myelomeningocele. *In* Swinyard, Chester A., ed. Decision Making and the Defective Newborn. Proceedings of a Conference on Spina Bifida and Ethics. Springfield, Ill.: Charles C. Thomas, 1978. p. 59-67. 16 refs. ISBN 0-398-03662-4. BE06952.
*extraordinary treatment; family problems; mentally retarded; morbidity; mortality; *newborns; physically handicapped; prognosis; prolongation of life; quality of life; *selection for treatment; *spina bifida; statistics

EXTRAORDINARY TREATMENT / TERMINALLY ILL

Ramsey, Paul. "Euthanasia" and dying well enough. *In his* Ethics at the Edges of Life: Medical and Legal Intersections. New Haven: Yale University Press, 1978. p. 145-188. 45 fn. ISBN 0-300-02137-2. BE06888.
active euthanasia; *adults; *allowing to die; chronically ill; coma; competence; congenital defects; death; *decision making; diagnosis; drugs; *euthanasia; *extraordinary treatment; human rights; legal aspects; medical ethics; moral obligations; newborns; pain; physicians; prognosis; quality of life; religious ethics; *selection for treatment; social interaction; spina bifida; *standards; terminal care; *terminally ill; theology; *third party consent; *treatment refusal; value of life; withholding treatment; McCormick, Richard A.; Veatch, Robert M.

EXTRAORDINARY TREATMENT *See under*

PROLONGATION OF LIFE / EXTRAORDINARY TREATMENT

PROLONGATION OF LIFE / EXTRAORDINARY TREATMENT / GENETIC DEFECTS

EYE DISEASES *See under*

HUMAN EXPERIMENTATION / INFANTS / EYE DISEASES

FAMILY MEMBERS *See under*

ORGAN DONATION / FAMILY MEMBERS

FAMILY PLANNING *See also* CONTRACEPTION

FAMILY PLANNING / DEVELOPING COUNTRIES / VALUES

Merrick, Thomas W.; Caplan, Arthur. International population programs: should they change local values? *Hastings Center Report* 7(5): 17-18, Oct 1977. Case study. BE06077.
contraception; cultural pluralism; *developing countries; *family planning; morality; religious beliefs; *values

FAMILY PLANNING / INDIA

Bose, Asit K. Abortion in India: a legal study. *Journal of the Indian Law Institute* 16(4): 535-548, Oct-Dec 1974. 19 fn. BE07061.
*abortion; *attitudes; *family planning; illegal abortion; *legislation; motivation; social impact; *India; *Medical Termination of Pregnancy Act

Datta-Ray, Sunanda. Sterilization backlash in India. *People* 4(3): 38-39, 1977. BE06065.
*family planning; involuntary sterilization; politics; public opinion; *public policy; *India

---

BE = bioethics accession number   fn. = footnotes   refs. = references

FAMILY PLANNING / NEW ZEALAND

Contraception, sterilisation and abortion in New Zealand. *New Zealand Medical Journal* 85(588): 441-445, 25 May 1977. Recommendations from the Report of the Royal Commission of Inquiry into Contraception, Sterilisation and Abortion (Government Printer, Wellington). BE05601.
> *abortion; adolescents; alternatives; community services; conscience; *contraception; counseling; education; employment; ethics committees; family planning; health personnel; information dissemination; *legal aspects; mentally handicapped; pregnant women; *public policy; review committees; sexuality; social workers; *sterilization; therapeutic abortion; *New Zealand

Contraception, sterilisation and abortion. *New Zealand Medical Journal* 85(588): 428-429, 25 May 1977. 1 ref. Editorial. BE05600.
> abortion; contraception; *family planning; legal aspects; sterilization; *New Zealand

FAMILY PLANNING / PHYSICIANS / ATTITUDES

Veatch, Robert M.; Draper, Thomas. The values of physicians. *In* Veatch, Robert M., ed. Population Policy and Ethics: The American Experience. New York: Irvington Publishers, distributed by Halsted Press, 1977. p. 377-408. A volume in the Irvington Population and Demography Series. A project of the Research Group on Ethics and Population, Institute of Society, Ethics and the Life Sciences. 42 refs. 5 fn. ISBN 0-470-15170-6. BE06871.
> abortion; *attitudes; confidentiality; *family planning; justice; organizational policies; *physicians; *population control; professional organizations; public policy; self determination; sterilization; survival; values

FAMILY PLANNING / PUBLIC POLICY / INDIA

Borders, William. India, eliminating coercion, makes sharp shift in birth-control policy. *New York Times,* 3 Oct 1977, p. 1+. BE06402.
> coercion; *family planning; population growth; *public policy; *India

FAMOUS PERSONS   *See under*

CONFIDENTIALITY / FAMOUS PERSONS / HEALTH

CONFIDENTIALITY / FAMOUS PERSONS / MORTALITY

CONFIDENTIALITY / MEDICAL RECORDS / FAMOUS PERSONS

FATHERS   *See under*

ABORTION / FATHERS / LEGAL RIGHTS

FEDERAL GOVERNMENT   *See under*

IMMUNIZATION / FEDERAL GOVERNMENT / LEGAL LIABILITY

FETAL DEVELOPMENT   *See under*

PERSONHOOD / FETAL DEVELOPMENT

PERSONHOOD / FETAL DEVELOPMENT / HUMAN RIGHTS

FETUSES   *See also* ABORTED FETUSES

FETUSES   *See under*

ABORTION / FETUSES / LEGAL RIGHTS

ABORTION / FETUSES / RIGHTS

HUMAN EXPERIMENTATION / FETUSES

HUMAN EXPERIMENTATION / FETUSES / CHRISTIAN ETHICS

HUMAN EXPERIMENTATION / FETUSES / LEGAL ASPECTS

HUMAN EXPERIMENTATION / FETUSES / MORALITY

HUMAN EXPERIMENTATION / FETUSES / RISKS AND BENEFITS

KILLING / FETUSES

KILLING / FETUSES / CRIMINAL LAW

PERSONHOOD / FETUSES

PERSONHOOD / FETUSES / LEGAL ASPECTS

PERSONHOOD / FETUSES / RIGHTS

VALUE OF LIFE / FETUSES

WRONGFUL DEATH / FETUSES

WRONGFUL DEATH / FETUSES / LEGAL ASPECTS

FINANCIAL SUPPORT   *See under*

ABORTION / FINANCIAL SUPPORT

ABORTION / FINANCIAL SUPPORT / ATTITUDES

ABORTION / FINANCIAL SUPPORT / CONSCIENCE

ABORTION / FINANCIAL SUPPORT / EQUAL PROTECTION

ABORTION / FINANCIAL SUPPORT / HYDE AMENDMENT

ABORTION / FINANCIAL SUPPORT / LEGAL ASPECTS

ABORTION / FINANCIAL SUPPORT / MEDICAID

ABORTION / FINANCIAL SUPPORT / MORALITY

ABORTION / FINANCIAL SUPPORT / PENNSYLVANIA

ABORTION / FINANCIAL SUPPORT / POLITICS

ABORTION / FINANCIAL SUPPORT / SOCIAL IMPACT

ABORTION / FINANCIAL SUPPORT / SUPREME COURT DECISIONS

BEHAVIOR CONTROL / FINANCIAL SUPPORT / LAW ENFORCEMENT ASSISTANCE ADMINISTRATION

BE = bioethics accession number   fn. = footnotes   refs. = references

FINANCIAL SUPPORT *See under (cont'd.)*

BIOMEDICAL RESEARCH / FINANCIAL SUPPORT

IN VITRO FERTILIZATION / FINANCIAL SUPPORT

FOOD AND DRUG ADMINISTRATION / LAETRILE

U.S. District Court, W.D. Oklahoma. Rutherford v. United States. 5 Dec 1977. *Federal Supplement* 438: 1287-1302. 28 fn. BE07125.
cancer; constitutional law; drugs; federal government; *government regulation; judicial action; law; *legal aspects; *legal rights; legislation; *patients; *privacy; toxicity; *Food and Drug Administration; Food, Drug, and Cosmetic Act; *Laetrile

FOOD AND DRUG ADMINISTRATION *See under*

GOVERNMENT REGULATION / FOOD AND DRUG ADMINISTRATION

FOOD *See under*

GENETIC INTERVENTION / FOOD

FORCE FEEDING / PRISONERS / GREAT BRITAIN

Zellick, Graham. The hunger-striking prisoner. *New Law Journal* 127(5820): 928-929, 22 Sep 1977. 13 fn. BE06341.
*force feeding; legal aspects; physicians; *prisoners; public policy; *Great Britain

FORT DETRICK *See under*

RECOMBINANT DNA RESEARCH / FORT DETRICK

RECOMBINANT DNA RESEARCH / PHYSICAL CONTAINMENT / FORT DETRICK

FRANCE *See under*

ORGAN TRANSPLANTATION / FRANCE

FREEDOM OF INFORMATION ACT *See under*

HUMAN EXPERIMENTATION / FREEDOM OF INFORMATION ACT

MEDICAL RECORDS / FREEDOM OF INFORMATION ACT / SOCIAL IMPACT

FREEDOM *See under*

HUMAN EXPERIMENTATION / PSYCHIATRY / FREEDOM

SCIENCE / CONSTITUTIONAL LAW / FREEDOM

SCIENCE / FREEDOM

FROZEN SEMEN *See under*

AIH / FROZEN SEMEN

FUTURE GENERATIONS *See under*

BIOMEDICAL TECHNOLOGIES / FUTURE GENERATIONS / CHRISTIAN ETHICS

MORAL OBLIGATIONS / FUTURE GENERATIONS / GENE POOL

GENE POOL *See under*

MORAL OBLIGATIONS / FUTURE GENERATIONS / GENE POOL

GENETIC COUNSELING

Murphy, Edmond A.; Chase, Gary A.; Rodriguez, Alejandro. Genetic intervention: some social, psychological, and philosophical aspects. *In* Cohen, Bernice H.; Lilienfeld, Abraham M.; Huang, P.C., eds. Genetic Issues in Public Health and Medicine. Springfield, Ill.: Charles C. Thomas, 1978. p. 358-398. 47 refs. ISBN 0-398-03659-4. BE06985.
amniocentesis; *eugenics; family planning; gene pool; *genetic counseling; genetic defects; human rights; involuntary sterilization; parents; reproduction; selective abortion

Riccardi, Vincent M. Ethical, moral, and legal aspects of clinical genetics. *In his* The Genetic Approach to Human Disease. New York: Oxford University Press, 1977. p. 247-257. 41 refs. BE06577.
confidentiality; disclosure; family members; *genetic counseling; genetic screening; informed consent; legal aspects; moral obligations; patients' rights; prenatal diagnosis; quality of life; selective abortion

Stine, Gerald J. Genetic counseling. *In his* Biosocial Genetics: Human Heredity and Social Issues. New York: Macmillan, 1977. p. 437-455. 19 refs. ISBN 0-02-416490-9. BE05705.
communication; decisiion making; disclosure; family problems; *genetic counseling; genetic defects; parents; psychological stress; selective abortion

Thorup, Oscar A., et al. Genetically Transmitted Disease. [Videorecording]. Charlottesville, Va.: University of Virginia Medical Center, Health Sciences Library, 1977. 1 cassette; 60 min.; sound; black and white; 3/4 in. Tape of the Medical Center Hour, 21 Sep 1977. Panel discussion moderated by Oscar A. Thorup. BE07053.
amniocentesis; case reports; coercion; decision making; economics; future generations; gene pool; *genetic counseling; *genetic defects; health care; goals; human rights; mandatory programs; moral obligations; obligations to society; parents; physicians; *prenatal diagnosis; public policy; reproduction; *selective abortion; self determination; sex determination; socioeconomic factors

Wojcik, Jan. Muted Consent: A Casebook in Modern

BE = bioethics accession number   fn. = footnotes   refs. = references

Medical Ethics. West Lafayette, Ind.: Purdue University, 1978. 164 p. 237 fn. Case study. ISBN 911198-50-4. BE06704.

> *abortion; *behavior control; *case reports; *determination of death; disclosure; drugs; electrical stimulation of the brain; ethics; eugenics; euthanasia; extraordinary treatment; fetuses; *genetic counseling; genetic defects; *genetic intervention; *genetic screening; health care; *human experimentation; *informed consent; involuntary commitment; mentally ill; operant conditioning; pregnant women; psychosurgery; quality of life; reproductive technologies; *resource allocation; rights; selective abortion; value of life; withholding treatment

GENETIC COUNSELING / ATTITUDES

Sorenson, James R.; Culbert, Arthur J. Genetic counselors and counseling orientations—unexamined topics in evaluation. *In* Lubs, Herbert A.; de la Cruz, Felix, eds. Genetic Counseling: A Monograph of the Nat'l. Inst. of Child Health and Human Development. New York: Raven Press, 1977. p. 131-156. 39 refs. ISBN 0-89004-150-4. BE05799.

> *attitudes; contraception; decision making; evaluation; *genetic counseling; genetic defects; goals; health personnel; investigators; moral obligations; physicians; referral and consultation; reproduction; sexuality

GENETIC COUNSELING / DOWN'S SYNDROME

Bergman, Garrett E.; Sorenson, John H. Bioethics in Genetic Counseling: Down's Syndrome/Amniocentesis and Abortion. [Slide]. Philadelphia: Medical College of Pa., Teaching Program in Human Values in Medicine, 1978. 54 slides; color; 2 x 2 in. Accompanied by 1 audiocassette (25 min.); program guide, including script. Available from Pennsylvania Medical Society, 20 Erford Road, Lemoyne, Pa. 17043. BE06903.

> amniocentesis; case reports; decision making; diagnosis; dignity; disclosure; *Down's syndrome; *genetic counseling; etiology; *moral obligations; incidence; obligations to society; parents; *physician patient relationship; *physicians; prognosis; quality of life; risks and benefits; selective abortion; values

GENETIC COUNSELING / HUNGARY

Hungary. Genetic counselling. *International Digest of Health Legislation* 28(3): 611-613, 1977. Directive No. 40204 of 1977 of the Ministry of Health. BE06463.

> amniocentesis; contraception; *genetic counseling; pregnant women; selective abortion; *Hungary

GENETIC COUNSELING / HUNTINGTON'S CHOREA / RISKS AND BENEFITS

MacKay, Charles R.; Shea, John M. Ethical considerations in research on Huntington's disease. *Clinical Research* 25(4): 241-247, Oct 1977. 8 refs. BE06348.

> carriers; diagnosis; *genetic counseling; *human experimentation; *Huntington's chorea; incidence; moral obligations; psychological stress; reproduction; *risks and benefits; selection of subjects

GENETIC COUNSELING / LEGAL ASPECTS

Dworkin, Roger. Legal issues in genetic counseling. *In* Smith, David H., ed. No Rush to Judgement: Essays on Medical Ethics. Bloomington, Ind.: Indiana University, Poynter Center, 1977. p. 44-67. 1 fn. BE05795.

> *abortion; competence; constitutional law; disclosure; family members; family relationship; fetuses; *genetic counseling; *genetic screening; informed consent; *legal aspects; legal obligations; legislation; legitimacy; mandatory programs; phenylketonuria; physicians; pregnant women; privacy *AID; rights; sickle cell anemia; state interest; sterilization; *Supreme Court decisions; Roe v. Wade

Reilly, Philip. Genetics, Law, and Social Policy. Cambridge, Mass.: Harvard University Press, 1977. 275 p. 297 refs. ISBN 0-674-34657-2. BE06519.

> AID; embryo transfer; eugenics; federal government; *genetic counseling; genetic defects; genetic fathers; *genetic screening; government regulation; *human rights; in vitro fertilization; involuntary sterilization; *legal aspects; *legislation; mandatory programs; medical records; mentally handicapped; *reproduction; *reproductive technologies; sickle cell anemia; *state government; Tay Sachs disease; voluntary programs; wrongful life; XYY karyotype

GENETIC COUNSELING / PHENYLKETONURIA

Bergman, Garrett E.; Sorenson, John H. Bioethics in Genetic Counseling: PKU and Cystic Fibrosis/Sterilization and Artificial Insemination. [Slide]. Philadelphia: Medical College of Pa., Teaching Program in Human Values in Medicine, 1978. 57 slides; color; 2 x 2 in. Accompanied by 1 audiocassette (25 min.); program guide, including script. Available from Pennsylvania Medical Society, 20 Erford Road, Lemoyne, Pa. 17043. BE06905.

> AID; carriers; case reports; communication; disclosure; *genetic counseling; males; moral obligations; motivation; obligations to society; parents; *phenylketonuria; physician patient relationship; physicians; *recessive genetic conditions; risks and benefits; voluntary sterilization

---

BE = bioethics accession number   fn. = footnotes   refs. = references

GENETIC COUNSELING / RECESSIVE GENETIC CONDITIONS

Bergman, Garrett E.; Sorenson, John H. Bioethics in Genetic Counseling: PKU and Cystic Fibrosis/Sterilization and Artificial Insemination. [Slide]. Philadelphia: Medical College of Pa., Teaching Program in Human Values in Medicine, 1978. 57 slides; color; 2 x 2 in. Accompanied by 1 audiocassette (25 min.); program guide, including script. Available from Pennsylvania Medical Society, 20 Erford Road, Lemoyne, Pa. 17043. BE06905.
   AID; carriers; case reports; communication; disclosure; *genetic counseling; males; moral obligations; motivation; obligations to society; parents; *phenylketonuria; physician patient relationship; physicians; *recessive genetic conditions; risks and benefits; voluntary sterilization

GENETIC COUNSELING / STATE GOVERNMENT

Reilly, Philip. Government support of genetic services. *Social Biology* 25(1): 23-32, Spring 1978. 29 refs. BE06926.
   confidentiality; federal government; *genetic screening; *genetic counseling; *legislation; *newborns; phenylketonuria; public policy; sickle cell anemia; *state government; Tay Sachs disease

GENETIC COUNSELING / TAY SACHS DISEASE

Bergman, Garrett E.; Sorenson, John H. Bioethics in Genetic Counseling: Tay Sachs Disease/Genetic Screening. [Slide]. Philadelphia: Medical College of Pa., Teaching Program in Human Values in Medicine, 1978. 40 slides; color; 2 x 2 in. Accompanied by 1 audiocassette (20 min.); program guide, including script. Available from Pennsylvania Medical Society, 20 Erford Road, Lemoyne, Pa. 17043. BE06904.
   adoption; alternatives; carriers; case reports; decision making; dignity; disclosure; *genetic counseling; *genetic screening; genetics; incidence; informed consent; Jews; patient participation; physician patient relationship; risks and benefits; *Tay Sachs disease; voluntary programs; voluntary sterilization

GENETIC DEFECTS   *See also* CONGENITAL DEFECTS

GENETIC DEFECTS

Malter, Susan. Genetic counseling...a responsibility of health-care professionals. *Nursing Forum* 16(1): 26-35, 1977. 10 refs. BE06231.
   eugenics; genetic counseling; *genetic defects; historical aspects; morality; quality of life

Milunsky, Aubrey. Know Your Genes. Boston:

Houghton Mifflin, 1977. 335 p. ISBN 0-395-25374-8. BE05885.
   AID; age; allowing to die; amniocentesis; cancer; carriers; chromosomal disorders; drugs; family members; genetic counseling; genetic screening; *genetic defects; heart diseases; in vitro fertilization; intelligence; mental health; mentally retarded; minority groups; newborns; obligations to society; prenatal diagnosis; selective abortion; sex linked defects; sex preselection; twinning; XYY karyotype

GENETIC DEFECTS   *See under*

   ALLOWING TO DIE / NEWBORNS / GENETIC DEFECTS
   AMNIOCENTESIS / SELECTIVE ABORTION / GENETIC DEFECTS
   CONFIDENTIALITY / MEDICAL RECORDS / GENETIC DEFECTS
   PATIENT CARE / CHILDREN / GENETIC DEFECTS
   PRENATAL DIAGNOSIS / SELECTIVE ABORTION / GENETIC DEFECTS
   PROLONGATION OF LIFE / CHILDREN / GENETIC DEFECTS
   PROLONGATION OF LIFE / EXTRAORDINARY TREATMENT / GENETIC DEFECTS
   REPRODUCTION / GENETIC DEFECTS / ATTITUDES
   SELECTIVE ABORTION / MORAL OBLIGATIONS / GENETIC DEFECTS
   WRONGFUL LIFE / GENETIC DEFECTS / LEGAL ASPECTS

GENETIC DIVERSITY

Bodmer, W.F. Social concern and biological advances. *Journal of the Royal Society of Arts* 125(5248): 180-194, Mar 1977. Discussion follows essay on p. 192-194. 5 refs. BE06293.
   aged; *biomedical technologies; communication; education; *genetic diversity; genetics; human equality; *intelligence; mass media; *population growth; *recombinant DNA research; science; social determinants; *social impact

GENETIC ENGINEERING   *See* GENETIC INTERVENTION, RECOMBINANT DNA RESEARCH

GENETIC FATHERS   *See under*
   AID / GENETIC FATHERS / LEGAL RIGHTS

GENETIC INTERVENTION   *See also* EUGENICS, GENETIC COUNSELING, GENETIC SCREENING, GENETICS, RECOMBINANT DNA RESEARCH, REPRODUCTIVE TECHNOLOGIES

GENETIC INTERVENTION

Cooke, Robert. Improving on Nature: The Brave New

BE = bioethics accession number   fn. = footnotes   refs. = references

World of Genetic Engineering. New York: Quadrangle/The New York Times Book Company, 1977. 248 p. ISBN 0-8129-0667-5. BE05699.

amniocentesis; *behavioral genetics; artificial genes; biological containment; biology; cancer; cloning; DNA therapy; embryo transfer; evolution; federal government; *food; genetic defects; *genetic intervention; genetic screening; genetics; government regulation; health hazards; in vitro fertilization; mutation; physical containment; prenatal diagnosis; *recombinant DNA research; risks and benefits; social impact

Goodfield, June. Playing God: Genetic Engineering and the Manipulation of Life. New York: Random House, 1977. 218 p. ISBN 0-394-40692-3. BE05879.

DNA therapy; decision making; federal government; genetic defects; *genetic intervention; government regulation; health hazards; historical aspects; international aspects; investigators; legal liability; morality; public opinion; public participation; *recombinant DNA research; reproductive technologies; *risks and benefits; science; self regulation; *social control; values; Asilomar Conference; NIH Guidelines

Packard, Vance. The People Shapers. Boston: Little, Brown, 1977. 398 p. 249 refs. ISBN 0-316-68750-2. BE06518.

AID; aggression; *behavior control; behavior disorders; cloning; brain; electrical stimulation of the brain; embryo transfer; eugenics; employment; frozen semen; *genetic intervention; host mothers; genetic screening; human characteristics; hypnosis; in vitro fertilization; intelligence; life extension; legal aspects; mass media; operant conditioning; organ transplantation; personality; privacy; psychoactive drugs; psychological stress; psychosurgery; selective abortion; sex determination; sex preselection; *social control; *social impact

Rifkin, Jeremy; Howard, Ted. Who should play God? *Progressive* 41(12): 16-22, Dec 1977. BE06147.

artificial insemination; cloning; embryo transfer; eugenics; *genetic intervention; in vitro fertilization; industry; psychoactive drugs; public policy; recombinant DNA research; social impact

Sinsheimer, Robert L. Genetic intervention and values: are all men created equal? *In* Ellison, Craig W., ed. Modifying Man: Implications and Ethics. Washington: University Press of America, 1978. p. 109-125. ISBN 0-8191-0302-0. BE06669.

behavioral genetics; cloning; genetic defects; genetic diversity; *genetic intervention; *human equality; *positive eugenics; selective abortion

GENETIC INTERVENTION / FOOD

Cooke, Robert. Improving on Nature: The Brave New World of Genetic Engineering. New York: Quadrangle/The New York Times Book Company, 1977. 248 p. ISBN 0-8129-0667-5. BE05699.

amniocentesis; *behavioral genetics; artificial genes; biological containment; biology; cancer; cloning; DNA therapy; embryo transfer; evolution; federal government; *food; genetic defects; *genetic intervention; genetic screening; genetics; government regulation; health hazards; in vitro fertilization; mutation; physical containment; prenatal diagnosis; *recombinant DNA research; risks and benefits; social impact

GENETIC INTERVENTION / LEGAL ASPECTS

Stine, Gerald J. Genetics, society, and the law. *In his* Biosocial Genetics: Human Heredity and Social Issues. New York: Macmillan, 1977. p. 503-524. 35 refs. ISBN 0-02-416490-9. BE05707.

AID children; amniocentesis; carriers; eugenics; family members; fetuses; genetic defects; genetic fathers; genetic identity; *genetic intervention; genetic screening; *genetics; human rights; *legal aspects; legal rights; legitimacy; mandatory programs; phenylketonuria; reproduction; selective abortion; sex preselection; sickle cell anemia; XYY karyotype

GENETIC INTERVENTION / RISKS AND BENEFITS

Beckwith, Jon. Recombinant DNA: does the fault lie within our genes? *Science for the People* 9(3): 14-17, May-Jun 1977. 19 refs. 1 fn. BE05604.

aggression; behavioral genetics; ecology; *genetic intervention; genetic screening; health hazards; hyperkinesis; industry; minority groups; *recombinant DNA research; *risks and benefits; *social impact; social problems

Stine, Gerald J. The future: promise or peril? *In his* Biosocial Genetics: Human Heredity and Social Issues. New York: Macmillan, 1977. p. 527-553. 22 refs. ISBN 0-02-416490-9. BE05708.

cloning; DNA therapy; *genetic intervention; investigators; moral obligations; positive eugenics; recombinant DNA research; *risks and benefits; self regulation; social control; social impact; technology

GENETIC INTERVENTION / SOCIAL IMPACT

Howard, Ted; Rifkin, Jeremy. Who Should Play God? The Artificial Creation of Life and What It Means for the Future of the Human Race. New York: Dell Publishing Company, 1977. 272 p. 526 fn. ISBN 0-440-19504-7. BE05882.

AID; behavioral genetics; biomedical technologies; cloning; DNA therapy; drugs; embryo transfer; *eugenics; genetic screening; *genetic intervention; government regulation; health hazards; historical aspects; in vitro fertilization; industry; involuntary sterilization; population control; recombinant DNA research; *reproductive technologies; risks and benefits; selective abortion; social control; *social impact

---

BE = bioethics accession number   fn. = footnotes   refs. = references

GENETIC MANIPULATION ADVISORY GROUP
See under

RECOMBINANT DNA RESEARCH / GENETIC
MANIPULATION ADVISORY GROUP

GENETIC SCREENING

Stine, Gerald J. Genetic screening. *In his* Biosocial Genetics: Human Heredity and Social Issues. New York: Macmillan, 1977. p. 456-471. 12 refs. ISBN 0-02-416490-9. BE05706.
amniocentesis; costs and benefits; disclosure; *genetic screening; newborns; risks and benefits; state government

Wojcik, Jan. Muted Consent: A Casebook in Modern Medical Ethics. West Lafayette, Ind.: Purdue University, 1978. 164 p. 237 fn. Case study. ISBN 911198-50-4. BE06704.
*abortion; *behavior control; *case reports; *determination of death; disclosure; drugs; electrical stimulation of the brain; ethics; eugenics; euthanasia; extraordinary treatment; fetuses; *genetic counseling; genetic defects; *genetic intervention; *genetic screening; health care; *human experimentation; *informed consent; involuntary commitment; mentally ill; operant conditioning; pregnant women; psychosurgery; quality of life; reproductive technologies; *resource allocation; rights; selective abortion; value of life; withholding treatment

GENETIC SCREENING / LEGISLATION

Reilly, Philip. Genetics, Law, and Social Policy. Cambridge, Mass.: Harvard University Press, 1977. 275 p. 297 refs. ISBN 0-674-34657-2. BE06519.
AID; embryo transfer; eugenics; federal government; *genetic counseling; genetic defects; genetic fathers; *genetic screening; government regulation; *human rights; in vitro fertilization; involuntary sterilization; *legal aspects; *legislation; mandatory programs; medical records; mentally handicapped; *reproduction; *reproductive technologies; sickle cell anemia; *state government; Tay Sachs disease; voluntary programs; wrongful life; XYY karyotype

GENETIC SCREENING / NEWBORNS

Reilly, Philip R. Mass neonatal screening in 1977. *American Journal of Human Genetics* 29(3): 302-304, May 1977. 8 refs. Letter. BE05626.
*genetic screening; *newborns; *public policy; *state government

GENETIC SCREENING / NEWBORNS / LEGISLATION

Reilly, Philip. Government support of genetic services. *Social Biology* 25(1): 23-32, Spring 1978. 29 refs. BE06926.
confidentiality; federal government; *genetic screening; *genetic counseling; *legislation; *newborns; phenylketonuria; public policy; sickle cell anemia; *state government; Tay Sachs disease

GENETIC SCREENING / NEWBORNS / PHENYLKETONURIA

Woodside, Gilbert. In defense of PKU screening. *Hospital Physician* 13(4): 11+, Apr 1977. BE05808.
*genetic screening; health care; intelligence; mentally retarded; *newborns; *phenylketonuria; prognosis; withholding treatment

GENETIC SCREENING / NEWBORNS / SICKLE CELL ANEMIA

Rowley, Peter T. Newborn screening for sickle-cell disease: benefits and burdens. *New York State Journal of Medicine* 78(1): 42-44, Jan 1978. 5 refs. BE06752.
case reports; genetic counseling; *genetic screening; *newborns; risks and benefits; *sickle cell anemia

GENETIC SCREENING / OBSTETRICS AND GYNECOLOGY / ATTITUDES

Gordis, Leon; Childs, Barton; Roseman, Myra G. Obstetricians' attitudes toward genetic screening. *American Journal of Public Health* 67(5): 469-471, May 1977. 6 refs. BE05615.
*attitudes; genetic defects; *genetic screening; *obstetrics and gynecology; physicians; selective abortion; *Tay Sachs disease

GENETIC SCREENING / PUBLIC POLICY / SICKLE CELL ANEMIA

Bowman, James E. Genetic screening programs and public policy. *Phylon* 38(2): 117-142, Jun 1977. 42 refs. BE05810.
adolescents; adults; behavioral genetics; *blacks; *carriers; children; costs and benefits; discrimination; diagnosis; employment; federal government; genetic counseling; genetic defects; *genetic screening; government agencies; incidence; intelligence; legal aspects; *legislation; mandatory programs; military personnel; prognosis; psychological stress; *public policy; *sickle cell anemia; *social impact; social problems; state government; sterilization; stigmatization; Greece; *United States

GENETIC SCREENING / SOCIAL IMPACT

Morgan, Graeme. Implications of genetic screening. *Medical Journal of Australia* 2(24): 789-790, 10 Dec 1977. 7 fn. Editorial. BE06374.
carriers; genetic defects; genetic counseling; *genetic screening; prenatal diagnosis; selective abortion; *social impact; XYY karyotype

---

BE = bioethics accession number   fn. = footnotes   refs. = references

GENETIC SCREENING / SPINA BIFIDA / COSTS AND BENEFITS

Yanchinski, Stephanie. Doctors split on screening pregnant women to detect spina bifida babies. *New Scientist* 77(1085): 68, 12 Jan 1978. BE06765.
amniocentesis; *costs and benefits; *genetic screening; incidence; pregnant women; selective abortion; *spina bifida

GENETIC SCREENING / TAY SACHS DISEASE

Bergman, Garrett E.; Sorenson, John H. Bioethics in Genetic Counseling: Tay Sachs Disease / Genetic Screening. [Slide]. Philadelphia: Medical College of Pa., Teaching Program in Human Values in Medicine, 1978. 40 slides; color; 2 x 2 in. Accompanied by 1 audiocassette (20 min.); program guide, including script. Available from Pennsylvania Medical Society, 20 Erford Road, Lemoyne, Pa. 17043. BE06904.
adoption; alternatives; carriers; case reports; decision making; dignity; disclosure; *genetic counseling; *genetic screening; genetics; incidence; informed consent; Jews; patient participation; physician patient relationship; risks and benefits; *Tay Sachs disease; voluntary programs; voluntary sterilization

GENETIC SCREENING / UNITED STATES

US skimps on genetic screening. *New Scientist* 77(1090): 412, 16 Feb 1978. BE06798.
costs and benefits; economics; federal government; financial support; genetic defects; *genetic screening; prenatal diagnosis; preventive medicine; *United States

GENETIC SCREENING / XYY KARYOTYPE

Kopelman, Loretta. Ethical controversies in medical research: the case of XYY screening. *Perspectives in Biology and Medicine* 21(2): 196-204, Winter 1978. 16 refs. BE06741.
behavioral research; disclosure; *genetic screening; informed consent; newborns; nontherapeutic research; parents; *risks and benefits; third party consent; *XYY karyotype

GENETIC SCREENING / XYY KARYOTYPE / RISKS AND BENEFITS

Hamerton, John L. Human population cytogenetics: dilemmas and problems. *American Journal of Human Genetics* 28(2): 107-122, Mar 1976. 42 refs. BE07097.
attitudes; behavior disorders; disclosure; *genetic screening; human characteristics; incidence; informed consent; institutionalized persons; parents; prisoners; *risks and benefits; socioeconomic factors; *XYY karyotype; Science for the People

GENETICS *See also* BEHAVIORAL GENETICS

GENETICS / LEGAL ASPECTS

Stine, Gerald J. Genetics, society, and the law. *In his* Biosocial Genetics: Human Heredity and Social Issues. New York: Macmillan, 1977. p. 503-524. 35 refs. ISBN 0-02-416490-9. BE05707.
AID children; amniocentesis; carriers; eugenics; family members; fetuses; genetic defects; genetic fathers; genetic identity; *genetic intervention; genetic screening; *genetics; human rights; *legal aspects; legal rights; legitimacy; mandatory programs; phenylketonuria; reproduction; selective abortion; sex preselection; sickle cell anemia; XYY karyotype

GENETICS *See under*

POPULATION CONTROL / GENETICS

GERMANY *See under*

EUGENICS / HISTORICAL ASPECTS / GERMANY

GOVERNMENT AGENCIES *See under*

HUMAN EXPERIMENTATION / PSYCHOACTIVE DRUGS / GOVERNMENT AGENCIES

RECOMBINANT DNA RESEARCH / GOVERNMENT AGENCIES / CONFLICT OF INTEREST

GOVERNMENT REGULATION *See also* JUDICIAL ACTION, LEGISLATION, SOCIAL CONTROL

GOVERNMENT REGULATION / FOOD AND DRUG ADMINISTRATION

Kennedy, Donald. Creative tension: FDA and medicine. *New England Journal of Medicine* 298(15): 846-850, 13 Apr 1978. 2 fn. BE06847.
biomedical research; *drugs; disclosure; federal government; government agencies; *government regulation; legislation; medical education; patient participation; physicians; risks and benefits; *Food and Drug Administration

GOVERNMENT REGULATION / LAETRILE

Clinite, Barbara J. Freedom of choice in medical treatment: reconsidering the efficacy requirement of the FDCA. *Loyola University Law Journal* 9(1): 205-226, Fall 1977. 150 fn. BE07033.
*cancer; *drugs; due process; federal government; government agencies; human experimentation; *government regulation; patient care; privacy; *self determination; standards; state interest; *Food and Drug Administration; *Laetrile

Hogan, Gypsy. Hands off Laetrile, judge orders FDA; appeal is expected. *Washington Post*, 6 Dec 1977, p. A9. BE06419.

BE = bioethics accession number   fn. = footnotes   refs. = references

cancer; drugs; *federal government; *government regulation; legal rights; *Laetrile

Spivak, Jonathan. Laetrile's message to the FDA. *Wall Street Journal,* 21 Jul 1977, p. 12. BE06057.
attitudes; authoritarianism; *cancer; *drugs; *federal government; *government agencies; *government regulation; physicians; public opinion; self determination; *Food and Drug Administration; *Laetrile

GOVERNMENT REGULATION / LAETRILE / LEGAL ASPECTS

U.S. District Court, W.D. Oklahoma. Rutherford v. United States. 5 Dec 1977. *Federal Supplement* 438: 1287-1302. 28 fn. BE07125.
cancer; constitutional law; drugs; federal government; *government regulation; judicial action; law; *legal aspects; *legal rights; legislation; *patients; *privacy; toxicity; *Food and Drug Administration; Food, Drug, and Cosmetic Act; *Laetrile

GOVERNMENT REGULATION / PSYCHOACTIVE DRUGS / CALIFORNIA

Armstrong, Barbara. The use of psychotropic drugs in state hospitals: a legal or medical decision? *Hospital and Community Psychiatry* 29(2): 118-121, Feb 1978. BE06770.
decision making; electroconvulsive therapy; *government regulation; *institutionalized persons; legal aspects; legislation; *mentally ill; *patient care; psychiatry; *psychoactive drugs; self regulation; *state government; *California

GOVERNMENT REGULATION / SCIENCE

Culliton, Barbara J. Science's restive public. *Daedalus* 107(2): 147-156, Spring 1978. 11 fn. BE06938.
federal government; *government regulation; fetuses; *human experimentation; investigators; *public participation; *recombinant DNA research; *science; self regulation

GOVERNMENT REGULATION *See under*
ABORTION / GOVERNMENT REGULATION
ABORTION / GOVERNMENT REGULATION / LEGAL ASPECTS
ABORTION / GOVERNMENT REGULATION / PLANNED PARENTHOOD OF CENTRAL MISSOURI V. DANFORTH
BEHAVIOR CONTROL / GOVERNMENT REGULATION
BEHAVIORAL RESEARCH / GOVERNMENT REGULATION
BEHAVIORAL RESEARCH / GOVERNMENT REGULATION / NORWAY
BEHAVIORAL RESEARCH / GOVERNMENT REGULATION / SOCIAL IMPACT
BIOMEDICAL RESEARCH / GOVERNMENT REGULATION / LEGAL ASPECTS

GOVERNMENT REGULATION *See under (cont'd.)*
CONTRACEPTION / GOVERNMENT REGULATION
ELECTROCONVULSIVE THERAPY / GOVERNMENT REGULATION / CALIFORNIA
HEALTH CARE DELIVERY / GOVERNMENT REGULATION / SOCIAL IMPACT
HUMAN EXPERIMENTATION / GOVERNMENT REGULATION
HUMAN EXPERIMENTATION / GOVERNMENT REGULATION / DHEW GUIDELINES
HUMAN EXPERIMENTATION / GOVERNMENT REGULATION / HISTORICAL ASPECTS
HUMAN EXPERIMENTATION / GOVERNMENT REGULATION / RISKS AND BENEFITS
INVOLUNTARY STERILIZATION / GOVERNMENT REGULATION
PSYCHOSURGERY / GOVERNMENT REGULATION
RECOMBINANT DNA RESEARCH / GOVERNMENT REGULATION
RECOMBINANT DNA RESEARCH / GOVERNMENT REGULATION / ATTITUDES
RECOMBINANT DNA RESEARCH / GOVERNMENT REGULATION / CALIFORNIA
RECOMBINANT DNA RESEARCH / GOVERNMENT REGULATION / CAMBRIDGE
RECOMBINANT DNA RESEARCH / GOVERNMENT REGULATION / CONTAINMENT
RECOMBINANT DNA RESEARCH / GOVERNMENT REGULATION / DENMARK
RECOMBINANT DNA RESEARCH / GOVERNMENT REGULATION / EUROPE
RECOMBINANT DNA RESEARCH / GOVERNMENT REGULATION / GREAT BRITAIN
RECOMBINANT DNA RESEARCH / GOVERNMENT REGULATION / HEALTH HAZARDS
RECOMBINANT DNA RESEARCH / GOVERNMENT REGULATION / LEGISLATION
RECOMBINANT DNA RESEARCH / GOVERNMENT REGULATION / NATIONAL ACADEMY OF SCIENCES
RECOMBINANT DNA RESEARCH / GOVERNMENT REGULATION / NIH GUIDELINES
RECOMBINANT DNA RESEARCH / GOVERNMENT REGULATION / POLITICS
RECOMBINANT DNA RESEARCH / GOVERNMENT REGULATION / PROGRAM DESCRIPTIONS
RECOMBINANT DNA RESEARCH / GOVERNMENT REGULATION / RISKS AND BENEFITS
RECOMBINANT DNA RESEARCH / GOVERNMENT REGULATION / WEST GERMANY
STERILIZATION / GOVERNMENT REGULATION
VOLUNTARY STERILIZATION / GOVERNMENT REGULATION

BE = bioethics accession number   fn. = footnotes   refs. = references

GREAT BRITAIN *See under*

ABORTION / ATTITUDES / GREAT BRITAIN

ABORTION / GREAT BRITAIN

ABORTION / LEGISLATION / GREAT BRITAIN

ABORTION / POLITICS / GREAT BRITAIN

CONFIDENTIALITY / MEDICAL RECORDS / GREAT BRITAIN

EUGENICS / HISTORICAL ASPECTS / GREAT BRITAIN

EUTHANASIA / LEGAL ASPECTS / GREAT BRITAIN

FORCE FEEDING / PRISONERS / GREAT BRITAIN

HUMAN EXPERIMENTATION / CHILDREN / GREAT BRITAIN

INFORMED CONSENT / CHILDREN / GREAT BRITAIN

PRENATAL DIAGNOSIS / SPINA BIFIDA / GREAT BRITAIN

PSYCHIATRY / HISTORICAL ASPECTS / GREAT BRITAIN

RECOMBINANT DNA RESEARCH / GOVERNMENT REGULATION / GREAT BRITAIN

RECOMBINANT DNA RESEARCH / GREAT BRITAIN

RECOMBINANT DNA RESEARCH / SOCIAL CONTROL / GREAT BRITAIN

HÄRING, BERNARD *See under*

BEHAVIOR CONTROL / HÄRING, BERNARD

HEALTH / TOTALITY / HÄRING, BERNARD

HANDICAPPED *See under*

JUSTIFIABLE KILLING / HANDICAPPED

VALUE OF LIFE / HANDICAPPED

HARE, R.M. *See under*

ABORTION / HARE, R.M.

PERSONHOOD / HUMAN RIGHTS / HARE, R.M.

HARVARD UNIVERSITY *See under*

RECOMBINANT DNA RESEARCH / HARVARD UNIVERSITY

HAWAII DECLARATION *See under*

CODES OF ETHICS / HAWAII DECLARATION

PATIENT CARE / PSYCHIATRY / HAWAII DECLARATION

HAWAII *See under*

ABORTION / ATTITUDES / HAWAII

ABORTION / LEGISLATION / HAWAII

DETERMINATION OF DEATH / HAWAII

INVOLUNTARY COMMITMENT / DANGEROUSNESS / HAWAII

HEALTH *See also* MENTAL HEALTH

HEALTH

Beauchamp, Tom L.; Walters, LeRoy, eds. Contemporary Issues in Bioethics. Encino, Calif.: Dickenson Publishing Company, 1978. 612 p. References and footnotes included. ISBN 0-8221-0200-5. BE06700.

*abortion; allowing to die; *behavior control; *bioethics; biomedical technologies; brain death; children; codes of ethics; confidentiality; decision making; deontological ethics; *determination of death; *euthanasia; fetuses; genetic intervention; *health; *health care; *human experimentation; *informed consent; involuntary commitment; justice; legal aspects; *medical ethics; natural law; *normative ethics; operant conditioning; patients' rights; personhood; psychosurgery; *resource allocation; rights; selection for treatment; treatment refusal; utilitarianism; Kaimowitz v. Department of Mental Health; Quinlan, Karen

Hellegers, André E. Biologic origins of bioethical problems. *In* Wynn, Ralph M., ed. Obstetrics and Gynecology Annual. Volume 6, 1977. New York: Appleton-Century-Crofts, 1977. p. 1-9. ISBN 0-8385-7181-6. BE06570.

*costs and benefits; deontological ethics; *health; health care delivery; *medicine; obligations to society; physician patient relationship; physicians; psychiatry; politics; quality of life; *resource allocation; *social control; teleological ethics; value of life

Newton, Lisa H. The healing of the person. *Connecticut Medicine* 41(10): 641-646, Oct 1977. 22 fn. BE06091.

decision making; dehumanization; paternalism; *health; patient advocacy; *patient care; *patient participation; *patients' rights; *physician patient relationship; physician's role; self determination

HEALTH CARE *See also* PATIENT CARE

HEALTH CARE DELIVERY / ADMINISTRATORS / ETHICS

Ethical Issues in Health Care Management: Proceedings of the Seventeenth Annual Symposium on Hospital Affairs, University of Chicago, 25-26 Apr 1975. Chicago: University of Chicago, Graduate Program in Hospital Administration, 1975. 67 p. Symposium conducted by the Graduate Program in Hospital Administration and Center for Health Administration Studies, Graduate School of Business, University of Chicago. 6 fn. BE07075.

*administrators; costs and benefits; *decision making; *ethics; *health care delivery; hospitals; institutional policies; methods; moral obligations; resource allocation; values

---

BE = bioethics accession number   fn. = footnotes   refs. = references

HEALTH CARE DELIVERY / ECONOMICS

Plant, Raymond. Gifts, exchanges and the political economy of health care. Part II: How should health care be distributed? *Journal of Medical Ethics* 4(1): 5-11, Mar 1978. 17 fn. BE06924.
  *economics; *health care delivery; *justice; state medicine; trust; values

HEALTH CARE DELIVERY / GOVERNMENT REGULATION / SOCIAL IMPACT

Mechanic, David. The growth of medical technology and bureaucracy: implications for medical care. *Milbank Memorial Fund Quarterly* 55(1): 61-78, Winter 1977. 21 refs. BE06287.
  biomedical technologies; decision making; economics; *government regulation; *health care delivery; health insurance; medical fees; medicine; *physician's role; physicians; resource allocation; self determination; *social impact; state medicine

HEALTH CARE DELIVERY / INTERNATIONAL ASPECTS

Levine, Carol. Ethics, justice, and international health. *Hastings Center Report* 7(2): 5-7, Apr 1977. Report from an Institute Conference on Ethical Issues in International Health, 11-12 Nov 1976. BE05564.
  decision making; developing countries; financial support; *health care delivery; human experimentation; *international aspects; justice; moral obligations; self determination

HEALTH CARE DELIVERY / JUSTICE

Plant, Raymond. Gifts, exchanges and the political economy of health care. Part II: How should health care be distributed? *Journal of Medical Ethics* 4(1): 5-11, Mar 1978. 17 fn. BE06924.
  *economics; *health care delivery; *justice; state medicine; trust; values

HEALTH CARE DELIVERY / MASS SCREENING / COSTS AND BENEFITS

Milio, Nancy. Ethics and the economics of community health services: the case of screening. *Linacre Quarterly* 44(4): 347-360, Nov 1977. 28 refs. 17 fn. BE06103.
  *costs and benefits; *health care delivery; *mass screening; preventive medicine; resource allocation

HEALTH CARE DELIVERY / MENTALLY ILL / COMMUNITY SERVICES

Klerman, Gerald L. Better but not well: social and ethical issues in the deinstitutionalization of the mentally ill. *Schizophrenia Bulletin* 3(4): 617-631, 1977. 38 refs. BE06280.
  biomedical technologies; *community services; dehumanization; *deinstitutionalized persons; *health care delivery; historical aspects; mental health; institutionalized persons; mental institutions; *mentally ill; *psychoactive drugs; psychotherapy; quality of life; schizophrenia; *social impact; state government

HEALTH CARE DELIVERY / MINORITY GROUPS

U.S. National Commission for the Protection of Human Subjects of Biomedical and Behavioral Research. National Commission for the Protection of Human Subjects of Biomedical and Behavioral Research. Transcript of the Meeting Proceedings (35th Meeting), 14-15 Oct 1977. Springfield, Va.: National Technical Information Service, 1977. 640 p. Report No. NCPHS/M-77/24. NTIS No. PB-275 903. Volume in two parts. BE06875.
  abortion on demand; adults; children; disadvantaged; ethical review; *ethics committees; federal government; financial support; government regulation; *health care delivery; homosexuals; *human experimentation; *informed consent; *institutionalized persons; mental health; *mentally handicapped; *minority groups; organizational policies; psychology; public participation; public policy; review committees; *risks and benefits; standards; sterilization; American Indians; Health Services Administration

HEALTH CARE DELIVERY / NURSES / COMMUNITY SERVICES

Aroskar, Mila A. Ethical dilemmas and community health nursing. *Linacre Quarterly* 44(4): 340-346, Nov 1977. 6 fn. BE06102.
  *community services; *health care delivery; minority groups; moral obligations; *nurses; obligations to society; patient care; resource allocation

HEALTH CARE DELIVERY / OBLIGATIONS OF SOCIETY

Rescher, Nicholas. Ethical issues regarding the delivery of health-care services. *Connecticut Medicine* 41(8): 501-506, Aug 1977. 6 fn. BE05973.
  abortion; *biomedical research; *costs and benefits; economics; future generations; *health care delivery; human equality; moral obligations; *obligations of society; resource allocation; self induced illness; state medicine

HEALTH CARE DELIVERY / PHYSICIAN'S ROLE

Mechanic, David. The growth of medical technology and bureaucracy: implications for medical care. *Milbank Memorial Fund Quarterly* 55(1): 61-78, Winter 1977. 21 refs. BE06287.
  biomedical technologies; decision making; economics; *government regulation; *health care delivery; health insurance; medical fees; medicine;

BE = bioethics accession number   fn. = footnotes   refs. = references

*physician's role; physicians; resource allocation; self determination; *social impact; state medicine

HEALTH CARE DELIVERY / RESOURCE ALLOCATION

Kirklin, John W.; Bridgers, William F.; Hearn, Thomas K. Panel discussion on: "The allocation of medical resources—who should decide and how". *Alabama Journal of Medical Sciences* 14(3): 316-321, Jul 1977. BE05821.
*decision making; economics; health care; *health care delivery; justice; legal aspects; physicians; public participation; *resource allocation; scarcity

Mechanic, David. Ethics, justice, and medical care systems. *In* Barber, Bernard, ed. Medical Ethics and Social Change. Philadelphia: American Academy of Political and Social Science, 1978. p. 74-85. Annals of the American Academy of Political and Social Science, Vol. 437, May 1978. 20 fn. ISBN 0-87761-226-9. BE07000.
economics; *health care delivery; justice; physician patient relationship; physicians; remuneration; *resource allocation; scarcity

HEALTH CARE DELIVERY / RESOURCE ALLOCATION / SWEDEN

Rexed, Bror; Juda, Daniel. Planning for scarcity in Sweden: an interview with Bror Rexed. *Hastings Center Report* 7(3): 5-7, Jun 1977. BE05662.
aged; community services; developing countries; economics; *health care delivery; medical education; physicians; public policy; *resource allocation; *scarcity; *Sweden

HEALTH CARE / ADOLESCENTS / LEGAL ASPECTS

Chabon, Robert S. The physician and parental consent. *Journal of Legal Medicine* 5(6): 33-37, Jun 1977. 8 refs. BE05641.
*adolescents; children; emergency care; *health care; *legal aspects; legislation; *parental consent; physicians; state government; treatment refusal

Hofmann, Adele D.; Becker, R.D.; Gabriel, H. Paul. Consent and confidentiality. *In their* The Hospitalized Adolescent: A Guide to Managing the Ill and Injured Youth. New York: Free Press, 1976. p. 211-232. 1 fn. ISBN 0-02-914790-5. BE07088.
*adolescents; *confidentiality; emergency care; *health care; hospitals; informed consent; institutional policies; *legal aspects; legal rights; *medical records; *parental consent

Holder, Angela R. The minor's right to consent to medical treatment. *Connecticut Medicine* 41(9): 579-582, Sep 1977. BE05779.
*adolescents; children; age; competence; disclosure; drug abuse; emergency care; *health care; *informed consent; legal aspects; legal rights; legis-

lation; parental consent; patient care; risks and benefits; state government; treatment refusal; venereal diseases

HEALTH CARE / ADOLESCENTS / LEGAL RIGHTS

Jamir, Vinson F. Children: health services for minors in Oklahoma—capacity to give self-consent to medical care and treatment. *Oklahoma Law Review* 30(2): 385-408, Spring 1977. 183 fn. BE05922.
abortion; *adolescents; contraception; disclosure; emergency care; *health care; health personnel; human experimentation; *informed consent; legal liability; *legal rights; legislation; medical fees; parental consent; physicians; state government; *Oklahoma

Mancini, Marguerite. Nursing, minors, and the law. *American Journal of Nursing* 78(1): 124+, Jan 1978. 2 fn. BE06745.
*adolescents; children; *health care; informed consent; *legal rights; nurses; parental consent

Pennsylvania. Court of Common Pleas, Adams County. In re Nancy F. 1 Dec 1977. *Pennsylvania District and County Reports, 3d Series,* 3: 612-617. BE07124.
*adolescents; dentistry; *health care; judicial action; *legal rights; *right to treatment; surgery; third party consent

HEALTH CARE / ADOLESCENTS / OKLAHOMA

Jamir, Vinson F. Children: health services for minors in Oklahoma—capacity to give self-consent to medical care and treatment. *Oklahoma Law Review* 30(2): 385-408, Spring 1977. 183 fn. BE05922.
abortion; *adolescents; contraception; disclosure; emergency care; *health care; health personnel; human experimentation; *informed consent; legal liability; *legal rights; legislation; medical fees; parental consent; physicians; state government; *Oklahoma

HEALTH CARE / AMERICAN HEART ASSOCIATION

American Heart Association. Committee on Ethics. The American Heart Association and community cardiovascular health: an ethical comment. Report of the Committee on Ethics of the American Heart Association. *Circulation* 57(3): 645A-649A, Mar 1978. BE06814.
biomedical research; coercion; community medicine; *community services; disadvantaged; *health care; health care delivery; heart diseases; human experimentation; informal social control; obligations to society; *organizational policies; *preventive medicine; *privacy; public advocacy; public

---

BE = bioethics accession number   fn. = footnotes   refs. = references

participation; self induced illness; *American Heart Association

HEALTH CARE / CHILDREN / LEGAL ASPECTS

Bennett, Robert. Allocation of child medical care decisionmaking authority. *In* Harvard Child Health Project. Developing a Better Health Care System for Children. Volume 3. Cambridge, Mass.: Ballinger Publishing Company, 1977. p. 231-275. 28 refs. 145 fn. BE06181.
adolescents; child neglect; *children; communicable diseases; decision making; fluoridation; genetic screening; *health care; immunization; *legal aspects; legal rights; legislation; parents; *parental consent; physician's role; religious beliefs; state government; state interest

HEALTH CARE / COSTS AND BENEFITS

Cohn, Victor. Harvard team challenges view that a human life is priceless. *Washington Post,* 24 May 1977, p. A12. BE06021.
*costs and benefits; decision making; *health care; *surgery

HEALTH CARE / DECISION MAKING

Burt, Robert A. The limits of law in regulating health care decisions. *Hastings Center Report* 7(6): 29-32, Dec 1977. BE06133.
conflict of interest; deception; *decision making; dehumanization; *health care; *judicial action; *law; legal aspects; parent child relationship; parental consent; patient participation; physician patient relationship; physicians

HEALTH CARE / LAW

Burt, Robert A. The limits of law in regulating health care decisions. *Hastings Center Report* 7(6): 29-32, Dec 1977. BE06133.
conflict of interest; deception; *decision making; dehumanization; *health care; *judicial action; *law; legal aspects; parent child relationship; parental consent; patient participation; physician patient relationship; physicians

HEALTH CARE / MENTAL HEALTH / MEDICAL ETHICS

Tancredi, Laurence R.; Slaby, Andrew E. Ethical Policy in Mental Health Care: The Goals of Psychiatric Intervention. New York: Prodist, 1977. 153 p. 234 refs. 30 fn. ISBN 0-88202-102-8. BE05710.
behavior control; compensation; conflict of interest; diagnosis; health care delivery; *health care; health personnel; human rights; informed consent; involuntary commitment; justice; *medical ethics; *mental health; mentally ill; moral obligations; patient care; patients' rights; physician patient relationship; psychiatry; psychotherapy; public

advocacy; quality of life; resource allocation; risks and benefits; stigmatization; treatment refusal; value of life

HEALTH CARE / MORAL OBLIGATIONS / SELF INDUCED ILLNESS

Crawford, Robert. You are dangerous to your health: the ideology and politics of victim blaming. *International Journal of Health Services* 7(4): 663-680, 1977. 52 refs. 1 fn. BE06474.
cancer; economics; education; employment; *health care; health care delivery; industrial medicine; *moral obligations; politics; preventive medicine; psychology; public policy; *self induced illness; social control; socioeconomic factors

HEALTH CARE / MORAL OBLIGATIONS / VALUES

McCormick, Richard A. Man's moral responsibility for health. *Catholic Hospital* 5(4): 6-9+, Jul-Aug 1977. BE05962.
allowing to die; Christian ethics; *health; *health care; health care delivery; love; *moral obligations; morality; physician patient relationship; quality of life; self determination; society; technology; *values

HEALTH CARE / OFFICE OF TECHNOLOGY ASSESSMENT

Daddario, Emilio Q. Technology assessment, the Congress, and health. *Connecticut Medicine* 41(1): 23-25, Jan 1977. BE06480.
*biomedical technologies; drugs; *health care; program descriptions; *technology assessment; *Office of Technology Assessment

HEALTH CARE / PARENTAL CONSENT / LEGAL ASPECTS

Bennett, Robert. Allocation of child medical care decisionmaking authority. *In* Harvard Child Health Project. Developing a Better Health Care System for Children. Volume 3. Cambridge, Mass.: Ballinger Publishing Company, 1977. p. 231-275. 28 refs. 145 fn. BE06181.
adolescents; child neglect; *children; communicable diseases; decision making; fluoridation; genetic screening; *health care; immunization; *legal aspects; legal rights; legislation; parents; *parental consent; physician's role; religious beliefs; state government; state interest

Chabon, Robert S. The physician and parental consent. *Journal of Legal Medicine* 5(6): 33-37, Jun 1977. 8 refs. BE05641.
*adolescents; children; emergency care; *health care; *legal aspects; legislation; *parental consent; physicians; state government; treatment refusal

Hofmann, Adele D.; Becker, R.D.; Gabriel, H. Paul.

BE = bioethics accession number   fn. = footnotes   refs. = references

Consent and confidentiality. *In their* The Hospitalized Adolescent: A Guide to Managing the Ill and Injured Youth. New York: Free Press, 1976. p. 211-232. 1 fn. ISBN 0-02-914790-5. BE07088.
   *adolescents; *confidentiality; emergency care; *health care; hospitals; informed consent; institutional policies; *legal aspects; legal rights; *medical records; *parental consent

MacKay, P.M. Consent to treatment—problems of minors. *In* Papers and Proceedings of the National Conference on Health and the Law, Ottawa, 23-25 Sep 1975. Toronto: Canadian Hospital Association, 1975. p. 122-128. BE07065.
   abortion; *adolescents; *children; contraception; *health care; informed consent; *legal aspects; organ transplantation; *parental consent; withholding treatment

HEALTH CARE / PRISONERS / MINNESOTA

U.S. District Court, D. Minnesota, Fourth Division. Hines v. Anderson. 27 May 1977. *Federal Supplement* 439: 12-24. BE07111.
   confidentiality; *health care; *legal rights; medical records; nutrition; patients' rights; *prisoners; standards; *Minnesota

HEALTH CARE / PUBLIC POLICY

Childress, James F. Priorities in the allocation of health care resources. *In* Smith, David H., ed. No Rush to Judgement: Essays on Medical Ethics. Bloomington, Ind.: Indiana University, Poynter Center, 1977. p. 129-148. 30 fn. BE05798.
   economics; freedom; health; *health care; human equality; *preventive medicine; *public policy; renal dialysis; *resource allocation; rights; scarcity; selection for treatment; self induced illness; social worth; socioeconomic factors; values

HEALTH CARE / PUBLIC POLICY / ECONOMICS

Hellegers, André E.; Wakin, Edward. Is the right-to-die wrong? *U.S. Catholic* 43(3): 13-17, Mar 1978. BE06815.
   allowing to die; attitudes to death; biomedical research; biomedical technologies; cancer; costs and benefits; death; decision making; *economics; extraordinary treatment; *health care; health insurance; heart diseases; life extension; morbidity; mortality; physician's role; preventive medicine; prolongation of life; *public policy; self induced illness; statistics; values

HEALTH CARE / RESOURCE ALLOCATION

Childress, James F. Priorities in the allocation of health care resources. *In* Smith, David H., ed. No Rush to Judgement: Essays on Medical Ethics. Bloomington, Ind.: Indiana University, Poynter Center, 1977. p. 129-148. 30 fn. BE05798.

economics; freedom; health; *health care; human equality; *preventive medicine; *public policy; renal dialysis; *resource allocation; rights; scarcity; selection for treatment; self induced illness; social worth; socioeconomic factors; values

Marty, Martin E. Physicians leave too much to the bedside encounter; economists stop everything before it starts. *American Medical News* 20(25): Impact 4+, 20 Jun 1977. BE05635.
   economics; *health care; random selection; *resource allocation; selection for treatment; theology; *value of life

HEALTH CARE / RESOURCE ALLOCATION / COSTS AND BENEFITS

Rhoads, Steven E. How much should we spend to save a life? *Public Interest* No. 51: 74-92, Spring 1978. 4 fn. BE06927.
   *costs and benefits; *economics; equal protection; *health care; indigents; occupational diseases; preventive medicine; public opinion; public policy; *resource allocation; scarcity; *value of life

HEALTH CARE / RESOURCE ALLOCATION / ECONOMICS

Welch, William J. Care of the sick can't be governed by bookkeeping and cost analyses. *American Medical News* 20(25): Impact 3, 20 Jun 1977. BE05634.
   costs and benefits; *economics; *health care; *resource allocation; value of life

Wicks, Elliot K. It's impossible to avoid a decision that saves some and condemns others. *American Medical News* 20(25): Impact 2-3, 20 Jun 1977. BE05633.
   biomedical technologies; costs and benefits; decision making; *economics; *health care; *resource allocation

HEALTH CARE / RIGHTS

Blackstone, William T. More on the right to health care. *Georgia Law Review* 11(3): 539-544, Spring 1977. 24 fn. BE05947.
   *health care; human rights; legal rights; morality; philosophy; politics; *rights; utilitarianism

Soble, Alan. Philosophical justifications, political activity, and adequate health care. *Georgia Law Review* 11(3): 525-538, Spring 1977. 24 fn. BE05946.
   *health care; *human rights; justice; *legal rights; moral obligations; philosophy; politics; *rights; utilitarianism

HEALTH CARE / VALUE OF LIFE

Marty, Martin E. Physicians leave too much to the bedside encounter; economists stop everything before it starts. *American Medical News* 20(25): Impact 4+, 20 Jun 1977. BE05635.
   economics; *health care; random selection; *re-

BE = bioethics accession number   fn. = footnotes   refs. = references

source allocation; selection for treatment; theology; *value of life

HEALTH CARE / VALUE OF LIFE / ECONOMICS

Card, W.I.; Mooney, G.H. What is the monetary value of a human life? *British Medical Journal* 2(6103): 1627-1629, 24-31 Dec 1977. 18 refs. BE06507.
costs and benefits; decision making; *economics; *health care; *resource allocation; social worth; *value of life; values

Rhoads, Steven E. How much should we spend to save a life? *Public Interest* No. 51: 74-92, Spring 1978. 4 fn. BE06927.
*costs and benefits; *economics; equal protection; *health care; indigents; occupational diseases; preventive medicine; public opinion; public policy; *resource allocation; scarcity; *value of life

HEALTH CARE   See under

RESOURCE ALLOCATION / HEALTH CARE

HEALTH HAZARDS   See under

BIOMEDICAL RESEARCH / SOCIAL CONTROL / HEALTH HAZARDS

MORAL OBLIGATIONS / INDUSTRIAL MEDICINE / HEALTH HAZARDS

RECOMBINANT DNA RESEARCH / CONTAINMENT / HEALTH HAZARDS

RECOMBINANT DNA RESEARCH / GOVERNMENT REGULATION / HEALTH HAZARDS

RECOMBINANT DNA RESEARCH / HEALTH HAZARDS

RECOMBINANT DNA RESEARCH / SOCIAL CONTROL / HEALTH HAZARDS

REPRODUCTION / INDUSTRIAL MEDICINE / HEALTH HAZARDS

HEALTH PERSONNEL   See under

MEDICAL ETHICS / HEALTH PERSONNEL / EDUCATION

PATIENT CARE / HEALTH PERSONNEL / STRIKES

HEALTH / TOTALITY / HÄRING, BERNARD

Soane, Brendan. The literature of medical ethics: Bernard Häring. *Journal of Medical Ethics* 3(2): 85-92, Jun 1977. 37 fn. BE05669.
AIH; *behavior control; biology; Christian ethics; contraception; genetic intervention; goals; *health; health care; medical ethics; medicine; *normality; operant conditioning; philosophy; psychology; Roman Catholic ethics; *self determination; *teleological ethics; *totality; values; voluntary sterilization; *Häring, Bernard

HEALTH   See under

CONFIDENTIALITY / FAMOUS PERSONS / HEALTH

HEARTS   See under

ORGAN TRANSPLANTATION / HEARTS

HEMODIALYSIS   See RENAL DIALYSIS

HISTORICAL ASPECTS   See under

ABORTION / HISTORICAL ASPECTS

ABORTION / ROMAN CATHOLIC ETHICS / HISTORICAL ASPECTS

EUGENICS / DISCRIMINATION / HISTORICAL ASPECTS

EUGENICS / HISTORICAL ASPECTS / GERMANY

EUGENICS / HISTORICAL ASPECTS / GREAT BRITAIN

EUGENICS / HISTORICAL ASPECTS / USSR

HUMAN EXPERIMENTATION / CONTROL GROUPS / HISTORICAL ASPECTS

HUMAN EXPERIMENTATION / GOVERNMENT REGULATION / HISTORICAL ASPECTS

HUMAN EXPERIMENTATION / HISTORICAL ASPECTS

HUMAN EXPERIMENTATION / INVESTIGATORS / HISTORICAL ASPECTS

HUMAN EXPERIMENTATION / POLITICS / HISTORICAL ASPECTS

HUMAN EXPERIMENTATION / RISKS AND BENEFITS / HISTORICAL ASPECTS

INFANTICIDE / HISTORICAL ASPECTS / ENGLAND

INFANTICIDE / HISTORICAL ASPECTS / UNITED STATES

INFANTICIDE / LAW / HISTORICAL ASPECTS

MEDICAL ETHICS / HISTORICAL ASPECTS

MEDICAL ETHICS / HISTORICAL ASPECTS / ISLAM

MEDICINE / HISTORICAL ASPECTS

PHYSICIAN PATIENT RELATIONSHIP / HISTORICAL ASPECTS

POPULATION CONTROL / PUBLIC POLICY / HISTORICAL ASPECTS

PSYCHIATRY / HISTORICAL ASPECTS / GREAT BRITAIN

PSYCHIATRY / METHODS / HISTORICAL ASPECTS

STERILIZATION / LEGISLATION / HISTORICAL ASPECTS

VALUE OF LIFE / MEDICINE / HISTORICAL ASPECTS

HOMOSEXUALS   See under

CONFIDENTIALITY / HOMOSEXUALS

BE = bioethics accession number   fn. = footnotes   refs. = references

HOSPICES *See under*
TERMINAL CARE / HOSPICES
TERMINAL CARE / HOSPICES / UNITED STATES

HOSPITALS *See also* HOSPICES, MENTAL INSTITUTIONS

HOSPITALS *See under*
ALLOWING TO DIE / HOSPITALS / CALIFORNIA NATURAL DEATH ACT
CONFIDENTIALITY / MEDICAL RECORDS / HOSPITALS
HUMAN EXPERIMENTATION / ADMINISTRATORS / HOSPITALS
HUMAN EXPERIMENTATION / ETHICS COMMITTEES / HOSPITALS
INFORMED CONSENT / HOSPITALS / LEGAL ASPECTS
INSTITUTIONAL POLICIES / HOSPITALS / MEDICAL RECORDS
INSTITUTIONAL POLICIES / HOSPITALS / TERMINALLY ILL
MEDICAL RECORDS / HOSPITALS / LEGAL ASPECTS
TERMINAL CARE / HOSPITALS

HUMAN CHARACTERISTICS *See under*
SOCIOBIOLOGY / HUMAN CHARACTERISTICS

HUMAN EXPERIMENTATION *See also* BEHAVIORAL RESEARCH, BIOMEDICAL RESEARCH

HUMAN EXPERIMENTATION

Beauchamp, Tom L.; Walters, LeRoy, eds. Contemporary Issues in Bioethics. Encino, Calif.: Dickenson Publishing Company, 1978. 612 p. References and footnotes included. ISBN 0-8221-0200-5. BE06700.
*abortion; allowing to die; *behavior control; *bioethics; biomedical technologies; brain death; children; codes of ethics; confidentiality; decision making; deontological ethics; *determination of death; *euthanasia; fetuses; genetic intervention; *health; *health care; *human experimentation; *informed consent; involuntary commitment; justice; legal aspects; *medical ethics; natural law; *normative ethics; operant conditioning; patients' rights; personhood; psychosurgery; *resource allocation; rights; selection for treatment; treatment refusal; utilitarianism; Kaimowitz v. Department of Mental Health; Quinlan, Karen

Childress, James F. Concerns of the community about experimentation on human subjects. *In* Mittler, Peter, ed. Research to Practice in Mental Retardation. Volume 1: Care and Intervention. Baltimore: University Park Press, 1977. p. 59-64. 8 refs. ISBN 0-8391-1122-3. BE06178.
compensation; *human experimentation; informed

consent; justice; nontherapeutic research; risks and benefits

Goldiamond, Israel. Protection of human subjects and patients. *In* Krapfl, Jon E.; Vargas, Ernest A., eds. Behaviorism and Ethics. Kalamazoo: Behaviordelia, 1977. p. 129-195. Commentary by Pamela Meadowcroft on p. 188-195. 41 refs. ISBN 0-914-47425-1. BE06546.
*behavior control; *coercion; consequences; ethical review; *human experimentation; goals; intention; informed consent; investigator subject relationship; *operant conditioning; remuneration; risks and benefits; self determination

Hunter, Thomas H. A philosophy of clinical research based on needs. *Pharos* 41(2): 26-28, Apr 1978. 5 refs. BE06845.
decision making; *human experimentation; legal aspects; rights; values

Hunter, Thomas H.; Marston, Robert; Fletcher, Joseph. Dilemmas of Clinical Trials. [Videorecording]. Charlottesville, Va.: University of Virginia Medical Center, Health Sciences Library, 1973. 1 cassette; 60 min.; sound; black and white; 3/4 in. Tape of the Medical Center Hour, 1 Oct 1973. Panel discussion moderated by Thomas H. Hunter. BE07042.
control groups; diabetes; federal government; government regulation; *human experimentation; morality; placebos; random selection; risks and benefits

Kupfer, Sherman. Experimentation and ethics. *Mount Sinai Journal of Medicine* 44(5): 648-656, Sep-Oct 1977. 9 refs. BE06334.
disclosure; ethics committees; *human experimentation; informed consent; normative ethics; physician's role; risks and benefits

Wecht, Cyril H. Human experimentation and clinical investigation—legal and ethical considerations. *In* Wecht, Cyril H., ed. Legal Medicine Annual 1976. New York: Appleton-Century-Crofts, 1977. p. 299-313. 23 fn. ISBN 0-8385-5654-X. BE06522.
*human experimentation; informed consent; nontherapeutic research; risks and benefits

Wojcik, Jan. Muted Consent: A Casebook in Modern Medical Ethics. West Lafayette, Ind.: Purdue University, 1978. 164 p. 237 fn. Case study. ISBN 911198-50-4. BE06704.
*abortion; *behavior control; *case reports; *determination of death; disclosure; drugs; electrical stimulation of the brain; ethics; eugenics; euthanasia; extraordinary treatment; fetuses; *genetic counseling; genetic defects; *genetic intervention; *genetic screening; health care; *human experimentation; *informed consent; involuntary commitment; mentally ill; operant conditioning; pregnant women; psychosurgery; quality of life; reproductive technologies; *resource allocation;

BE = bioethics accession number   fn. = footnotes   refs. = references

rights; selective abortion; value of life; withholding treatment

HUMAN EXPERIMENTATION / ADMINISTRATORS / HOSPITALS

Wielk, Carol A. Human experimentation: issues before the hospital administrator. *Hospital and Health Services Administration* 22(3): 4-25, Summer 1977. 6 refs. 90 fn. BE05939.
*administrators; codes of ethics; *ethics committees; federal government; government regulation; *hospitals; *human experimentation; informed consent; institutional policies; mentally ill; prisoners

HUMAN EXPERIMENTATION / ALCOHOL ABUSE

Goldberg, Martin, et al. Ethics in alcoholism research. *American Journal of Psychiatry* 134(12): 1447, Dec 1977. Letter. BE06138.
*alcohol abuse; ethics committees; *human experimentation

HUMAN EXPERIMENTATION / ALCOHOL ABUSE / RISKS AND BENEFITS

Cowan, Dale H. Ethical considerations of metabolic research in alcoholic patients. *Alcoholism: Clinical and Experimental Research* 1(3): 193-198, Jul 1977. 14 refs. BE05723.
*alcohol abuse; coercion; competence; *human experimentation; hospitals; *informed consent; investigator subject relationship; investigators; moral obligations; nontherapeutic research; *patients; physician patient relationship; physician's role; physicians; research subjects; rights; *risks and benefits; *selection of subjects; socioeconomic factors; therapeutic research

HUMAN EXPERIMENTATION / CADAVERS / INJURIES

Cadaver tests target of ban. *American Medical News* 21(7): 1+, 13 Feb 1978. BE06804.
*cadavers; government agencies; *human experimentation; *injuries; research institutes; universities; Department of Transportation; Wayne State University

The quick, the dead, and the cadaver population. *Science* 199(4336): 1420, 31 Mar 1978. BE06805.
*cadavers; government agencies; *human experimentation; *injuries; morality; universities; Department of Transportation; Moss, John E.

HUMAN EXPERIMENTATION / CALIFORNIA

California. Legislature. Senate. Human Experimentation Act of 1976. Senate Bill No. 2051, 25 Mar 1976. 4 p. By Senator Robbins, et al. Died in Senate Health and Welfare Committee, 1975-76 Ses-

sion. BE06599.
government regulation; *human experimentation; informed consent; state government; *California

HUMAN EXPERIMENTATION / CANCER / COMMON GOOD

Krant, Melvin J.; Cohen, Joseph L.; Rosenbaum, Charles. Moral dilemmas in clinical cancer experimentation. *Medical and Pediatric Oncology* 3(2): 141-147, 1977. 12 refs. BE06479.
*cancer; coercion; *common good; drugs; *human experimentation; nontherapeutic research; obligations to society; physician patient relationship; physician's role; risks and benefits; terminally ill; values

HUMAN EXPERIMENTATION / CENTRAL INTELLIGENCE AGENCY

Horrock, Nicholas M. C.I.A. data show 14-year project on controlling human behavior. *New York Times,* 21 Jul 1977, p. A1+. BE06056.
behavior control; drugs; *federal government; *government agencies; *human experimentation; *Central Intelligence Agency

U.S. Congress. Senate. Project MKULTRA, the CIA's Program of Research in Behavioral Modification. Joint Hearing. Washington: U.S. Government Printing Office, 1977. 171 p. Joint Hearing, 95th Cong., 1st Sess., before the Select Committee on Intelligence and the Committee on Human Resources, Subcommittee on Health and Scientific Research, 3 Aug 1977. References and footnotes included. BE06693.
behavior control; *federal government; *government agencies; *human experimentation; LSD; *psychoactive drugs; *Central Intelligence Agency; *Project MKULTRA

U.S. National Commission for the Protection of Human Subjects of Biomedical and Behavioral Research. National Commission for the Protection of Human Subjects of Biomedical and Behavioral Research. Transcript of the Meeting Proceedings (32d Meeting), 8-9 Jul 1977. Springfield, Va.: National Technical Information Service, 1977. 326 p. Report No. NCPHS/M-77/20. NTIS No. PB-270 857. BE06652.
*children; epidemiology; *ethics committees; genetics; government agencies; *human experimentation; infants; informed consent; *institutionalized persons; *mentally handicapped; mentally ill; mentally retarded; research subjects; risks and benefits; self determination; *Central Intelligence Agency

HUMAN EXPERIMENTATION / CHILDREN

Abrams, Natalie. Medical experimentation: the consent of prisoners and children. *In* Spicker, Stuart F.; Engelhardt, H. Tristram, eds. Philosophical Medical

---

BE = bioethics accession number   fn. = footnotes   refs. = references

Ethics: Its Nature and Significance. Proceedings. Boston: D. Reidel, 1977. p. 111-124. Proceedings of the Third Trans-Disciplinary Symposium on Philosophy and Medicine, Farmington, Connecticut, 11-13 Dec 1975. 6 refs. ISBN 90-277-0772-3. BE05867.
    *children; *coercion; *human experimentation; incentives; informed consent; morality; *parental consent; *prisoners; risks and benefits; self determination

Bartholome, William G. Central themes in the debate over involvement of infants and children in biomedical research: a critical examination. *In* Van Eys, Jan, ed. Research on Children: Medical Imperatives, Ethical Quandaries, and Legal Constraints. Baltimore: University Park Press, 1978. p. 69-76. Proceedings of a conference held at Houston, Texas, 29-30 Apr 1977, sponsored by the Univ. of Texas System Cancer Center, the Inst. of Religion, and the Univ. of Texas Health Science Center. 10 refs. ISBN 0-8391-1191-6. BE06990.
    *children; *human experimentation; infants; informed consent; *parental consent

Cavanaugh, Paul F. Dilemmas of experimentation on children. *Linacre Quarterly* 44(3): 273-277, Aug 1977. 1 ref. BE05835.
    *children; *human experimentation; nontherapeutic research; normality; risks and benefits; therapeutic research; value of life

Cohn, Victor. Panel: gain child's assent to research. *Washington Post,* 14 May 1977, p. A9. BE06020.
    *children; *human experimentation; *informed consent; National Commission for Protection of Human Subjects

Cooke, Robert E. An ethical and procedural basis for research on children. *Journal of Pediatrics* 90(4): 681-682, Apr 1977. 3 refs. Editorial. BE05547.
    *children; ethics committees; *human experimentation; informed consent; *nontherapeutic research; parental consent

Fried, Charles. Children as subjects for medical experimentation. *In* Van Eys, Jan, ed. Research on Children: Medical Imperatives, Ethical Quandaries, and Legal Constraints. Baltimore: University Park Press, 1978. p. 107-115. Proceedings of a conference held at Houston, Texas, 29-30 Apr 1977, sponsored by the Univ. of Texas System Cancer Center, the Inst. of Religion, and the Univ. of Texas Health Science Center. ISBN 0-8391-1191-6. BE06993.
    adolescents; *human experimentation; *children; informed consent; legal rights; random selection; risks and benefits; self determination

Lockhart, Jean D. Pediatric drug testing: is it at risk? *Hastings Center Report* 7(3): 8-10, Jun 1977. 1 fn. BE05653.
    case reports; *children; *drugs; ethical review; *human experimentation; human rights; informed

consent; minority groups; pediatrics; politics; risks and benefits; social impact

Ramsey, Paul. Children as research subjects: a reply. *Hastings Center Report* 7(2): 40-42, Apr 1977. 9 fn. BE05576.
    *children; *human experimentation; *moral obligations; nontherapeutic research; parental consent; risks and benefits; McCormick, Richard A.

Ramsey, Paul; Bartholome, William G. Ordinary risks of childhood. *Hastings Center Report* 7(2): 4, Apr 1977. Letter. BE05533.
    *children; *human experimentation; morality; nontherapeutic research; pediatrics; *risks and benefits

Richie, Ellen. Biomedical research: for the patient or on the patient? *In* Van Eys, Jan, ed. The Truly Cured Child: The New Challenge in Pediatric Cancer Care. Baltimore: University Park Press, 1977. p. 25-37. Proceedings of a workshop, Department of Pediatrics, M.D. Anderson Hospital and Tumor Institute, Houston, Texas, 13-14 Mar 1976. 16 refs. ISBN 0-8391-1108-8. BE05703.
    *children; *human experimentation; informed consent; investigator subject relationship; nontherapeutic research; physician patient relationship; risks and benefits; therapeutic research

Ryan, Kenneth J.; Silverman, William A. Research involving children. [Title provided]. *Scientific American* 237(6): 6-7, Dec 1977. Letter. BE06148.
    *children; federal government; government regulation; *human experimentation; risks and benefits

U.S. Department of Health, Education, and Welfare. Research involving children: report and recommendations of the National Commission for the Protection of Human Subjects of Biomedical and Behavioral Research. *Federal Register* 43(9): 2084-2114, 13 Jan 1978. 81 fn. BE06913.
    *children; ethical review; ethics committees; *human experimentation; *informed consent; legal aspects; moral obligations; nontherapeutic research; parental consent; risks and benefits; therapeutic research; National Commission for Protection of Human Subjects

U.S. National Commission for the Protection of Human Subjects of Biomedical and Behavioral Research. Research Involving Children: Report and Recommendations. Washington: U.S. Government Printing Office, 1977. 154 p. DHEW Publication No. (OS)77-0004. 77 fn. BE06193.
    adolescents; attitudes; behavioral research; *children; ethicists; ethics committees; federal government; government agencies; *human experimentation; investigators; *informed consent; legal aspects; minority groups; moral obligations; nontherapeutic research; parental consent; public

---

opinion; *risks and benefits; social impact; therapeutic research

U.S. National Commission for the Protection of Human Subjects of Biomedical and Behavioral Research. National Commission for the Protection of Human Subjects of Biomedical and Behavioral Research. Transcript of the Meeting Proceedings (26th Meeting), 7-8 Jan 1977. Springfield, Va.: National Technical Information Service, 1977. 345 p. Report No. NCPHS/M-77/06. NTIS No. PB-263 280. BE06647.
*children; *ethics committees; federal government; financial support; government regulation; *human experimentation; informed consent; nontherapeutic research; *parental consent; *psychosurgery; risks and benefits; therapeutic research; treatment refusal

U.S. National Commission for the Protection of Human Subjects of Biomedical and Behavioral Research. National Commission for the Protection of Human Subjects of Biomedical and Behavioral Research. Transcript of the Meeting Proceedings (28th Meeting), 11-12 Mar 1977. Springfield, Va.: National Technical Information Service, 1977. 467 p. Report No. NCPHS/M-77/10. NTIS No. PB-265 847. BE06648.
*children; confidentiality; disclosure; ethics committees; federal government; financial support; *human experimentation; informed consent; *parental consent; risks and benefits

U.S. National Commission for the Protection of Human Subjects of Biomedical and Behavioral Research. National Commission for the Protection of Human Subjects of Biomedical and Behavioral Research. Transcript of the Meeting Proceedings (31st Meeting), 10-11 Jun 1977. Springfield, Va.: National Technical Information Service, 1977. 336 p. Report No. NCPHS/M-77/18. NTIS No. PB-269 377. BE06651.
*children; compensation; ethical review; *ethics committees; *evaluation; government agencies; *human experimentation; informed consent; institutionalized persons; legal aspects; mentally handicapped; research subjects; rights; risks and benefits; social control; Institutional Review Boards

U.S. National Commission for the Protection of Human Subjects of Biomedical and Behavioral Research. National Commission for the Protection of Human Subjects of Biomedical and Behavioral Research. Transcript of the Meeting Proceedings (32d Meeting), 8-9 Jul 1977. Springfield, Va.: National Technical Information Service, 1977. 326 p. Report No. NCPHS/M-77/20. NTIS No. PB-270 857. BE06652.
*children; epidemiology; *ethics committees; genetics; government agencies; *human experimentation; infants; informed consent; *institutionalized

persons; *mentally handicapped; mentally ill; mentally retarded; research subjects; risks and benefits; self determination; *Central Intelligence Agency

U.S. National Commission for the Protection of Human Subjects of Biomedical and Behavioral Research. National Commission for the Protection of Human Subjects of Biomedical and Behavioral Research. Transcript of the Meeting Proceedings (30th Meeting), 13-14 May 1977. Springfield, Va.: National Technical Information Service, 1977. 346 p. Report No. NCPHS/M-77/15. NTIS No. PB-268 100. BE06650.
*children; *compensation; ethical review; *ethics committees; government agencies; *informed consent; *human experimentation; institutionalized persons; legal guardians; nontherapeutic research; parental consent; *research subjects; *risks and benefits; suffering; therapeutic research; third party consent; Institutional Review Boards

Van Eys, Jan, ed. Research on Children: Medical Imperatives, Ethical Quandaries, and Legal Constraints. Baltimore: University Park Press, 1978. 152 p. Proceedings of a conference held at Houston, Texas, 29-30 Apr 1977, sponsored by the Univ. of Texas System Cancer Center, the Inst. of Religion, and the Univ. of Texas Health Science Center. 101 refs. 6 fn. ISBN 0-8391-1191-6. BE06987.
adolescents; cancer; *children; drugs; federal government; government regulation; human development; *human experimentation; *informed consent; legal aspects; *parental consent; patient care; psychological stress; risks and benefits

HUMAN EXPERIMENTATION / CHILDREN / CHRISTIAN ETHICS

Tripp, John. Ethical problems in paediatrics. *In* Vale, J.A., ed. Medicine and the Christian Mind. London: Christian Medical Fellowship Publications, 1975. p. 106-113. 11 refs. ISBN 85111-956-5. BE07073.
*allowing to die; *Christian ethics; *children; *congenital defects; family problems; *human experimentation; *newborns; physicians; withholding treatment

HUMAN EXPERIMENTATION / CHILDREN / COMMUNICATION

Berryman, Jerome. Discussing the ethics of research on children. *In* Van Eys, Jan, ed. Research on Children: Medical Imperatives, Ethical Quandaries, and Legal Constraints. Baltimore: University Park Press, 1978. p. 85-101. Proceedings of a conference held at Houston, Texas, 29-30 Apr 1977, sponsored by the Univ. of Texas System Cancer Center, the Inst. of Religion, and the Univ. of Texas Health Science Center. 16 refs. ISBN 0-8391-1191-6. BE06992.
cancer; *children; *communication; decision mak-

BE = bioethics accession number   fn. = footnotes   refs. = references

ing; ethical analysis; *human experimentation; moral obligations; religion; trust

HUMAN EXPERIMENTATION / CHILDREN / DRUGS

Shirkey, Harry C. Therapeutic orphans—who speaks for children? *In* Shirkey, Harry C., ed. Pediatric Therapy. St. Louis: C.V. Mosby, 1975. p. 62-65. 25 refs. 2 fn. ISBN 0-8016-4600-6. BE07077.
　*children; drug industry; *drugs; federal government; government agencies; government regulation; *human experimentation; pediatrics; Food and Drug Administration

HUMAN EXPERIMENTATION / CHILDREN / GREAT BRITAIN

Skegg, P.D.G. English law relating to experimentation on children. *Lancet* 2(8041): 754-755, 8 Oct 1977. 9 refs. BE06093.
　*children; *human experimentation; *informed consent; *legal aspects; nontherapeutic research; parental consent; *Great Britain

Soothill, J.F. Research on infants. *Lancet* 2(8051): 1278-1279, 17 Dec 1977. 9 fn. Letter. BE06258.
　*children; *human experimentation; standards; Department of Health and Social Security; *Great Britain; Medical Research Council

HUMAN EXPERIMENTATION / CHILDREN / LEGAL ASPECTS

Valid parental consent. *Lancet* 1(8026): 1346-1347, 25 Jun 1977. BE05673.
　*children; *human experimentation; *legal aspects; *nontherapeutic research; *parental consent

HUMAN EXPERIMENTATION / CHILDREN / LEGAL RIGHTS

U.S. National Commission for the Protection of Human Subjects of Biomedical and Behavioral Research. Legal issues. *In its* Research Involving Children: Report and Recommendations. Washington: U.S. Government Printing Office, 1977. p. 73-90. 37 fn. BE06194.
　adolescents; behavioral research; *children; health care; *human experimentation; *legal rights; *informed consent; legislation; nontherapeutic research; organ donation; parental consent; risks and benefits; state government; therapeutic research

HUMAN EXPERIMENTATION / CHILDREN / MENTALLY HANDICAPPED

Solnit, Albert J., et al. Responses to "Changes in the parent-child relationship...". *Journal of Autism and Childhood Schizophrenia* 7(1): 94-108, Mar 1977. 3 refs. 7 fn. BE05910.
　adolescents; behavioral research; *children; *decision making; ethical review; hospitals; *human ex-

perimentation; institutionalized persons; *involuntary commitment; *legal aspects; legal rights; mental institutions; *mentally handicapped; parent child relationship; *parental consent; patient care; right to treatment

HUMAN EXPERIMENTATION / CHILDREN / NATIONAL COMMISSION FOR PROTECTION OF HUMAN SUBJECTS

U.S. National Commission for the Protection of Human Subjects of Biomedical and Behavioral Research. National Commission for the Protection of Human Subjects of Biomedical and Behavioral Research. Transcript of the Meeting Proceedings (33d Meeting), 12-13 Aug 1977. Springfield, Va.: National Technical Information Service, 1977. 335 p. Report No. NCPHS/M-77/21. NTIS No. PB-271 917. BE06873.
　*children; conflict of interest; consent forms; dignity; ethical review; ethics; *ethics committees; health care delivery; *human experimentation; informed consent; injuries; *institutionalized persons; investigators; *mentally handicapped; *mentally ill; *mentally retarded; nontherapeutic research; research subjects; risks and benefits; social impact; therapeutic research; Central Intelligence Agency; *National Commission for Protection of Human Subjects

HUMAN EXPERIMENTATION / CHILDREN / RISKS AND BENEFITS

Ramsey, Paul. Ethical dimensions of experimental research on children. *In* Van Eys, Jan, ed. Research on Children: Medical Imperatives, Ethical Quandaries, and Legal Constraints. Baltimore: University Park Press, 1978. p. 57-68. Proceedings of a conference held at Houston, Texas, 29-30 Apr 1977, sponsored by the Univ. of Texas System Cancer Center, the Inst. of Religion, and the Univ. of Texas Health Science Center. 17 refs. 5 fn. ISBN 0-8391-1191-6. BE06989.
　*children; compensation; *human experimentation; moral obligations; nontherapeutic research; *parental consent; *risks and benefits

U.S. National Commission for the Protection of Human Subjects of Biomedical and Behavioral Research. Ethical issues. *In its* Research Involving Children: Report and Recommendations. Washington: U.S. Government Printing Office, 1977. p. 91-121. 40 fn. BE06195.
　*children; ethicists; *human experimentation; informed consent; *moral obligations; morality; nontherapeutic research; obligations to society; parental consent; *risks and benefits; self determination; therapeutic research

U.S. National Commission for the Protection of Human Subjects of Biomedical and Behavioral Research. National Commission for the Protection of

---

BE = bioethics accession number　　　　fn. = footnotes　　　　refs. = references

Human Subjects of Biomedical and Behavioral Research. Transcript of the Meeting Proceedings (29th Meeting), 8-9 Apr 1977. Springfield, Va.: National Technical Information Service, 1977. 382 p. Report No. NCPHS/M-77/14. NTIS No. PB-267 013. BE06649.
*children; *human experimentation; immunization; nontherapeutic research; *risks and benefits; therapeutic research

U.S. National Commission for the Protection of Human Subjects of Biomedical and Behavioral Research. National Commission for the Protection of Human Subjects of Biomedical and Behavioral Research. Transcript of the Meeting Proceedings (30th Meeting), 13-14 May 1977. Springfield, Va.: National Technical Information Service, 1977. 346 p. Report No. NCPHS/M-77/15. NTIS No. PB-268 100. BE06650.
*children; *compensation; ethical review; *ethics committees; government agencies; *informed consent; *human experimentation; institutionalized persons; legal guardians; nontherapeutic research; parental consent; *research subjects; *risks and benefits; suffering; therapeutic research; third party consent; Institutional Review Boards

HUMAN EXPERIMENTATION / CHILDREN / STANDARDS

American Academy of Pediatrics. Committee on Drugs. Guidelines for the ethical conduct of studies to evaluate drugs in pediatric populations. Pediatrics 60(1): 91-101, Jul 1977. 1 ref. BE05718.
adolescents; age; attitudes; *children; disclosure; compensation; *drugs; ethical review; *ethics committees; evaluation; human development; *human experimentation; infants; *informed consent; institutionalized persons; investigator subject relationship; *investigators; mentally retarded; nontherapeutic research; pediatrics; placebos; psychological stress; remuneration; research design; *risks and benefits; selection of subjects; socioeconomic factors; *standards; therapeutic research; third party consent

HUMAN EXPERIMENTATION / CHILDREN / VALUES

Vaux, Kenneth. Lazarus revisited: moral and spiritual aspects of experimental therapeutics with children. In Van Eys, Jan, ed. Research on Children: Medical Imperatives, Ethical Quandaries, and Legal Constraints. Baltimore: University Park Press, 1978. p. 77-84. Proceedings of a conference held at Houston, Texas, 29-30 Apr 1977, sponsored by the Univ. of Texas System Cancer Center, the Inst. of Religion, and the Univ. of Texas Health Science Center. 4 refs. ISBN 0-8391-1191-6. BE06991.
allowing to die; *children; *human experimentation; informed consent; random selection; *values

HUMAN EXPERIMENTATION / CHRISTIAN ETHICS

Reece, Robert D. Christian ethics and human experimentation. In Hollis, Harry N., comp. A Matter of Life and Death: Christian Perspectives. Nashville: Broadman Press, 1977. p. 80-98. 5 fn. ISBN 0-8054-6118-3. BE06532.
*Christian ethics; hearts; *human experimentation; informed consent; investigators; moral obligations; organ transplantation; physicians; public policy; risks and benefits; syphilis

HUMAN EXPERIMENTATION / COERCION

Abrams, Natalie. Medical experimentation: the consent of prisoners and children. In Spicker, Stuart F.; Engelhardt, H. Tristram, eds. Philosophical Medical Ethics: Its Nature and Significance. Proceedings. Boston: D. Reidel, 1977. p. 111-124. Proceedings of the Third Trans-Disciplinary Symposium on Philosophy and Medicine, Farmington, Connecticut, 11-13 Dec 1975. 6 refs. ISBN 90-277-0772-3. BE05867.
*children; *coercion; *human experimentation; incentives; informed consent; morality; *parental consent; *prisoners; risks and benefits; self determination

Cohen, Carl. Medical experimentation on prisoners. Perspectives in Biology and Medicine 21(3): 357-372, Spring 1978. 6 refs. 4 fn. BE06811.
*coercion; costs and benefits; *human experimentation; incentives; *informed consent; institutionalized persons; moral obligations; paternalism; *prisoners; remuneration; socioeconomic factors

HUMAN EXPERIMENTATION / COMPENSATION

Annas, George J.; Glantz, Leonard H.; Katz, Barbara F. Compensation for harm: an additional protection for human subjects. In their Informed Consent to Human Experimentation: The Subject's Dilemma. Cambridge, Mass.: Ballinger Publishing Company, 1977. p. 257-277. 78 fn. ISBN 0-88410-147-9. BE06860.
*compensation; federal government; costs and benefits; *human experimentation; informed consent; insurance; legal liability; obligations of society

HUMAN EXPERIMENTATION / COMPENSATION / LEGAL ASPECTS

Curran, William J. Legal liability in clinical investigations. New England Journal of Medicine 298(14): 778-779, 6 Apr 1978. 5 refs. BE06838.
*compensation; *human experimentation; informed consent; *legal aspects; legal liability; prisoners

---

BE = bioethics accession number   fn. = footnotes   refs. = references

HUMAN EXPERIMENTATION / CONFIDENTI-ALITY / SEXUALITY

Kelty, Miriam F. Ethical issues and requirements for sex research with humans: confidentiality. *In* Masters, William H.; Johnson, Virginia E.; Kolodny, Robert C., eds. Ethical Issues in Sex Therapy and Research. Boston: Little, Brown, 1977. p. 84-118. Includes discussion by Richard Green and others on p. 106-118. 36 refs. ISBN 0-316-549-835. BE06162.
> *confidentiality; data bases; disclosure; education; federal government; government agencies; government regulation; *human experimentation; journalism; *legal aspects; legal rights; *privacy; *sexuality; XYY karyotype

HUMAN EXPERIMENTATION / CONTROL GROUPS / HISTORICAL ASPECTS

Armstrong, David. Clinical sense and clinical science. *Social Science and Medicine* 11(11-13): 599-601, Sep 1977. 28 refs. BE06494.
> attitudes; authoritarianism; *control groups; *historical aspects; *human experimentation; physicians

HUMAN EXPERIMENTATION / DEPARTMENT OF THE ARMY / LSD

Army moving slowly with LSD followup. *New York Times,* 19 Oct 1977, p. A17. BE06390.
> federal government; government agencies; *human experimentation; *LSD; military personnel; *Department of the Army

Cannon, Lou. Army gave private LSD in 3-month interrogation in 1961. *Washington Post,* 8 Oct 1977, p. A4. BE06389.
> federal government; government agencies; *human experimentation; *LSD; military personnel; *Department of the Army

Carlsen, William. Army data describe LSD test on soldier. *New York Times,* 7 Oct 1977, p. A13. BE06388.
> federal government; government agencies; *human experimentation; *LSD; military personnel; *Department of the Army

HUMAN EXPERIMENTATION / DEPARTMENT OF THE NAVY

Navy issues research guidelines. *U.S. Medicine* 12(10): 1+, 15 May 1976. BE06205.
> ethics committees; federal government; government agencies; human experimentation; informed consent; prisoners; *Department of the Navy

HUMAN EXPERIMENTATION / DRUG INDUSTRY / VOLUNTEERS

Butler, J.K. Is it ethical to conduct volunteer studies within the pharmaceutical industry? *Lancet* 1(8068): 816-818, 15 Apr 1978. 8 fn. BE06837.
> *drug industry; *employment; ethics committees; *human experimentation; informed consent; investigators; motivation; program descriptions; remuneration; *volunteers

HUMAN EXPERIMENTATION / DRUGS

American Academy of Pediatrics. Committee on Drugs. Guidelines for the ethical conduct of studies to evaluate drugs in pediatric populations. *Pediatrics* 60(1): 91-101, Jul 1977. 1 ref. BE05718.
> adolescents; age; attitudes; *children; disclosure; compensation; *drugs; ethical review; *ethics committees; evaluation; human development; *human experimentation; infants; *informed consent; institutionalized persons; investigator subject relationship; *investigators; mentally retarded; nontherapeutic research; pediatrics; placebos; psychological stress; remuneration; research design; *risks and benefits; selection of subjects; socioeconomic factors; *standards; therapeutic research; third party consent

Arnold, John D. Ethical considerations in selecting subject populations for drug research. *In* Barber, Bernard, ed. Medical Ethics and Social Change. Philadelphia: American Academy of Political and Social Science, 1978. p. 111-115. Annals of the American Academy of Political and Social Science, Vol. 437, May 1978. ISBN 0-87761-226-9. BE07003.
> compensation; *drugs; *human experimentation; risks and benefits; *selection of subjects

Curran, William J., ed. Rights and Responsibilities in Drug Research. Washington: Medicine in the Public Interest, 1977. 84 p. Report of an interdisciplinary conference sponsored by Medicine in the Public Interest, Dedham, Mass., 18-20 Mar 1976. Annotated Bibliography: p. 76-84. BE06537.
> carcinogens; compensation; *disclosure; *drugs; ethics committees; federal government; government agencies; government regulation; *human experimentation; industry; *informed consent; investigators; *legal aspects; legal liability; *moral obligations; *sponsoring agencies; Food and Drug Administration

Vere, D.W. Ethics of clinical trials. *In* Good, C.S., ed. The Principles and Practice of Clinical Trials. Edinburgh: Churchill Livingstone, 1976. p. 3-12. Distributed by Longman, New York. 13 refs. ISBN 0-443-01525-2. BE06665.
> *drugs; goals; government regulation; *human experimentation; informed consent; placebos

HUMAN EXPERIMENTATION / DRUGS / CENTRAL INTELLIGENCE AGENCY

Scientists back CIA behaviour control. *New Scientist* 75(1064): 340, 11 Aug 1977. BE05766.
> *behavior control; *drugs; *federal government;

BE = bioethics accession number   fn. = footnotes   refs. = references

*government agencies; *human experimentation; *Central Intelligence Agency

HUMAN EXPERIMENTATION / EPIDEMIOLOGY

Susser, Mervyn; Stein, Zena; Kline, Jennie. Ethics in epidemiology. *In* Barber, Bernard, ed. Medical Ethics and Social Change. Philadelphia: American Academy of Political and Social Science, 1978. p. 128-141. Annals of the American Academy of Political and Social Science, Vol. 437, May 1978. 21 fn. ISBN 0-87761-226-9. BE07005.
  confidentiality; control groups; *epidemiology; *human experimentation; medical records; methods; nutrition; research design; risks and benefits; standards; withholding treatment

HUMAN EXPERIMENTATION / EPIDEMIOLOGY / LEGAL LIABILITY

Berger, Pamela M.; Stallones, Reuel A. Legal liability and epidemiologic research. *American Journal of Epidemiology* 106(3): 177-183, Sep 1977. 32 refs. BE05984.
  compensation; confidentiality; disclosure; *epidemiology; federal government; government regulation; *human experimentation; informed consent; injuries; *investigators; *legal liability; medical records; risks and benefits

HUMAN EXPERIMENTATION / ETHICAL REVIEW

Applications for ethical approval: a report by the Working Group in Current Medical/Ethical Problems, Northern Regional Health Authority. *Lancet* 1(8055): 87-89, 14 Jan 1978. 1 fn. BE06735.
  *ethical review; *ethics committees; evaluation; *human experimentation; statistics

HUMAN EXPERIMENTATION / ETHICS COMMITTEES

American Academy of Pediatrics. Committee on Drugs. Guidelines for the ethical conduct of studies to evaluate drugs in pediatric populations. *Pediatrics* 60(1): 91-101, Jul 1977. 1 ref. BE05718.
  adolescents; age; attitudes; *children; disclosure; compensation; *drugs; ethical review; *ethics committees; evaluation; human development; *human experimentation; infants; *informed consent; institutionalized persons; investigator subject relationship; *investigators; mentally retarded; nontherapeutic research; pediatrics; placebos; psychological stress; remuneration; research design; *risks and benefits; selection of subjects; socioeconomic factors; *standards; therapeutic research; third party consent

Gray, Bradford H. The functions of human subjects review committees. *American Journal of Psychiatry* 134(8): 907-910, Aug 1977. 12 refs. BE06241.

attitudes; ethical review; *ethics committees; evaluation; *human experimentation; informed consent; investigators; *politics; *public participation; risks and benefits; social impact; universities

Local institutional review boards. *Journal of Medical Education* 52(7): 604-606, Jul 1977. Editorial. BE06492.
  costs and benefits; ethical review; *ethics committees; *human experimentation; program descriptions

Turnbull, H.R. Consent procedures: a conceptual approach to protection without overprotection. *In* Mittler, Peter, ed. Research to Practice in Mental Retardation. Volume 1: Care and Intervention. Baltimore: University Park Press, 1977. p. 65-70. ISBN 0-8391-1122-3. BE06179.
  *ethics committees; *human experimentation; *informed consent; legal aspects; self determination

U.S. National Commission for the Protection of Human Subjects of Biomedical and Behavioral Research. National Commission for the Protection of Human Subjects of Biomedical and Behavioral Research. Transcript of the Meeting Proceedings (32d Meeting), 8-9 Jul 1977. Springfield, Va.: National Technical Information Service, 1977. 326 p. Report No. NCPHS/M-77/20. NTIS No. PB-270 857. BE06652.
  *children; epidemiology; *ethics committees; genetics; government agencies; *human experimentation; infants; informed consent; *institutionalized persons; *mentally handicapped; mentally ill; mentally retarded; research subjects; risks and benefits; self determination; *Central Intelligence Agency

U.S. National Commission for the Protection of Human Subjects of Biomedical and Behavioral Research. National Commission for the Protection of Human Subjects of Biomedical and Behavioral Research. Transcript of the Meeting Proceedings (31st Meeting), 10-11 Jun 1977. Springfield, Va.: National Technical Information Service, 1977. 336 p. Report No. NCPHS/M-77/18. NTIS No. PB-269 377. BE06651.
  *children; compensation; ethical review; *ethics committees; *evaluation; government agencies; *human experimentation; informed consent; institutionalized persons; legal aspects; mentally handicapped; research subjects; rights; risks and benefits; social control; Institutional Review Boards

U.S. National Commission for the Protection of Human Subjects of Biomedical and Behavioral Research. National Commission for the Protection of Human Subjects of Biomedical and Behavioral Research. Transcript of the Meeting Proceedings (33d Meeting), 12-13 Aug 1977. Springfield, Va.: National Technical Information Service, 1977. 335 p. Report No. NCPHS/M-77/21. NTIS No. PB-271 917. BE06873.

BE = bioethics accession number   fn. = footnotes   refs. = references

*children; conflict of interest; consent forms; dignity; ethical review; ethics; *ethics committees; health care delivery; *human experimentation; informed consent; injuries; *institutionalized persons; investigators; *mentally handicapped; *mentally ill; *mentally retarded; nontherapeutic research; research subjects; risks and benefits; social impact; therapeutic research; Central Intelligence Agency; *National Commission for Protection of Human Subjects

U.S. National Commission for the Protection of Human Subjects of Biomedical and Behavioral Research. National Commission for the Protection of Human Subjects of Biomedical and Behavioral Research. Transcript of the Meeting Proceedings (34th Meeting), 9-10 Sep 1977. Springfield, Va.: National Technical Information Service, 1977. 414 p. Report No. NCPHS/M-77/22. NTIS No. PB-272 887. BE06874.
   adults; children; ethical review; *ethics committees; federal government; financial support; *government regulation; *human experimentation; *informed consent; *institutionalized persons; judicial action; legal aspects; legal guardians; *mentally handicapped; mentally retarded; mentally ill; public policy; risks and benefits; selection of subjects; *standards; DHEW Guidelines; National Commission for Protection of Human Subjects

U.S. National Commission for the Protection of Human Subjects of Biomedical and Behavioral Research. National Commission for the Protection of Human Subjects of Biomedical and Behavioral Research. Transcript of the Meeting Proceedings (35th Meeting), 14-15 Oct 1977. Springfield, Va.: National Technical Information Service, 1977. 640 p. Report No. NCPHS/M-77/24. NTIS No. PB-275 903. Volume in two parts. BE06875.
   abortion on demand; adults; children; disadvantaged; ethical review; *ethics committees; federal government; financial support; government regulation; *health care delivery; homosexuals; *human experimentation; *informed consent; *institutionalized persons; mental health; *mentally handicapped; *minority groups; organizational policies; psychology; public participation; public policy; review committees; *risks and benefits; standards; sterilization; American Indians; Health Services Administration

HUMAN EXPERIMENTATION / ETHICS COMMITTEES / DENMARK

Riis, Povl. Planning of scientific-ethical committees. *British Medical Journal* 2(6080): 173-174, 16 Jul 1977. BE05732.
   ethical review; evaluation; *ethics committees; *human experimentation; legislation; program descriptions; research design; *Denmark

HUMAN EXPERIMENTATION / ETHICS COMMITTEES / HOSPITALS

Wielk, Carol A. Human experimentation: issues before the hospital administrator. *Hospital and Health Services Administration* 22(3): 4-25, Summer 1977. 6 refs. 90 fn. BE05939.
   *administrators; codes of ethics; *ethics committees; federal government; government regulation; *hospitals; *human experimentation; informed consent; institutional policies; mentally ill; prisoners

HUMAN EXPERIMENTATION / FETUSES

Friedman, Jane M. The federal fetal experimentation regulations: an establishment clause analysis. *Minnesota Law Review* 61(6): 961-1005, Jun 1977. 202 fn. BE05934.
   abortion; Christian ethics; ethicists; ethics committees; evolution; *federal government; *fetuses; *government regulation; *human experimentation; medicine; morality; *personhood; *religion; social impact; value of life; wedge argument; DHEW Guidelines; National Commission for Protection of Human Subjects

Hunter, Thomas H., et al. Fetal Research. [Videorecording]. Charlottesville, Va.: University of Virginia Medical Center, Health Sciences Library, 1976. 1 cassette; 60 min.; sound; black and white; 3/4 in. Tape of the Medical Center Hour, 4 Feb 1976. Panel discussion moderated by Thomas H. Hunter. BE07048.
   aborted fetuses; *fetuses; government regulation; *human experimentation; nontherapeutic research; legal aspects; *personhood; *risks and benefits; Supreme Court decisions; therapeutic research; value of life; viability; *National Commission for Protection of Human Subjects

Hunter, Thomas H., et al. Research Using "Live" Human Fetuses: When Is It Justifiable? [Videorecording]. Charlottesville, Va.: University of Virginia Medical Center, Health Sciences Library, 1974. 1 cassette; 60 min.; sound; black and white; 3/4 in. Tape of the Medical Center Hour, 1 Apr 1974. Panel discussion moderated by Thomas H. Hunter. BE07046.
   *aborted fetuses; abortion; attitudes; federal government; *fetuses; government regulation; *human experimentation; investigators; informed consent; parental consent; quality of life; religious beliefs; Supreme Court decisions; value of life

HUMAN EXPERIMENTATION / FETUSES / CHRISTIAN ETHICS

Quinn, James J. Christianity views fetal research. *Linacre Quarterly* 45(1): 55-63, Feb 1978. 8 fn. BE06793.
   aborted fetuses; abortion; *Christian ethics; *fetuses; future generations; *human experimenta-

tion; human rights; informed consent; justice; *moral obligations; natural law; *nontherapeutic research; parental consent; personhood; philosophy; pregnant women; third party consent; utilitarianism; Fletcher, Joseph; McCormick, Richard A.; Ramsey, Paul; Walters, LeRoy

HUMAN EXPERIMENTATION / FETUSES / LEGAL ASPECTS

Annas, George J.; Glantz, Leonard H.; Katz, Barbara F. Fetal research: the limited role of informed consent in protecting the unborn. *In their* Informed Consent to Human Experimentation: The Subject's Dilemma. Cambridge, Mass.: Ballinger Publishing Company, 1977. p. 195-213. 77 fn. ISBN 0-88410-147-9. BE06858.
aborted fetuses; abortion; *fetuses; *human experimentation; *legal aspects; legislation; nontherapeutic research; personhood; state government; therapeutic research

HUMAN EXPERIMENTATION / FETUSES / MORALITY

Humber, James M. Abortion, fetal research, and the law. *Social Theory and Practice* 4(2): 127-147, Spring 1977. 28 fn. BE06297.
aborted fetuses; *abortion; *alternatives; *artificial organs; constitutional amendments; *embryo transfer; *fetuses; government regulation; *human experimentation; informed consent; *legal aspects; legislation; *morality; nontherapeutic research; pain; parental consent; *placentas; rights; selective abortion; state government; therapeutic research; utilitarianism

HUMAN EXPERIMENTATION / FETUSES / RISKS AND BENEFITS

Markey, Kathleen. Federal regulation of fetal research: toward a public policy founded on ethical reasoning. *University of Miami Law Review* 31(3): 675-696, Spring 1977. 115 fn. BE05951.
aborted fetuses; drugs; abortion; *federal government; *fetuses; *government regulation; *human experimentation; legal rights; nontherapeutic research; parental consent; *risks and benefits; therapeutic research; viability

HUMAN EXPERIMENTATION / FREEDOM OF INFORMATION ACT

U.S. National Commission for the Protection of Human Subjects of Biomedical and Behavioral Research. Disclosure of Research Information under the Freedom of Information Act. Report and Recommendations. Washington: U.S. Department of Health, Education, and Welfare, 1977. 42 p. DHEW Publication No. (OS)77-0003. BE06173.
attitudes; biomedical research; confidentiality; *disclosure; ethics committees; federal government; financial support; government regulation; *human experimentation; informed consent; investigators; legal aspects; *peer review; privacy; professional organizations; public policy; risks and benefits; *Freedom of Information Act

HUMAN EXPERIMENTATION / GOVERNMENT REGULATION

Friedman, Jane M. The federal fetal experimentation regulations: an establishment clause analysis. *Minnesota Law Review* 61(6): 961-1005, Jun 1977. 202 fn. BE05934.
abortion; Christian ethics; ethicists; ethics committees; evolution; *federal government; *fetuses; *government regulation; *human experimentation; medicine; morality; *personhood; *religion; social impact; value of life; wedge argument; DHEW Guidelines; National Commission for Protection of Human Subjects

Greenberg, Daniel S. Getting a grip on 'rogue science'. *Washington Post,* 9 Aug 1977, p. A19. BE06392.
federal government; government agencies; *government regulation; *human experimentation; Central Intelligence Agency

Swazey, Judith P. Protecting the "animal of necessity": limits to inquiry in clinical investigation. *Daedalus* 107(2): 129-145, Spring 1978. 46 fn. BE06937.
biomedical research; codes of ethics; drugs; ethics committees; *federal government; government agencies; *government regulation; *human experimentation; informed consent; investigators; legal aspects; *self regulation; social control

U.S. Congress. Senate. A bill to amend the Public Health Service Act to establish the President's Commission for the Protection of Human Subjects of Biomedical and Behavioral Research, and for other purposes. S. 1893, 95th Cong., 1st Sess., 19 Jul 1977. 33 p. By Edward Kennedy, et al. Referred to the Committee on Human Resources. BE06471.
behavioral research; bioethical issues; confidentiality; ethical review; *ethics committees; evaluation; federal government; financial support; goals; government agencies; government regulation; health care delivery; *human experimentation; *informed consent; legal aspects; *program descriptions; records; research subjects; *President's Commission for Protection of Human Subjects

U.S. Energy Research and Development Administration. Title 10—Energy, Chapter III, Part 745—Protection of human subjects. *Federal Register* 42(115): 30492-30493, 15 Jun 1977. BE07105.
federal government; government agencies; *government regulation; *human experimentation; *Energy Research and Development Administration

U.S. Energy Research and Development Administration. Title 10—Energy, Chapter III, Part 745—Protection of human subjects. *Federal Register* 41(231):

BE = bioethics accession number   fn. = footnotes   refs. = references

52434-52438, 30 Nov 1976. BE07104.
confidentiality; ethical review; ethics committees; federal government; financial support; government agencies; *government regulation; *human experimentation; informed consent; *Energy Research and Development Administration

U.S. National Commission for the Protection of Human Subjects of Biomedical and Behavioral Research. National Commission for the Protection of Human Subjects of Biomedical and Behavioral Research. Transcript of the Meeting Proceedings (34th Meeting), 9-10 Sep 1977. Springfield, Va.: National Technical Information Service, 1977. 414 p. Report No. NCPHS/M-77/22. NTIS No. PB-272 887. BE06874.
adults; children; ethical review; *ethics committees; federal government; financial support; *government regulation; *human experimentation; *informed consent; *institutionalized persons; judicial action; legal aspects; legal guardians; *mentally handicapped; mentally retarded; mentally ill; public policy; risks and benefits; selection of subjects; *standards; DHEW Guidelines; National Commission for Protection of Human Subjects

HUMAN EXPERIMENTATION / GOVERNMENT REGULATION / DHEW GUIDELINES

Standridge, Linda W. Experimentation on humans in biomedical research: implications for the industry of recent legislation and cases. *Women Lawyers Journal* 63(3): 88-95, Summer 1977. BE05936.
case reports; compensation; ethics committees; *federal government; fetuses; *government regulation; *human experimentation; informed consent; legislation; psychosurgery; state government; *DHEW Guidelines; National Commission for Protection of Human Subjects

HUMAN EXPERIMENTATION / GOVERNMENT REGULATION / HISTORICAL ASPECTS

Zegel, Vikki A. Biomedical Ethics: Human Experimentation. Issue Brief Number IB74095. Washington: Library of Congress, Congressional Research Service, 20 Mar 1977. 9 p. Congressional Research Service Major Issues System. Originally published 23 Jul 1974; updated 20 Mar 1977. Available only through Members of Congress. 35 refs. BE06571.
*government regulation; *historical aspects; *human experimentation; legislation; *politics; National Commission for Protection of Human Subjects

HUMAN EXPERIMENTATION / GOVERNMENT REGULATION / RISKS AND BENEFITS

Markey, Kathleen. Federal regulation of fetal research: toward a public policy founded on ethical reasoning. *University of Miami Law Review* 31(3): 675-696, Spring 1977. 115 fn. BE05951.

aborted fetuses; drugs; abortion; *federal government; *fetuses; *government regulation; *human experimentation; legal rights; nontherapeutic research; parental consent; *risks and benefits; therapeutic research; viability

HUMAN EXPERIMENTATION / HISTORICAL ASPECTS

Musto, David F. Freedom of inquiry and subjects' rights: historical perspective. *American Journal of Psychiatry* 134(8): 893-896, Aug 1977. 6 refs. BE06237.
ancient history; attitudes; costs and benefits; dehumanization; freedom; *historical aspects; *human experimentation; human rights; investigators; literature; research subjects; science; Eighteenth Century; Nineteenth Century; Seventeenth Century; Twentieth Century

HUMAN EXPERIMENTATION / HUNTINGTON'S CHOREA

MacKay, Charles R.; Shea, John M. Ethical considerations in research on Huntington's disease. *Clinical Research* 25(4): 241-247, Oct 1977. 8 refs. BE06348.
carriers; diagnosis; *genetic counseling; *human experimentation; *Huntington's chorea; incidence; moral obligations; psychological stress; reproduction; *risks and benefits; selection of subjects

HUMAN EXPERIMENTATION / INFANTS

Gairdner, Douglas; Dodge, John A.; Evans, John. Research on infants. *Lancet* 1(8016): 852, 16 Apr 1977. 3 refs. Letter. BE05557.
children; *human experimentation; *infants; nontherapeutic research; risks and benefits; *venepuncture

Pratt, Helenor. Research on infants. *Lancet* 1(8020): 1052, 14 May 1977. 7 refs. Letter. BE05624.
*human experimentation; *infants; risks and benefits; venepuncture

HUMAN EXPERIMENTATION / INFANTS / EYE DISEASES

Silverman, William A. The lesson of retrolental fibroplasia. *Scientific American* 236(6): 100-107, Jun 1977. BE05667.
control groups; *eye diseases; *human experimentation; *infants; informed consent; morbidity; mortality; *newborns; nontherapeutic research; *prematurity; random selection; risks and benefits; values

HUMAN EXPERIMENTATION / INFORMED CONSENT

American Academy of Pediatrics. Committee on Drugs. Guidelines for the ethical conduct of studies

---

BE = bioethics accession number   fn. = footnotes   refs. = references

to evaluate drugs in pediatric populations. *Pediatrics* 60(1): 91-101, Jul 1977. 1 ref. BE05718.
adolescents; age; attitudes; *children; disclosure; compensation; *drugs; ethical review; *ethics committees; evaluation; human development; *human experimentation; infants; *informed consent; institutionalized persons; investigator subject relationship; *investigators; mentally retarded; nontherapeutic research; pediatrics; placebos; psychological stress; remuneration; research design; *risks and benefits; selection of subjects; socioeconomic factors; *standards; therapeutic research; third party consent

Annas, George J.; Glantz, Leonard H.; Katz, Barbara F. Law of informed consent in human experimentation: institutionalized mentally infirm. *In* U.S. National Commission for Protection of Human Subjects.... Research Involving Those Institutionalized as Mentally Infirm: Appendix. Washington: U.S. Government Printing Office, 1978. p. 3.1-3.75. DHEW Publication No. (OS) 78-0007. 198 fn. BE06898.
behavior control; *human experimentation; competence; *informed consent; *institutionalized persons; involuntary sterilization; legal aspects; *legal rights; *mentally handicapped; mentally ill; mentally retarded; nontherapeutic research; psychosurgery; third party consent; treatment refusal

Cohen, Carl. Medical experimentation on prisoners. *Perspectives in Biology and Medicine* 21(3): 357-372, Spring 1978. 6 refs. 4 fn. BE06811.
*coercion; costs and benefits; *human experimentation; incentives; *informed consent; institutionalized persons; moral obligations; paternalism; *prisoners; remuneration; socioeconomic factors

Curran, William J., ed. Rights and Responsibilities in Drug Research. Washington: Medicine in the Public Interest, 1977. 84 p. Report of an interdisciplinary conference sponsored by Medicine in the Public Interest, Dedham, Mass., 18-20 Mar 1976. Annotated bibliography: p. 76-84. BE06537.
carcinogens; compensation; *disclosure; *drugs; ethics committees; federal government; government agencies; government regulation; *human experimentation; industry; *informed consent; investigators; *legal aspects; legal liability; *moral obligations; *sponsoring agencies; Food and Drug Administration

De Wet, B.S. Clinical trials, children and the law. *South African Medical Journal* 51(16): 528, 16 Apr 1977. 2 refs. Letter. BE05550.
children; drugs; *human experimentation; *informed consent; parental consent

Donagan, Alan. Informed consent in therapy and experimentation. *Journal of Medicine and Philosophy* 2(4): 307-329, Dec 1977. 41 refs. 2 fn. BE06136.
Christian ethics; disclosure; *human experimentation; *informed consent; patient care; *physician

patient relationship; risks and benefits; utilitarianism

Goldstein, Joseph. On the right of the "institutionalized mentally infirm" to consent to or refuse to participate as subjects in biomedical and behavioral research. *In* U.S. National Commission for Protection of Human Subjects.... Research Involving Those Institutionalized as Mentally Infirm: Appendix. Washington: U.S. Government Printing Office, 1978. p. 2.1-2.39. DHEW Publication No. (OS) 78-0007. 4 fn. BE06897.
behavioral research; *coercion; *competence; dangerousness; disclosure; *human experimentation; *informed consent; *institutionalized persons; involuntary commitment; legal aspects; *mentally handicapped; self determination

Jackson, Jacquelyne J. Informed consent: ethical issues in behavioral research. *In* U.S. National Commission for Protection of Human Subjects.... Research Involving Those Institutionalized as Mentally Infirm: Appendix. Washington: U.S. Government Printing Office, 1978. p. 5.1-5.18. DHEW Publication No. (OS) 78-0007. Paper prepared for the National Minority Conference on Human Experimentation, 6-8 Jan 1976. 3 refs. BE06900.
*behavioral research; *blacks; federal government; government regulation; *human experimentation; *informed consent

Martindale, David. The ethics of experimentation. *New Physician* 26(7): 25-26, Jul 1977. BE06320.
children; *human experimentation; *informed consent; mentally handicapped; prisoners

Morgenbesser, Sidney. Experimentation and consent: a note. *In* Spicker, Stuart F.; Engelhardt, H. Tristram, eds. Philosophical Medical Ethics: Its Nature and Significance. Proceedings. Boston: D. Reidel, 1977. p. 97-110. Proceedings of the Third Trans-Disciplinary Symposium on Philosophy and Medicine, Farmington, Connecticut, 11-13 Dec 1975. 40 refs. 7 fn. ISBN 90-277-0772-3. BE05866.
drugs; ethics committees; *human experimentation; *informed consent; moral obligations; physicians

Sharpe, Gilbert. Consent and experimentation. *Canadian Medical Association Journal* 118(2): 194+, 21 Jan 1978. BE06911.
coercion; *disclosure; *human experimentation; *informed consent; legal aspects; patient care; physician's role; *risks and benefits

Turnbull, H.R. Consent procedures: a conceptual approach to protection without overprotection. *In* Mittler, Peter, ed. Research to Practice in Mental Retardation. Volume 1: Care and Intervention. Baltimore: University Park Press, 1977. p. 65-70. ISBN 0-8391-1122-3. BE06179.

BE = bioethics accession number   fn. = footnotes   refs. = references

*ethics committees; *human experimentation; *informed consent; legal aspects; self determination

U.S. Department of Health, Education, and Welfare. Research involving children: report and recommendations of the National Commission for the Protection of Human Subjects of Biomedical and Behavioral Research. *Federal Register* 43(9): 2084-2114, 13 Jan 1978. 81 fn. BE06913.

*children; ethical review; ethics committees; *human experimentation; *informed consent; legal aspects; moral obligations; nontherapeutic research; parental consent; risks and benefits; therapeutic research; National Commission for Protection of Human Subjects

U.S. National Commission for the Protection of Human Subjects of Biomedical and Behavioral Research. National Commission for the Protection of Human Subjects of Biomedical and Behavioral Research. Transcript of the Meeting Proceedings (37th Meeting), 9-10 Dec 1977. Springfield, Va.: National Technical Information Service, 1977. 360 p. NTIS No. PB-275 748. BE06877.

disclosure; ethical review; ethics committees; *human experimentation; *informed consent; *institutionalized persons; *mentally handicapped; risks and benefits

U.S. National Commission for the Protection of Human Subjects of Biomedical and Behavioral Research. Research Involving Those Institutionalized as Mentally Infirm: Appendix. Washington: U.S. Government Printing Office, 1978. 320 p. DHEW Publication No. (OS) 78-0007. References and footnotes included. BE06895.

behavior control; behavioral research; competence; decision making; *human experimentation; *informed consent; *institutionalized persons; involuntary commitment; legal rights; *mentally handicapped; minority groups; risks and benefits

HUMAN EXPERIMENTATION / INFORMED CONSENT / ALCOHOL ABUSE

Cowan, Dale H. Ethical considerations of metabolic research in alcoholic patients. *Alcoholism: Clinical and Experimental Research* 1(3): 193-198, Jul 1977. 14 refs. BE05723.

*alcohol abuse; coercion; competence; *human experimentation; hospitals; *informed consent; investigator subject relationship; investigators; moral obligations; nontherapeutic research; *patients; physician patient relationship; physician's role; physicians; research subjects; rights; *risks and benefits; *selection of subjects; socioeconomic factors; therapeutic research

HUMAN EXPERIMENTATION / INFORMED CONSENT / COMPREHENSION

Williams, Roger L., et al. The use of a test to determine that consent is informed. *Military Medicine* 141(7):

542-545, Jul 1977. 11 refs. BE05825.

*comprehension; disclosure; evaluation; *human experimentation; *informed consent; *prisoners

HUMAN EXPERIMENTATION / INFORMED CONSENT / DIGNITY

Lebacqz, Karen; Levine, Robert J. Respect for persons and informed consent to participate in research. *Clinical Research* 25(3): 101-107, Apr 1977. 27 refs. 8 fn. BE05806.

adults; children; codes of ethics; coercion; competence; comprehension; *dignity; disclosure; *human experimentation; *informed consent; injuries; institutionalized persons; moral obligations; physician patient relationship; research subjects; risks and benefits; self determination; third party consent; Nuremberg Code

HUMAN EXPERIMENTATION / INFORMED CONSENT / LEGAL ASPECTS

Annas, George J.; Glantz, Leonard H.; Katz, Barbara F. The current status of the law of informed consent to human experimentation. *In their* Informed Consent to Human Experimentation: The Subject's Dilemma. Cambridge, Mass.: Ballinger Publishing Company, 1977. p. 27-61. 144 fn. ISBN 0-88410-147-9. BE06857.

battery; consent forms; decision making; disclosure; federal government; government regulation; *human experimentation; *informed consent; *legal aspects; legislation; negligence; *physician patient relationship; risks and benefits; self determination; state government

Annas, George J.; Glantz, Leonard H.; Katz, Barbara F. Informed Consent to Human Experimentation: The Subject's Dilemma. Cambridge, Mass.: Ballinger Publishing Company, 1977. 333 p. References and footnotes included. ISBN 0-88410-147-9. BE06855.

children; compensation; fetuses; government regulation; *informed consent; *human experimentation; institutionalized persons; *legal aspects; mentally handicapped; parental consent; prisoners; psychosurgery

Annas, George J.; Glantz, Leonard H.; Katz, Barbara F. Origins of the law of informed consent to human experimentation. *In their* Informed Consent to Human Experimentation: The Subject's Dilemma. Cambridge, Mass.: Ballinger Publishing Company, 1977. p. 1-25. 68 fn. ISBN 0-88410-147-9. BE06856.

artificial organs; attitudes; codes of ethics; disclosure; historical aspects; *human experimentation; *informed consent; *legal aspects; legal liability; nontherapeutic research; organ transplantation; physicians; risks and benefits; Nuremberg Code

---

BE = bioethics accession number   fn. = footnotes   refs. = references

HUMAN EXPERIMENTATION / INFORMED CONSENT / LEGAL RIGHTS

U.S. National Commission for the Protection of Human Subjects of Biomedical and Behavioral Research. Legal issues. *In its* Research Involving Children: Report and Recommendations. Washington: U.S. Government Printing Office, 1977. p. 73-90. 37 fn. BE06194.

adolescents; behavioral research; *children; health care; *human experimentation; *legal rights; *informed consent; legislation; nontherapeutic research; organ donation; parental consent; risks and benefits; state government; therapeutic research

HUMAN EXPERIMENTATION / INFORMED CONSENT / RISKS AND BENEFITS

Gardner, E. Clinton. Ethical issues in the testing of new drugs in man. *Journal of Drug Issues* 7(3): 275-286, Summer 1977. 29 refs. 4 fn. BE05676.

disclosure; *drugs; ethicists; ethics committees; federal government; government regulation; hepatitis; *human experimentation; *informed consent; nontherapeutic research; obligations to society; *risks and benefits; third party consent; DHEW Guidelines

Gordon-Smith, Ian. Research on patients. *In* Vale, J.A., ed. Medicine and the Christian Mind. London: Christian Medical Fellowship Publications, 1975. p. 126-133. 6 refs. ISBN 85111-956-5. BE07074.

Christian ethics; disclosure; *human experimentation; *informed consent; investigators; motivation; *risks and benefits

Herbert, Victor. Acquiring new information while retaining old ethics. *Science* 198(4318): 690-693, 18 Nov 1977. 28 fn. BE06115.

carcinogens; case reports; disclosure; government regulation; *human experimentation; *informed consent; recombinant DNA research; *risks and benefits

U.S. National Commission for the Protection of Human Subjects of Biomedical and Behavioral Research. Research Involving Children: Report and Recommendations. Washington: U.S. Government Printing Office, 1977. 154 p. DHEW Publication No. (OS)77-0004. 77 fn. BE06193.

adolescents; attitudes; behavioral research; *children; ethicists; ethics committees; federal government; government agencies; *human experimentation; investigators; *informed consent; legal aspects; minority groups; moral obligations; nontherapeutic research; parental consent; public opinion; *risks and benefits; social impact; therapeutic research

HUMAN EXPERIMENTATION / INFORMED CONSENT / SEXUALITY

Kolodny, Robert C. Ethical requirements for sex research in humans: informed consent and general principles. *In* Masters, William H.; Johnson, Virginia E.; Kolodny, Robert C., eds. Ethical Issues in Sex Therapy and Research. Boston: Little, Brown, 1977. p. 52-83. Includes discussion by Ruth Macklin and others on p. 69-83. 9 refs. ISBN 0-316-549-835. BE06161.

adolescents; attitudes; children; confidentiality; disclosure; ethics committees; *human experimentation; *informed consent; morality; peer review; prisoners; research design; research subjects; *sexuality

HUMAN EXPERIMENTATION / INSTITUTIONALIZED PERSONS / MENTALLY HANDICAPPED

Annas, George J.; Glantz, Leonard H.; Katz, Barbara F. Law of informed consent in human experimentation: institutionalized mentally infirm. *In* U.S. National Commission for Protection of Human Subjects.... Research Involving Those Institutionalized as Mentally Infirm: Appendix. Washington: U.S. Government Printing Office, 1978. p. 3.1-3.75. DHEW Publication No. (OS) 78-0007. 198 fn. BE06898.

behavior control; *human experimentation; competence; *informed consent; *institutionalized persons; involuntary sterilization; legal aspects; *legal rights; *mentally handicapped; mentally ill; mentally retarded; nontherapeutic research; psychosurgery; third party consent; treatment refusal

Goldstein, Joseph. On the right of the "institutionalized mentally infirm" to consent to or refuse to participate as subjects in biomedical and behavioral research. *In* U.S. National Commission for Protection of Human Subjects.... Research Involving Those Institutionalized as Mentally Infirm: Appendix. Washington: U.S. Government Printing Office, 1978. p. 2.1-2.39. DHEW Publication No. (OS) 78-0007. 4 fn. BE06897.

behavioral research; *coercion; *competence; dangerousness; disclosure; *human experimentation; *informed consent; *institutionalized persons; involuntary commitment; legal aspects; *mentally handicapped; self determination

Tannenbaum, Arnold S.; Cooke, Robert A. University of Michigan. Survey Research Center. Report on the mentally infirm. *In* U.S. National Commission for Protection of Human Subjects.... Research Involving Those Institutionalized as Mentally Infirm: Appendix. Washington: U.S. Government Printing Office, 1978. p. 1.1-1.117. DHEW Publication No. (OS) 78-0007. 6 fn. BE06896.

attitudes; *behavioral research; consent forms; ethical review; ethics committees; *human experimentation; informed consent; *institutionalized persons; investigators; *mentally handicapped; nontherapeutic research; research subjects; risks

---

BE = bioethics accession number   fn. = footnotes   refs. = references

and benefits; selection of subjects; *statistics; therapeutic research; third party consent

U.S. Department of Health, Education, and Welfare. Protection of human subjects: research involving those institutionalized as mentally infirm; report and recommendations for public comment. *Federal Register* 43(53): 11328-11358, 17 Mar 1978. 5 refs. 56 fn. BE06932.

    beneficence; competence; ethical review; ethics committees; *human experimentation; informed consent; *institutionalized persons; justice; legal guardians; legal rights; *mentally handicapped; mentally ill; mentally retarded; risks and benefits; self determination; treatment refusal

U.S. National Commission for the Protection of Human Subjects of Biomedical and Behavioral Research. National Commission for the Protection of Human Subjects of Biomedical and Behavioral Research. Transcript of the Meeting Proceedings (32d Meeting), 8-9 Jul 1977. Springfield, Va.: National Technical Information Service, 1977. 326 p. Report No. NCPHS/M-77/20. NTIS No. PB-270 857. BE06652.

    *children; epidemiology; *ethics committees; genetics; government agencies; *human experimentation; infants; informed consent; *institutionalized persons; *mentally handicapped; mentally ill; mentally retarded; research subjects; risks and benefits; self determination; *Central Intelligence Agency

U.S. National Commission for the Protection of Human Subjects of Biomedical and Behavioral Research. National Commission for the Protection of Human Subjects of Biomedical and Behavioral Research. Transcript of the Meeting Proceedings (34th Meeting), 9-10 Sep 1977. Springfield, Va.: National Technical Information Service, 1977. 414 p. Report No. NCPHS/M-77/22. NTIS No. PB-272 887. BE06874.

    adults; children; ethical review; *ethics committees; federal government; financial support; *government regulation; *human experimentation; *informed consent; *institutionalized persons; judicial action; legal aspects; legal guardians; *mentally handicapped; mentally retarded; mentally ill; public policy; risks and benefits; selection of subjects; *standards; DHEW Guidelines; National Commission for Protection of Human Subjects

U.S. National Commission for the Protection of Human Subjects of Biomedical and Behavioral Research. National Commission for the Protection of Human Subjects of Biomedical and Behavioral Research. Transcript of the Meeting Proceedings (33d Meeting), 12-13 Aug 1977. Springfield, Va.: National Technical Information Service, 1977. 335 p. Report No. NCPHS/M-77/21. NTIS No. PB-271 917. BE06873.

    *children; conflict of interest; consent forms; dignity; ethical review; ethics; *ethics committees; health care delivery; *human experimentation; in-

formed consent; injuries; *institutionalized persons; investigators; *mentally handicapped; *mentally ill; *mentally retarded; nontherapeutic research; research subjects; risks and benefits; social impact; therapeutic research; Central Intelligence Agency; *National Commission for Protection of Human Subjects

U.S. National Commission for the Protection of Human Subjects of Biomedical and Behavioral Research. National Commission for the Protection of Human Subjects of Biomedical and Behavioral Research. Transcript of the Meeting Proceedings (37th Meeting), 9-10 Dec 1977. Springfield, Va.: National Technical Information Service, 1977. 360 p. NTIS No. PB-275 748. BE06877.

    disclosure; ethical review; ethics committees; *human experimentation; *informed consent; *institutionalized persons; *mentally handicapped; risks and benefits

U.S. National Commission for the Protection of Human Subjects of Biomedical and Behavioral Research. National Commission for the Protection of Human Subjects of Biomedical and Behavioral Research. Transcript of the Meeting Proceedings (36th Meeting), 11-12 Nov 1977. Springfield, Va.: National Technical Information Service, 1977. 450 p. Report No. NCPHS/M-77/25. NTIS No. PB-275 399. BE06876.

    children; ethical review; ethics committees; *human experimentation; informed consent; *institutionalized persons; *mentally handicapped; nontherapeutic research; risks and benefits; therapeutic research

U.S. National Commission for the Protection of Human Subjects of Biomedical and Behavioral Research. National Commission for the Protection of Human Subjects of Biomedical and Behavioral Research. Transcript of the Meeting Proceedings (35th Meeting), 14-15 Oct 1977. Springfield, Va.: National Technical Information Service, 1977. 640 p. Report No. NCPHS/M-77/24. NTIS No. PB-275 903. Volume in two parts. BE06875.

    abortion on demand; adults; children; disadvantaged; ethical review; *ethics committees; federal government; financial support; government regulation; *health care delivery; homosexuals; *human experimentation; *informed consent; *institutionalized persons; mental health; *mentally handicapped; *minority groups; organizational policies; psychology; public participation; public policy; review committees; *risks and benefits; standards; sterilization; American Indians; Health Services Administration

U.S. National Commission for the Protection of Human Subjects of Biomedical and Behavioral Research. Research Involving Those Institutionalized as Mentally Infirm: Report and Recommendations. Washington: U.S. Government Printing Office, 1978. 125 p. DHEW Publication No. (OS) 78-

---

BE = bioethics accession number   fn. = footnotes   refs. = references

0006. References and footnotes included. BE06894.

> ethical review; ethics committees; *human experimentation; informed consent; *institutionalized persons; legal aspects; *mentally handicapped; mentally ill; mentally retarded; nontherapeutic research; risks and benefits; therapeutic research; third party consent; treatment refusal

U.S. National Commission for the Protection of Human Subjects of Biomedical and Behavioral Research. Research Involving Those Institutionalized as Mentally Infirm: Appendix. Washington: U.S. Government Printing Office, 1978. 320 p. DHEW Publication No. (OS) 78-0007. References and footnotes included. BE06895.

> behavior control; behavioral research; competence; decision making; *human experimentation; *informed consent; *institutionalized persons; involuntary commitment; legal rights; *mentally handicapped; minority groups; risks and benefits

HUMAN EXPERIMENTATION / INVESTIGATORS / HISTORICAL ASPECTS

Reiser, Stanley J. Human experimentation and the convergence of medical research and patient care. *In* Barber, Bernard, ed. Medical Ethics and Social Change. Philadelphia: American Academy of Political and Social Science, 1978. p. 8-18. Annals of the American Academy of Political and Social Science, Vol. 437, May 1978. 45 fn. ISBN 0-87761-226-9. BE06996.

> *historical aspects; *human experimentation; investigator subject relationship; *investigators; medical ethics; patient care

HUMAN EXPERIMENTATION / LAETRILE

Lipsett, Mortimer B.; Fletcher, John C. Ethics of Laetrile clinical trials. *New England Journal of Medicine* 297(21): 1183-1184, 24 Nov 1977. Letter. BE06098.

> cancer; drugs; *human experimentation; medical ethics; *Laetrile

Moertel, Charles G. Clinical evaluation of Laetrile: two perspectives. A trial of Laetrile now. *New England Journal of Medicine* 298(4): 218-219, 26 Jan 1978. BE06721.

> cancer; drugs; evaluation; *human experimentation; judicial action; legislation; medical ethics; patient care; public opinion; self determination; volunteers; *Laetrile; National Cancer Institute

Relman, Arnold S. Laetrilomania—again. *New England Journal of Medicine* 298(4): 215-216, 26 Jan 1978. 5 refs. Editorial. BE06719.

> cancer; drugs; *evaluation; *human experimentation; medical records; *Laetrile; National Cancer Institute

HUMAN EXPERIMENTATION / LEGAL ASPECTS

Cooper, Cindy. The test culture: medical experimentation on prisoners. *New England Journal on Prison Law* 2(2): 261-313, Spring 1976. 254 fn. BE06635.

> coercion; contracts; disclosure; drugs; due process; *human experimentation; incentives; investigators; informed consent; *legal aspects; legal liability; legal rights; legislation; *prisoners; privacy; risks and benefits; torts

HUMAN EXPERIMENTATION / LSD

Thomas, Jo. C.I.A. sought to spray drug on partygoers. *New York Times,* 21 Sep 1977, p. A11. BE06386.

> *behavior control; federal government; government agencies; *human experimentation; *LSD; *Central Intelligence Agency

Treaster, Joseph B. Researchers say that students were among 200 who took LSD in tests financed by C.I.A. in early '50's. *New York Times,* 9 Aug 1977, p. 21. BE06383.

> *behavior control; federal government; government agencies; *human experimentation; *LSD; students; *Central Intelligence Agency

HUMAN EXPERIMENTATION / MENTAL HEALTH

Harper, Mary S. Ethical issues on mental health research from a minority perspective. *In* U.S. National Commission for Protection of Human Subjects.... Research Involving Those Institutionalized as Mentally Infirm: Appendix. Washington: U.S. Government Printing Office, 1978. p. 6.1-6.41. DHEW Publication No. (OS) 78-0007. Paper prepared for the National Minority Conference on Human Experimentation, 6-8 Jan 1976. 9 refs. 81 fn. BE06901.

> blacks; discrimination; health personnel; *human experimentation; investigators; *mental health; *minority groups; social problems

HUMAN EXPERIMENTATION / MICHIGAN

Michigan. Protection of human subjects in biomedical experimentation: provisions applicable to mental patients. *International Digest of Health Legislation* 28(2): 395-397, 1977. Administrative rules of the State Department of Mental Health, Aug 1975. BE06467.

> ethics committees; *human experimentation; informed consent; *Michigan

HUMAN EXPERIMENTATION / MILITARY PERSONNEL

U.S. Congress. House. A bill to limit use of prison inmates in medical research. H.R. 7051, 95th Cong., 1st Sess., 10 May 1977. 7 p. By Robert W. Kasten-

---

BE = bioethics accession number   fn. = footnotes   refs. = references

meier. Referred to the Committee on the Judiciary. BE06469.

> *human experimentation; *military personnel; *prisoners

HUMAN EXPERIMENTATION / MINORITY GROUPS

Harper, Mary S. Ethical issues on mental health research from a minority perspective. *In* U.S. National Commission for Protection of Human Subjects.... Research Involving Those Institutionalized as Mentally Infirm: Appendix. Washington: U.S. Government Printing Office, 1978. p. 6.1-6.41. DHEW Publication No. (OS) 78-0007. Paper prepared for the National Minority Conference on Human Experimentation, 6-8 Jan 1976. 9 refs. 81 fn. BE06901.

> blacks; discrimination; health personnel; *human experimentation; investigators; *mental health; *minority groups; social problems

HUMAN EXPERIMENTATION / MORAL OBLIGATIONS / RISKS AND BENEFITS

U.S. National Commission for the Protection of Human Subjects of Biomedical and Behavioral Research. Ethical issues. *In its* Research Involving Children: Report and Recommendations. Washington: U.S. Government Printing Office, 1977. p. 91-121. 40 fn. BE06195.

> *children; ethicists; *human experimentation; informed consent; *moral obligations; morality; nontherapeutic research; obligations to society; parental consent; *risks and benefits; self determination; therapeutic research

HUMAN EXPERIMENTATION / NATIONAL COMMISSION FOR PROTECTION OF HUMAN SUBJECTS

Human experimentation rules debated. *Science News* 111(15): 230-231, 9 Apr 1977. BE05560.

> children; drugs; federal government; government regulation; *human experimentation; mentally ill; prisoners; *National Commission for Protection of Human Subjects

HUMAN EXPERIMENTATION / NEWBORNS

Cross, K.W. Experimentation on children. *Lancet* 2(8043): 866, 22 Oct 1977. Letter. BE06086.

> children; *human experimentation; *newborns; parental consent; Great Britain

HUMAN EXPERIMENTATION / NURSES / LEGAL ASPECTS

Creighton, Helen. Legal concerns of nursing research. *Nursing Research* 26(5): 337-341, Sep-Oct 1977. 32 refs. BE06332.

> codes of ethics; drugs; federal government; govern-

ment regulation; *human experimentation; informed consent; investigators; *legal aspects; legal liability; *nurses

HUMAN EXPERIMENTATION / NURSING ETHICS

Armiger, Bernadette. Ethics of nursing research: profile, principles, perspective. *Nursing Research* 26(5): 330-336, Sep-Oct 1977. 72 refs. 1 fn. BE06329.

> codes of ethics; ethics committees; *human experimentation; human rights; informed consent; international aspects; investigators; nurses; *nursing ethics; values

Kratz, Charlotte. The ethics of research. *Nursing Mirror* 145(3): 17-20, 21 Jul 1977. 8 refs. BE05815.

> *codes of ethics; ethics committees; information dissemination; *human experimentation; investigator subject relationship; investigators; moral obligations; nurses; *nursing ethics; patient care; patients; professional organizations; professional patient relationship; research subjects; sponsoring agencies

Royal College of Nursing of the United Kingdom. Ethics Related to Research in Nursing. London: Royal College of Nursing of the United Kingdom, 1977. 6 refs. BE05803.

> confidentiality; *human experimentation; informed consent; investigators; moral obligations; nurses; *nursing ethics; patient care; sponsoring organizations; technical expertise

HUMAN EXPERIMENTATION / PARENTAL CONSENT

Abrams, Natalie. Medical experimentation: the consent of prisoners and children. *In* Spicker, Stuart F.; Engelhardt, H. Tristram, eds. Philosophical Medical Ethics: Its Nature and Significance. Proceedings. Boston: D. Reidel, 1977. p. 111-124. Proceedings of the Third Trans-Disciplinary Symposium on Philosophy and Medicine, Farmington, Connecticut, 11-13 Dec 1975. 6 refs. ISBN 90-277-0772-3. BE05867.

> *children; *coercion; *human experimentation; incentives; informed consent; morality; *parental consent; *prisoners; risks and benefits; self determination

Bartholome, William G. Central themes in the debate over involvement of infants and children in biomedical research: a critical examination. *In* Van Eys, Jan, ed. Research on Children: Medical Imperatives, Ethical Quandaries, and Legal Constraints. Baltimore: University Park Press, 1978. p. 69-76. Proceedings of a conference held at Houston, Texas, 29-30 Apr 1977, sponsored by the Univ. of Texas System Cancer Center, the Inst. of Religion, and the Univ. of Texas Health Science Center. 10 refs. ISBN 0-8391-1191-6. BE06990.

BE = bioethics accession number   fn. = footnotes   refs. = references

*children; *human experimentation; infants; informed consent; *parental consent

Ramsey, Paul. Ethical dimensions of experimental research on children. *In* Van Eys, Jan, ed. Research on Children: Medical Imperatives, Ethical Quandaries, and Legal Constraints. Baltimore: University Park Press, 1978. p. 57-68. Proceedings of a conference held at Houston, Texas, 29-30 Apr 1977, sponsored by the Univ. of Texas System Cancer Center, the Inst. of Religion, and the Univ. of Texas Health Science Center. 17 refs. 5 fn. ISBN 0-8391-1191-6. BE06989.
    *children; compensation; *human experimentation; moral obligations; nontherapeutic research; *parental consent; *risks and benefits

U.S. National Commission for the Protection of Human Subjects of Biomedical and Behavioral Research. National Commission for the Protection of Human Subjects of Biomedical and Behavioral Research. Transcript of the Meeting Proceedings (28th Meeting), 11-12 Mar 1977. Springfield, Va.: National Technical Information Service, 1977. 467 p. Report No. NCPHS/M-77/10. NTIS No. PB-265 847. BE06648.
    *children; confidentiality; disclosure; ethics committees; federal government; financial support; *human experimentation; informed consent; *parental consent; risks and benefits

Van Eys, Jan, ed. Research on Children: Medical Imperatives, Ethical Quandaries, and Legal Constraints. Baltimore: University Park Press, 1978. 152 p. Proceedings of a conference held at Houston, Texas, 29-30 Apr 1977, sponsored by the Univ. of Texas System Cancer Center, the Inst. of Religion, and the Univ. of Texas Health Science Center. 101 refs. 6 fn. ISBN 0-8391-1191-6. BE06987.
    adolescents; cancer; *children; drugs; federal government; government regulation; human development; *human experimentation; *informed consent; legal aspects; *parental consent; patient care; psychological stress; risks and benefits

HUMAN EXPERIMENTATION / PARENTAL CONSENT / LEGAL ASPECTS

Solnit, Albert J., et al. Responses to "Changes in the parent-child relationship...". *Journal of Autism and Childhood Schizophrenia* 7(1): 94-108, Mar 1977. 3 refs. 7 fn. BE05910.
    adolescents; behavioral research; *children; *decision making; ethical review; hospitals; *human experimentation; institutionalized persons; *involuntary commitment; *legal aspects; legal rights; mental institutions; *mentally handicapped; parent child relationship; *parental consent; patient care; right to treatment

Valid parental consent. *Lancet* 1(8026): 1346-1347, 25 Jun 1977. BE05673.

*children; *human experimentation; *legal aspects; *nontherapeutic research; *parental consent

HUMAN EXPERIMENTATION / PLACEBOS

Vrhovac, B. Placebo and its importance in medicine. *International Journal of Clinical Pharmacology* 15(4): 161-165, Apr 1977. 31 refs. BE05586.
    control groups; drugs; evaluation; *human experimentation; physician patient relationship; *placebos

HUMAN EXPERIMENTATION / POLITICS / HISTORICAL ASPECTS

Zegel, Vikki A. Biomedical Ethics: Human Experimentation. Issue Brief Number IB74095. Washington: Library of Congress, Congressional Research Service, 20 Mar 1977. 9 p. Congressional Research Service Major Issues System. Originally published 23 Jul 1974; updated 20 Mar 1977. Available only through Members of Congress. 35 refs. BE06571.
    *government regulation; *historical aspects; *human experimentation; legislation; *politics; National Commission for Protection of Human Subjects

HUMAN EXPERIMENTATION / PRESIDENT'S COMMISSION FOR PROTECTION OF HUMAN SUBJECTS

Kennedy, Edward M. S. 1893—a bill to amend the Public Health Service Act to establish the President's Commission for the Protection of Human Subjects.... *Congressional Record (Daily Edition)* 123(122): S12371-S12377, 19 Jul 1977. BE05960.
    ethical review; ethics committees; federal government; government agencies; government regulation; *human experimentation; *President's Commission for Protection of Human Subjects

U.S. Congress. Senate. President's Commission for the Protection of Human Subjects of Biomedical and Behavioral Research Act of 1978. S. 2579, 95th Cong., 2d Sess., 23 Feb 1978. 17 p. By Edward Kennedy, et al. Referred to the Committee on Human Resources. BE06718.
    *behavioral research; confidentiality; determination of death; ethics committees; federal government; genetic counseling; government agencies; *government regulation; *human experimentation; informed consent; privacy; *program descriptions; records; *President's Commission for Protection of Human Subjects

U.S. Congress. Senate. Committee on Human Resources. President's Commission for the Protection of Human Subjects of Biomedical and Behavioral Research Act of 1978. Report. Washington: U.S. Congress, Senate, 1978. 50 p. Senate Report No. 95-852, to accompany S. 2579, 95th Cong., 2d Sess. BE06902.

BE = bioethics accession number   fn. = footnotes   refs. = references

drugs; federal government; government agencies; *human experimentation; informed consent; program descriptions; risks and benefits; *President's Commission for Protection of Human Subjects

HUMAN EXPERIMENTATION / PRISONERS

Abrams, Natalie. Medical experimentation: the consent of prisoners and children. *In* Spicker, Stuart F.; Engelhardt, H. Tristram, eds. Philosophical Medical Ethics: Its Nature and Significance. Proceedings. Boston: D. Reidel, 1977. p. 111-124. Proceedings of the Third Trans-Disciplinary Symposium on Philosophy and Medicine, Farmington, Connecticut, 11-13 Dec 1975. 6 refs. ISBN 90-277-0772-3. BE05867.
　　*children; *coercion; *human experimentation; incentives; informed consent; morality; *parental consent; *prisoners; risks and benefits; self determination

Branson, Roy. Why use prisoners for drug testing? *Washington Post,* 14 Jul 1977, p. A27. BE06055.
　　*drugs; *federal government; *government regulation; *human experimentation; incentives; *prisoners; National Commission for Protection of Human Subjects

California. Biomedical and behavioral research. *California Penal Code (Deering),* Title 2.1, Sects. 3500-3524. Approved 1 Oct 1977. 6 p. BE06656.
　　*behavioral research; drugs; ethics committees; *human experimentation; informed consent; legal liability; *prisoners; *California

Cohen, Carl. Medical experimentation on prisoners. *Perspectives in Biology and Medicine* 21(3): 357-372, Spring 1978. 6 refs. 4 fn. BE06811.
　　*coercion; costs and benefits; *human experimentation; incentives; *informed consent; institutionalized persons; moral obligations; paternalism; *prisoners; remuneration; socioeconomic factors

U.S. Congress. House. A bill to limit use of prison inmates in medical research. H.R. 7051, 95th Cong., 1st Sess., 10 May 1977. 7 p. By Robert W. Kastenmeier. Referred to the Committee on the Judiciary. BE06469.
　　*human experimentation; *military personnel; *prisoners

U.S. Department of Health, Education, and Welfare. Protection of human subjects: proposed regulations on research involving prisoners. *Federal Register* 43(3): 1050-1053, 5 Jan 1978. BE06759.
　　ethics committees; *federal government; *government regulation; *human experimentation; informed consent; *prisoners; risks and benefits

Williams, Roger L., et al. The use of a test to determine that consent is informed. *Military Medicine* 141(7): 542-545, Jul 1977. 11 refs. BE05825.
　　*comprehension; disclosure; evaluation; *human experimentation; *informed consent; *prisoners

HUMAN EXPERIMENTATION / PRISONERS / CODES OF ETHICS

Bloomberg, Seth A.; Wilkins, Leslie T. Ethics of research involving human subjects in criminal justice. *Crime and Delinquency* 23(4): 435-444, Oct 1977. 26 fn. BE06085.
　　*codes of ethics; confidentiality; *human experimentation; informed consent; peer review; *prisoners; risks and benefits

HUMAN EXPERIMENTATION / PRIVACY

Scheinberg, I. Herbert. Protection of privacy of experimental subjects. *New England Journal of Medicine* 297(1): 64-65, 7 Jul 1977. 2 refs. Letter. BE05717.
　　*human experimentation; investigators; moral obligations; *privacy; *research subjects

HUMAN EXPERIMENTATION / PSYCHIATRY

Mufson, Michael; Dorwart, Robert A.; Eisenberg, Leon. Political torture and overpopulation. *New England Journal of Medicine* 297(1): 63-64, 7 Jul 1977. 12 refs. Letter. BE05716.
　　behavioral research; drugs; *human experimentation; mentally ill; military personnel; moral obligations; physician's role; population growth; *psychiatry; social determinants; social problems; Central Intelligence Agency

HUMAN EXPERIMENTATION / PSYCHIATRY / FREEDOM

Reiss, David. Freedom of inquiry and subjects' rights: an introduction. *American Journal of Psychiatry* 134(8): 891-892, Aug 1977. BE06236.
　　behavioral genetics; behavioral research; ethics committees; *freedom; *human experimentation; investigators; *legal rights; politics; *psychiatry; *research subjects

HUMAN EXPERIMENTATION / PSYCHOACTIVE DRUGS

Cole, Jonathan O. Research barriers in psychopharmacology. *American Journal of Psychiatry* 134(8): 896-898, Aug 1977. BE06238.
　　ethics committees; government regulation; *human experimentation; informal social control; *informed consent; mentally ill; mentally retarded; psychiatry; *psychoactive drugs

Crewdson, John M. Abuses in testing of drugs by C.I.A. to be panel focus. *New York Times,* 20 Sep 1977, p. 1+. BE06385.
　　*behavior control; federal government; government agencies; *human experimentation; program descriptions; *psychoactive drugs; *Central Intelligence Agency

Horrock, Nicholas M. Drugs tested by C.I.A. on mental patients. *New York Times,* 3 Aug 1977, p. A1+.

BE = bioethics accession number   fn. = footnotes   refs. = references

BE06379.
*behavior control; federal government; government agencies; *human experimentation; LSD; mentally ill; *psychoactive drugs; *Central Intelligence Agency

HUMAN EXPERIMENTATION / PSYCHOACTIVE DRUGS / GOVERNMENT AGENCIES

Szasz, Thomas. Patriotic poisoners: psychiatrists, physicians and the CIA drug program. *Humanist* 36(6): 5-7, Nov/Dec 1976. 21 fn. BE06646.
federal government; *government agencies; *human experimentation; LSD; physician's role; *psychiatry; *psychoactive drugs; *Central Intelligence Agency; *Department of the Army

HUMAN EXPERIMENTATION / RANDOM SELECTION

Kolata, Gina B. Clinical trials: methods and ethics are debated. *Science* 198(4322): 1127-1131, 16 Dec 1977. BE06141.
control groups; heart diseases; *human experimentation; preventive medicine; *random selection; *risks and benefits

Randomised clinical trials. *British Medical Journal* 1(6071): 1238-1239, 14 May 1977. 9 refs. BE05625.
*human experimentation; methods; *random selection; research design

HUMAN EXPERIMENTATION / RELIGION

Friedman, Jane M. The federal fetal experimentation regulations: an establishment clause analysis. *Minnesota Law Review* 61(6): 961-1005, Jun 1977. 202 fn. BE05934.
abortion; Christian ethics; ethicists; ethics committees; evolution; *federal government; *fetuses; *government regulation; *human experimentation; medicine; morality; *personhood; *religion; social impact; value of life; wedge argument; DHEW Guidelines; National Commission for Protection of Human Subjects

HUMAN EXPERIMENTATION / RESEARCH DESIGN

Miké, Valerie; Good, Robert A. Old problems, new challenges. *Science* 198(4318): 677-678, 18 Nov 1977. 12 fn. BE06117.
epidemiology; ethics committees; evaluation; *human experimentation; medical ethics; random selection; *research design

HUMAN EXPERIMENTATION / RESEARCH SUBJECTS / COMPENSATION

U.S. National Commission for the Protection of Human Subjects of Biomedical and Behavioral Research. National Commission for the Protection of Human Subjects of Biomedical and Behavioral Research. Transcript of the Meeting Proceedings (30th Meeting), 13-14 May 1977. Springfield, Va.: National Technical Information Service, 1977. 346 p. Report No. NCPHS/M-77/15. NTIS No. PB-268 100. BE06650.
*children; *compensation; ethical review; *ethics committees; government agencies; *informed consent; *human experimentation; institutionalized persons; legal guardians; nontherapeutic research; parental consent; *research subjects; *risks and benefits; suffering; therapeutic research; third party consent; Institutional Review Boards

HUMAN EXPERIMENTATION / RESEARCH SUBJECTS / LEGAL RIGHTS

Reiss, David. Freedom of inquiry and subjects' rights: an introduction. *American Journal of Psychiatry* 134(8): 891-892, Aug 1977. BE06236.
behavioral genetics; behavioral research; ethics committees; *freedom; *human experimentation; investigators; *legal rights; politics; *psychiatry; *research subjects

HUMAN EXPERIMENTATION / RIGHTS

Kobler, Arthur L. Civil liberties. *In* Wolman, Benjamin B., ed. International Encyclopedia of Psychiatry, Psychology, Psychoanalysis, and Neurology. Volume 3. New York: Van Nostrand Reinhold, 1977. p. 151-154. 4 refs. ISBN 0-918228-01-8. BE06564.
*behavior control; *confidentiality; *human experimentation; human rights; legal rights; medical records; operant conditioning; *rights; self determination; social control

HUMAN EXPERIMENTATION / RISKS AND BENEFITS / HISTORICAL ASPECTS

Geison, Gerald L. Pasteur's work on rabies: reexamining the ethical issues. *Hastings Center Report* 8(2): 26-33, Apr 1978. 47 fn. BE06842.
animal experimentation; *historical aspects; *human experimentation; immunization; methods; *risks and benefits; science; *Pasteur, Louis

HUMAN EXPERIMENTATION / SELECTION OF SUBJECTS

Arnold, John D. Ethical considerations in selecting subject populations for drug research. *In* Barber, Bernard, ed. Medical Ethics and Social Change. Philadelphia: American Academy of Political and Social Science, 1978. p. 111-115. Annals of the American Academy of Political and Social Science, Vol. 437, May 1978. ISBN 0-87761-226-9. BE07003.
compensation; *drugs; *human experimentation; risks and benefits; *selection of subjects

---

BE = bioethics accession number   fn. = footnotes   refs. = references

HUMAN EXPERIMENTATION / SELECTION OF SUBJECTS / ALCOHOL ABUSE

Cowan, Dale H. Ethical considerations of metabolic research in alcoholic patients. *Alcoholism: Clinical and Experimental Research* 1(3): 193-198, Jul 1977. 14 refs. BE05723.
   *alcohol abuse; coercion; competence; *human experimentation; hospitals; *informed consent; investigator subject relationship; investigators; moral obligations; nontherapeutic research; *patients; physician patient relationship; physician's role; physicians; research subjects; rights; *risks and benefits; *selection of subjects; socioeconomic factors; therapeutic research

HUMAN EXPERIMENTATION / SELF REGULATION

Swazey, Judith P. Protecting the "animal of necessity": limits to inquiry in clinical investigation. *Daedalus* 107(2): 129-145, Spring 1978. 46 fn. BE06937.
   biomedical research; codes of ethics; drugs; ethics committees; *federal government; government agencies; *government regulation; *human experimentation; informed consent; investigators; legal aspects; *self regulation; social control

HUMAN EXPERIMENTATION / SEXUALITY

Masters, William H.; Johnson, Virginia E.; Kolodny, Robert C., eds. Ethical Issues in Sex Therapy and Research. Boston: Little, Brown, 1977. 227 p. Reproductive Biology Research Foundation Conference, 22-23 Jan 1976. 146 fn. ISBN 0-316-549-835. BE06160.
   behavioral research; confidentiality; data bases; *human experimentation; informed consent; patient care; physician patient relationship; privacy; religious ethics; research design; *sexuality

Money, John. Issues and attitudes in research and treatment of variant forms of human sexual behavior. *In* Masters, William H.; Johnson, Virginia E.; Kolodny, Robert C., eds. Ethical Issues in Sex Therapy and Research. Boston: Little, Brown, 1977. p. 119-142. Includes discussion by H. Tristram Engelhardt, Jr. and others on p. 132-142. 14 refs. ISBN 0-316-549-835. BE06163.
   decision making; homosexuals; *human experimentation; legal aspects; morality; physicians; sex offenses; *sexuality; XYY karyotype

HUMAN EXPERIMENTATION / SOCIAL CONTROL / INTERNATIONAL ASPECTS

Ioirysh, A.I. Law and the new potentials of biology. *Soviet Law and Government* 16(1): 40-54, Summer 1977. 12 fn. BE05678.
   biological warfare; codes of ethics; *human experimentation; *international aspects; investigators; legal aspects; *recombinant DNA research; *risks

and benefits; self regulation; *social control; therapeutic research; Nuremberg Code

HUMAN EXPERIMENTATION / SOCIAL CONTROL / SOCIAL IMPACT

Eisenberg, Leon. The social imperatives of medical research. *Science* 198(4322): 1105-1110, 16 Dec 1977. 43 fn. BE06137.
   *biomedical research; control groups; developing countries; diagnosis; historical aspects; *human experimentation; methods; morbidity; mortality; patient care; physicians; risks and benefits; *social control; *social impact; surgery; thalassemia

HUMAN EXPERIMENTATION / STATE UNIVERSITY OF NEW YORK AT ALBANY

Morehouse, Ward. Human research: questions raised. *Christian Science Monitor (Eastern Edition),* 23 Nov 1977, p. 1+. BE06458.
   ethics committees; *human experimentation; psychology; *universities; *State University of New York at Albany

Sheppard, Nathaniel. State U. experiments assailed by officials. *New York Times,* 24 Sep 1977, p. 45. BE06398.
   behavioral research; ethical review; *human experimentation; informed consent; psychology; universities; *State University of New York at Albany

Smith, R. Jeffrey. Electroshock experiment at Albany violates ethics guidelines. *Science* 198(4315): 383-386, 28 Oct 1977. BE06094.
   coercion; ethics committees; federal government; financial support; *government regulation; *human experimentation; informed consent; psychology; state government; students; *universities; *State University of New York at Albany

Smith, R. Jeffrey. SUNY at Albany admits research violations. *Science* 198(4318): 708, 18 Nov 1977. BE06123.
   government agencies; government regulation; *human experimentation; state government; *universities; *State University of New York at Albany

Studies on humans raise controversy. *New York Times,* 9 Oct 1977, sect. 1, p. 43. BE06404.
   ethics committees; *human experimentation; informed consent; universities; *State University of New York at Albany

Tedeschi, James T.; Gallup, Gordon G. Human subjects research. *Science* 198(4322): 1099-1100, 16 Dec 1977. Letter. BE06152.
   ethics committees; *human experimentation; informed consent; legislation; psychology; state government; *universities; *State University of New York at Albany

---

BE = bioethics accession number   fn. = footnotes   refs. = references

HUMAN EXPERIMENTATION / STATISTICS / METHODS

Byar, David P. Sound advice for conducting clinical trials. *New England Journal of Medicine* 297(10): 553-554, 8 Sep 1977. 3 refs. Editorial. BE05774.
*human experimentation; *methods; random selection; *statistics

HUMAN EXPERIMENTATION / USSR

U.S. Congress. Senate. Committee on the Judiciary. Subcommittee to Investigate the Administration of the Internal Security Act.... Humans Used as Guinea Pigs in the Soviet Union. Hearing. Washington: U.S. Government Printing Office, 1976. 48 p. Hearing, 94th Cong., 2d Sess., 30 Mar 1976. BE06610.
*case reports; dissent; drugs; government agencies; *human experimentation; human rights; journalism; *Markish, Luba; *USSR

HUMAN EXPERIMENTATION / VOLUNTEERS / MOTIVATION

Novak, Ervin; Seckman, Clarence E.; Stewart, Raymond D. Motivations for volunteering as research subjects. *Journal of Clinical Pharmacology* 17(7): 365-371, Jul 1977. 7 refs. BE05822.
*drugs; employment; *human experimentation; *motivation; prisoners; students; *volunteers

HUMAN REPRODUCTION   *See* REPRODUCTION

HUMAN RIGHTS   *See also* LEGAL RIGHTS, PATIENTS' RIGHTS, RIGHTS, WOMEN'S RIGHTS

HUMAN RIGHTS   *See under*

BEHAVIOR CONTROL / HUMAN RIGHTS

INVOLUNTARY COMMITMENT / MENTALLY ILL / HUMAN RIGHTS

PERSONHOOD / FETAL DEVELOPMENT / HUMAN RIGHTS

PERSONHOOD / HUMAN RIGHTS / HARE, R.M.

REPRODUCTION / HUMAN RIGHTS

TREATMENT REFUSAL / HUMAN RIGHTS

HUMAN TISSUE ACT 1961   *See under*

ORGAN TRANSPLANTATION / CADAVERS / HUMAN TISSUE ACT 1961

HUMANISM   *See under*

ALLOWING TO DIE / HUMANISM

JUDICIAL ACTION / HUMANISM

MEDICAL ETHICS / HUMANISM

HUMANNESS   *See* PERSONHOOD

HUNGARY   *See under*

GENETIC COUNSELING / HUNGARY

HUNTINGTON'S CHOREA   *See under*

GENETIC COUNSELING / HUNTINGTON'S CHOREA / RISKS AND BENEFITS

HUMAN EXPERIMENTATION / HUNTINGTON'S CHOREA

HYDE AMENDMENT   *See under*

ABORTION / FINANCIAL SUPPORT / HYDE AMENDMENT

ABORTION / HYDE AMENDMENT

HYPERKINESIS   *See under*

BEHAVIOR CONTROL / CHILDREN / HYPERKINESIS

HYPERTENSION   *See under*

RESOURCE ALLOCATION / HYPERTENSION / COSTS AND BENEFITS

ICELAND   *See under*

ABORTION / ICELAND

STERILIZATION / ICELAND

ILLINOIS   *See under*

CONFIDENTIALITY / MEDICAL RECORDS / ILLINOIS

IMMUNIZATION / CHILDREN / COMPENSATION

Teff, Harvey. Compensating vaccine-damaged children. *New Law Journal* 127(5819): 904-905, 15 Sep 1977. 13 fn. BE06339.
*children; *compensation; disclosure; *immunization; industry; *legal liability; negligence; physicians; risks and benefits

IMMUNIZATION / COMPENSATION

Franklin, Marc A.; Mais, Joseph E. Tort law and mass immunization programs: lessons from the polio and flu episodes. *California Law Review* 65(4): 754-775, Jul 1977. 83 fn. BE05958.
alternatives; *compensation; disclosure; drugs; *federal government; *immunization; industry; *influenza; *legal liability; *poliomyelitis; legislation; public health; risks and benefits; state government; *torts; Reyes v. Wyeth Laboratories

IMMUNIZATION / DISCLOSURE / RISKS AND BENEFITS

Moore, William E. Duty to warn extended to bystander in close contact with polio vaccinee. *Mercer Law Review* 29(2): 643-647, Winter 1978. 27 fn. BE06746.

BE = bioethics accession number   fn. = footnotes   refs. = references

compensation; *disclosure; *drug industry; *immunization; injuries; *legal obligations; poliomyelitis; *risks and benefits; Givens v. Lederle

IMMUNIZATION / FEDERAL GOVERNMENT / LEGAL LIABILITY

U.S. District Court, M.D. Tennessee, Northeastern Division. Wolfe v. Merrill National Laboratories, Inc. 13 Jun 1977. *Federal Supplement* 433: 231-238. 5 fn. BE06276.
  compensation; drugs; *federal government; *immunization; industry; *influenza; informed consent; *legal liability; negligence; voluntary programs

IMMUNIZATION / INFLUENZA / LEGAL LIABILITY

Franklin, Marc A.; Mais, Joseph E. Tort law and mass immunization programs: lessons from the polio and flu episodes. *California Law Review* 65(4): 754-775, Jul 1977. 83 fn. BE05958.
  alternatives; *compensation; disclosure; drugs; *federal government; *immunization; industry; *influenza; *legal liability; *poliomyelitis; legislation; public health; risks and benefits; state government; *torts; Reyes v. Wyeth Laboratories

IMMUNIZATION / INFORMED CONSENT / LEGAL ASPECTS

Woolley, Andrea P. Informed consent to immunization: the risks and benefits of individual autonomy. *California Law Review* 65(6): 1286-1314, Dec 1977. 127 fn. BE06854.
  *consent forms; *disclosure; *immunization; influenza; *informed consent; *legal aspects; mandatory programs; physician's role; *risks and benefits; self determination; standards

IMMUNIZATION / LEGAL LIABILITY

Trout, Monroe E. Immunizations—a societal dilemma. *New York Law School Law Review* 22(4): 943-960, 1977. 86 fn. BE06070.
  common good; disclosure; drugs; federal government; *immunization; industry; influenza; *legal liability; legislation; poliomyelitis; public health; risks and benefits; torts

IMMUNIZATION / MANDATORY PROGRAMS

Altman, Lawrence K. Carter's immunization plan. *New York Times,* 11 Apr 1977, p. 14. BE06012.
  *immunization; informed consent; legislation; *mandatory programs; public policy; state government

IMMUNIZATION / POLIOMYELITIS / RISKS AND BENEFITS

U.S. Court of Appeals, Fifth Circuit. Givens v. Lederle. 8 Aug 1977. *Federal Reporter, 2d Series,* 556: 1341-1346. 2 fn. BE06274.
  children; compensation; *disclosure; drugs; *immunization; industry; parents; physicians; *poliomyelitis; *risks and benefits

IMMUNIZATION / POLIOMYELITIS / TORTS

Franklin, Marc A.; Mais, Joseph E. Tort law and mass immunization programs: lessons from the polio and flu episodes. *California Law Review* 65(4): 754-775, Jul 1977. 83 fn. BE05958.
  alternatives; *compensation; disclosure; drugs; *federal government; *immunization; industry; *influenza; *legal liability; *poliomyelitis; legislation; public health; risks and benefits; state government; *torts; Reyes v. Wyeth Laboratories

IMMUNIZATION / PUBLIC POLICY

Immunization experts foresee problems. *Medical World News* 18(9): 31-32, 2 May 1977. BE05619.
  communicable diseases; *immunization; informed consent; legal liability; *public policy; risks and benefits

IMMUNIZATION / PUBLIC POLICY / POLIOMYELITIS

Nightingale, Elena O. Recommendations for a national policy on poliomyelitis vaccination. *New England Journal of Medicine* 297(5): 249-253, 4 Aug 1977. 7 refs. 13 fn. BE05762.
  alternatives; biomedical research; *immunization; incidence; informed consent; *poliomyelitis; *public policy; *risks and benefits

IMMUNIZATION / RISKS AND BENEFITS

Gillie, Oliver. Whooping cough: the facts parents have not been told. *Sunday Times (London),* 6 Mar 1977, p. 6. BE06007.
  communicable diseases; *immunization; morbidity; mortality; public policy; *risks and benefits; Great Britain

Grady, George F.; Wetterlow, Leslie H. Pertussis vaccine: reasonable doubt? *New England Journal of Medicine* 298(17): 966-967, 27 Apr 1978. 11 refs. Editorial. BE06843.
  compensation; *immunization; morbidity; *risks and benefits

Karzon, David T. Immunization on public trial. *New England Journal of Medicine* 297(5): 275-277, 4 Aug 1977. 2 refs. Editorial. BE05760.
  human experimentation; *immunization; morbidity; public policy; *risks and benefits

Ombudsman upholds parents' complaint. *Nursing Times* 73(44): 1692, 3 Nov 1977. 1 ref. BE06362.
  compensation; *disclosure; health personnel; *immunization; *risks and benefits; Great Britain

---

BE = bioethics accession number   fn. = footnotes   refs. = references

IN RE QUINLAN

Beresford, H. Richard. The Quinlan decision: problems and legislative alternatives. *Annals of Neurology* 2(1): 74-81, Jul 1977. 26 fn. BE05819.
*allowing to die; brain pathology; coma; decision making; determination of death; ethics committees; family members; judicial action; killing; legislation; living wills; physicians; terminally ill; withholding treatment; *In re Quinlan

IN RE QUINLAN   *See under*

ALLOWING TO DIE / IN RE QUINLAN

ALLOWING TO DIE / NEW JERSEY SUPREME COURT / IN RE QUINLAN

IN VITRO FERTILIZATION   *See also* EMBRYO TRANSFER

IN VITRO FERTILIZATION

British awaiting birth of a baby conceived in laboratory process. *New York Times,* 12 Jul 1978, p. A1+. BE07010.
embryo transfer; *in vitro fertilization; Great Britain

Cohn, Victor. Test-tube baby reported near; British press in bidding war for exclusive story. *Washington Post,* 12 Jul 1978, p. A1+. BE07011.
embryo transfer; *in vitro fertilization; mass media; Great Britain

Cohn, Victor. U.S. scientists cautious on baby-implant effort. *Washington Post,* 27 Jul 1978, p. A19. BE07023.
embryo transfer; ethics committees; federal government; fertility; financial support; *in vitro fertilization; risks and benefits

Curtin, Leah L.; Petrick, Joseph A. Reproductive manipulation: technical advances, options, and ethical ramifications. *Nursing Forum* 16(1): 6-25, 1977. 53 fn. BE06230.
artificial organs; attitudes; embryo transfer; host mothers; *in vitro fertilization; placentas; risks and benefits

Sullivan, Walter. New era in reproduction seen in British laboratory's embryo. *New York Times,* 15 Jul 1978, p. 1+. BE07012.
congenital defects; embryo transfer; *in vitro fertilization; methods; Great Britain

IN VITRO FERTILIZATION / BROWN, LOUISE

The first test-tube baby. *Time* 112(5): 58-59+, 31 Jul 1978. BE07026.
attitudes; embryo transfer; *in vitro fertilization; journalism; mass media; methods; risks and benefits; *Brown, Louise

Gwynne, Peter, et al. All about that baby. *Newsweek* 92(6): 66-72, 7 Aug 1978. BE07029.

animal experimentation; congenital defects; embryo transfer; fertility; *in vitro fertilization; methods; newborns; physician's role; products of in vitro fertilization; reproductive technologies; *Brown, Louise; Edwards, Robert G.; Steptoe, Patrick C.

Nossiter, Bernard. Test tube baby 'well': doctors predict more successes. *Washington Post,* 27 Jul 1978, p. A1+. BE07022.
embryo transfer; *in vitro fertilization; newborns; *Brown, Louise; Edwards, Robert G.; Steptoe, Patrick C.

Reed, Roy. Scientists praise British birth as triumph: early insertion of embryo into womb is linked to successful gestation. *New York Times,* 27 Jul 1978, p. A1+. BE07019.
embryo transfer; *in vitro fertilization; journalism; newborns; *Brown, Louise

Sullivan, Walter. Scientists praise British birth as triumph: doctors' success in conception in the laboratory intensifies the debate over reproductive control. *New York Times,* 27 Jul 1978, p. A1+. BE07020.
congenital defects; embryo transfer; *in vitro fertilization; methods; products of in vitro fertilization; *Brown, Louise; Edwards, Robert G.

Weintraub, Richard M. First test-tube baby born in British hospital. *Washington Post,* 26 Jul 1978, p. A1+. BE07016.
embryo transfer; females; *in vitro fertilization; newborns; products of in vitro fertilization; *Brown, Louise; Great Britain; Steptoe, Patrick C.

Woman gives birth to baby conceived outside the body. *New York Times,* 26 Jul 1978, p. A1+. BE07017.
*in vitro fertilization; methods; newborns; *Brown, Louise

IN VITRO FERTILIZATION / EMBRYO TRANSFER

Milunsky, Aubrey. Test-tube babies—a reality? *In his* Know Your Genes. Boston: Houghton Mifflin, 1977. p. 298-304. ISBN 0-395-25374-8. BE05888.
*embryo transfer; host mothers; *in vitro fertilization; legal aspects; ovum donors

Restak, Richard M. Can there be new forms of life before birth? *New York Times,* 16 Jul 1978, sect. 4, p. 8. BE07013.
congenital defects; *embryo transfer; *in vitro fertilization; social impact

Sullivan, Walter. Implants of monkeys may explain success with human embryo. *New York Times,* 25 Jul 1978, p. A1+. BE07015.
animal experimentation; *embryo transfer; *in vitro fertilization; Edwards, Robert G.; Great Britain; Steptoe, Patrick C.

BE = bioethics accession number   fn. = footnotes   refs. = references

IN VITRO FERTILIZATION / FINANCIAL SUPPORT

Cohn, Victor. HEW urged to support test-tube fertilization. *Washington Post,* 5 Aug 1978, p. A8. BE07028.
*federal government; *financial support; government regulation; *in vitro fertilization

IN VITRO FERTILIZATION / MORALITY

Hellegers, André E.; McCormick, Richard A. Unanswered questions on test tube life. *America* 139(4): 74-78, 12-19 Aug 1978. BE07030.
AID; AIH; abortion; *artificial insemination; biomedical technologies; embryo transfer; *in vitro fertilization; marital relationship; *morality; newborns; products of in vitro fertilization; reproduction; resource allocation; risks and benefits; *Roman Catholic ethics; social impact

IN VITRO FERTILIZATION / PHYSICIANS / LEGAL LIABILITY

Conrad, Miriam. Trial opening into loss of test-tube embryo. *Washington Post,* 17 Jul 1978, p. A11. BE07014.
*in vitro fertilization; *legal liability; *physicians; Del Zio, Doris

Footlick, Jerrold K.; Agrest, Susan. Test-tube bereavement. *Newsweek* 92(5): 70, 31 Jul 1978. BE07027.
*in vitro fertilization; *legal liability; married persons; *physicians; products of in vitro fertilization; psychological stress; Del Zio, Doris

IN VITRO FERTILIZATION / RELIGIOUS BELIEFS

Hyer, Marjorie. Theologians react cautiously to test-tube baby process. *Washington Post,* 28 Jul 1978, p. A3. BE07024.
*in vitro fertilization; products of in vitro fertilization; *religious beliefs; risks and benefits

Vecsey, George. Religious leaders differ on implant. *New York Times,* 27 Jul 1978, p. A16. BE07021.
AID; *attitudes; embryo transfer; *in vitro fertilization; *religious beliefs; Roman Catholic ethics

INCENTIVES *See under*

POPULATION CONTROL / INCENTIVES
POPULATION CONTROL / JUSTICE / INCENTIVES
POPULATION CONTROL / PUBLIC POLICY / INCENTIVES

INDIA *See under*

ABORTION / CONSCIENCE / INDIA
ABORTION / LEGISLATION / INDIA
ABORTION / SOCIAL IMPACT / INDIA
FAMILY PLANNING / INDIA

INDIA *See under (cont'd.)*

FAMILY PLANNING / PUBLIC POLICY / INDIA
INVOLUNTARY STERILIZATION / INDIA
INVOLUNTARY STERILIZATION / PUBLIC POLICY / INDIA
INVOLUNTARY STERILIZATION / ROMAN CATHOLIC ETHICS / INDIA
POPULATION CONTROL / INDIA
POPULATION CONTROL / PUBLIC POLICY / INDIA

INDIGENTS *See under*

INVOLUNTARY STERILIZATION / INDIGENTS

INDUSTRIAL MEDICINE *See under*

CODES OF ETHICS / INDUSTRIAL MEDICINE
CONFIDENTIALITY / MEDICAL RECORDS / INDUSTRIAL MEDICINE
MORAL OBLIGATIONS / INDUSTRIAL MEDICINE / HEALTH HAZARDS
REPRODUCTION / INDUSTRIAL MEDICINE / HEALTH HAZARDS

INDUSTRY *See under*

BIOMEDICAL TECHNOLOGIES / INDUSTRY / SOCIAL IMPACT
RECOMBINANT DNA RESEARCH / INDUSTRY
RECOMBINANT DNA RESEARCH / INDUSTRY / LEGAL ASPECTS

INFANTICIDE / HISTORICAL ASPECTS / ENGLAND

Damme, Catherine J. Infanticide: the worth of an infant under law. *Medical History* 22(1): 1-24, Jan 1978. 112 fn. BE06728.
abortion; allowing to die; *attitudes; congenital defects; criminal law; fetuses; health care; *historical aspects; *infanticide; *infants; judicial action; killing; *law; law enforcement; legal liability; legislation; *newborns; parental consent; parents; personhood; physicians; pregnant women; privacy; rights; state interest; Supreme Court decisions; treatment refusal; *value of life; *England; *United States

INFANTICIDE / HISTORICAL ASPECTS / UNITED STATES

Damme, Catherine J. Infanticide: the worth of an infant under law. *Medical History* 22(1): 1-24, Jan 1978. 112 fn. BE06728.
abortion; allowing to die; *attitudes; congenital defects; criminal law; fetuses; health care; *historical aspects; *infanticide; *infants; judicial action; killing; *law; law enforcement; legal liability; legislation; *newborns; parental consent; parents; personhood; physicians; pregnant women; privacy; rights; state interest; Supreme Court decisions;

treatment refusal; *value of life; *England; *United States

INFANTICIDE / LAW / HISTORICAL ASPECTS

Damme, Catherine J. Infanticide: the worth of an infant under law. *Medical History* 22(1): 1-24, Jan 1978. 112 fn. BE06728.
abortion; allowing to die; *attitudes; congenital defects; criminal law; fetuses; health care; *historical aspects; *infanticide; *infants; judicial action; killing; *law; law enforcement; legal liability; legislation; *newborns; parental consent; parents; personhood; physicians; pregnant women; privacy; rights; state interest; Supreme Court decisions; treatment refusal; *value of life; *England; *United States

INFANTS   See also CHILDREN, NEWBORNS

INFANTS   See under

HUMAN EXPERIMENTATION / INFANTS

HUMAN EXPERIMENTATION / INFANTS / EYE DISEASES

INFLUENZA   See under

IMMUNIZATION / INFLUENZA / LEGAL LIABILITY

INFORMAL SOCIAL CONTROL   See under

ABORTION / INFORMAL SOCIAL CONTROL / ROMAN CATHOLICISM

BEHAVIORAL GENETICS / INFORMAL SOCIAL CONTROL

INFORMED CONSENT   See also PARENTAL CONSENT, SPOUSAL CONSENT, THIRD PARTY CONSENT, TREATMENT REFUSAL

INFORMED CONSENT

Beauchamp, Tom L.; Walters, LeRoy, eds. Contemporary Issues in Bioethics. Encino, Calif.: Dickenson Publishing Company, 1978. 612 p. References and footnotes included. ISBN 0-8221-0200-5. BE06700.
*abortion; allowing to die; *behavior control; *bioethics; biomedical technologies; brain death; children; codes of ethics; confidentiality; decision making; deontological ethics; *determination of death; *euthanasia; fetuses; genetic intervention; *health; *health care; *human experimentation; *informed consent; involuntary commitment; justice; legal aspects; *medical ethics; natural law; *normative ethics; operant conditioning; patients' rights; personhood; psychosurgery; *resource allocation; rights; selection for treatment; treatment refusal; utilitarianism; Kaimowitz v. Department of Mental Health; Quinlan, Karen

Gray, Bradford H. Complexities of informed consent.

In Barber, Bernard, ed. Medical Ethics and Social Change. Philadelphia: American Academy of Political and Social Science, 1978. p. 37-48. Annals of the American Academy of Political and Social Science, Vol. 437, May 1978. 33 fn. ISBN 0-87761-226-9. BE06998.
attitudes; federal government; government regulation; human experimentation; immunization; *informed consent; legal aspects; paternalism; patient care; physicians; risks and benefits; self determination

Katz, Ronald L. Informed consent: *is it bad medicine?* *Western Journal of Medicine* 126(5): 426-428, May 1977. 2 refs. BE05621.
anesthesia; *disclosure; *informed consent; patients; recall; risks and benefits; surgery

MacKay, P.M., et al. Panel discussion on consent to treatment. *In* Papers and Proceedings of the National Conference on Health and the Law, Ottawa, 23-25 Sep 1975. Toronto: Canadian Hospital Association, 1975. p. 158-180. BE07068.
abortion; adolescents; case reports; disclosure; emergency care; *informed consent; legal aspects; parental consent; physician's role; spousal consent; sterilization; surgery

INFORMED CONSENT / ADOLESCENTS / CANADA

Rozovsky, Lorne E. Can a minor consent to treatment? *Dimensions in Health Service* 54(5): 10-12, May 1977. 13 fn. BE05628.
*adolescents; competence; *informed consent; *legislation; *patient care; surgery; *Canada

INFORMED CONSENT / ADOLESCENTS / LEGAL ASPECTS

Holder, Angela R. The minor's right to consent to medical treatment. *Connecticut Medicine* 41(9): 579-582, Sep 1977. BE05779.
*adolescents; children; age; competence; disclosure; drug abuse; emergency care; *health care; *informed consent; legal aspects; legal rights; legislation; parental consent; patient care; risks and benefits; state government; treatment refusal; venereal diseases

INFORMED CONSENT / ADOLESCENTS / LEGAL RIGHTS

Jamir, Vinson F. Children: health services for minors in Oklahoma—capacity to give self-consent to medical care and treatment. *Oklahoma Law Review* 30(2): 385-408, Spring 1977. 183 fn. BE05922.
abortion; *adolescents; contraception; disclosure; emergency care; *health care; health personnel; human experimentation; *informed consent; legal liability; *legal rights; legislation; medical fees;

BE = bioethics accession number   fn. = footnotes   refs. = references

parental consent; physicians; state government; *Oklahoma

INFORMED CONSENT / ANESTHESIA

Kucera, William R. What the nurse anesthetist should know about the law's view of informed consent. *Journal of the American Association of Nurse Anesthetists* 45(3): 309-311, Jun 1977. BE05935.

*anesthesia; disclosure; *informed consent; legal liability; *nurses; risks and benefits

INFORMED CONSENT / CHILDREN

American Academy of Pediatrics. Committee on Drugs. Guidelines for the ethical conduct of studies to evaluate drugs in pediatric populations. *Pediatrics* 60(1): 91-101, Jul 1977. 1 ref. BE05718.

adolescents; age; attitudes; *children; disclosure; compensation; *drugs; ethical review; *ethics committees; evaluation; human development; *human experimentation; infants; *informed consent; institutionalized persons; investigator subject relationship; *investigators; mentally retarded; nontherapeutic research; pediatrics; placebos; psychological stress; remuneration; research design; *risks and benefits; selection of subjects; socioeconomic factors; *standards; therapeutic research; third party consent

Holt, John. The right of children to informed consent. *In* Van Eys, Jan, ed. Research on Children: Medical Imperatives, Ethical Quandaries, and Legal Constraints. Baltimore: University Park Press, 1978. p. 5-16. Proceedings of a conference held at Houston, Texas, 29-30 Apr 1977, sponsored by the Univ. of Texas System Cancer Center, the Inst. of Religion, and the Univ. of Texas Health Science Center. 6 refs. ISBN 0-8391-1191-6. BE06988.

attitudes to death; *children; death; decision making; disclosure; *informed consent; parents; *patient care

U.S. Department of Health, Education, and Welfare. Research involving children: report and recommendations of the National Commission for the Protection of Human Subjects of Biomedical and Behavioral Research. *Federal Register* 43(9): 2084-2114, 13 Jan 1978. 81 fn. BE06913.

*children; ethical review; ethics committees; *human experimentation; *informed consent; legal aspects; moral obligations; nontherapeutic research; parental consent; risks and benefits; therapeutic research; National Commission for Protection of Human Subjects

INFORMED CONSENT / CHILDREN / GREAT BRITAIN

Skegg, P.D.G. English law relating to experimentation on children. *Lancet* 2(8041): 754-755, 8 Oct 1977. 9 refs. BE06093.

*children; *human experimentation; *informed consent; *legal aspects; nontherapeutic research; parental consent; *Great Britain

INFORMED CONSENT / COMMUNICATION / CONSENT FORMS

Vaccarino, James M. Consent, informed consent and the consent form. *New England Journal of Medicine* 298(8): 455, 23 Feb 1978. 2 refs. Editorial. BE06799.

*communication; *consent forms; disclosure; *informed consent; legal obligations; physician patient relationship; physicians; risks and benefits

INFORMED CONSENT / COMPETENCE

Sharpe, Gilbert. Consent and competency. *Canadian Medical Association Journal* 117(10): 1215-1216, 19 Nov 1977. BE06503.

age; *competence; diagnosis; family members; hospitals; *informed consent; legal aspects; legal guardians; mentally ill; physician's role; physicians; third party consent

INFORMED CONSENT / CONSENT FORMS

Fine, Arthur. Informed consent in California: latent liability without 'negligence'. *Western Journal of Medicine* 127(2): 158-163, Aug 1977. 6 refs. 5 fn. BE06322.

alternatives; *disclosure; *consent forms; *informed consent; *legal liability; malpractice; negligence; *physicians; recall; *risks and benefits

Levine, Robert J. Informed consent to participate in research: Part II. *Bioethics Digest* 1(12): 1-16, Apr 1977. 42 fn. BE05566.

*consent forms; disclosure; *ethics committees; federal government; government regulation; health personnel; human experimentation; *informed consent; investigators; *standards; DHEW Guidelines

INFORMED CONSENT / ETHICS COMMITTEES

Levine, Robert J. Informed consent to participate in research: Part II. *Bioethics Digest* 1(12): 1-16, Apr 1977. 42 fn. BE05566.

*consent forms; disclosure; *ethics committees; federal government; government regulation; health personnel; human experimentation; *informed consent; investigators; *standards; DHEW Guidelines

INFORMED CONSENT / EXPERT TESTIMONY / MALPRACTICE

Florida. District Court of Appeal, Second District. Thomas v. Berrios. 29 Jun 1977. *Southern Reporter, 2d Series,* 348: 905-910. BE06260.

disclosure; *expert testimony; *informed consent; *malpractice; negligence; physicians; surgery

---

BE = bioethics accession number   fn. = footnotes   refs. = references

INFORMED CONSENT / HOSPITALS / LEGAL ASPECTS

Horty, John F. Hospitals seen liable in consent case ruling. *Modern Healthcare* 7(7): 74+, Jul 1977. BE05727.
> blood transfusions; disclosure; hepatitis; *hospitals; *informed consent; *legal aspects; legal liability; legal obligations; legislation; physicians; Georgia

INFORMED CONSENT / INSTITUTIONALIZED PERSONS / MENTALLY HANDICAPPED

Annas, George J.; Glantz, Leonard H.; Katz, Barbara F. Law of informed consent in human experimentation: institutionalized mentally infirm. *In* U.S. National Commission for Protection of Human Subjects.... Research Involving Those Institutionalized as Mentally Infirm: Appendix. Washington: U.S. Government Printing Office, 1978. p. 3.1-3.75. DHEW Publication No. (OS) 78-0007. 198 fn. BE06898.
> behavior control; *human experimentation; competence; *informed consent; *institutionalized persons; involuntary sterilization; legal aspects; *legal rights; *mentally handicapped; mentally ill; mentally retarded; nontherapeutic research; psychosurgery; third party consent; treatment refusal

Goldstein, Joseph. On the right of the "institutionalized mentally infirm" to consent to or refuse to participate as subjects in biomedical and behavioral research. *In* U.S. National Commission for Protection of Human Subjects.... Research Involving Those Institutionalized as Mentally Infirm: Appendix. Washington: U.S. Government Printing Office, 1978. p. 2.1-2.39. DHEW Publication No. (OS) 78-0007. 4 fn. BE06897.
> behavioral research; *coercion; *competence; dangerousness; disclosure; *human experimentation; *informed consent; *institutionalized persons; involuntary commitment; legal aspects; *mentally handicapped; self determination

U.S. National Commission for the Protection of Human Subjects of Biomedical and Behavioral Research. National Commission for the Protection of Human Subjects of Biomedical and Behavioral Research. Transcript of the Meeting Proceedings (37th Meeting), 9-10 Dec 1977. Springfield, Va.: National Technical Information Service, 1977. 360 p. NTIS No. PB-275 748. BE06877.
> disclosure; ethical review; ethics committees; *human experimentation; *informed consent; *institutionalized persons; *mentally handicapped; risks and benefits

U.S. National Commission for the Protection of Human Subjects of Biomedical and Behavioral Research. Research Involving Those Institutionalized as Mentally Infirm: Appendix. Washington: U.S. Government Printing Office, 1978. 320 p. DHEW

Publication No. (OS) 78-0007. References and footnotes included. BE06895.
> behavior control; behavioral research; competence; decision making; *human experimentation; *informed consent; *institutionalized persons; involuntary commitment; legal rights; *mentally handicapped; minority groups; risks and benefits

INFORMED CONSENT / LEGAL ASPECTS

Beaty, Gene R.; Knapp, Thomas. Informed consent to medical treatment. *Air Force Law Review* 19(1): 63-75, Spring 1977. 42 fn. BE05917.
> *disclosure; emergency care; *informed consent; *legal aspects; military personnel; patients; physicians; recall; risks and benefits; third party consent

Creighton, Helen. The right to refuse treatment. *Supervisor Nurse* 8(2): 13-16, Feb 1977. 25 fn. BE05906.
> adults; blood transfusions; children; competence; *disclosure; *informed consent; *legal aspects; legal obligations; patients; physicians; religious beliefs; risks and benefits; self determination; *treatment refusal

Edelman, Alvin; Edelman, Leon F. Informed consent: the patient's dilemma and the doctor's problem. *International Surgery* 62(4): 220-224, Apr 1977. 30 fn. BE05553.
> battery; decision making; *disclosure; *informed consent; *legal aspects; *legal obligations; patients; *physicians; risks and benefits; self determination

Fierstein, Ronald K. Who's afraid of informed consent? An affirmative approach to the medical malpractice crisis. *Brooklyn Law Review* 44(2): 241-284, Winter 1978. 157 fn. BE06733.
> consent forms; *informed consent; disclosure; *legal aspects; legal liability; malpractice; negligence; patients; physicians; risks and benefits; self determination; standards

Foster, Henry W. Why bother with informed consent? *Journal of Medical Education* 53(2): 154-155, Feb 1978. Editorial. BE06777.
> disclosure; *informed consent; *legal aspects; physician's role

Grad, Frank P. Legal aspects of informed consent. *In* Swinyard, Chester A., ed. Decision Making and the Defective Newborn. Proceedings of a Conference on Spina Bifida and Ethics. Springfield, Ill.: Charles C. Thomas, 1978. p. 435-445. ISBN 0-398-03662-4. BE06960.
> children; *disclosure; extraordinary treatment; fetuses; human experimentation; *informed consent; *legal aspects; legal obligations; mentally handicapped; *newborns; organ donation; *parental consent; *patient care; physicians; selection for treatment; *spina bifida; third party consent; treatment refusal; withholding treatment

Hoffman, P.Browning, et al. Between Doctor and Pa-

BE = bioethics accession number   fn. = footnotes   refs. = references

tient: How Informed Must Consent Be? [Videorecording]. Charlottesville, Va.: University of Virginia Medical Center, Health Sciences Library, 1976. 1 cassette; 60 min.; sound; black and white; 3/4 in. Tape of the Medical Center Hour, 5 May 1976. Panel discussion moderated by Browning P. Hoffman. BE07049.

consent forms; *disclosure; *informed consent; *legal aspects; legal liability; patients; physician patient relationship; physicians; recall; risks and benefits; torts

Katz, Jay. Informed consent—a fairy tale? Law's vision. *University of Pittsburgh Law Review* 39(2): 137-174, Winter 1977. 113 fn. BE06526.

battery; decision making; *disclosure; *informed consent; *judicial action; *law; *legal aspects; legal liability; legal obligations; malpractice; negligence; patient participation; patients; physicians; *risks and benefits; *self determination; standards

Linden, A.M. The law of consent to treatment. *In* Papers and Proceedings of the National Conference on Health and the Law, Ottawa, 23-25 Sep 1975. Toronto: Canadian Hospital Association, 1975. p. 139-157. 43 fn. BE07067.

battery; disclosure; *informed consent; *legal aspects; negligence; parental consent; physicians

Martin, Andrew J. Consent to treatment. *Nursing Times* 73(22): 810-811, 2 Jun 1977. BE05655.

coercion; electroconvulsive therapy; emergency care; *informed consent; *legal aspects; mentally handicapped; parental consent; patient care

Sharpe, Gilbert. Consent to medical treatment. *Canadian Medical Association Journal* 117(6): 692-694+, 17 Sep 1977. BE06495.

battery; emergency care; *informed consent; *legal aspects; *legal liability; negligence; *physicians; surgery; torts; treatment refusal

INFORMED CONSENT / LEGAL LIABILITY

Morison, C. David. Informed consent or contractual absolution? The legitimacy of contracts removing liability for negligence in the delivery of medical care. *Journal of the Tennessee Medical Association* 71(4): 293-294, Apr 1978. BE06942.

*informed consent; *legal liability; *negligence; *physicians

INFORMED CONSENT / LEGAL LIABILITY / CANADA

Sharpe, Gilbert. Recent Canadian court decisions on consent. *Canadian Medical Association Journal* 117(12): 1421-1423, 17 Dec 1977. BE06586.

battery; *informed consent; disclosure; *legal liability; negligence; physicians; risks and benefits; *Canada

INFORMED CONSENT / LEGAL OBLIGATIONS

Ontario. High Court of Justice. Kelly v. Hazlett. 29 Jul 1976. *Dominion Law Reports, 3d Series,* 75: 536-567. BE06625.

*battery; *disclosure; *informed consent; *legal aspects; *legal obligations; *negligence; *physicians; *risks and benefits; standards; surgery; torts; Canada

INFORMED CONSENT / MALPRACTICE

Annas, George J. Avoiding malpractice suits through the use of informed consent. *In* Wecht, Cyril H., ed. Legal Medicine Annual 1977. New York: Appleton-Century-Crofts, 1977. p. 219-246. 37 refs. ISBN 0-8385-5655-8. BE06511.

attitudes; decision making; *disclosure; *informed consent; legal aspects; *malpractice; patients' rights; physicians

INFORMED CONSENT / PHYSICIAN PATIENT RELATIONSHIP

Donagan, Alan. Informed consent in therapy and experimentation. *Journal of Medicine and Philosophy* 2(4): 307-329, Dec 1977. 41 refs. 2 fn. BE06136.

Christian ethics; disclosure; *human experimentation; *informed consent; patient care; *physician patient relationship; risks and benefits; utilitarianism

INFORMED CONSENT / PHYSICIAN PATIENT RELATIONSHIP / LEGAL ASPECTS

Annas, George J.; Glantz, Leonard H.; Katz, Barbara F. The current status of the law of informed consent to human experimentation. *In their* Informed Consent to Human Experimentation: The Subject's Dilemma. Cambridge, Mass.: Ballinger Publishing Company, 1977. p. 27-61. 144 fn. ISBN 0-88410-147-9. BE06857.

battery; consent forms; decision making; disclosure; federal government; government regulation; *human experimentation; *informed consent; *legal aspects; legislation; negligence; *physician patient relationship; risks and benefits; self determination; state government

INFORMED CONSENT / PHYSICIANS

Maryland. Court of Appeals. Sard v. Hardy. 9 Nov 1977. *Atlantic Reporter, 2d Series,* 379: 1014-1027. 6 fn. BE07121.

alternatives; *disclosure; females; *informed consent; law; *legal aspects; legal liability; *legal obligations; legal rights; patients; *physicians; *risks and benefits; *standards; surgery; *voluntary sterilization

Quimby, Charles W. Informed consent: a dialogue. *Radiology* 123(3): 805-806, Jun 1977. BE05636.

*disclosure; *informed consent; *legal obligations;

BE = bioethics accession number   fn. = footnotes   refs. = references

*physicians; radiology; risks and benefits; standards

INFORMED CONSENT / PHYSICIANS / LEGAL ASPECTS

James, A. Everette; Johson, Burton A.; Hall, Donald J. Informed consent: some newer aspects and their relation to the specialty of radiology. *Radiology* 123(3): 809-813, Jun 1977. 12 refs. BE05638.
*disclosure; *informed consent; *legal aspects; legal liability; legal obligations; malpractice; negligence; *physicians; radiology; risks and benefits

INFORMED CONSENT / PHYSICIANS / LEGAL LIABILITY

Fine, Arthur. Informed consent in California: latent liability without 'negligence'. *Western Journal of Medicine* 127(2): 158-163, Aug 1977. 6 refs. 5 fn. BE06322.
alternatives; *disclosure; *consent forms; *informed consent; *legal liability; malpractice; negligence; *physicians; recall; *risks and benefits

Harney, David M. Authority for treatment. *In his* Medical Malpractice: 1977 Pocket Supplement. Indianapolis: Allen Smith, 1977. p. 11-23. Footnotes included. BE05881.
autopsies; battery; *disclosure; expert testimony; *informed consent; *legal liability; organ transplantation; *physicians; risks and benefits; surgery; treatment refusal; withholding treatment

INFORMED CONSENT / PRISONERS

Jobson, K. Consent to treatment—problems of prisoners. *In* Papers and Proceedings of the National Conference on Health and the Law, Ottawa, 23-25 Sep 1975. Toronto: Canadian Hospital Association, 1975. p. 134-137. BE07066.
coercion; incentives; *informed consent; *patient care; *prisoners

INFORMED CONSENT / PRISONERS / CALIFORNIA

California. Legislature. Assembly. An act...relating to drug therapy for prisoners. Assembly Bill No. 16, 1977-78 Regular Session, 7 Dec 1976. 13 p. By Assemblyman Alatorre, et al. Vetoed by Governor Edmund G. Brown, Jr., 28 Sep 1977. BE06594.
*informed consent; mentally ill; *prisoners; psychiatry; *psychoactive drugs; psychosurgery; *California

INFORMED CONSENT / PSYCHIATRY / LEGAL ASPECTS

Rada, Richard T. Informed consent in the care of psychiatric patients. *Journal of the National Association of Private Psychiatric Hospitals* 8(2): 9-12, Summer 1976. 16 refs. BE07099.

disclosure; historical aspects; human experimentation; *informed consent; *legal aspects; *psychiatry; risks and benefits

INFORMED CONSENT / PSYCHOACTIVE DRUGS / RISKS AND BENEFITS

Ayd, Frank J. Ethical and legal dilemmas posed by tardive dyskinesia. *Medical-Moral Newsletter* 14(8): 29-32, Oct 1977. BE06083.
central nervous system diseases; disclosure; *informed consent; legal aspects; *mentally ill; physicians; *psychoactive drugs; *risks and benefits; schizophrenia

INFORMED CONSENT / PSYCHOLOGICAL STRESS / LEGAL ASPECTS

Annas, George J.; Meisel, Alan; Horne, Rikki. Legal aspects of informed consent. *New England Journal of Medicine* 297(4): 228, 28 Jul 1977. 12 refs. Letter. BE05719.
*disclosure; *informed consent; *legal aspects; legal liability; *patients; physicians; *psychological stress; risks and benefits

INFORMED CONSENT / RADIOLOGY

Allen, Robert W. Informed consent: a medical decision (II). *Radiology* 123(3): 807, Jun 1977. 2 refs. BE05637.
*disclosure; *informed consent; legal aspects; moral obligations; physicians; *radiology

INFORMED CONSENT / RADIOLOGY / RISKS AND BENEFITS

Allen, Robert W.; Ochsner, Seymour. Informed consent for urography? *American Journal of Roentgenology* 129(2): 358-359, Aug 1977. 3 refs. Letter. BE05833.
*disclosure; *informed consent; legal aspects; *radiology; *risks and benefits

INFORMED CONSENT / RISKS AND BENEFITS / LEGAL ASPECTS

Ontario. High Court of Justice. Kelly v. Hazlett. 29 Jul 1976. *Dominion Law Reports, 3d Series,* 75: 536-567. BE06625.
*battery; *disclosure; *informed consent; *legal aspects; *legal obligations; *negligence; *physicians; *risks and benefits; standards; surgery; torts; Canada

INFORMED CONSENT / STANDARDS

Levine, Robert J. Informed consent to participate in research. *Bioethics Digest* 1(11): 1-13, Mar 1977. 49 fn. BE05565.
alternatives; behavioral research; comprehension; contracts; deception; *disclosure; ethics committees; human experimentation; *informed consent;

BE = bioethics accession number   fn. = footnotes   refs. = references

legal aspects; patient care; patients; research subjects; risks and benefits; self determination; *standards; third party consent

Levine, Robert J. Informed consent to participate in research: Part II. *Bioethics Digest* 1(12): 1-16, Apr 1977. 42 fn. BE05566.
*consent forms; disclosure; *ethics committees; federal government; government regulation; health personnel; human experimentation; *informed consent; investigators; *standards; DHEW Guidelines

INFORMED CONSENT / SURGERY / ALTERNATIVES

Washington. Court of Appeals, Division One. Archer v. Galbraith. 15 Aug 1977. *Pacific Reporter, 2d Series,* 567: 1155-1161. 2 fn. BE06272.
*alternatives; disclosure; *informed consent; legal obligations; malpractice; physicians; risks and benefits; *surgery

INFORMED CONSENT / SURGERY / RISKS AND BENEFITS

Massachusetts. Supreme Judicial Court, Norfolk. Schroeder v. Lawrence. 15 Feb 1977. *North Eastern Reporter, 2d Series,* 359: 1301-1303. 4 fn. BE06262.
disclosure; *informed consent; negligence; *risks and benefits; *surgery

INFORMED CONSENT   *See under*

BEHAVIORAL RESEARCH / INFORMED CONSENT

BEHAVIORAL RESEARCH / INFORMED CONSENT / EVALUATION

BEHAVIORAL RESEARCH / INFORMED CONSENT / RISKS AND BENEFITS

BEHAVIORAL RESEARCH / INFORMED CONSENT / SOCIAL IMPACT

ELECTROCONVULSIVE THERAPY / INFORMED CONSENT

HUMAN EXPERIMENTATION / INFORMED CONSENT

HUMAN EXPERIMENTATION / INFORMED CONSENT / ALCOHOL ABUSE

HUMAN EXPERIMENTATION / INFORMED CONSENT / COMPREHENSION

HUMAN EXPERIMENTATION / INFORMED CONSENT / DIGNITY

HUMAN EXPERIMENTATION / INFORMED CONSENT / LEGAL ASPECTS

HUMAN EXPERIMENTATION / INFORMED CONSENT / LEGAL RIGHTS

HUMAN EXPERIMENTATION / INFORMED CONSENT / RISKS AND BENEFITS

HUMAN EXPERIMENTATION / INFORMED CONSENT / SEXUALITY

INFORMED CONSENT   *See under (cont'd.)*

IMMUNIZATION / INFORMED CONSENT / LEGAL ASPECTS

NEGATIVE REINFORCEMENT / INFORMED CONSENT

PSYCHOTHERAPY / INFORMED CONSENT / CONSENT FORMS

STERILIZATION / INFORMED CONSENT / LEGAL ASPECTS

VOLUNTARY STERILIZATION / INFORMED CONSENT / LEGAL ASPECTS

INJURIES   *See under*

HUMAN EXPERIMENTATION / CADAVERS / INJURIES

INSTITUTE OF SOCIETY, ETHICS AND THE LIFE SCIENCES   *See under*

BIOETHICS / INSTITUTE OF SOCIETY, ETHICS AND THE LIFE SCIENCES

INSTITUTIONAL POLICIES / HOSPITALS / MEDICAL RECORDS

Melum, Mara M. Balancing information and privacy. *Hospital Progress* 58(7): 68-69+, Jul 1977. 6 fn. BE05730.
biomedical research; confidentiality; *disclosure; hospitals; *informed consent; institutional policies; *medical records; patient access; patients; privacy; professional organizations; *standards; statistics; Minnesota; Minnesota Hospital Association

INSTITUTIONAL POLICIES / HOSPITALS / TERMINALLY ILL

Shannon, Thomas A. Caring for the dying patient: what guidance from the guidelines? *Hastings Center Report* 7(3): 28-30, Jun 1977. BE05666.
*allowing to die; competence; *decision making; economics; ethics committees; *hospitals; family members; *institutional policies; patients' rights; physicians; prognosis; resuscitation; self determination; standards; *terminally ill; treatment refusal; *withholding treatment

INSTITUTIONAL POLICIES   *See under*

ABORTION / INSTITUTIONAL POLICIES

ABORTION / INSTITUTIONAL POLICIES / SUPREME COURT DECISIONS

VOLUNTARY STERILIZATION / INSTITUTIONAL POLICIES

INSTITUTIONALIZED PERSONS / LEGAL RIGHTS

U.S. Congress. Senate. Committee on the Judiciary. Subcommittee on the Constitution. Civil Rights of Institutionalized Persons. Hearings. Washington: U.S. Government Printing Office,

---

BE = bioethics accession number   fn. = footnotes   refs. = references

1977. 1138 p. Hearings before the Subcommittee, 95th Cong., 1st Sess., on S. 1393, a bill...to redress deprivations of constitutional...rights of institutionalized persons; 17, 22, 23, 30 Jun and 1 Jul 1977. BE06538.
> adults; case reports; children; federal government; health care; *institutionalized persons; *legal aspects; *legal rights; *legislation; *mental institutions; *mentally ill; *mentally retarded; organizational policies; patient advocacy; patient care; *prisoners; professional organizations; *public hospitals; residential facilities; standards; state government; Bronx Psychiatric Center; Caswell Training School; Department of Justice; Fairview State Hospital; Willowbrook State School

INSTITUTIONALIZED PERSONS / MENTALLY RETARDED / LEGAL RIGHTS

Friedman, Paul R. Rights of mentally retarded persons in institutions. *In his* The Rights of Mentally Retarded Persons: The Basic ACLU Guide for the Mentally Retarded Persons' Rights. New York: Avon Books, 1976. p. 57-95. 55 fn. ISBN 0-380-00868-8. BE06200.
> compensation; due process; employment; equal protection; freedom; *institutionalized persons; *legal rights; mental institutions; *mentally retarded; right to treatment; sexuality; standards; treatment refusal; Eighth Amendment

Logan, Harold J. Retarded patients' rights expanded. *Washington Post,* 25 Dec 1977, p. A13. BE06430.
> *institutionalized persons; involuntary commitment; *legal rights; *mental institutions; *mentally retarded

INSTITUTIONALIZED PERSONS   *See under*

ALLOWING TO DIE / INSTITUTIONALIZED PERSONS / MENTALLY RETARDED

BEHAVIOR CONTROL / INSTITUTIONALIZED PERSONS / LEGAL RIGHTS

BEHAVIORAL RESEARCH / INSTITUTIONALIZED PERSONS / MENTALLY HANDICAPPED

COERCION / INSTITUTIONALIZED PERSONS / MENTALLY ILL

COERCION / INSTITUTIONALIZED PERSONS / PENNSYLVANIA

CONFIDENTIALITY / INSTITUTIONALIZED PERSONS

ELECTROCONVULSIVE THERAPY / INSTITUTIONALIZED PERSONS / MENTALLY ILL

HUMAN EXPERIMENTATION / INSTITUTIONALIZED PERSONS / MENTALLY HANDICAPPED

INFORMED CONSENT / INSTITUTIONALIZED PERSONS / MENTALLY HANDICAPPED

PATIENT CARE / ETHICS COMMITTEES / INSTITUTIONALIZED PERSONS

PATIENT CARE / INSTITUTIONALIZED PERSONS / MODEL LEGISLATION

INSTITUTIONALIZED PERSONS   *See under* (*cont'd.*)

PATIENT CARE / PSYCHOACTIVE DRUGS / INSTITUTIONALIZED PERSONS

REVIEW COMMITTEES / INSTITUTIONALIZED PERSONS / PROGRAM DESCRIPTIONS

RIGHT TO TREATMENT / INSTITUTIONALIZED PERSONS

RIGHT TO TREATMENT / INSTITUTIONALIZED PERSONS / MENTALLY ILL

RIGHT TO TREATMENT / INSTITUTIONALIZED PERSONS / O'CONNOR V. DONALDSON

RIGHT TO TREATMENT / INSTITUTIONALIZED PERSONS / PENNSYLVANIA

TREATMENT REFUSAL / INSTITUTIONALIZED PERSONS / CALIFORNIA

TREATMENT REFUSAL / INSTITUTIONALIZED PERSONS / LEGAL RIGHTS

TREATMENT REFUSAL / INSTITUTIONALIZED PERSONS / MENTALLY ILL

INSURANCE   *See under*

CONFIDENTIALITY / PHYSICIAN PATIENT RELATIONSHIP / INSURANCE

INTELLIGENCE   *See under*

BEHAVIORAL GENETICS / INTELLIGENCE / MORALITY

INTENSIVE CARE UNITS   *See under*

BIOETHICAL ISSUES / INTENSIVE CARE UNITS

PATIENT CARE / INTENSIVE CARE UNITS

RESUSCITATION / INTENSIVE CARE UNITS / COSTS AND BENEFITS

SELECTION FOR TREATMENT / NEWBORNS / INTENSIVE CARE UNITS

INTERNATIONAL ASPECTS   *See under*

ABORTION / INTERNATIONAL ASPECTS

ABORTION / LEGISLATION / INTERNATIONAL ASPECTS

HEALTH CARE DELIVERY / INTERNATIONAL ASPECTS

HUMAN EXPERIMENTATION / SOCIAL CONTROL / INTERNATIONAL ASPECTS

RECOMBINANT DNA RESEARCH / SOCIAL CONTROL / INTERNATIONAL ASPECTS

INVESTIGATORS   *See under*

HUMAN EXPERIMENTATION / INVESTIGATORS / HISTORICAL ASPECTS

MEDICAL ETHICS / INVESTIGATORS

RECOMBINANT DNA RESEARCH / INVESTIGATORS / ATTITUDES

RECOMBINANT DNA RESEARCH / INVESTIGATORS / POLITICAL ACTIVITY

BE = bioethics accession number   fn. = footnotes   refs. = references

INVOLUNTARY COMMITMENT

Weisstub, David N. Involuntary commitment. *Canadian Psychiatric Association Journal* 22(4): 177-179, Jun 1977. 4 fn. BE05811.
 dangerousness; *involuntary commitment; legal aspects; legislation; *mentally ill; psychiatry; Canada; *Donaldson, Kenneth

INVOLUNTARY COMMITMENT / ADOLESCENTS / DUE PROCESS

California. Supreme Court, In Bank. In re Roger S. 15 Sep 1977. *Pacific Reporter, 2d Series,* 569: 1286-1302. 12 fn. BE07116.
 *adolescents; *due process; equal protection; *involuntary commitment; legal rights; mentally ill; parent child relationship; parental consent; voluntary admission

Curtiss, Zelda. Constitutional law—fourteenth amendment—due process—deprivation of children's rights—civil commitment. *Duquesne Law Review* 15(2): 337-347, Winter 1976-1977. 55 fn. BE07094.
 *adolescents; *due process; *involuntary commitment; legal rights; mentally ill; parent child relationship; *parental consent

INVOLUNTARY COMMITMENT / ADOLESCENTS / LEGAL ASPECTS

Miller, Derek; Burt, Robert A. On children's rights and therapeutic institutions. *In* Feinstein, Sherman C.; Giovacchini, Peter L., eds. Adolescent Psychiatry: Developmental and Clinical Studies. Volume 5. New York: Jason Aronson, 1977. p. 39-53. 16 refs. ISBN 0-87668-258-1. BE06576.
 *adolescents; *involuntary commitment; *legal aspects; legal rights; *mentally ill; parent child relationship; parental consent; physician patient relationship; psychiatry; psychological stress; residential facilities

INVOLUNTARY COMMITMENT / CASE REPORTS / USSR

Bloch, Sidney; Reddaway, Peter. Psychiatric Terror: How Soviet Psychiatry Is Used to Suppress Dissent. New York: Basic Books, 1977. 510 p. 613 fn. ISBN 0-465-06488-4. BE05876.
 *case reports; *dissent; historical aspects; international aspects; *involuntary commitment; legal aspects; mental health; mental institutions; mentally ill; professional organizations; psychiatric diagnosis; *psychiatry; religious beliefs; schizophrenia; *USSR

INVOLUNTARY COMMITMENT / CHILDREN / LEGAL RIGHTS

Can parents "volunteer" their children into mental institutions: an analysis of *Kremens v. Bartley* and *Par-*

ham v. J.L. and J.R. Mental Retardation and the Law : 1-8, Jan 1978. BE06724.
 *adolescents; *children; decision making; due process; *involuntary commitment; *legal rights; legislation; mental institutions; mentally retarded; *mentally ill; parent child relationship; parental consent; parents; private hospitals; public hospitals; *Supreme Court decisions; Georgia; *Kremens v. Bartley; *Parham v. J.L. and J.R.; *Pennsylvania

INVOLUNTARY COMMITMENT / CHILDREN / MENTALLY HANDICAPPED

Solnit, Albert J., et al. Responses to "Changes in the parent-child relationship...". *Journal of Autism and Childhood Schizophrenia* 7(1): 94-108, Mar 1977. 3 refs. 7 fn. BE05910.
 adolescents; behavioral research; *children; *decision making; ethical review; hospitals; *human experimentation; institutionalized persons; *involuntary commitment; *legal aspects; legal rights; mental institutions; *mentally handicapped; parent child relationship; *parental consent; patient care; right to treatment

INVOLUNTARY COMMITMENT / DANGEROUSNESS / HAWAII

Curran, William J. Psychiatric emergency commitments in Hawaii: tests of dangerousness. *New England Journal of Medicine* 298(5): 265-266, 2 Feb 1978. 5 refs. BE06774.
 *dangerousness; *involuntary commitment; judicial action; legislation; *mentally ill; standards; state government; *Hawaii

INVOLUNTARY COMMITMENT / DANGEROUSNESS / LEGAL ASPECTS

Groethe, Reed. Overt dangerous behavior as a constitutional requirement for involuntary civil commitment of the mentally ill. *University of Chicago Law Review* 44(3): 562-593, Spring 1977. 135 fn. BE06295.
 *dangerousness; *due process; *involuntary commitment; *legal aspects; legislation; mentally ill; prognosis; state government; *Supreme Court decisions; O'Connor v. Donaldson

INVOLUNTARY COMMITMENT / DISSENT / USSR

Bloch, Sidney; Reddaway, Peter. Psychiatric Terror: How Soviet Psychiatry Is Used to Suppress Dissent. New York: Basic Books, 1977. 510 p. 613 fn. ISBN 0-465-06488-4. BE05876.
 *case reports; *dissent; historical aspects; international aspects; *involuntary commitment; legal aspects; mental health; mental institutions; mentally ill; professional organizations; psychiatric diagnosis; *psychiatry; religious beliefs; schizophrenia; *USSR

BE = bioethics accession number   fn. = footnotes   refs. = references

Bloch, Sidney; Reddaway, Peter. Your disease is dissent! *New Scientist* 75(1061): 149-151, 21 Jul 1977. 1 fn. BE05722.
case reports; *dissent; *involuntary commitment; mental institutions; politics; professional organizations; psychiatric diagnosis; *psychiatry; schizophrenia; *USSR

Rich, Vera. Heading for Honolulu. *Nature* 268(5621): 578-579, 18 Aug 1977. BE05839.
*attitudes; *dissent; *involuntary commitment; politics; professional ethics; *professional organizations; psychiatry; *USSR; *World Psychiatric Association

Segal, Boris M. Soviet approaches to involuntary hospitalisation. *International Journal of Social Psychiatry* 23(2): 94-102, Summer 1977. 4 fn. BE06310.
case reports; *dissent; government regulation; *involuntary commitment; law enforcement; political activity; psychiatric diagnosis; *psychiatry; *USSR

Working Group on the Internment of Dissenters in Mental Hospitals. The Political Abuse of Psychiatry in the Soviet Union. London: Working Group on the Internment of Dissenters in Mental Hospitals, 1977. 16 p. Available from: Margrit Wreschner, 60 Riverside Drive, New York, NY. 10024; or, Hon. Secretary, 13 Armitage Rd., Golders Green, London NW11 8QT. 48 fn. BE06005.
*dissent; human rights; *involuntary commitment; malpractice; political activity; psychiatric diagnosis; *psychiatry; public opinion; schizophrenia; *USSR

INVOLUNTARY COMMITMENT / MENTALLY ILL

Lipp, Martin R. Competency, refusal of treatment, and committability. *In his* Respectful Treatment: The Human Side of Medical Care. Hagerstown, Md.: Harper and Row, Medical Department, 1977. 61-73. 11 refs. ISBN 0-06-141550-2. BE06679.
adults; children; *competence; dangerousness; duration of commitment; decision making; emergency care; *informed consent; *involuntary commitment; legal aspects; legal guardians; *mentally ill; parental consent; patient care; physician patient relationship; psychiatric diagnosis; *treatment refusal

Monahan, John. John Stuart Mill on the liberty of the mentally ill: a historical note. *American Journal of Psychiatry* 134(12): 1428-1429, Dec 1977. 4 refs. BE06145.
*involuntary commitment; *mentally ill; *self determination; Mill, John Stuart

Roth, Loren H. Involuntary civil commitment: the right to treatment and the right to refuse treatment. *Psychiatric Annals* 7(5): 50-51+, May 1977. 55 refs. BE05809.
coercion; dangerousness; competence; drugs; due

process; economics; informed consent; *institutionalized persons; *involuntary commitment; judicial action; *legal aspects; legal rights; legislation; mental institutions; *mentally ill; mentally retarded; operant conditioning; patients; psychiatry; review committees; *right to treatment; standards; state government; *treatment refusal

INVOLUNTARY COMMITMENT / MENTALLY ILL / HUMAN RIGHTS

Hoffman, P. Browning. Living with your rights off. *Bulletin of the American Academy of Psychiatry and the Law* 5(1): 68-74, 1977. 17 fn. BE05895.
case reports; dangerousness; *decision making; *human rights; *involuntary commitment; legal aspects; *mentally ill; prognosis; *psychiatry; right to treatment

INVOLUNTARY COMMITMENT / MENTALLY ILL / LEGAL ASPECTS

Martin, Andrew. A right to treatment. *Nursing Times* 73(42): 1620-1621, 20 Oct 1977. BE06349.
informed consent; *institutionalized persons; *involuntary commitment; *legal aspects; legal rights; *mentally ill; *right to treatment

INVOLUNTARY COMMITMENT / MENTALLY ILL / LEGAL RIGHTS

Troland, Mary B. Involuntary commitment of the mentally ill. *Montana Law Review* 38(2): 307-325, Summer 1977. 117 fn. BE05953.
dangerousness; due process; *involuntary commitment; law; *legal rights; legislation; *mentally ill; prognosis; psychiatric diagnosis; *standards; state government; Montana

INVOLUNTARY COMMITMENT / MENTALLY ILL / O'CONNOR V. DONALDSON

Szasz, Thomas S. Psychiatric Slavery. New York: Free Press, 1977. 159 p. 240 fn. Case study. ISBN 0-02-931600-6. BE05709.
dangerousness; *institutionalized persons; *involuntary commitment; law; *legal aspects; *mentally ill; professional organizations; psychiatric diagnosis; psychiatry; *right to treatment; Supreme Court decisions; treatment refusal; American Civil Liberties Union; American Psychiatric Association; *Donaldson, Kenneth; *Mental Health Law Project; *O'Connor v. Donaldson

INVOLUNTARY COMMITMENT / PARENTAL CONSENT

Curtiss, Zelda. Constitutional law—fourteenth amendment—due process—deprivation of children's rights—civil commitment. *Duquesne Law Review* 15(2): 337-347, Winter 1976-1977. 55 fn. BE07094.
*adolescents; *due process; *involuntary commit-

BE = bioethics accession number   fn. = footnotes   refs. = references

ment; legal rights; mentally ill; parent child relationship; *parental consent

INVOLUNTARY COMMITMENT / PARENTAL CONSENT / LEGAL ASPECTS

Solnit, Albert J., et al. Responses to "Changes in the parent-child relationship...". *Journal of Autism and Childhood Schizophrenia* 7(1): 94-108, Mar 1977. 3 refs. 7 fn. BE05910.
   adolescents; behavioral research; *children; *decision making; ethical review; hospitals; *human experimentation; institutionalized persons; *involuntary commitment; *legal aspects; legal rights; mental institutions; *mentally handicapped; parent child relationship; *parental consent; patient care; right to treatment

INVOLUNTARY COMMITMENT / POLITICS

Greenberg, Joel. The ethics of psychiatry: who is sick? *Science News* 112(21): 346-347, 19 Nov 1977. BE06593.
   dangerousness; dissent; *involuntary commitment; physician's role; *politics; professional organizations; psychiatry; public policy; social control; *United States; *USSR

INVOLUNTARY COMMITMENT / PSYCHIATRIC DIAGNOSIS / USSR

Bukovsky, Vladimir K. Gen. Svetlichny: 'We will let him rot in the insane asylum'. *New York Times,* 3 May 1977, p. 41. BE06216.
   dissent; *involuntary commitment; *psychiatric diagnosis; psychiatry; *USSR

INVOLUNTARY COMMITMENT / STANDARDS / PENNSYLVANIA

Meyers, Barton A. Standards for involuntary civil commitment in Pennsylvania. *University of Pittsburgh Law Review* 38(3): 535-549, Spring 1977. 64 fn. BE05925.
   dangerousness; *involuntary commitment; legal aspects; legislation; *mentally ill; *psychiatric diagnosis; psychiatry; *standards; state government; state interest; Mental Health Procedures Act; *Pennsylvania

INVOLUNTARY COMMITMENT / WORLD PSYCHIATRIC ASSOCIATION / USSR

Benson, Bruce. Psychiatrist group condemns abuses charged to Soviets. *Washington Post,* 2 Sep 1977, p. A16. BE06394.
   *dissent; *involuntary commitment; professional organizations; *psychiatry; *USSR; *World Psychiatric Association

Rich, Vera. Heading for Honolulu. *Nature* 268(5621): 578-579, 18 Aug 1977. BE05839.
   *attitudes; *dissent; *involuntary commitment;

politics; professional ethics; *professional organizations; psychiatry; *USSR; *World Psychiatric Association

Soviets finally condemned for psychiatric malpractices... will South Africa be next? *New Scientist* 75(1068): 571, 8 Sep 1977. BE05785.
   blacks; discrimination; *dissent; *involuntary commitment; professional organizations; *psychiatry; public policy; South Africa; *USSR; *World Psychiatric Association

INVOLUNTARY STERILIZATION / ADOLESCENTS / MENTALLY RETARDED

New York. Surrogate's Court, Nassau County. Application of A.D. 28 Apr 1977. *New York Supplement, 2d Series,* 394: 139-141. BE06271.
   *adolescents; *involuntary sterilization; legal rights; *mentally retarded; reproduction

INVOLUNTARY STERILIZATION / AMERICAN INDIANS

Harley, Richard M. Indian women plan to sue U.S. in sterilization cases. *Christian Science Monitor (Eastern Edition),* 27 May 1977, p. 6. BE06442.
   informed consent; *involuntary sterilization; *American Indians

INVOLUNTARY STERILIZATION / GOVERNMENT REGULATION

Brozan, Nadine. The volatile issue of sterilization abuse: a tangle of accusations and remedies. *New York Times,* 9 Dec 1977, p. B10. BE06422.
   federal government; females; *government regulation; indigents; informed consent; *involuntary sterilization; legal aspects; *legal rights

INVOLUNTARY STERILIZATION / INDIA

Hanlon, Joseph; Agarwal, Anil. Mass sterilisation at gunpoint. *New Scientist* 74(1050): 268-270, 5 May 1977. BE05617.
   family planning; government regulation; incentives; indigents; *involuntary sterilization; population control; public opinion; public policy; socioeconomic factors; *India

The issue that inflamed India. *Time* 109(14): 38-39, 4 Apr 1977. BE05568.
   *involuntary sterilization; public opinion; public policy; social impact; *India

Jamali, Naseem Z. Compulsory sterilization. *Hastings Center Report* 7(2): 4, Apr 1977. Letter. BE05534.
   human rights; *involuntary sterilization; socioeconomic factors; *India

Kramer, Barry. Resistance to India's sterilization policy grows amid reports that zealous officials use force. *Wall Street Journal (Eastern Edition),* 4 Feb 1977, p. 30. BE06435.

BE = bioethics accession number   fn. = footnotes   refs. = references

*involuntary sterilization; public opinion; public policy; *India

Simons, Lewis M. Compulsory sterilization provokes fear, contempt. *Washington Post,* 4 Jul 1977, p. A1+. BE06048.
coercion; family planning; government regulation; *involuntary sterilization; population control; public opinion; values; *India

Sinha, Arun. Compulsory sterilization. *Dissent* 24(1): 94-95, Winter 1977. BE06482.
family planning; incentives; incidence; *involuntary sterilization; public policy; *India

INVOLUNTARY STERILIZATION / INDIGENTS

U.S. Court of Appeals, Fourth Circuit. Walker v. Pierce. 26 Jul 1977. *Federal Reporter, 2d Series,* 560: 609-615. 5 fn. BE06473.
discrimination; *indigents; *involuntary sterilization; *obstetrics and gynecology; physicians; socioeconomic factors; state action; Medicaid

INVOLUNTARY STERILIZATION / JUDICIAL ACTION / LEGAL LIABILITY

U.S. Supreme Court. Stump v. Sparkman. 28 Mar 1978. *United States Reports* 435: 349-370. 21 fn. BE07032.
adolescents; *involuntary sterilization; *judicial action; *legal liability; mentally retarded; parental consent

INVOLUNTARY STERILIZATION / MENTALLY HANDICAPPED / LEGAL ASPECTS

Burgdorf, Robert L.; Burgdorf, Marcia P. The wicked witch is almost dead: *Buck v. Bell* and the sterilization of handicapped persons. *Temple Law Quarterly* 50(4): 995-1034, 1977. 271 fn. BE06850.
due process; equal protection; *involuntary sterilization; *legal aspects; legislation; *mentally handicapped; *mentally retarded; negative eugenics; state government; *Supreme Court decisions; *Buck v. Bell

INVOLUNTARY STERILIZATION / MENTALLY RETARDED

Schumacher, Edward. The right to bear children: a test case in Maine. *Washington Post,* 23 Jan 1977, p. C3. BE06212.
adults; family members; *involuntary sterilization; *mentally retarded; physician's role; social workers

INVOLUNTARY STERILIZATION / MENTALLY RETARDED / LEGAL ASPECTS

Chernus, Roy O. Substantive due process—compulsory sterilization of the mentally deficient. *New York Law School Law Review* 23(1): 151-158, 1977. 42 fn. BE06278.

due process; equal protection; *involuntary sterilization; *legal aspects; legislation; *mentally retarded; negative eugenics; state government; *In re Sterilization of Moore

INVOLUNTARY STERILIZATION / PUBLIC POLICY / INDIA

Gulhati, Kaval. Compulsory sterilization: a new dimension in India's population policy. *Draper World Population Fund Report* No. 3: 26-29, Autumn-Winter 1976. BE06644.
coercion; economics; family planning; *involuntary sterilization; *public policy; socioeconomic factors; *India

INVOLUNTARY STERILIZATION / ROMAN CATHOLIC ETHICS / INDIA

Catholic Bishops of India. A pastoral letter on sterilization. *Catholic Mind* 75(1313): 4-6, May 1977. BE05605.
coercion; family planning; government regulation; *involuntary sterilization; population control; *Roman Catholic ethics; *India

INVOLUNTARY STERILIZATION / WALKER V. PIERCE

Curran, William J. The freedom of medical practice, sterilization, and economic medical philosophy. *New England Journal of Medicine* 298(1): 32-33, 5 Jan 1978. 4 refs. BE06727.
*coercion; discrimination; federal government; freedom; financial support; indigents; informed consent; *involuntary sterilization; legal aspects; *legal rights; *physicians; pregnant women; socioeconomic factors; *sterilization; Medicaid; *Walker v. Pierce

IRELAND *See under*

ISLAM *See under*

ISRAEL *See under*

ITALY *See under*

JEHOVAH'S WITNESSES *See under*

JEHOVAH'S WITNESSES See under (cont'd.)
NESSES / LEGAL RIGHTS

TREATMENT REFUSAL / JEHOVAH'S WIT-
NESSES / LEUKEMIA

JEWISH ETHICS   See under

ABORTION / VALUE OF LIFE / JEWISH ETHICS

ALLOWING TO DIE / TERMINALLY ILL / JEWISH
ETHICS

DETERMINATION OF DEATH / JEWISH ETHICS

EXTRAORDINARY TREATMENT / JEWISH ETH-
ICS

LIVING WILLS / JEWISH ETHICS

PATIENT CARE / PHYSICIANS / JEWISH ETHICS

PATIENT CARE / SPINA BIFIDA / JEWISH ETHICS

VALUE OF LIFE / JEWISH ETHICS

JONES V. SAIKEWICZ

Brant, Jonathan.   The right to die in peace: substituted
consent and the mentally incompetent.   *Suffolk University Law Review* 11(4): 959-973, Spring 1977.   105
fn.   BE06234.
aged; *allowing to die; children; *competence; *decision making; judicial action; *legal aspects; legal
guardians; legal rights; leukemia; *mentally retarded; organ donation; parental consent; privacy;
quality of life; state interest; suffering; terminally
ill; *third party consent; treatment refusal; *withholding treatment; *Jones v. Saikewicz

Schultz, Stephen; Swartz, William; Appelbaum, Judith
C.   Deciding right-to-die cases involving incompetent
patients: Jones v. Saikewicz.   *Suffolk University Law
Review*   11(4): 936-958, Spring 1977.   95   fn.
BE06233.
*allowing to die; competence; constitutional law;
*decision making; institutionalized persons; *legal
aspects; legal guardians; legal rights; leukemia;
*mentally retarded; physician's role; privacy; qual-
ity of life; right to treatment; state courts; *third
party consent; value of life; *withholding treatment; *Jones v. Saikewicz

JOURNALISM   See under

BIOMEDICAL RESEARCH / JOURNALISM

JUDICIAL ACTION / HUMANISM

Csank, James F.   The right to a natural death.   *Human
Life Review*   4(1): 44-54, Winter 1978.   19   fn.
BE06725.
abortion; *allowing to die; *coma; decision making; extraordinary treatment; *humanism; *judicial action; law; legal guardians; physician's role;
privacy; state interest; terminally ill; third party
consent; withholding treatment; *In re Quinlan;
New Jersey Supreme Court

JUDICIAL ACTION   See under

ALLOWING TO DIE / JUDICIAL ACTION

ALLOWING TO DIE / JUDICIAL ACTION / MAS-
SACHUSETTS SUPREME JUDICIAL COURT

ALLOWING TO DIE / MENTALLY RETARDED /
JUDICIAL ACTION

ALLOWING TO DIE / QUALITY OF LIFE / JUDI-
CIAL ACTION

ALLOWING TO DIE / TERMINALLY ILL / JUDI-
CIAL ACTION

ALLOWING TO DIE / THIRD PARTY CONSENT /
JUDICIAL ACTION

INVOLUNTARY STERILIZATION / JUDICIAL AC-
TION / LEGAL LIABILITY

PROFESSIONAL ORGANIZATIONS / JUDICIAL
ACTION / STANDARDS

TREATMENT REFUSAL / JUDICIAL ACTION

TREATMENT REFUSAL / SURGERY / JUDICIAL
ACTION

JUSTICE   See under

HEALTH CARE DELIVERY / JUSTICE

PHYSICIANS / JUSTICE / STRIKES

POPULATION CONTROL / JUSTICE

POPULATION CONTROL / JUSTICE / INCEN-
TIVES

JUSTIFIABLE KILLING / HANDICAPPED

Hostler, John.   The right to life.   *Journal of Medical
Ethics* 3(3): 143-145, Sep 1977.   5 refs.   BE05982.
allowing to die; *handicapped; *justifiable killing;
killing; moral obligations; personhood; quality of
life; *value of life

KIDNEY DISEASES   See under

ALLOWING TO DIE / KIDNEY DISEASES

RESOURCE ALLOCATION / KIDNEY DISEASES

KIDNEYS   See under

ORGAN DONORS / KIDNEYS / SOCIAL ADJUST-
MENT

ORGAN TRANSPLANTATION / KIDNEYS / RISKS
AND BENEFITS

ORGAN TRANSPLANTATION / KIDNEYS / SCAR-
CITY

KILLING   See also JUSTIFIABLE KILLING

KILLING

Glover, Jonathan.   Causing Death and Saving Lives.
New York: Penguin Books, 1977.   328 p.   67 fn.
Bibliography: p. 299-324.   ISBN 0-1402-2003-8.
BE06513.
*abortion; active euthanasia; *allowing to die; cap-
ital punishment; consequences; *decision making;
double effect; infanticide; involuntary euthanasia;

---

*killing; morality; newborns; paternalism; personhood; prolongation of life; quality of life; self determination; suicide; utilitarianism; value of life; voluntary euthanasia; war; wedge argument; women's rights

KILLING / BRAIN DEATH / LAW ENFORCEMENT

Michigan. Court of Appeals. People v. Vanderford. 9 Aug 1977. *North Western Reporter, 2d Series,* 258: 502-504. BE07113.
allowing to die; *brain death; *killing; *law enforcement; withholding treatment

KILLING / DECISION MAKING

McMullin, Ernan. Introduction. *In* McMullin, Ernan, ed. Death and Decision. Boulder, Colo.: Westview Press, 1978. p. 1-13. American Association for the Advancement of Science Selected Symposium 18. ISBN 0-89158-152-9. BE06706.
allowing to die; case studies; death; *decision making; euthanasia; *killing; prolongation of life; utilitarianism; value of life

KILLING / FETUSES

New Jersey. Superior Court, Law Division. State v. Anderson. 14 Jul 1975. *Atlantic Reporter, 2d Series,* 343: 505-509. 2 fn. BE07080.
*fetuses; *killing; law enforcement; legal rights; pregnant women; prematurity

KILLING / FETUSES / CRIMINAL LAW

California. Court of Appeal, Fifth District. People v. Apodaca. 2 Feb 1978. *California Reporter,* 142: 830-840. 6 fn. BE06712.
*criminal law; *fetuses; *killing; *viability; *California

KILLING / MORAL OBLIGATIONS

Russell, Bruce. On the relative strictness of negative and positive duties. *American Philosophical Quarterly* 14(2): 87-97, Apr 1977. 28 fn. BE06300.
*allowing to die; alternatives; double effect; *killing; *moral obligations; philosophy

KILLING / MORAL OBLIGATIONS / RIGHTS

Williams, Peter C. Rights and the alleged right of innocents to be killed. *Ethics* 87(4): 383-394, Jul 1977. 29 fn. BE05739.
active euthanasia; allowing to die; consequences; ethics; *euthanasia; family members; *human rights; *killing; law; legal rights; *moral obligations; physicians; *rights; suffering; terminally ill

KILLING / NORMATIVE ETHICS

Hare, R.M. Medical ethics: can the moral philosopher

help? *In* Spicker, Stuart F.; Engelhardt, H. Tristram, eds. Philosophical Medical Ethics: Its Nature and Significance. Proceedings. Boston: D. Reidel, 1977. p. 49-61. Proceedings of the Third Trans-Disciplinary Symposium on Philosophy and Medicine, Farmington, Connecticut, 11-13 Dec 1975. 1 fn. ISBN 90-277-0772-3. BE05862.
abortion; deontological ethics; *killing; *medical ethics; medicine; morality; *normative ethics; philosophy; physicians; *utilitarianism; wedge argument

KILLING / VALUE OF LIFE

Rachels, James. Medical ethics and the rule against killing: comments on Professor Hare's paper. *In* Spicker, Stuart F.; Engelhardt, H. Tristram, eds. Philosophical Medical Ethics: Its Nature and Significance. Proceedings. Boston: D. Reidel, 1977. p. 63-69. Proceedings of the Third Trans-Disciplinary Symposium on Philosophy and Medicine, Farmington, Connecticut, 11-13 Dec 1975. 1 ref. ISBN 90-277-0772-3. BE05863.
euthanasia; *killing; utilitarianism; *value of life; *wedge argument

KILLING / WEDGE ARGUMENT

Rachels, James. Medical ethics and the rule against killing: comments on Professor Hare's paper. *In* Spicker, Stuart F.; Engelhardt, H. Tristram, eds. Philosophical Medical Ethics: Its Nature and Significance. Proceedings. Boston: D. Reidel, 1977. p. 63-69. Proceedings of the Third Trans-Disciplinary Symposium on Philosophy and Medicine, Farmington, Connecticut, 11-13 Dec 1975. 1 ref. ISBN 90-277-0772-3. BE05863.
euthanasia; *killing; utilitarianism; *value of life; *wedge argument

KILLING *See under*

ALLOWING TO DIE / KILLING

ORGAN TRANSPLANTATION / KILLING

LAETRILE / EVALUATION

Newell, Guy R. Clinical evaluation of Laetrile: two perspectives. Why the National Cancer Institute chooses a case-record review of Laetrile. *New England Journal of Medicine* 298(4): 216-218, 26 Jan 1978. 2 refs. BE06720.
cancer; drugs; *evaluation; human experimentation; *medical records; Food and Drug Administration; *Laetrile; *National Cancer Institute

Relman, Arnold S. Laetrilomania—again. *New England Journal of Medicine* 298(4): 215-216, 26 Jan 1978. 5 refs. Editorial. BE06719.
cancer; drugs; *evaluation; *human experimenta-

BE = bioethics accession number   fn. = footnotes   refs. = references

tion; medical records; *Laetrile; National Cancer Institute

LAETRILE *See under*

FOOD AND DRUG ADMINISTRATION / LAETRILE

GOVERNMENT REGULATION / LAETRILE

GOVERNMENT REGULATION / LAETRILE / LEGAL ASPECTS

HUMAN EXPERIMENTATION / LAETRILE

PATIENT CARE / LAETRILE / CANCER

PATIENT CARE / LAETRILE / LEGAL ASPECTS

PATIENT CARE / LAETRILE / LEGAL RIGHTS

PATIENT CARE / LAETRILE / LEGISLATION

PATIENT CARE / LAETRILE / NEW YORK ACADEMY OF MEDICINE

PATIENT CARE / LAETRILE / RIGHTS

PATIENT CARE / LAETRILE / RISKS AND BENEFITS

PRIVACY / LAETRILE / LEGAL RIGHTS

SELF DETERMINATION / LAETRILE / CANCER

LAW ENFORCEMENT ASSISTANCE ADMINISTRATION *See under*

BEHAVIOR CONTROL / FINANCIAL SUPPORT / LAW ENFORCEMENT ASSISTANCE ADMINISTRATION

LAW ENFORCEMENT *See under*

ABORTION / LAW ENFORCEMENT / SOCIAL IMPACT

ALLOWING TO DIE / BRAIN DEATH / LAW ENFORCEMENT

KILLING / BRAIN DEATH / LAW ENFORCEMENT

LAW *See under*

BIOETHICAL ISSUES / LAW / CANADA

HEALTH CARE / LAW

INFANTICIDE / LAW / HISTORICAL ASPECTS

MEDICINE / LAW

PSYCHIATRY / LAW

RECOMBINANT DNA RESEARCH / TECHNOLOGY ASSESSMENT / LAW

LEGAL ASPECTS *See under*

ABORTION / ADOLESCENTS / LEGAL ASPECTS

ABORTION / ALTERNATIVES / LEGAL ASPECTS

ABORTION / FINANCIAL SUPPORT / LEGAL ASPECTS

ABORTION / GOVERNMENT REGULATION / LEGAL ASPECTS

ABORTION / LEGAL ASPECTS

ABORTION / LEGAL ASPECTS / NEW ZEALAND

ABORTION / LEGAL ASPECTS / UNITED STATES

ABORTION / PARENTAL CONSENT / LEGAL AS-

LEGAL ASPECTS *See under (cont'd.)*
PECTS

ABORTION / PUBLIC HOSPITALS / LEGAL ASPECTS

ABORTION / SPOUSAL CONSENT / LEGAL ASPECTS

ABORTION / VIABILITY / LEGAL ASPECTS

AID / LEGAL ASPECTS

ALLOWING TO DIE / BRAIN PATHOLOGY / LEGAL ASPECTS

ALLOWING TO DIE / COMA / LEGAL ASPECTS

ALLOWING TO DIE / LEGAL ASPECTS

ALLOWING TO DIE / LIVING WILLS / LEGAL ASPECTS

ALLOWING TO DIE / MENTALLY RETARDED / LEGAL ASPECTS

ALLOWING TO DIE / PHYSICIANS / LEGAL ASPECTS

ALLOWING TO DIE / TERMINALLY ILL / LEGAL ASPECTS

BEHAVIOR CONTROL / COERCION / LEGAL ASPECTS

BEHAVIOR CONTROL / PSYCHOACTIVE DRUGS / LEGAL ASPECTS

BIOETHICAL ISSUES / MEDICAL RECORDS / LEGAL ASPECTS

BIOMEDICAL RESEARCH / GOVERNMENT REGULATION / LEGAL ASPECTS

BRAIN DEATH / LEGAL ASPECTS

CHILD NEGLECT / DIAGNOSIS / LEGAL ASPECTS

CONFIDENTIALITY / DANGEROUSNESS / LEGAL ASPECTS

CONFIDENTIALITY / MEDICAL RECORDS / LEGAL ASPECTS

CONFIDENTIALITY / PRIVACY / LEGAL ASPECTS

CONFIDENTIALITY / PSRO'S / LEGAL ASPECTS

CONFIDENTIALITY / PSYCHIATRY / LEGAL ASPECTS

CONFIDENTIALITY / PSYCHOTHERAPY / LEGAL ASPECTS

CONTRACEPTION / ADOLESCENTS / LEGAL ASPECTS

DETERMINATION OF DEATH / LEGAL ASPECTS

DISCLOSURE / LEGAL ASPECTS

EUTHANASIA / LEGAL ASPECTS

EUTHANASIA / LEGAL ASPECTS / GREAT BRITAIN

GENETIC COUNSELING / LEGAL ASPECTS

GENETIC INTERVENTION / LEGAL ASPECTS

GENETICS / LEGAL ASPECTS

GOVERNMENT REGULATION / LAETRILE / LEGAL ASPECTS

HEALTH CARE / ADOLESCENTS / LEGAL ASPECTS

HEALTH CARE / CHILDREN / LEGAL ASPECTS

BE = bioethics accession number   fn. = footnotes   refs. = references

LEGAL ASPECTS *See under (cont'd.)*

HEALTH CARE / PARENTAL CONSENT / LEGAL ASPECTS

HUMAN EXPERIMENTATION / CHILDREN / LEGAL ASPECTS

HUMAN EXPERIMENTATION / COMPENSATION / LEGAL ASPECTS

HUMAN EXPERIMENTATION / FETUSES / LEGAL ASPECTS

HUMAN EXPERIMENTATION / INFORMED CONSENT / LEGAL ASPECTS

HUMAN EXPERIMENTATION / LEGAL ASPECTS

HUMAN EXPERIMENTATION / NURSES / LEGAL ASPECTS

HUMAN EXPERIMENTATION / PARENTAL CONSENT / LEGAL ASPECTS

IMMUNIZATION / INFORMED CONSENT / LEGAL ASPECTS

INFORMED CONSENT / ADOLESCENTS / LEGAL ASPECTS

INFORMED CONSENT / HOSPITALS / LEGAL ASPECTS

INFORMED CONSENT / LEGAL ASPECTS

INFORMED CONSENT / PHYSICIAN PATIENT RELATIONSHIP / LEGAL ASPECTS

INFORMED CONSENT / PHYSICIANS / LEGAL ASPECTS

INFORMED CONSENT / PSYCHIATRY / LEGAL ASPECTS

INFORMED CONSENT / PSYCHOLOGICAL STRESS / LEGAL ASPECTS

INFORMED CONSENT / RISKS AND BENEFITS / LEGAL ASPECTS

INVOLUNTARY COMMITMENT / ADOLESCENTS / LEGAL ASPECTS

INVOLUNTARY COMMITMENT / DANGEROUSNESS / LEGAL ASPECTS

INVOLUNTARY COMMITMENT / MENTALLY ILL / LEGAL ASPECTS

INVOLUNTARY COMMITMENT / PARENTAL CONSENT / LEGAL ASPECTS

INVOLUNTARY STERILIZATION / MENTALLY HANDICAPPED / LEGAL ASPECTS

INVOLUNTARY STERILIZATION / MENTALLY RETARDED / LEGAL ASPECTS

MEDICAL RECORDS / HOSPITALS / LEGAL ASPECTS

OPERANT CONDITIONING / PRISONERS / LEGAL ASPECTS

ORGAN DONATION / CHILDREN / LEGAL ASPECTS

ORGAN TRANSPLANTATION / LEGAL ASPECTS / AUSTRALIA

PARENTAL CONSENT / ADOLESCENTS / LEGAL ASPECTS

PARENTAL CONSENT / PARENT CHILD RELATIONSHIP / LEGAL ASPECTS

LEGAL ASPECTS *See under (cont'd.)*

PARENTAL CONSENT / SPINA BIFIDA / LEGAL ASPECTS

PATIENT ACCESS / MEDICAL RECORDS / LEGAL ASPECTS

PATIENT CARE / LAETRILE / LEGAL ASPECTS

PATIENT CARE / NURSES / LEGAL ASPECTS

PATIENT CARE / PHYSICIANS / LEGAL ASPECTS

PATIENT CARE / SPINA BIFIDA / LEGAL ASPECTS

PERSONHOOD / FETUSES / LEGAL ASPECTS

PHYSICIAN PATIENT RELATIONSHIP / LEGAL ASPECTS

POPULATION CONTROL / DEVELOPING COUNTRIES / LEGAL ASPECTS

POPULATION CONTROL / LEGAL ASPECTS

POPULATION CONTROL / PUBLIC POLICY / LEGAL ASPECTS

PRECONCEPTION INJURIES / LEGAL ASPECTS

PRENATAL DIAGNOSIS / MORALITY / LEGAL ASPECTS

PRENATAL INJURIES / LEGAL ASPECTS

PRIVILEGED COMMUNICATION / TARASOFF V. REGENTS / LEGAL ASPECTS

PSYCHIATRY / LEGAL ASPECTS

PSYCHOSURGERY / LEGAL ASPECTS

PSYCHOSURGERY / LEGAL ASPECTS / NEW SOUTH WALES

PSYCHOSURGERY / LEGAL ASPECTS / UNITED STATES

RECOMBINANT DNA RESEARCH / INDUSTRY / LEGAL ASPECTS

STERILIZATION / INFORMED CONSENT / LEGAL ASPECTS

STERILIZATION / LEGAL ASPECTS

STERILIZATION / MENTALLY RETARDED / LEGAL ASPECTS

SUICIDE / LEGAL ASPECTS

TRANSPLANTATION / PARENTAL CONSENT / LEGAL ASPECTS

TRANSSEXUALISM / LEGAL ASPECTS

TREATMENT REFUSAL / LEGAL ASPECTS

TREATMENT REFUSAL / MENTALLY ILL / LEGAL ASPECTS

TREATMENT REFUSAL / SPINA BIFIDA / LEGAL ASPECTS

VOLUNTARY EUTHANASIA / LEGAL ASPECTS / AUSTRALIA

VOLUNTARY STERILIZATION / INFORMED CONSENT / LEGAL ASPECTS

VOLUNTARY STERILIZATION / LEGAL ASPECTS

VOLUNTARY STERILIZATION / SPOUSAL CONSENT / LEGAL ASPECTS

WRONGFUL DEATH / FETUSES / LEGAL ASPECTS

WRONGFUL DEATH / VIABILITY / LEGAL ASPECTS

BE = bioethics accession number   fn. = footnotes   refs. = references

LEGAL ASPECTS *See under (cont'd.)*

WRONGFUL LIFE / COMPENSATION / LEGAL ASPECTS

WRONGFUL LIFE / GENETIC DEFECTS / LEGAL ASPECTS

WRONGFUL LIFE / LEGAL ASPECTS

LEGAL GUARDIANS *See under*

ALLOWING TO DIE / LEGAL GUARDIANS

LEGAL LIABILITY / BRAIN PATHOLOGY / CRIMINAL LAW

Florida. District Court of Appeal, Third District. Tunsil v. State. 26 Oct 1976. *Southern Reporter, 2d Series,* 338: 874-876. BE06626.

*brain pathology; coma; *criminal law; *killing; law enforcement; *legal liability; withholding treatment

LEGAL LIABILITY *See under*

ALLOWING TO DIE / BRAIN DEATH / LEGAL LIABILITY

ALLOWING TO DIE / SPINA BIFIDA / LEGAL LIABILITY

CONFIDENTIALITY / PSYCHOTHERAPY / LEGAL LIABILITY

HUMAN EXPERIMENTATION / EPIDEMIOLOGY / LEGAL LIABILITY

IMMUNIZATION / FEDERAL GOVERNMENT / LEGAL LIABILITY

IMMUNIZATION / INFLUENZA / LEGAL LIABILITY

IMMUNIZATION / LEGAL LIABILITY

IN VITRO FERTILIZATION / PHYSICIANS / LEGAL LIABILITY

INFORMED CONSENT / LEGAL LIABILITY

INFORMED CONSENT / LEGAL LIABILITY / CANADA

INFORMED CONSENT / PHYSICIANS / LEGAL LIABILITY

INVOLUNTARY STERILIZATION / JUDICIAL ACTION / LEGAL LIABILITY

PATIENT CARE / PHYSICIANS / LEGAL LIABILITY

PHYSICIANS / CONGENITAL DEFECTS / LEGAL LIABILITY

PSYCHIATRY / DEINSTITUTIONALIZED PERSONS / LEGAL LIABILITY

RECOMBINANT DNA RESEARCH / TORTS / LEGAL LIABILITY

RESUSCITATION / LEGAL LIABILITY

SUICIDE / PSYCHIATRY / LEGAL LIABILITY

VOLUNTARY STERILIZATION / UNWANTED CHILDREN / LEGAL LIABILITY

WRONGFUL LIFE / PHYSICIANS / LEGAL LIABILITY

LEGAL OBLIGATIONS / PHYSICIANS / ABORTED FETUSES

Robertson, John A. After *Edelin* : little guidance. *Hastings Center Report* 7(3): 15-17+, Jun 1977. 1 fn. BE05663.

*aborted fetuses; abortion; judicial action; killing; legal liability; *legal obligations; *patient care; *physicians; viability; *Edelin, Kenneth

LEGAL OBLIGATIONS / PHYSICIANS / DEINSTITUTIONALIZED PERSONS

U.S. Court of Appeals, Fourth Circuit. Semler v. Psychiatric Institute of Washington, D.C. 17 Feb 1976. *Federal Reporter, 2d Series,* 538: 121-127. 2 fn. BE06627.

*dangerousness; killing; *deinstitutionalized persons; legal liability; *legal obligations; mental institutions; *mentally ill; negligence; obligations to society; *physicians; psychiatry; torts; Virginia

LEGAL OBLIGATIONS *See under*

EMERGENCY CARE / LEGAL OBLIGATIONS

INFORMED CONSENT / LEGAL OBLIGATIONS

PATIENT CARE / LEGAL OBLIGATIONS

LEGAL RIGHTS *See also* HUMAN RIGHTS

LEGAL RIGHTS *See under*

ABORTION / ADOLESCENTS / LEGAL RIGHTS

ABORTION / FATHERS / LEGAL RIGHTS

ABORTION / FETUSES / LEGAL RIGHTS

ABORTION / MENTALLY RETARDED / LEGAL RIGHTS

ABORTION / PARENTAL CONSENT / LEGAL RIGHTS

ABORTION / PREGNANT WOMEN / LEGAL RIGHTS

ABORTION / SPOUSAL CONSENT / LEGAL RIGHTS

AID / GENETIC FATHERS / LEGAL RIGHTS

AUTOPSIES / LEGAL RIGHTS / MARYLAND

BEHAVIOR CONTROL / INSTITUTIONALIZED PERSONS / LEGAL RIGHTS

BEHAVIOR CONTROL / PRISONERS / LEGAL RIGHTS

CHILDREN / LEGAL RIGHTS

CONTRACEPTION / ADOLESCENTS / LEGAL RIGHTS

CONTRACEPTION / PARENTAL CONSENT / LEGAL RIGHTS

HEALTH CARE / ADOLESCENTS / LEGAL RIGHTS

HUMAN EXPERIMENTATION / CHILDREN / LEGAL RIGHTS

HUMAN EXPERIMENTATION / INFORMED CONSENT / LEGAL RIGHTS

BE = bioethics accession number   fn. = footnotes   refs. = references

LEGAL RIGHTS *See under (cont'd.)*

    HUMAN EXPERIMENTATION / RESEARCH SUBJECTS / LEGAL RIGHTS

    INFORMED CONSENT / ADOLESCENTS / LEGAL RIGHTS

    INSTITUTIONALIZED PERSONS / LEGAL RIGHTS

    INSTITUTIONALIZED PERSONS / MENTALLY RETARDED / LEGAL RIGHTS

    INVOLUNTARY COMMITMENT / CHILDREN / LEGAL RIGHTS

    INVOLUNTARY COMMITMENT / MENTALLY ILL / LEGAL RIGHTS

    PATIENT ACCESS / MEDICAL RECORDS / LEGAL RIGHTS

    PATIENT CARE / LAETRILE / LEGAL RIGHTS

    PRENATAL INJURIES / COMPENSATION / LEGAL RIGHTS

    PRIVACY / LAETRILE / LEGAL RIGHTS

    PSYCHIATRY / LEGAL RIGHTS

    PSYCHOSURGERY / PRISONERS / LEGAL RIGHTS

    SOCIAL SCIENCES / LEGAL RIGHTS

    STERILIZATION / PHYSICIANS / LEGAL RIGHTS

    THIRD PARTY CONSENT / MENTALLY RETARDED / LEGAL RIGHTS

    TREATMENT REFUSAL / CRITICALLY ILL / LEGAL RIGHTS

    TREATMENT REFUSAL / INSTITUTIONALIZED PERSONS / LEGAL RIGHTS

    TREATMENT REFUSAL / JEHOVAH'S WITNESSES / LEGAL RIGHTS

    TREATMENT REFUSAL / LEGAL RIGHTS

    TREATMENT REFUSAL / MENTALLY ILL / LEGAL RIGHTS

    TREATMENT REFUSAL / SELF DETERMINATION / LEGAL RIGHTS

    TREATMENT REFUSAL / TERMINALLY ILL / LEGAL RIGHTS

LEGISLATION *See also* JUDICIAL ACTION, MODEL LEGISLATION

LEGISLATION *See under*

    ABORTION / LEGISLATION / ATTITUDES

    ABORTION / LEGISLATION / CANADA

    ABORTION / LEGISLATION / CONSCIENCE

    ABORTION / LEGISLATION / EUROPE

    ABORTION / LEGISLATION / GREAT BRITAIN

    ABORTION / LEGISLATION / HAWAII

    ABORTION / LEGISLATION / INDIA

    ABORTION / LEGISLATION / INTERNATIONAL ASPECTS

    ABORTION / LEGISLATION / ISRAEL

    ABORTION / LEGISLATION / ITALY

    ALLOWING TO DIE / LEGISLATION

    ALLOWING TO DIE / LEGISLATION / SOCIAL IM-

LEGISLATION *See under (cont'd.)*

    PACT

    ALLOWING TO DIE / LEGISLATION / SOUTH CAROLINA

    ALLOWING TO DIE / LEGISLATION / TENNESSEE

    ALLOWING TO DIE / LIVING WILLS / LEGISLATION

    DETERMINATION OF DEATH / LEGISLATION / UNITED STATES

    ELECTROCONVULSIVE THERAPY / LEGISLATION / CALIFORNIA

    GENETIC SCREENING / LEGISLATION

    GENETIC SCREENING / NEWBORNS / LEGISLATION

    LIVING WILLS / LEGISLATION

    LIVING WILLS / LEGISLATION / TENNESSEE

    PATIENT CARE / LAETRILE / LEGISLATION

    PRIVILEGED COMMUNICATION / LEGISLATION / CALIFORNIA

    PSYCHOSURGERY / LEGISLATION / NEW SOUTH WALES

    RECOMBINANT DNA RESEARCH / GOVERNMENT REGULATION / LEGISLATION

    RECOMBINANT DNA RESEARCH / LEGISLATION

    RECOMBINANT DNA RESEARCH / LEGISLATION / POLITICAL ACTIVITY

    RECOMBINANT DNA RESEARCH / LEGISLATION / U.S. CONGRESS

    RECOMBINANT DNA RESEARCH / ORGANIZATIONAL POLICIES / LEGISLATION

    RECOMBINANT DNA RESEARCH / REVIEW COMMITTEES / LEGISLATION

    RENAL DIALYSIS / LEGISLATION

    STERILIZATION / LEGISLATION / HISTORICAL ASPECTS

    TREATMENT REFUSAL / LEGISLATION / TENNESSEE

LEGITIMACY *See under*

    AID CHILDREN / LEGITIMACY

LEUKEMIA *See under*

    PATIENT CARE / CHILDREN / LEUKEMIA

    PATIENT CARE / LEUKEMIA / RISKS AND BENEFITS

    TREATMENT REFUSAL / JEHOVAH'S WITNESSES / LEUKEMIA

LIFE EXTENSION

Engelhardt, H. Tristram. Treating aging: restructuring the human condition. *In* Neugarten, Bernice L.; Havighurst, Robert J., eds. Extending the Human Life Span: Social Policy and Social Ethics. Washington: U.S. Government Printing Office, 1977. p. 33-39. Report prepared for the National Science Founda-

tion, RANN-Research Applications Directorate. Report No. NSF/RA 770123. 16 fn. BE06699.
*aged; biomedical research; *attitudes to death; contraception; *death; developing countries; euthanasia; *health; *life extension; *moral obligations; normality; population growth; quality of life; *resource allocation; social impact; suicide; values

Goddard, James L. Extension of the life span: a national goal? *In* Neugarten, Bernice L.; Havighurst, Robert J., eds. Extending the Human Life Span: Social Policy and Social Ethics. Washington: U.S. Government Printing Office, 1977. p. 19-26. Report prepared for the National Science Foundation, RANN-Research Applications Directorate. Report No. NSF/RA 770123. 15 fn. BE06697.
aged; biomedical research; drugs; family; goals; government regulation; human experimentation; *life extension; methods; population growth; public policy; resource allocation; risks and benefits; *social impact

Gustafson, James M. Extension of the active life: ethical issues. *In* Neugarten, Bernice L.; Havighurst, Robert J., eds. Extending the Human Life Span: Social Policy and Social Ethics. Washington: U.S. Government Printing Office, 1977. p. 27-32. Report prepared for the National Science Foundation, RANN-Research Applications Directorate. Report No. NSF/RA 770123. BE06698.
*common good; consequences; costs and benefits; *human rights; justice; *life extension; public policy; quality of life; resource allocation; rights; *self determination; utilitarianism; values

Lasch, Christopher. Aging in a culture without a future. *Hastings Center Report* 7(4): 42-44, Aug 1977. BE05761.
*age; aged; *attitudes; biomedical technologies; future generations; *life extension; medicine; reproduction; social worth; socioeconomic factors; values

Miller, Albert J.; Acri, Michael J. Science. *In their* Death: A Bibliographical Guide. Metuchen, N.J.: Scarecrow Press, 1977. p. 205-210. Bibliography. ISBN 0-8108-1025-5. BE06187.
*cryonic suspension; *death; determination of death; *life extension; mortality; *science

LIFE EXTENSION / SOCIAL CONTROL / SOCIAL IMPACT

Morison, Robert S. Misgivings about life-extending technologies. *Daedalus* 107(2): 211-226, Spring 1978. 13 fn. BE06940.
biomedical research; genetic intervention; justice; *life extension; psychological stress; science; sex determination; *social control; *social impact

LITERATURE *See under*
CONFIDENTIALITY / PSYCHIATRY / LITERATURE

LIVING WILLS / JEWISH ETHICS

Cohn, Hillel. Natural death—humane, just and Jewish. *Sh'ma* 7(132): 99-101, 15 Apr 1977. BE05539.
*allowing to die; decision making; *Jewish ethics; *living wills; prolongation of life; *terminally ill; withholding treatment

Siegel, Seymour. Jewish law permits natural death. *Sh'ma* 7(132): 96-97, 15 Apr 1977. BE05537.
*allowing to die; *Jewish ethics; legislation; *living wills; state government; *terminally ill; withholding treatment; California Natural Death Act

LIVING WILLS / LEGISLATION

Raible, Jane A. The right to refuse treatment and natural death legislation. *Medicolegal News* 5(4): 6-8+, Fall 1977. 9 fn. BE06245.
*adults; *allowing to die; aged; attitudes; brain death; children; competence; decision making; legal aspects; legal rights; *legislation; *living wills; medical records; physicians; public opinion; pregnant women; religion; state government; terminal care; *terminally ill; treatment refusal; withholding treatment

LIVING WILLS / LEGISLATION / TENNESSEE

McKenney, Edward J. Death and dying in Tennessee. *Memphis State University Law Review* 7(4): 503-554, Summer 1977. 343 fn. BE06307.
*allowing to die; *brain death; *coma; *determination of death; euthanasia; *extraordinary treatment; family members; judicial action; *legal aspects; legal liability; *legislation; *living wills; organ transplantation; physician patient relationship; physician's role; privacy; state government; rights; state interest; terminally ill; third party consent; treatment refusal; *withholding treatment; *Dockery v. Dockery; In re Quinlan; *Tennessee; *Tennessee Natural Death Act

LIVING WILLS *See under*
ALLOWING TO DIE / LIVING WILLS

ALLOWING TO DIE / LIVING WILLS / ARKANSAS

ALLOWING TO DIE / LIVING WILLS / LEGAL ASPECTS

ALLOWING TO DIE / LIVING WILLS / LEGISLATION

ALLOWING TO DIE / LIVING WILLS / NEVADA

ALLOWING TO DIE / LIVING WILLS / NORTH CAROLINA

ALLOWING TO DIE / LIVING WILLS / TEXAS

ALLOWING TO DIE / TERMINALLY ILL / LIVING WILLS

BE = bioethics accession number   fn. = footnotes   refs. = references

LIVING WILLS *See under (cont'd.)*
TERMINAL CARE / MEDICAL EDUCATION / LIVING WILLS
TREATMENT REFUSAL / LIVING WILLS

LSD *See under*
HUMAN EXPERIMENTATION / DEPARTMENT OF THE ARMY / LSD
HUMAN EXPERIMENTATION / LSD

MALPRACTICE *See under*
INFORMED CONSENT / EXPERT TESTIMONY / MALPRACTICE
INFORMED CONSENT / MALPRACTICE
PHYSICIAN PATIENT RELATIONSHIP / MALPRACTICE

MANDATORY PROGRAMS *See under*
ABORTION / MANDATORY PROGRAMS
IMMUNIZATION / MANDATORY PROGRAMS

MARYLAND *See under*
AUTOPSIES / LEGAL RIGHTS / MARYLAND
RECOMBINANT DNA RESEARCH / MARYLAND

MASS MEDIA *See under*
CARCINOGENS / MASS MEDIA / SOCIAL IMPACT

MASS SCREENING / CANCER

Sutnick, Alton I., et al. Ethical issues in investigation of screening strategies. *Medical and Pediatric Oncology* 3(2): 133-136, 1977. 6 refs. BE06477.
*cancer; control groups; *mass screening; random selection; risks and benefits

MASS SCREENING / PREGNANT WOMEN / SPINA BIFIDA

Check, William. Mass screening for open spina bifida needs careful consideration. *Journal of the American Medical Association* 238(14): 1441-1443+, 3 Oct 1977. BE05849.
*amniocentesis; fetuses; costs and benefits; incidence; informed consent; *mass screening; *pregnant women; *spina bifida; statistics; Great Britain; United States

Screening for spina bifida. *Journal of Medical Ethics* 4(1): 3-4, Mar 1978. Editorial. BE06929.
amniocentesis; *mass screening; *pregnant women; prenatal diagnosis; risks and benefits; selective abortion; social impact; *spina bifida; Great Britain

MASS SCREENING / SPINA BIFIDA

Prevention of spina bifida—the parents' choice. *Nature* 271(5646): 595, 16 Feb 1978. Editorial. BE06792.
allowing to die; decision making; handicapped; incidence; *mass screening; methods; parents; pregnant women; *prenatal diagnosis; selective abortion; *spina bifida; Great Britain

MASS SCREENING / SPINA BIFIDA / COSTS AND BENEFITS

Screening for neural-tube defects. *Lancet* 1(8026): 1345-1346, 25 Jun 1977. 9 fn. BE05664.
amniocentesis; *costs and benefits; *mass screening; pregnant women; *prenatal diagnosis; *selective abortion; *spina bifida

MASS SCREENING *See under*
HEALTH CARE DELIVERY / MASS SCREENING / COSTS AND BENEFITS

MASSACHUSETTS SUPREME JUDICIAL COURT *See under*
ALLOWING TO DIE / JUDICIAL ACTION / MASSACHUSETTS SUPREME JUDICIAL COURT

MEDICAID *See under*
ABORTION / DISCRIMINATION / MEDICAID
ABORTION / FINANCIAL SUPPORT / MEDICAID
ABORTION / MEDICAID
VOLUNTARY STERILIZATION / MEDICAID

MEDICAL EDUCATION / ETHICS

Davidson, G.P.; Roberts, F.J. Relational ethics in medicine. *New Zealand Medical Journal* 86(598): 388-391, 26 Oct 1977. 13 refs. BE06345.
dehumanization; deontological ethics; *ethics; evaluation; *medical education; situational ethics; *social interaction; teaching methods; utilitarianism

MEDICAL EDUCATION *See under*
DEATH / MEDICAL EDUCATION
MEDICAL ETHICS / MEDICAL EDUCATION
MEDICAL ETHICS / MEDICAL EDUCATION / AUSTRALIA
MEDICAL ETHICS / MEDICAL EDUCATION / IRELAND
MEDICAL ETHICS / MEDICAL EDUCATION / SOUTHAMPTON UNIVERSITY MEDICAL SCHOOL
TERMINAL CARE / MEDICAL EDUCATION / LIVING WILLS

MEDICAL ETHICS *See also* BIOETHICS, CODES OF ETHICS, NURSING ETHICS, PROFESSIONAL ETHICS

MEDICAL ETHICS

American Medical Association. Judicial Council. Opinions and Reports of the Judicial Council, Including the Principles of Medical Ethics and Rules of the

Judicial Council. Chicago: American Medical Association, 1977. 52 p. BE05875.
> bioethical issues; codes of ethics; mass media; *medical ethics; medical etiquette; medical fees; medical records; moral obligations; professional ethics

Barber, Bernard, ed. Medical Ethics and Social Change. Philadelphia: American Academy of Political and Social Science, 1978. 201 p. Annals of the American Academy of Political and Social Science, Vol. 437, May 1978. 303 fn. ISBN 0-87761-226-9. BE06994.
> abortion; allowing to die; *bioethical issues; critically ill; decision making; disclosure; drugs; epidemiology; health care delivery; historical aspects; human experimentation; informed consent; justice; legal aspects; *medical ethics; medical education; physicians; preventive medicine; public policy; resource allocation; selective abortion; self induced illness; social impact; withholding treatment

Beauchamp, Tom L.; Walters, LeRoy, eds. Contemporary Issues in Bioethics. Encino, Calif.: Dickenson Publishing Company, 1978. 612 p. References and footnotes included. ISBN 0-8221-0200-5. BE06700.
> *abortion; allowing to die; *behavior control; *bioethics; biomedical technologies; brain death; children; codes of ethics; confidentiality; decision making; deontological ethics; *determination of death; *euthanasia; fetuses; genetic intervention; *health; *health care; *human experimentation; *informed consent; involuntary commitment; justice; legal aspects; *medical ethics; natural law; *normative ethics; operant conditioning; patients' rights; personhood; psychosurgery; *resource allocation; rights; selection for treatment; treatment refusal; utilitarianism; Kaimowitz v. Department of Mental Health; Quinlan, Karen

Burns, Chester R., ed. Legacies in Ethics and Medicine. New York: Science History Publications, 1977. 326 p. References and footnotes included. ISBN 0-88202-166-4. BE05683.
> *codes of ethics; historical aspects; *medical ethics; moral obligations; physicians

Campbell, A.V. Medical ethics today. *Scottish Medical Journal* 22(4): 248-250, Oct 1977. 15 refs. BE06496.
> bioethics; decision making; interdisciplinary communication; *medical ethics; physicians; professional organizations

Engelhardt, H. Tristram; Spicker, Stuart F. Introduction to medical ethics. *In* Spicker, Stuart F.; Engelhardt, H. Tristram, eds. Philosophical Medical Ethics: Its Nature and Significance. Proceedings. Boston: D. Reidel, 1977. p. 3-17. Proceedings of the Third Trans-Disciplinary Symposium on Philosophy and Medicine, Farmington, Connecticut, 11-13 Dec 1975. 36 refs. ISBN 90-277-0772-3. BE05859.

bioethical issues; *medical ethics; morality; philosophy

Jonsen, Albert R. Do no harm. *Annals of Internal Medicine* 88(6): 827-832, Jun 1978. 28 fn. BE06974.
> codes of ethics; double effect; injuries; *medical ethics; medicine; moral obligations; patient care; physician's role; risks and benefits

Lasagna, Louis; Veatch, Robert M.; Kurtz, Paul. Medical Ethics. [Videorecording]. Amherst, N.Y.: American Humanist Association, 1977. 1 cassette; 30 min.; sound; color; 3/4 in. Ethics in America Series presented by the American Humanist Association. Produced at WNED-TV, Buffalo, New York. BE07041.
> decision making; disclosure; informed consent; *medical ethics; *paternalism; *patient care; *patient participation; *physician patient relationship; physicians; risks and benefits; self determination; values

Murray, David S. Medical ethics. *New Humanist* 91(10): 265-266, Feb 1976. BE06634.
> *medical ethics; medical fees; moral obligations; obligations to society; physician's role; physicians; state medicine

Sloan, Sherwin H. Ethics in ophthalmology. *Transactions of the American Academy of Ophthalmology and Otolaryngology* 83(5): OP849-OP852, Sep-Oct 1977. BE06588.
> codes of ethics; eye diseases; *medical ethics

Smith, David H., ed. No Rush to Judgement: Essays on Medical Ethics. Bloomington, Ind.: Indiana University, Poynter Center, 1977. 149 p. 23 refs. 136 fn. BE05792.
> *abortion; active euthanasia; *allowing to die; congenital defects; decision making; ethics; *euthanasia; genetic counseling; *genetic intervention; *health care; legal aspects; *medical ethics; medicine; newborns; public policy; *resource allocation; selective abortion; third party consent

Taylor, Nancy K., comp. Bibliography of Society, Ethics and the Life Sciences: Supplement for 1977-78. Hastings-on-Hudson, N.Y.: Institute of Society, Ethics and the Life Sciences, 1977. 26 p. Full bibliography compiled by Sharmon Sollitto and Robert M. Veatch. Bibliography. BE06688.
> behavior control; *bioethical issues; bioethics; death; euthanasia; genetic intervention; health care delivery; human experimentation; *medical ethics; population control

Veatch, Robert M. Medical ethics. *Journal of the American Medical Association* 239(6): 514-515, 6 Feb 1978. 10 refs. BE06801.
> bioethical issues; codes of ethics; education; ethics; government regulation; human experimentation; *interdisciplinary communication; *medical edu-

---

BE = bioethics accession number   fn. = footnotes   refs. = references

cation; *medical ethics; medicine; nursing ethics; public policy; social impact

Walters, Hugh. Medical ethics. *New Humanist* 92(1): 30-31, May-Jun 1976. BE06637.
*medical ethics; medical fees; *medicine; moral obligations; physicians; *politics; state medicine

Watson, Charles G. Can and should the ethics of a cottage industry survive? Ethics for young surgeons. *Archives of Surgery* 112(9): 1111-1114, Sep 1977. 4 refs. BE05992.
economics; *medical ethics; moral obligations; obligations to society; physician patient relationship; physicians

MEDICAL ETHICS / CHRISTIAN ETHICS

Crouch, Muriel. A basis for medical ethics. *In* Vale, J.A., ed. Medicine and the Christian Mind. London: Christian Medical Fellowship Publications, 1975. p. 17-25. 20 fn. ISBN 85111-956-5. BE07070.
*Christian ethics; humanism; *medical ethics; *physicians

MEDICAL ETHICS / EDUCATION

Veatch, Robert M. Medicine, biology, and ethics. *Hastings Center Report* 7(6): Special Suppl. 2-3, Dec 1977. BE06154.
bioethics; *education; ethicists; health personnel; *medical education; *medical ethics; physicians

MEDICAL ETHICS / HEALTH PERSONNEL / EDUCATION

Purtilo, Ruth B. Ethics teaching in allied health fields. *Hastings Center Report* 8(2): 14-16, Apr 1978. 2 fn. BE06945.
*education; *health personnel; *medical ethics

MEDICAL ETHICS / HISTORICAL ASPECTS

Burns, Chester R. American medical ethics: some historical roots. *In* Spicker, Stuart F.; Engelhardt, H. Tristram, eds. Philosophical Medical Ethics: Its Nature and Significance. Proceedings. Boston: D. Reidel, 1977. p. 21-26. Proceedings of the Third Trans-Disciplinary Symposium on Philosophy and Medicine, Farmington, Connecticut, 11-13 Dec 1975. 15 refs. 1 fn. ISBN 90-277-0772-3. BE05860.
Christian ethics; *historical aspects; *medical ethics; philosophy; physicians; professional ethics

MEDICAL ETHICS / HISTORICAL ASPECTS / ISLAM

Levey, Martin. Medical deontology in ninth century Islam. *In* Burns, Chester R., ed. Legacies in Ethics and Medicine. New York: Science History Publications, 1977. p. 129-144. 54 fn. ISBN 0-88202-

166-4. BE05684.
*historical aspects; Islamic ethics; *medical ethics; moral obligations; physician patient relationship; physicians; professional ethics; *Islam; Ninth Century

MEDICAL ETHICS / HUMANISM

Angrist, Alfred A., et al. Symposium: humanism and medical ethics. *New York State Journal of Medicine* 77(9): 1448-1462, Aug 1977. 22 refs. BE05834.
altruism; biomedical technologies; compensation; curriculum; dehumanization; hospitals; *humanism; *medical education; *medical ethics; medical fees; *moral obligations; patient care; *physician patient relationship; physicians; professional ethics; self determination; teaching methods; *values

MEDICAL ETHICS / INVESTIGATORS

Lally, John J. The making of the compassionate physician-investigator. *In* Barber, Bernard, ed. Medical Ethics and Social Change. Philadelphia: American Academy of Political and Social Science, 1978. p. 86-98. Annals of the American Academy of Political and Social Science, Vol. 437, May 1978. 38 fn. ISBN 0-87761-226-9. BE07001.
attitudes; human experimentation; curriculum; *investigators; *medical education; *medical ethics; physicians; social impact; values

MEDICAL ETHICS / MEDICAL EDUCATION

Pellegrino, Edmund D. Ethics and the moment of clinical truth. *Journal of the American Medical Association* 239(10): 960-961, 6 Mar 1978. 4 refs. Editorial. BE06822.
decision making; ethicists; *ethics; interdisciplinary communication; *medical education; *medical ethics; *medicine; patient care; physicians

Schmeck, Harold M. Ethics, medicine's painful research. *New York Times,* 24 Apr 1977, sect. 4, p. 6. BE06018.
*medical education; *medical ethics

MEDICAL ETHICS / MEDICAL EDUCATION / AUSTRALIA

The teaching of medical ethics. *Medical Journal of Australia* 1(24): 871-872, 11 Jun 1977. 4 fn. Editorial. BE05672.
*medical education; *medical ethics; medical etiquette; *Australia

MEDICAL ETHICS / MEDICAL EDUCATION / IRELAND

Clarke, Dolores D. The teaching of medical ethics: University College, Cork, Ireland. *Journal of Medical Ethics* 4(1): 36-39, Mar 1978. 3 refs. BE06919.
bioethical issues; decision making; legal aspects; legislation; *medical education; *medical ethics;

BE = bioethics accession number   fn. = footnotes   refs. = references

physicians; Roman Catholic ethics; teaching methods; technical expertise; values; *Ireland; University College, Cork

MEDICAL ETHICS / MEDICAL EDUCATION / SOUTHAMPTON UNIVERSITY MEDICAL SCHOOL

Dennis, K.J.; Hall, M.R.P. The teaching of medical ethics at Southampton University Medical School. *Journal of Medical Ethics* 3(4): 183-185, Dec 1977. 8 refs. BE06252.
> curriculum; law; *medical education; *medical ethics; teaching methods; England; *Southampton University Medical School

MEDICAL ETHICS / MEDICAL STAFF / EDUCATION

Levine, Melvin D.; Scott, Lee; Curran, William J. Ethics rounds in a children's medical center: evaluation of a hospital-based program for continuing education in medical ethics. *Pediatrics* 60(2): 202-208, Aug 1977. 8 refs. BE06325.
> evaluation; case studies; children; decision making; disclosure; *education; hospitals; interdisciplinary communication; *medical ethics; *medical staff; patient care; physicians; *program descriptions

MEDICAL ETHICS / NORMATIVE ETHICS

Hare, R.M. Medical ethics: can the moral philosopher help? *In* Spicker, Stuart F.; Engelhardt, H. Tristram, eds. Philosophical Medical Ethics: Its Nature and Significance. Proceedings. Boston: D. Reidel, 1977. p. 49-61. Proceedings of the Third Trans-Disciplinary Symposium on Philosophy and Medicine, Farmington, Connecticut, 11-13 Dec 1975. 1 fn. ISBN 90-277-0772-3. BE05862.
> abortion; deontological ethics; *killing; *medical ethics; medicine; morality; *normative ethics; philosophy; physicians; *utilitarianism; wedge argument

MEDICAL ETHICS / NURSING EDUCATION

Aroskar, Mila A.; Veatch, Robert M. Ethics teaching in nursing schools. *Hastings Center Report* 7(4): 23-26, Aug 1977. BE05748.
> curriculum; decision making; interdisciplinary communication; *medical ethics; nurses; *nursing education; nursing ethics

Steinfels, Margaret O. Ethics, education, and nursing practice. *Hastings Center Report* 7(4): 20-21, Aug 1977. BE05746.
> curriculum; *medical ethics; nurses; *nursing education; professional ethics; professional organizations

MEDICAL ETHICS / NURSING EDUCATION / CURRICULUM

Aroskar, Mila A. Ethics in the nursing curriculum. *Nursing Outlook* 25(4): 260-264, Apr 1977. 6 fn. BE05926.
> *curriculum; interdisciplinary communication; *medical ethics; nurses; *nursing education; nursing ethics; teaching methods

MEDICAL ETHICS / PHILOSOPHY

Spicker, Stuart F.; Engelhardt, H. Tristram, eds. Philosophical Medical Ethics: Its Nature and Significance. Proceedings. Boston: D. Reidel, 1977. 252 p. Proceedings of the Third Trans-Disciplinary Symposium on Philosophy and Medicine, Farmington, Connecticut, 11-13 Dec 1975. References and footnotes included. ISBN 90-277-0772-3. BE05858.
> abortion; *bioethical issues; children; cultural pluralism; euthanasia; evolution; goals; human experimentation; informed consent; *medical ethics; medicine; moral obligations; morality; parental consent; personhood; *philosophy; physician patient relationship; physician's role; prisoners; rights; self determination; suicide; utilitarianism; value of life; voluntary euthanasia; wedge argument

MEDICAL ETHICS / PHYSICIANS / EDUCATION

Singer, Peter. Can ethics be taught in a hospital? *Pediatrics* 60(2): 253-255, Aug 1977. 1 ref. BE05974.
> *education; ethicists; ethics; *hospitals; medical education; *medical ethics; *physicians

MEDICAL ETHICS / PSYCHIATRY

Jellinek, Michael; Parmelee, Dean. Is there a role for medical ethics in postgraduate psychiatry courses? *American Journal of Psychiatry* 134(12): 1438-1439, Dec 1977. 2 refs. BE06139.
> *education; *medical ethics; *psychiatry

MEDICAL ETHICS / SOCIAL IMPACT

Barber, Bernard. Perspectives on medical ethics and social change. *In* Barber, Bernard, ed. Medical Ethics and Social Change. Philadelphia: American Academy of Political and Social Science, 1978. p. 1-7. Annals of the American Academy of Political and Social Science, Vol. 437, May 1978. 4 fn. ISBN 0-87761-226-9. BE06995.
> authoritarianism; *medical ethics; interdisciplinary communication; patient participation; physicians; *social impact; socioeconomic factors; values

MEDICAL ETHICS / STUDENTS / EVALUATION

Stevens, Carey; Firth, S.T. Responses to moral

BE = bioethics accession number   fn. = footnotes   refs. = references

dilemmas in medical students and psychiatric residents. *Canadian Psychiatric Association Journal* 22(8): 441-445, Dec 1977. 16 refs. BE06589.
codes of ethics; *decision making; *evaluation; medical education; *medical ethics; psychiatry; *students

MEDICAL ETHICS / TEACHING METHODS

Siegler, Mark. A legacy of Osler: teaching clinical ethics at the bedside. *Journal of the American Medical Association* 239(10): 951-956, 6 Mar 1978. 22 refs. BE06827.
bioethical issues; decision making; ethicists; hospitals; human experimentation; investigators; *medical education; *medical ethics; normative ethics; patient care; physician's role; *physicians; students; *teaching methods; technical expertise; universities; *Osler, William

MEDICAL ETHICS / WORLD HEALTH ORGANIZATION

Gellhorn, Alfred. WHO and medical ethics. *WHO Chronicle* 31(5): 194-195, May 1977. 1 fn. BE05612.
biomedical technologies; health; health personnel; human experimentation; human rights; *medical ethics; torture; *World Health Organization

MEDICAL ETHICS   *See under*

HEALTH CARE / MENTAL HEALTH / MEDICAL ETHICS
TORTURE / MEDICAL ETHICS

MEDICAL RECORDS / FREEDOM OF INFORMATION ACT / SOCIAL IMPACT

Shires, David B.; Duncan, Gaylen. Freedom of information legislation and medical records. *Canadian Medical Association Journal* 118(4): 343-344, 18 Feb 1978. 1 ref. Editorial. BE06796.
confidentiality; disclosure; famous persons; federal government; law enforcement; legal obligations; legal rights; *legislation; *medical records; mental health; patients; physicians; *social impact; *Canada; *Freedom of Information Act

MEDICAL RECORDS / HOSPITALS / LEGAL ASPECTS

Hayt, Emanuel. Medicolegal Aspects of Hospital Records. Second Edition. Berwyn, Ill.: Physicians' Record Company, 1977. 519 p. Footnotes included. ISBN 0-917036-13-1. BE06678.
abortion; adults; artificial insemination; autopsies; blood transfusions; children; *confidentiality; consent forms; contraception; data bases; disclosure; famous persons; federal government; *hospitals; *informed consent; *legal aspects; legal obligations; legal rights; legislation; mass media; *medical records; medical staff; organ donation; organ

transplantation; patients; physicians; privacy; *privileged communication; radiology; review committees; state government; sterilization

MEDICAL RECORDS   *See under*

BIOETHICAL ISSUES / MEDICAL RECORDS / LEGAL ASPECTS
CONFIDENTIALITY / MEDICAL RECORDS
CONFIDENTIALITY / MEDICAL RECORDS / ADOLESCENTS
CONFIDENTIALITY / MEDICAL RECORDS / ALCOHOL ABUSE
CONFIDENTIALITY / MEDICAL RECORDS / ATTITUDES
CONFIDENTIALITY / MEDICAL RECORDS / DRUG ABUSE
CONFIDENTIALITY / MEDICAL RECORDS / DRUGS
CONFIDENTIALITY / MEDICAL RECORDS / EPIDEMIOLOGY
CONFIDENTIALITY / MEDICAL RECORDS / FAMOUS PERSONS
CONFIDENTIALITY / MEDICAL RECORDS / GENETIC DEFECTS
CONFIDENTIALITY / MEDICAL RECORDS / GREAT BRITAIN
CONFIDENTIALITY / MEDICAL RECORDS / HOSPITALS
CONFIDENTIALITY / MEDICAL RECORDS / ILLINOIS
CONFIDENTIALITY / MEDICAL RECORDS / INDUSTRIAL MEDICINE
CONFIDENTIALITY / MEDICAL RECORDS / LEGAL ASPECTS
CONFIDENTIALITY / MEDICAL RECORDS / MENTAL HEALTH
CONFIDENTIALITY / MEDICAL RECORDS / NEW YORK
CONFIDENTIALITY / MEDICAL RECORDS / PRIVACY PROTECTION STUDY COMMISSION
CONFIDENTIALITY / MEDICAL RECORDS / PSYCHIATRY
CONFIDENTIALITY / MEDICAL RECORDS / SOCIAL IMPACT
DISCLOSURE / CLERGY / MEDICAL RECORDS
INSTITUTIONAL POLICIES / HOSPITALS / MEDICAL RECORDS
PATIENT ACCESS / MEDICAL RECORDS / LEGAL ASPECTS
PATIENT ACCESS / MEDICAL RECORDS / LEGAL RIGHTS
PATIENT ACCESS / MEDICAL RECORDS / STANDARDS

MEDICAL RESEARCH COUNCIL   *See under*

RECOMBINANT DNA RESEARCH / MEDICAL RESEARCH COUNCIL

BE = bioethics accession number   fn. = footnotes   refs. = references

MEDICAL STAFF *See under*

ABORTION / MEDICAL STAFF / CONSCIENCE

MEDICAL ETHICS / MEDICAL STAFF / EDUCATION

MEDICINE

Pellegrino, Edmund D. Profession, patient, compassion, consent: meditations on medical philology. *Connecticut Medicine* 42(3): 175-178, Mar 1978. 7 refs. BE06823.
　　codes of ethics; *informed consent; medical ethics; *medicine; moral obligations; *patients; physician patient relationship; physicians

MEDICINE / CHRISTIAN ETHICS

Vale, J.A., ed. Medicine and the Christian Mind. London: Christian Medical Fellowship Publications, 1975. 142 p. References and footnotes included. ISBN 85111-956-5. BE07069.
　　abortion; aged; *Christian ethics; congenital defects; death; euthanasia; human experimentation; medical education; medical ethics; *medicine; newborns; physician patient relationship; psychiatry; suffering

MEDICINE / DECISION MAKING

Davis, Alan; Horobin, Gordon. The problem of priorities. *Journal of Medical Ethics* 3(3): 107-109, Sep 1977. 12 refs. BE05978.
　　*decision making; ecology; economics; health care; hospitals; *medicine; self induced illness; social control

MEDICINE / ETHICS

Pellegrino, Edmund D. Ethics and the moment of clinical truth. *Journal of the American Medical Association* 239(10): 960-961, 6 Mar 1978. 4 refs. Editorial. BE06822.
　　decision making; ethicists; *ethics; interdisciplinary communication; *medical education; *medical ethics; *medicine; patient care; physicians

Spicker, Stuart F. Medicine's influence on ethics: reflections on the putative moral role of medicine. *In* Spicker, Stuart F.; Engelhardt, H. Tristram, eds. Philosophical Medical Ethics: Its Nature and Significance. Proceedings. Boston: D. Reidel, 1977. p. 143-151. Proceedings of the Third Trans-Disciplinary Symposium on Philosophy and Medicine, Farmington, Connecticut, 11-13 Dec 1975. 11 refs. 2 fn. ISBN 90-277-0772-3. BE05869.
　　*ethics; goals; health; *medicine; moral obligations; physicians

MEDICINE / ETHICS / ARISTOTLE

Owens, Joseph. Aristotelian ethics, medicine, and the changing nature of man. *In* Spicker, Stuart F.; En-

gelhardt, H. Tristram, eds. Philosophical Medical Ethics: Its Nature and Significance. Proceedings. Boston: D. Reidel, 1977. p. 127-142. Proceedings of the Third Trans-Disciplinary Symposium on Philosophy and Medicine, Farmington, Connecticut, 11-13 Dec 1975. 37 refs. 19 fn. ISBN 90-277-0772-3. BE05868.
　　*ethics; human characteristics; *medicine; morality; *Aristotle

MEDICINE / HISTORICAL ASPECTS

Burns, Chester R. Richard Clarke Cabot (1868-1939) and reformation in American medical ethics. *Bulletin of the History of Medicine* 51(3): 353-368, Fall 1977. 50 fn. BE06000.
　　autopsies; case studies; diagnosis; *historical aspects; hospitals; *medicine; moral obligations; patient care; physicians; professional ethics; social workers; *Cabot, Richard C.

MEDICINE / LAW

Burns, Chester R. Professional ethics and the development of American law as applied to medicine. *In* Burns, Chester R., ed. Legacies in Law and Medicine. New York: Science History Publications, 1977. p. 299-310. 16 fn. ISBN 0-88202-164-8. BE05877.
　　government regulation; *law; legislation; medical education; malpractice; *medicine; obligations to society; physicians; *professional ethics; public health

MEDICINE / MORALITY

Hauerwas, Stanley. Medicine as a tragic profession. *In* Smith, David H., ed. No Rush to Judgement: Essays on Medical Ethics. Bloomington, Ind.: Indiana University, Poynter Center, 1977. p. 93-128. 42 fn. BE05797.
　　authoritarianism; injuries; human experimentation; medical ethics; *medicine; moral obligations; *morality; patient care; patients; physician patient relationship; physician's role; physicians; self determination; society; trust; value of life; *values; Illich, Ivan; Thomas, Lewis

MacIntyre, Alasdair. Can medicine dispense with a theological perspective on human nature? *In* Engelhardt, H. Tristram; Callahan, Daniel, eds. Knowledge, Value and Belief. Hastings-on-Hudson, N.Y.: Institute of Society, Ethics and the Life Sciences, 1977. p. 25-43. Foundations of Ethics and Its Relationship to Science: Volume 2. 10 fn. ISBN 0-916558-02-9. BE06176.
　　historical aspects; *medicine; moral obligations; *morality; national socialism; philosophy; *teleological ethics; theology

---

BE = bioethics accession number　　　fn. = footnotes　　　refs. = references

MEDICINE / POLITICS

Walters, Hugh. Medical ethics. *New Humanist* 92(1): 30-31, May-Jun 1976. BE06637.
 *medical ethics; medical fees; *medicine; moral obligations; physicians; *politics; state medicine

MEDICINE / PROFESSIONAL ETHICS

Ethical codes and professional conduct. *Journal of Medical Ethics* 3(3): 105-106, Sep 1977. 2 refs. Editorial. BE05977.
 codes of ethics; *medicine; nurses; physicians; *professional ethics; self regulation

Toulmin, Stephen. The meaning of professionalism: doctors' ethics and biomedical science. *In* Engelhardt, H. Tristram; Callahan, Daniel, eds. Knowledge, Value and Belief. Hastings-on-Hudson, N.Y.: Institute of Society, Ethics and the Life Sciences, 1977. p. 254-278. Foundations of Ethics and Its Relationship to Science: Volume 2. 9 fn. ISBN 0-916558-02-9. BE06177.
 confidentiality; goals; *conflict of interest; human experimentation; government regulation; law; legal obligations; medical ethics; *medicine; *moral obligations; obligations to society; *physician patient relationship; physicians; professional competence; *professional ethics; professional organizations; *science; values

MEDICINE / TELEOLOGICAL ETHICS

MacIntyre, Alasdair. Can medicine dispense with a theological perspective on human nature? *In* Engelhardt, H. Tristram; Callahan, Daniel, eds. Knowledge, Value and Belief. Hastings-on-Hudson, N.Y.: Institute of Society, Ethics and the Life Sciences, 1977. p. 25-43. Foundations of Ethics and Its Relationship to Science: Volume 2. 10 fn. ISBN 0-916558-02-9. BE06176.
 historical aspects; *medicine; moral obligations; *morality; national socialism; philosophy; *teleological ethics; theology

MEDICINE   *See under*

MORAL OBLIGATIONS / MEDICINE

RESOURCE ALLOCATION / MEDICINE / COSTS AND BENEFITS

SOCIAL CONTROL / MEDICINE

VALUE OF LIFE / MEDICINE / HISTORICAL ASPECTS

MENTAL HEALTH   *See under*

CONFIDENTIALITY / MEDICAL RECORDS / MENTAL HEALTH

HEALTH CARE / MENTAL HEALTH / MEDICAL ETHICS

HUMAN EXPERIMENTATION / MENTAL HEALTH

MENTALLY HANDICAPPED   *See under*

BEHAVIORAL RESEARCH / INSTITUTIONALIZED PERSONS / MENTALLY HANDICAPPED

HUMAN EXPERIMENTATION / CHILDREN / MENTALLY HANDICAPPED

HUMAN EXPERIMENTATION / INSTITUTIONALIZED PERSONS / MENTALLY HANDICAPPED

INFORMED CONSENT / INSTITUTIONALIZED PERSONS / MENTALLY HANDICAPPED

INVOLUNTARY COMMITMENT / CHILDREN / MENTALLY HANDICAPPED

INVOLUNTARY STERILIZATION / MENTALLY HANDICAPPED / LEGAL ASPECTS

RIGHT TO TREATMENT / MENTALLY HANDICAPPED

MENTALLY ILL   *See under*

COERCION / INSTITUTIONALIZED PERSONS / MENTALLY ILL

DEINSTITUTIONALIZED PERSONS / MENTALLY ILL / SOCIAL IMPACT

ELECTROCONVULSIVE THERAPY / INSTITUTIONALIZED PERSONS / MENTALLY ILL

HEALTH CARE DELIVERY / MENTALLY ILL / COMMUNITY SERVICES

INVOLUNTARY COMMITMENT / MENTALLY ILL

INVOLUNTARY COMMITMENT / MENTALLY ILL / HUMAN RIGHTS

INVOLUNTARY COMMITMENT / MENTALLY ILL / LEGAL ASPECTS

INVOLUNTARY COMMITMENT / MENTALLY ILL / LEGAL RIGHTS

INVOLUNTARY COMMITMENT / MENTALLY ILL / O'CONNOR V. DONALDSON

OPERANT CONDITIONING / AGED / MENTALLY ILL

RIGHT TO TREATMENT / INSTITUTIONALIZED PERSONS / MENTALLY ILL

RIGHT TO TREATMENT / MENTALLY ILL / NURSING HOMES

SELF DETERMINATION / MENTALLY ILL

TREATMENT REFUSAL / INSTITUTIONALIZED PERSONS / MENTALLY ILL

TREATMENT REFUSAL / MENTALLY ILL / LEGAL ASPECTS

TREATMENT REFUSAL / MENTALLY ILL / LEGAL RIGHTS

VOLUNTARY ADMISSION / ADOLESCENTS / MENTALLY ILL

MENTALLY RETARDED   *See under*

ABORTION / MENTALLY RETARDED / LEGAL RIGHTS

ALLOWING TO DIE / INSTITUTIONALIZED PERSONS / MENTALLY RETARDED

ALLOWING TO DIE / MENTALLY RETARDED / COMPETENCE

BE = bioethics accession number   fn. = footnotes   refs. = references

MENTALLY RETARDED *See under (cont'd.)*

ALLOWING TO DIE / MENTALLY RETARDED / JUDICIAL ACTION

ALLOWING TO DIE / MENTALLY RETARDED / LEGAL ASPECTS

INSTITUTIONALIZED PERSONS / MENTALLY RETARDED / LEGAL RIGHTS

INVOLUNTARY STERILIZATION / ADOLESCENTS / MENTALLY RETARDED

INVOLUNTARY STERILIZATION / MENTALLY RETARDED

INVOLUNTARY STERILIZATION / MENTALLY RETARDED / LEGAL ASPECTS

NEGATIVE REINFORCEMENT / MENTALLY RETARDED / CONSENT FORMS

PATIENT CARE / MENTALLY RETARDED / WILLOWBROOK STATE SCHOOL

STERILIZATION / MENTALLY RETARDED / LEGAL ASPECTS

THIRD PARTY CONSENT / MENTALLY RETARDED

THIRD PARTY CONSENT / MENTALLY RETARDED / LEGAL RIGHTS

MERCY KILLING *See* EUTHANASIA

METHODS *See under*

BEHAVIORAL RESEARCH / DECEPTION / METHODS

HUMAN EXPERIMENTATION / STATISTICS / METHODS

PSYCHIATRY / METHODS / HISTORICAL ASPECTS

MICHIGAN *See under*

HUMAN EXPERIMENTATION / MICHIGAN

MILITARY PERSONNEL *See under*

HUMAN EXPERIMENTATION / MILITARY PERSONNEL

MINNESOTA *See under*

HEALTH CARE / PRISONERS / MINNESOTA

MINORITY GROUPS *See under*

HEALTH CARE DELIVERY / MINORITY GROUPS

HUMAN EXPERIMENTATION / MINORITY GROUPS

STERILIZATION / MINORITY GROUPS

MODEL LEGISLATION *See under*

PATIENT CARE / INSTITUTIONALIZED PERSONS / MODEL LEGISLATION

TREATMENT REFUSAL / MODEL LEGISLATION

MORAL OBLIGATIONS / ADMINISTRATORS / NURSING HOMES

Lavelle, Michael J. An economic analysis of resource allocation for care of the aged. *Hospital Progress* 58(9): 99-103, Sep 1977. BE05782.
   *administrators; *aged; attitudes; children; decision making; *economics; federal government; financial support; law; *moral obligations; morality; *nursing homes; patient care; personhood; philosophy; renal dialysis; *resource allocation; selection for treatment; terminal care

MORAL OBLIGATIONS / FUTURE GENERATIONS / GENE POOL

Gastonguay, Paul R. Human genetics: a model of responsibility. *Ethics in Science and Medicine* 4(3-4): 119-134, 1977. 7 refs. BE06067.
   artificial insemination; carriers; evolution; *future generations; *gene pool; *genetic defects; genetics; genetic screening; legal liability; *moral obligations; obligations of society; parents; preventive medicine; reproduction; selective abortion; social interaction; survival

MORAL OBLIGATIONS / INDUSTRIAL MEDICINE / HEALTH HAZARDS

Morton, William E. The responsibility to report occupational health risks. *Journal of Occupational Medicine* 19(4): 258-260, Apr 1977. 23 refs. BE05807.
   *communication; employment; ethical review; financial support; federal government; *health hazards; *industrial medicine; industry; medical ethics;* moral obligations; occupational diseases; *physicians

MORAL OBLIGATIONS / MEDICINE

Twiss, Sumner B. The problem of moral responsibility in medicine. *Journal of Medicine and Philosophy* 2(4): 330-375, Dec 1977. 38 refs. BE06153.
   administrators; biomedical research; hospitals; government regulation; human experimentation; investigator subject relationship; legal liability; *medicine; *moral obligations; nontherapeutic research; physician patient relationship; physician's role; public health; resource allocation

MORAL OBLIGATIONS / PHYSICIAN PATIENT RELATIONSHIP

Angrist, Alfred A., et al. Symposium: humanism and medical ethics. *New York State Journal of Medicine* 77(9): 1448-1462, Aug 1977. 22 refs. BE05834.
   altruism; biomedical technologies; compensation; curriculum; dehumanization; hospitals; *humanism; *medical education; *medical ethics; medical fees; *moral obligations; patient care; *physician patient relationship; physicians; professional ethics; self determination; teaching methods; *values

---

BE = bioethics accession number   fn. = footnotes   refs. = references

MORAL OBLIGATIONS / PHYSICIAN PATIENT RELATIONSHIP / CONFLICT OF INTEREST

Toulmin, Stephen. The meaning of professionalism: doctors' ethics and biomedical science. *In* Engelhardt, H. Tristram; Callahan, Daniel, eds. Knowledge, Value and Belief. Hastings-on-Hudson, N.Y.: Institute of Society, Ethics and the Life Sciences, 1977. p. 254-278. Foundations of Ethics and Its Relationship to Science: Volume 2. 9 fn. ISBN 0-916558-02-9. BE06177.
   confidentiality; goals; *conflict of interest; human experimentation; government regulation; law; legal obligations; medical ethics; *medicine; *moral obligations; obligations to society; *physician patient relationship; physicians; professional competence; *professional ethics; professional organizations; *science; values

MORAL OBLIGATIONS / PHYSICIAN PATIENT RELATIONSHIP / SEXUALITY

Hill, Denis. The General Medical Council: frame of reference or arbiter of morals? *Journal of Medical Ethics* 3(3): 110-114, Sep 1977. 4 refs. BE05979.
   attitudes; education; *moral obligations; morality; peer review; *physician patient relationship; physicians; professional ethics; *sexuality; *General Medical Council; Great Britain

MORAL OBLIGATIONS / PHYSICIANS

Childress, James F. Citizen and physician: harmonious or conflicting responsibilities? *Journal of Medicine and Philosophy* 2(4): 401-409, Dec 1977. 3 refs. 3 fn. BE06135.
   justice; *moral obligations; obligations to society; *patient care; *physicians; resource allocation; utilitarianism

Jonsen, Albert R. Do no harm: axiom of medical ethics. *In* Spicker, Stuart F.; Engelhardt, H. Tristram, eds. Philosophical Medical Ethics: Its Nature and Significance. Proceedings. Boston: D. Reidel, 1977. p. 27-46. Proceedings of the Third Trans-Disciplinary Symposium on Philosophy and Medicine, Farmington, Connecticut, 11-13 Dec 1975. Discussion by Louis Lasagna on p. 43-46. 25 refs. 1 fn. ISBN 90-277-0772-3. BE05861.
   codes of ethics; double effect; injuries; medical ethics; medicine; *moral obligations; *patient care; physician's role; *physicians; risks and benefits; standards

Jonsen, Albert R.; Jameton, Andrew L. Social and political responsibilities of physicians. *Journal of Medicine and Philosophy* 2(4): 376-400, Dec 1977. 35 refs. 2 fn. BE06140.
   diagnosis; health care; justice; *moral obligations; *obligations to society; *patient care; *physicians; political activity; resource allocation

Pellegrino, Edmund D. Moral agency and professional

ethics: some notes on transformation of the physician-patient encounter. *In* Spicker, Stuart F.; Engelhardt, H. Tristram, eds. Philosophical Medical Ethics: Its Nature and Significance. Proceedings. Boston: D. Reidel, 1977. p. 213-220. Proceedings of the Third Trans-Disciplinary Symposium on Philosophy and Medicine, Farmington, Connecticut, 11-13 Dec 1975. 9 refs. 1 fn. ISBN 90-277-0772-3. BE05872.
   authoritarianism; medical ethics; *moral obligations; physician patient relationship; *physicians; self determination; values

Spicker, Stuart F., et al. The physician as moral agent: round table discussion. *In* Spicker, Stuart F.; Engelhardt, H. Tristram, eds. Philosophical Medical Ethics: Its Nature and Significance. Proceedings. Boston: D. Reidel, 1977. p. 221-244. Proceedings of the Third Trans-Disciplinary Symposium on Philosophy and Medicine, Farmington, Connecticut, 11-13 Dec 1975. 15 refs. 1 fn. ISBN 90-277-0772-3. BE05873.
   legal aspects; medical ethics; medicine; *moral obligations; obligations to society; physician patient relationship; *physicians; resource allocation

MORAL OBLIGATIONS / PHYSICIANS / STRIKES

Dworkin, Gerald. Strikes and the National Health Service: some legal and ethical issues. *Journal of Medical Ethics* 3(2): 76-82, Jun 1977. 12 refs. BE05647.
   criminal law; employment; health personnel; injuries; legal liability; *moral obligations; patient care; *physicians; social impact; state medicine; *strikes; Great Britain; National Health Service

Hunter, Thomas H., et al. Has the Physician the Right to Strike? [Videorecording]. Charlottesville, Va.: University of Virginia Medical Center, Health Sciences Library, 1975. 1 cassette; 60 min.; sound; black and white; 3/4 in. Tape of the Medicine and Society Conference, 11 Nov 1975. Panel discussion moderated by Thomas H. Hunter. BE07047.
   alternatives; costs and benefits; employment; federal government; hospitals; legal aspects; legislation; *moral obligations; morality; obligations to society; patient care; *physicians; rights; *strikes

---

BE = bioethics accession number   fn. = footnotes   refs. = references

MORAL OBLIGATIONS *See under (cont'd.)*

KILLING / MORAL OBLIGATIONS / RIGHTS

PATIENT CARE / MORAL OBLIGATIONS

PROLONGATION OF LIFE / MORAL OBLIGATIONS

RECOMBINANT DNA RESEARCH / MORAL OBLIGATIONS

SELECTIVE ABORTION / MORAL OBLIGATIONS / GENETIC DEFECTS

TORTURE / MORAL OBLIGATIONS / PHYSICIAN'S ROLE

MORAL POLICY *See under*

RESOURCE ALLOCATION / MORAL POLICY / SCARCITY

SELECTION FOR TREATMENT / MORAL POLICY / SPINA BIFIDA

MORALITY *See under*

ABORTION / FINANCIAL SUPPORT / MORALITY

ABORTION / MORALITY

ABORTION / MORALITY / ATTITUDES

ACTIVE EUTHANASIA / MORALITY

ALLOWING TO DIE / MORALITY

BEHAVIORAL GENETICS / INTELLIGENCE / MORALITY

HUMAN EXPERIMENTATION / FETUSES / MORALITY

IN VITRO FERTILIZATION / MORALITY

MEDICINE / MORALITY

NURSES / STRIKES / MORALITY

PRENATAL DIAGNOSIS / MORALITY / LEGAL ASPECTS

REPRODUCTIVE TECHNOLOGIES / MORALITY

SOCIOBIOLOGY / MORALITY

SUICIDE / MORALITY

VOLUNTARY EUTHANASIA / MORALITY

MORALS *See* ETHICS

MORGENTALER V. THE QUEEN *See under*

ABORTION / MORGENTALER V. THE QUEEN

MORTALITY *See under*

CONFIDENTIALITY / FAMOUS PERSONS / MORTALITY

MOTIVATION *See under*

HUMAN EXPERIMENTATION / VOLUNTEERS / MOTIVATION

NATIONAL ACADEMY OF SCIENCES *See under*

RECOMBINANT DNA RESEARCH / GOVERN-

NATIONAL ACADEMY OF SCIENCES *See under (cont'd.)*

MENT REGULATION / NATIONAL ACADEMY OF SCIENCES

NATIONAL COMMISSION FOR PROTECTION OF HUMAN SUBJECTS *See under*

HUMAN EXPERIMENTATION / CHILDREN / NATIONAL COMMISSION FOR PROTECTION OF HUMAN SUBJECTS

HUMAN EXPERIMENTATION / NATIONAL COMMISSION FOR PROTECTION OF HUMAN SUBJECTS

PSYCHOSURGERY / NATIONAL COMMISSION FOR PROTECTION OF HUMAN SUBJECTS

PSYCHOSURGERY / NATIONAL COMMISSION FOR PROTECTION OF HUMAN SUBJECTS / EVALUATION

NATIONAL HEALTH INSURANCE / COSTS AND BENEFITS

Bayles, Michael D. National health insurance and non-covered services. *Journal of Health Politics, Policy and Law* 2(3): 335-348, Fall 1977. 27 fn. BE06342.
abortion; altruism; *costs and benefits; decision making; extraordinary treatment; health care delivery; *national health insurance; politics; prognosis; review committees

NATIONAL INSTITUTES OF HEALTH *See under*

RECOMBINANT DNA RESEARCH / NATIONAL INSTITUTES OF HEALTH

NEGATIVE REINFORCEMENT / INFORMED CONSENT

Cook, J. William; Altman, Karl; Haavik, Sarah. Consent for aversive treatment: a model form. *Mental Retardation* 16(1): 47-51, Feb 1978. 9 refs. BE06915.
children; *consent forms; ethics committees; *informed consent; legal aspects; *mentally retarded; *negative reinforcement; standards

NEGATIVE REINFORCEMENT / MENTALLY RETARDED / CONSENT FORMS

Cook, J. William; Altman, Karl; Haavik, Sarah. Consent for aversive treatment: a model form. *Mental Retardation* 16(1): 47-51, Feb 1978. 9 refs. BE06915.
children; *consent forms; ethics committees; *informed consent; legal aspects; *mentally retarded; *negative reinforcement; standards

NEGATIVE REINFORCEMENT / PUNISHMENT

Logan, Daniel L.; Turnage, John R. Ethical

BE = bioethics accession number   fn. = footnotes   refs. = references

considerations in the use of faradic aversion therapy. *Behavioral Engineering* 3(1): 29-34, Summer 1975. 38 refs. BE06607.

adults; children; dangerousness; injuries; *methods; *negative reinforcement; operant conditioning; positive reinforcement; *punishment; risks and benefits

NEGLIGENCE *See under*

ABORTION / PHYSICIANS / NEGLIGENCE

NETHERLANDS *See under*

RECOMBINANT DNA RESEARCH / NETHERLANDS

NEVADA *See under*

ALLOWING TO DIE / LIVING WILLS / NEVADA

NEW JERSEY SUPREME COURT *See under*

ALLOWING TO DIE / NEW JERSEY SUPREME COURT / IN RE QUINLAN

NEW SOUTH WALES *See under*

PSYCHOSURGERY / ETHICS COMMITTEES / NEW SOUTH WALES
PSYCHOSURGERY / EVALUATION / NEW SOUTH WALES
PSYCHOSURGERY / LEGAL ASPECTS / NEW SOUTH WALES
PSYCHOSURGERY / LEGISLATION / NEW SOUTH WALES

NEW YORK ACADEMY OF MEDICINE *See under*

PATIENT CARE / LAETRILE / NEW YORK ACADEMY OF MEDICINE

NEW YORK *See under*

CONFIDENTIALITY / MEDICAL RECORDS / NEW YORK

NEW ZEALAND *See under*

ABORTION / LEGAL ASPECTS / NEW ZEALAND
ABORTION / NEW ZEALAND
CONTRACEPTION / NEW ZEALAND
FAMILY PLANNING / NEW ZEALAND
PRIVILEGED COMMUNICATION / PHYSICIANS / NEW ZEALAND
STERILIZATION / NEW ZEALAND
VOLUNTARY STERILIZATION / NEW ZEALAND

NEWBORNS *See also* CHILDREN, INFANTS

NEWBORNS / CONGENITAL DEFECTS / PHYSICIAN'S ROLE

Shaw, Anthony. The ethics of proxy consent. *In* Swin-yard, Chester A., ed. Decision Making and the Defective Newborn. Proceedings of a Conference on Spina Bifida and Ethics. Springfield, Ill.: Charles C. Thomas, 1978. p. 589-597. 11 refs. ISBN 0-398-03662-4. BE06964.

adolescents; *allowing to die; blood transfusions; children; *congenital defects; *decision making; family relationship; informed consent; legal aspects; *newborns; parental consent; *patient care; *physician's role; quality of life; society; *third party consent; treatment refusal; withholding treatment

NEWBORNS / SPINA BIFIDA / ATTITUDES

Fletcher, John C. Spina bifida with myelomeningocele: a case study in attitudes towards defective newborns. *In* Swinyard, Chester A., ed. Decision Making and the Defective Newborn. Proceedings of a Conference on Spina Bifida and Ethics. Springfield, Ill.: Charles C. Thomas, 1978. p. 281-303. 39 refs. ISBN 0-398-03662-4. BE06956.

active euthanasia; allowing to die; *attitudes; attitudes to death; congenital defects; decision making; extraordinary treatment; family problems; handicapped; *moral policy; *newborns; obligations of society; patient care; parents; physicians; prognosis; *quality of life; resource allocation; *selection for treatment; *spina bifida; *value of life; values

NEWBORNS / SPINA BIFIDA / DECISION MAKING

Duff, Raymond S. A physician's role in the decision-making process: a physician's experience. *In* Swinyard, Chester A., ed. Decision Making and the Defective Newborn. Proceedings of a Conference on Spina Bifida and Ethics. Springfield, Ill.: Charles C. Thomas, 1978. p. 194-219. 18 refs. Case study. ISBN 0-398-03662-4. BE06954.

*allowing to die; alternatives; biomedical technologies; *decision making; disclosure; ethics; extraordinary treatment; family members; family problems; *handicapped; hospitals; informed consent; institutional policies; killing; legal aspects; medical staff; medicine; *newborns; parents; *physician's role; patient care; prognosis; physicians; psychological stress; public opinion; quality of life; risks and benefits; social impact; *spina bifida; terminally ill; withholding treatment

Fost, Norman. How decisions are made: a physician's view. *In* Swinyard, Chester A., ed. Decision Making and the Defective Newborn. Proceedings of a Conference on Spina Bifida and Ethics. Springfield, Ill.: Charles C. Thomas, 1978. p. 220-230. 16 refs. ISBN 0-398-03662-4. BE06955.

allowing to die; *decision making; ethical analysis; ethics; extraordinary treatment; family problems; handicapped; informed consent; legal aspects; *newborns; parents; *patient care; *physician's

---

BE = bioethics accession number   fn. = footnotes   refs. = references

role; physicians; prognosis; quality of life; selection for treatment; *spina bifida; withholding treatment

Swinyard, Chester A., ed. Decision Making and the Defective Newborn. Proceedings of a Conference on Spina Bifida and Ethics. Springfield, Ill.: Charles C. Thomas, 1978. 649 p. References and footnotes included. ISBN 0-398-03662-4. BE06951.

*allowing to die; attitudes; congenital defects; *decision making; ethics; extraordinary treatment; informed consent; legal aspects; medical ethics; *newborns; parental consent; *patient care; physician's role; *quality of life; *selection for treatment; socioeconomic factors; *spina bifida; withholding treatment

Veatch, Robert M. Abnormal newborns and the physician's role: models of physician decision making. *In* Swinyard, Chester A., ed. Decision Making and the Defective Newborn. Proceedings of a Conference on Spina Bifida and Ethics. Springfield, Ill.: Charles C. Thomas, 1978. p. 174-193. 37 fn. ISBN 0-398-03662-4. BE06953.

allowing to die; blood transfusions; congenital defects; *contracts; *decision making; extraordinary treatment; judicial action; killing; legal aspects; legal obligations; medical ethics; *moral obligations; *newborns; *parental consent; physician patient relationship; *physician's role; physicians; prognosis; *prolongation of life; *quality of life; *selection for treatment; *spina bifida; technical expertise; *treatment refusal; values; withholding treatment; Hippocratic Oath

NEWBORNS / SPINA BIFIDA / PHYSICIAN'S ROLE

Veatch, Robert M. Abnormal newborns and the physician's role: models of physician decision making. *In* Swinyard, Chester A., ed. Decision Making and the Defective Newborn. Proceedings of a Conference on Spina Bifida and Ethics. Springfield, Ill.: Charles C. Thomas, 1978. p. 174-193. 37 fn. ISBN 0-398-03662-4. BE06953.

allowing to die; blood transfusions; congenital defects; *contracts; *decision making; extraordinary treatment; judicial action; killing; legal aspects; legal obligations; medical ethics; *moral obligations; *newborns; *parental consent; physician patient relationship; *physician's role; physicians; prognosis; *prolongation of life; *quality of life; *selection for treatment; *spina bifida; technical expertise; *treatment refusal; values; withholding treatment; Hippocratic Oath

NEWBORNS   *See under*

ACTIVE EUTHANASIA / NEWBORNS / SPINA BIFIDA

ALLOWING TO DIE / NEWBORNS

ALLOWING TO DIE / NEWBORNS / CHRISTIAN ETHICS

ALLOWING TO DIE / NEWBORNS / CONGENI-

NEWBORNS   *See under (cont'd.)*

TAL DEFECTS

ALLOWING TO DIE / NEWBORNS / DOWN'S SYNDROME

ALLOWING TO DIE / NEWBORNS / GENETIC DEFECTS

ALLOWING TO DIE / NEWBORNS / SPINA BIFIDA

EXTRAORDINARY TREATMENT / NEWBORNS / SPINA BIFIDA

GENETIC SCREENING / NEWBORNS

GENETIC SCREENING / NEWBORNS / LEGISLATION

GENETIC SCREENING / NEWBORNS / PHENYLKETONURIA

GENETIC SCREENING / NEWBORNS / SICKLE CELL ANEMIA

HUMAN EXPERIMENTATION / NEWBORNS

PATIENT CARE / NEWBORNS / CONGENITAL DEFECTS

SELECTION FOR TREATMENT / NEWBORNS / INTENSIVE CARE UNITS

SELECTION FOR TREATMENT / NEWBORNS / SPINA BIFIDA

TREATMENT REFUSAL / NEWBORNS / CONGENITAL DEFECTS

WITHHOLDING TREATMENT / NEWBORNS / CONGENITAL DEFECTS

NIH GUIDELINES   *See under*

RECOMBINANT DNA RESEARCH / GOVERNMENT REGULATION / NIH GUIDELINES

RECOMBINANT DNA RESEARCH / NIH GUIDELINES

NOLAN, KIERAN   *See under*

EUTHANASIA / ROMAN CATHOLIC ETHICS / NOLAN, KIERAN

NORMATIVE ETHICS   *See under*

KILLING / NORMATIVE ETHICS

MEDICAL ETHICS / NORMATIVE ETHICS

SELECTION FOR TREATMENT / SPINA BIFIDA / NORMATIVE ETHICS

NORTH CAROLINA   *See under*

ALLOWING TO DIE / LIVING WILLS / NORTH CAROLINA

BRAIN DEATH / NORTH CAROLINA

NORWAY   *See under*

BEHAVIORAL RESEARCH / GOVERNMENT REGULATION / NORWAY

CONFIDENTIALITY / BEHAVIORAL RESEARCH / NORWAY

BE = bioethics accession number   fn. = footnotes   refs. = references

NUCLEAR ENERGY *See under*
>   RECOMBINANT DNA RESEARCH / NUCLEAR
>   ENERGY

NURSES / STRIKES / MORALITY

Jarvis, Peter. Nursing and the ethics of withdrawing
professional services. *Nursing Mirror* 146(7): 30-31,
16 Feb 1978. BE06780.
>   *morality; *nurses; patient care; *strikes

NURSES *See under*
>   ATTITUDES TO DEATH / NURSES / NURSING
>   EDUCATION
>   BIOETHICAL ISSUES / NURSES
>   ETHICS / NURSES
>   HEALTH CARE DELIVERY / NURSES / COMMU-
>   NITY SERVICES
>   HUMAN EXPERIMENTATION / NURSES / LEGAL
>   ASPECTS
>   PATIENT CARE / NURSES / LEGAL ASPECTS
>   RESUSCITATION / NURSES
>   RESUSCITATION / NURSES / CHRONICALLY ILL
>   TERMINAL CARE / NURSES

NURSING EDUCATION *See under*
>   ATTITUDES TO DEATH / NURSES / NURSING
>   EDUCATION
>   ETHICS / NURSING EDUCATION
>   MEDICAL ETHICS / NURSING EDUCATION
>   MEDICAL ETHICS / NURSING EDUCATION /
>   CURRICULUM

NURSING ETHICS

Churchill, Larry. Ethical issues of a profession in tran-
sition. *American Journal of Nursing* 77(5): 873-875,
May 1977. 1 fn. BE05595.
>   nurses; *nursing ethics; physicians; professional
>   competence; social dominance

Davis, Ann J. Ethical dilemmas and nursing practice.
Linacre Quarterly 44(4): 302-311, Nov 1977. 9 refs.
BE06099.
>   codes of ethics; disclosure; discrimination; females;
>   hospitals; nurses; *nursing ethics; physician nurse
>   relationship; professional organizations

Jameton, Andrew. The nurse: when roles and rules conflict. *Hastings Center Report* 7(4): 22-23, Aug 1977.
BE05747.
>   bioethics; decision making; nurses; *nursing ethics;
>   physicians

Maurice, Shirley; Warrick, Louise. Ethics and morals
in nursing. *MCN: American Journal of Maternal
Child Nursing* 2(6): 343-347, Nov/Dec 1977. 15 fn.
BE06502.
>   codes of ethics; ethics; moral obligations; nurses;
>   *nursing ethics

The Nurse's Dilemma: Ethical Considerations in Nurs-
ing Practice. Geneva: International Council of
Nurses, 1977. 114 p. Bibliography: p. 111-114.
BE06003.
>   abortion; codes of ethics; decision making; confi-
>   dentiality; disclosure; family planning; malprac-
>   tice; nurses; *nursing ethics; patient care;
>   physicians; professional ethics; religious beliefs;
>   values; withholding treatment

NURSING ETHICS / PSYCHIATRY

Mellor, Peter D. Moral dilemmas in psychiatric nursing. *Nursing Mirror* 145(3): 20-22, 21 Jul 1977.
BE05816.
>   electroconvulsive therapy; freedom; informed con-
>   sent; institutionalized persons; mentally ill; moral
>   obligations; nurses; *nursing ethics; patient care;
>   patients' rights; professional patient relationship;
>   *psychiatry

NURSING ETHICS *See under*
>   CODES OF ETHICS / NURSING ETHICS
>   HUMAN EXPERIMENTATION / NURSING ETH-
>   ICS
>   PATIENT CARE / NURSING ETHICS

NURSING HOMES *See under*
>   MORAL OBLIGATIONS / ADMINISTRATORS /
>   NURSING HOMES
>   RESOURCE ALLOCATION / AGED / NURSING
>   HOMES
>   RIGHT TO TREATMENT / MENTALLY ILL /
>   NURSING HOMES

O'CONNOR V. DONALDSON *See under*
>   INVOLUNTARY COMMITMENT / MENTALLY
>   ILL / O'CONNOR V. DONALDSON
>   RIGHT TO TREATMENT / INSTITUTIONALIZED
>   PERSONS / O'CONNOR V. DONALDSON

OBLIGATIONS OF SOCIETY *See under*
>   HEALTH CARE DELIVERY / OBLIGATIONS OF
>   SOCIETY

OBLIGATIONS TO SOCIETY / PHYSICIANS

Hunter, Thomas H., et al. The Physician as Double
Agent. [Videorecording]. Charlottesville, Va.: Uni-
versity of Virginia Medical Center, Health Sciences
Library, 1977. 1 cassette; 60 min.; sound; color; 3/4
in. Tape of the Medical Center Hour, 5 Jan 1977.
Panel discussion moderated by Thomas H. Hunter.
BE07052.
>   case reports; *confidentiality; conflict of interest;
>   dangerousness; decision making; disclosure; em-
>   ployment; informed consent; legal aspects; legal
>   obligations; medical education; mentally ill; moral
>   obligations; *obligations to society; *physician pa-
>   tient relationship; *physicians; psychiatric diagno-

BE = bioethics accession number   fn. = footnotes   refs. = references

sis; psychiatry; students; teaching methods; universities

Jonsen, Albert R.; Jameton, Andrew L. Social and political responsibilities of physicians. *Journal of Medicine and Philosophy* 2(4): 376-400, Dec 1977. 35 refs. 2 fn. BE06140.
diagnosis; health care; justice; *moral obligations; *obligations to society; *patient care; *physicians; political activity; resource allocation

OBSTETRICS AND GYNECOLOGY *See under*
ABORTION / OBSTETRICS AND GYNECOLOGY / ATTITUDES
GENETIC SCREENING / OBSTETRICS AND GYNECOLOGY / ATTITUDES

OCCUPATIONAL DISEASES *See under*
CONFIDENTIALITY / OCCUPATIONAL DISEASES

OFFICE OF TECHNOLOGY ASSESSMENT *See under*
HEALTH CARE / OFFICE OF TECHNOLOGY ASSESSMENT

OKLAHOMA *See under*
HEALTH CARE / ADOLESCENTS / OKLAHOMA

OPERANT CONDITIONING *See also* NEGATIVE REINFORCEMENT, POSITIVE REINFORCEMENT

OPERANT CONDITIONING

Goldiamond, Israel. Protection of human subjects and patients. *In* Krapfl, Jon E.; Vargas, Ernest A., eds. Behaviorism and Ethics. Kalamazoo: Behaviordelia, 1977. p. 129-195. Commentary by Pamela Meadowcroft on p. 188-195. 41 refs. ISBN 0-914-47425-1. BE06546.
*behavior control; *coercion; consequences; ethical review; *human experimentation; goals; intention; informed consent; investigator subject relationship; *operant conditioning; remuneration; risks and benefits; self determination

Roos, P. Ethical use of behavior modification techniques. *In* Mittler, Peter, ed. Research to Practice in Mental Retardation. Volume 1: Care and Intervention. Baltimore: University Park Press, 1977. p. 71-77. 11 refs. ISBN 0-8391-1122-3. BE06180.
ethics committees; *goals; *negative reinforcement; *operant conditioning

OPERANT CONDITIONING / AGED / MENTALLY ILL

Harris, Sandra L., et al. Behavior modification therapy with elderly demented patients: implementation and ethical considerations. *Journal of Chronic Diseases* 30(3): 129-134, Mar 1977. 18 refs. BE05913.
*aged; coercion; goals; institutionalized persons;

*mentally ill; *operant conditioning; paternalism; risks and benefits; self determination; values

OPERANT CONDITIONING / CLONCE V. RICHARDSON

Brenner, Edgar H. Behavior modification. *In* Bernstein, Sidney, ed. Criminal Defense Techniques: Volume 4. New York: Matthew Bender, 1976. p. 73-1—73-51. 222 fn. BE06609.
adolescents; adults; behavioral research; due process; ethics committees; federal government; government regulation; human experimentation; informed consent; institutionalized persons; involuntary programs; judicial action; legal rights; *legal aspects; legislation; mentally ill; *operant conditioning; patient care; *positive reinforcement; *prisoners; *program descriptions; right to treatment; state government; *Clonce v. Richardson; National Commission for Protection of Human Subjects; National Research Act; *Project START

OPERANT CONDITIONING / PATIENT CARE / CONTRACTS

Atthowe, John M. Legal and ethical accountability in everyday practice. *Behavioral Engineering* 3(2): 35-38, Fall 1975. 7 refs. BE06608.
behavior disorders; *contracts; disclosure; health personnel; human experimentation; *informed consent; institutionalized persons; legal aspects; legal obligations; mentally ill; moral obligations; *operant conditioning; *patient care; patient participation; peer review; psychology; records

OPERANT CONDITIONING / PRISONERS

Chavkin, Samuel. The Mind Stealers: Psychosurgery and Mind Control. Boston: Houghton Mifflin, 1978. 228 p. 400 fn. ISBN 0-395-26381-6. BE06977.
*adolescents; aggression; *behavior control; behavioral genetics; brain pathology; children; *electrical stimulation of the brain; evaluation; human experimentation; imprisonment; law enforcement; institutionalized persons; legal aspects; legal rights; *operant conditioning; *prisoners; prognosis; psychoactive drugs; *psychosurgery; socioeconomic factors; *violence; Project START

OPERANT CONDITIONING / PRISONERS / LEGAL ASPECTS

Brenner, Edgar H. Behavior modification. *In* Bernstein, Sidney, ed. Criminal Defense Techniques: Volume 4. New York: Matthew Bender, 1976. p. 73-1—73-51. 222 fn. BE06609.
adolescents; adults; behavioral research; due process; ethics committees; federal government; government regulation; human experimentation; informed consent; institutionalized persons; involuntary programs; judicial action; legal rights; *le-

---

BE = bioethics accession number   fn. = footnotes   refs. = references

gal aspects; legislation; mentally ill; *oper-ant conditioning; patient care; *positive rein-forcement; *prisoners; *program descriptions; right to treatment; state government; *Clonce v. Richardson; National Commission for Protection of Human Subjects; National Research Act; *Project START

OPERANT CONDITIONING / PROJECT START

Brenner, Edgar H. Behavior modification. *In* Bernstein, Sidney, ed. Criminal Defense Techniques: Volume 4. New York: Matthew Bender, 1976. p. 73-1—73-51. 222 fn. BE06609.
adolescents; adults; behavioral research; due process; ethics committees; federal government; government regulation; human experimentation; informed consent; institutionalized persons; involuntary programs; judicial action; legal rights; *legal aspects; legislation; mentally ill; *operant conditioning; patient care; *positive reinforcement; *prisoners; *program descriptions; right to treatment; state government; *Clonce v. Richardson; National Commission for Protection of Human Subjects; National Research Act; *Project START

OPERANT CONDITIONING / VALUES

London, Perry. Behavior control, values and the future. *In* Ellison, Craig W., ed. Modifying Man: Implications and Ethics. Washington: University Press of America, 1978. p. 189-208. 8 fn. ISBN 0-8191-0302-0. BE06671.
*behavior control; behavior disorders; coercion; common good; drugs; freedom; homosexuals; human rights; mentally ill; *operant conditioning; politics; psychotherapy; *self determination; *values

ORGAN DONATION

Simmons, Roberta G.; Klein, Susan D.; Simmons, Richard L. Gift of Life: The Social and Psychological Impact of Organ Transplantation. New York: John Wiley, 1977. 526 p. Bibliography: p. 485-507. ISBN 0-471-79197-0. BE06686.
adolescents; *cadavers; case reports; children; chronically ill; costs and benefits; *decision making; *family members; family relationship; financial support; health care delivery; kidney diseases; *kidneys; motivation; *organ donation; *organ donors; *organ transplantation; *psychological stress; quality of life; *social adjustment; *socioeconomic factors; standards; statistics; *transplant recipients

ORGAN DONATION / CADAVERS

Simmons, Roberta G.; Klein, Susan D.; Simmons, Richard L. Transplantation and changing norms: cultural lag and ethical ambiguities. *In their* Gift of Life: The Social and Psychological Impact of Organ

Transplantation. New York: John Wiley, 1977. p. 9-44. References included. ISBN 0-471-79197-0. BE06687.
adolescents; aged; attitudes; brain death; *cadavers; *case reports; chronically ill; decision making; determination of death; donor cards; *family members; federal government; financial support; *informed consent; *kidneys; legal aspects; medical staff; *organ donors; *organ donation; *organ transplantation; *physician's role; physicians; public opinion; remuneration; resource allocation; scarcity; *selection for treatment; *socioeconomic factors; standards; *transplant recipients; volunteers; withholding treatment

ORGAN DONATION / CHILDREN

Cohen, Cynthia B., et al. Children who donate kidneys. *New England Journal of Medicine* 296(23): 1361-1362, 9 Jun 1977. 5 refs. Letter. BE05644.
adults; *children; coercion; *informed consent; international aspects; legal aspects; *organ donation; parental consent; risks and benefits; siblings; twinning

ORGAN DONATION / CHILDREN / LEGAL ASPECTS

Clothier, C.M. The law and the juvenile donor. *Lancet* 1(8026): 1356-1357, 25 Jun 1977. 5 fn. BE05642.
*children; decision making; *informed consent; *legal aspects; motivation; *organ donation; parental consent; surgery; twinning

ORGAN DONATION / FAMILY MEMBERS

Needleman, Lionel. Valuing other people's lives. *Manchester School of Economics and Social Studies* 44(4): 309-342, Dec 1976. 57 refs. 14 fn. BE06208.
economics; *family members; kidneys; motivation; *organ donation; organ transplantation; parents; *prolongation of life; *risks and benefits; statistics; treatment refusal; *value of life

Simmons, Roberta G.; Klein, Susan D.; Simmons, Richard L. Transplantation and changing norms: cultural lag and ethical ambiguities. *In their* Gift of Life: The Social and Psychological Impact of Organ Transplantation. New York: John Wiley, 1977. p. 9-44. References included. ISBN 0-471-79197-0. BE06687.
adolescents; aged; attitudes; brain death; *cadavers; *case reports; chronically ill; decision making; determination of death; donor cards; *family members; federal government; financial support; *informed consent; *kidneys; legal aspects; medical staff; *organ donors; *organ donation; *organ transplantation; *physician's role; physicians; public opinion; remuneration; resource allocation; scarcity; *selection for treatment; *socioeconomic

---

BE = bioethics accession number   fn. = footnotes   refs. = references

factors; standards; *transplant recipients; volunteers; withholding treatment

ORGAN DONATION / PSYCHOLOGICAL STRESS

Adams, Virginia. Organ transplant controversy: a gift of life or lifetime I.O.U.? *New York Times,* 9 Jul 1977, p. 18. BE06052.
> family members; kidneys; motivation; *organ donation; *psychological stress

ORGAN DONATION / REMUNERATION

Brams, Marvin. Transplantable human organs: should their sale be authorized by state statutes? *American Journal of Law and Medicine* 3(2): 183-195, Summer 1977. 27 fn. BE05812.
> altruism; cadavers; costs and benefits; economics; kidneys; legislation; *organ donation; organ donors; organ transplantation; *remuneration; scarcity; transplantation

ORGAN DONATION / UNIFORM ANATOMICAL GIFT ACT

Weissman, Steven I. Why the Uniform Anatomical Gift Act has failed. *Trusts and Estates* 116(4): 264-267+, Apr 1977. 62 fn. BE05589.
> blood donation; *body parts and fluids; cadavers; family members; informed consent; kidneys; mandatory programs; model legislation; *organ donation; organ transplantation; patients; property rights; public opinion; remuneration; *scarcity; self determination; *Uniform Anatomical Gift Act

ORGAN DONORS

Bernstein, Dorothy M. The organ donor. *Journal of the American Medical Association* 237(24): 2643-2644, 13 Jun 1977. 10 refs. BE05639.
> adolescents; family members; evaluation; kidneys; motivation; *organ donors; psychological stress; risks and benefits; self concept

ORGAN DONORS / KIDNEYS / SOCIAL ADJUSTMENT

Marshall, John R.; Fellner, Carl H. Kidney donors revisited. *American Journal of Psychiatry* 134(5): 575-576, May 1977. 6 refs. BE05623.
> attitudes; family members; *kidneys; *organ donors; psychological stress; *social adjustment

ORGAN TRANSPLANTATION

Anderson, Norman. The prolongation of life, transplant surgery, euthanasia and suicide. *In his* Issues of Life and Death: Abortion, Birth Control, Capital Punishment, Euthanasia. Downers Grove, Ill.: InterVarsity Press, 1977. p. 85-107. 19 fn. ISBN 0-87784-721-5. BE05791.
> abortion; active euthanasia; adults; *allowing to

die; brain death; congenital defects; decision making; determination of death; *euthanasia; fetuses; hearts; infanticide; *kidneys; legal aspects; morality; *newborns; organ donors; *organ transplantation; physicians; prenatal diagnosis; prolongation of life; suffering; suicide; *terminally ill; voluntary euthanasia

Argentina. Removal of human tissues and organs for therapeutic or scientific purposes. *International Digest of Health Legislation* 28(4): 889-894, 1977. Law No. 21541 of 21 Mar 1977. BE07128.
> age; cadavers; government regulation; determination of death; health facilities; informed consent; organ donors; *organ transplantation; physicians; third party consent; transplantation

Duncan, George E. Christian ethics and biomedical issues: a physician's perspective. *In* Hollis, Harry N., comp. A Matter of Life and Death: Christian Perspectives. Nashville: Broadman Press, 1977. p. 50-55. 1 fn. ISBN 0-8054-6118-3. BE06530.
> determination of death; organ donors; *organ transplantation; resource allocation

Simmons, Roberta G.; Klein, Susan D.; Simmons, Richard L. Gift of Life: The Social and Psychological Impact of Organ Transplantation. New York: John Wiley, 1977. 526 p. Bibliography: p. 485-507. ISBN 0-471-79197-0. BE06686.
> adolescents; *cadavers; case reports; children; chronically ill; costs and benefits; *decision making; *family members; family relationship; financial support; health care delivery; kidney diseases; *kidneys; motivation; *organ donation; *organ donors; *organ transplantation; *psychological stress; quality of life; *social adjustment; *socioeconomic factors; standards; statistics; *transplant recipients

ORGAN TRANSPLANTATION / AUSTRALIA

Human tissue transplants. *British Medical Journal* 1(6107): 195-196, 28 Jan 1978. 5 refs. Editorial. BE06737.
> determination of death; legislation; organ donors; *organ transplantation; *Australia

Law Reform Commission. Human Tissue Transplants. Canberra: Australian Government Publishing Service, 1977. 154 p. Law Reform Commission Report No. 7. 513 fn. Bibliography: p. 144-150. ISBN 0-642-03018-9. BE06172.
> autopsies; blood donation; brain death; cadavers; children; determination of death; economics; international aspects; legal liability; legislation; organ donation; organ donors; *organ transplantation; physicians; privacy; public opinion; transplant recipients; *transplantation; *Australia

ORGAN TRANSPLANTATION / BRAIN DEATH

Pappworth, M.H. Coroners and transplants. *Times (London),* 16 May 1977, p. 15. Letter. BE06444.

BE = bioethics accession number   fn. = footnotes   refs. = references

*brain death; *health personnel; legal aspects; *organ transplantation

Sells, Robert A. Coroners and transplants. *Times (London),* 6 Jun 1977, p. 9. Letter. BE06445.
*brain death; *health personnel; *organ transplantation

ORGAN TRANSPLANTATION / CADAVERS

Coroners and transplants. *British Medical Journal* 1(6073): 1418, 28 May 1977. 13 refs. BE05609.
*cadavers; determination of death; family members; legal aspects; *organ transplantation; third party consent

ORGAN TRANSPLANTATION / CADAVERS / HUMAN TISSUE ACT 1961

Skegg, P.D.G. Liability for the unauthorized removal of cadaveric transplant material: some further comments. *Medicine, Science, and the Law* 17(2): 123-126, Apr 1977. 4 refs. BE05581.
body parts and fluids; *cadavers; *legal liability; legislation; negligence; *organ transplantation; property rights; *Human Tissue Act 1961

ORGAN TRANSPLANTATION / CADAVERS / PORTUGAL

Portugal. Removal of human organs or tissues for therapeutic purposes. *International Digest of Health Legislation* 28(4): 1041, 1977. Decree-Law No. 553 of 13 Jun 1976. BE07130.
*cadavers; determination of death; organ donation; *organ transplantation; physicians; *Portugal

ORGAN TRANSPLANTATION / EAST GERMANY

German Democratic Republic. Removal of human organs and tissues for therapeutic purposes. *International Digest of Health Legislation* 28(3): 511-514, 1977. Ordinance of 4 Jul 1975. 1 fn. BE06462.
cadavers; compensation; determination of death; informed consent; organ donors; *organ transplantation; *East Germany

ORGAN TRANSPLANTATION / FRANCE

France. Removal of human organs for therapeutic or scientific purposes. *International Digest of Health Legislation* 28(2): 271-272, 1977. Translation of Law No. 76-1181, 22 Dec 1976. BE06461.
cadavers; informed consent; *organ transplantation; *France

ORGAN TRANSPLANTATION / HEARTS

Colen, B.D. Heart transplants decline decade after 1st operation. *Washington Post,* 5 Sep 1977, p. A1+. BE06395.

attitudes; incidence; *hearts; mortality; *organ transplantation; physicians; resource allocation

Schmeck, Harold M. Decade after first heart transplant: fanfare dead but at least 85 live. *New York Times,* 2 Dec 1977, p. A20. BE06415.
evaluation; *hearts; mortality; *organ transplantation

ORGAN TRANSPLANTATION / ITALY

Italy. Removal of organs and tissues for therapeutic purposes. *International Digest of Health Legislation* 28(3): 621-627, 1977. Law No. 644 of 2 Dec 1975. 3 fn. BE06466.
body parts and fluids; cadavers; determination of death; government agencies; legal liability; *organ transplantation; *Italy

ORGAN TRANSPLANTATION / KIDNEYS / RISKS AND BENEFITS

Levine, Carol. Dialysis or transplant: values and choices. *Hastings Center Report* 8(2): 8-10, Apr 1978. BE06848.
*decision making; economics; *kidneys; morbidity; mortality; *organ transplantation; psychological stress; *renal dialysis; *risks and benefits

ORGAN TRANSPLANTATION / KIDNEYS / SCARCITY

Machalaba, Daniel. New breed of sleuth is chasing ambulances to ease kidney supply. *Wall Street Journal (Eastern Edition),* 14 Mar 1977, p. 1+. BE06438.
brain death; health personnel; *kidneys; organ donation; *organ transplantation; *scarcity

ORGAN TRANSPLANTATION / KILLING

Singer, Peter. Utility and the survival lottery. *Philosophy* 52(200): 218-222, Apr 1977. 6 fn. BE05580.
costs and benefits; *killing; organ donation; *organ transplantation; *random selection; *risks and benefits; selection of subjects; self induced illness; utilitarianism

ORGAN TRANSPLANTATION / LEGAL ASPECTS / AUSTRALIA

Stewart, J.H. The law and transplantation. *Medical Journal of Australia* 2(21): 689-691, 19 Nov 1977. 5 fn. Editorial. BE06365.
*brain death; cadavers; determination of death; informed consent; *legal aspects; legislation; *organ donation; *organ transplantation; remuneration; *Australia

ORGAN TRANSPLANTATION / PERSONHOOD / BRAIN

Shaffer, Jerome A. Personal identity: the implications of brain bisection and brain transplants. *Journal of*

BE = bioethics accession number   fn. = footnotes   refs. = references

Medicine and Philosophy 2(2): 147-161, Jun 1977. 6 refs. BE05665.
> *brain; *organ transplantation; personality; *personhood; philosophy; psychology; surgery

ORGAN TRANSPLANTATION / PHYSICIAN'S ROLE

Simmons, Roberta G.; Klein, Susan D.; Simmons, Richard L. Transplantation and changing norms: cultural lag and ethical ambiguities. *In their* Gift of Life: The Social and Psychological Impact of Organ Transplantation. New York: John Wiley, 1977. p. 9-44. References included. ISBN 0-471-79197-0. BE06687.
> adolescents; aged; attitudes; brain death; *cadavers; *case reports; chronically ill; decision making; determination of death; donor cards; *family members; federal government; financial support; *informed consent; *kidneys; legal aspects; medical staff; *organ donors; *organ donation; *organ transplantation; *physician's role; physicians; public opinion; remuneration; resource allocation; scarcity; *selection for treatment; *socioeconomic factors; standards; *transplant recipients; volunteers; withholding treatment

ORGAN TRANSPLANTATION / RANDOM SELECTION / RISKS AND BENEFITS

Singer, Peter. Utility and the survival lottery. *Philosophy* 52(200): 218-222, Apr 1977. 6 fn. BE05580.
> costs and benefits; *killing; organ donation; *organ transplantation; *random selection; *risks and benefits; selection of subjects; self induced illness; utilitarianism

ORGAN TRANSPLANTATION / SELECTION FOR TREATMENT

Simmons, Roberta G.; Klein, Susan D.; Simmons, Richard L. Transplantation and changing norms: cultural lag and ethical ambiguities. *In their* Gift of Life: The Social and Psychological Impact of Organ Transplantation. New York: John Wiley, 1977. p. 9-44. References included. ISBN 0-471-79197-0. BE06687.
> adolescents; aged; attitudes; brain death; *cadavers; *case reports; chronically ill; decision making; determination of death; donor cards; *family members; federal government; financial support; *informed consent; *kidneys; legal aspects; medical staff; *organ donors; *organ donation; *organ transplantation; *physician's role; physicians; public opinion; remuneration; resource allocation; scarcity; *selection for treatment; *socioeconomic factors; standards; *transplant recipients; volunteers; withholding treatment

ORGANIZATIONAL POLICIES / WORLD PSYCHIATRIC ASSOCIATION / USSR

Reich, Walter. Soviet psychiatry on trial. *Commentary* 65(1): 40-48, Jan 1978. 1 fn. BE06749.
> case reports; *dissent; involuntary commitment; *organizational policies; politics; *professional organizations; *psychiatric diagnosis; *psychiatry; schizophrenia; *USSR; *World Psychiatric Association

ORGANIZATIONAL POLICIES   See under
> RECOMBINANT DNA RESEARCH / ORGANIZATIONAL POLICIES / LEGISLATION

PARENT CHILD RELATIONSHIP   See under
> PARENTAL CONSENT / PARENT CHILD RELATIONSHIP / LEGAL ASPECTS

PARENTAL CONSENT / ADOLESCENTS / LEGAL ASPECTS

Sharpe, Gilbert. Valid consent: determining the minor's ability to make decisions. *Canadian Medical Association Journal* 117(8): 934+, 22 Oct 1977. BE06498.
> *adolescents; children; emergency care; human experimentation; informed consent; *legal aspects; legislation; *parental consent; *patient care; physician's role; *Canada

PARENTAL CONSENT / CHILDREN

Hellegers, André E. 'Incompetence' and consent. *Ob. Gyn. News* 12(21): 20-21, 1 Nov 1977. BE06114.
> *children; competence; informed consent; *mentally ill; mentally retarded; *parental consent; *third party consent

PARENTAL CONSENT / PARENT CHILD RELATIONSHIP / LEGAL ASPECTS

Beyer, Henry A. Changes in the parent-child legal relationship—what they mean to the clinician and researcher. *Journal of Autism and Childhood Schizophrenia* 7(1): 84-94, Mar 1977. 56 fn. BE05909.
> adolescents; children; human experimentation; involuntary commitment; *legal aspects; mental health; mentally retarded; organ donation; *parent child relationship; *parental consent

PARENTAL CONSENT / SPINA BIFIDA / LEGAL ASPECTS

Grad, Frank P. Legal aspects of informed consent. *In* Swinyard, Chester A., ed. Decision Making and the Defective Newborn. Proceedings of a Conference on Spina Bifida and Ethics. Springfield, Ill.: Charles C. Thomas, 1978. p. 435-445. ISBN 0-398-03662-4. BE06960.
> children; *disclosure; extraordinary treatment; fetuses; human experimentation; *informed con-

---

BE = bioethics accession number   fn. = footnotes   refs. = references

sent; *legal aspects; legal obligations; mentally handicapped; *newborns; organ donation; *parental consent; *patient care; physicians; selection for treatment; *spina bifida; third party consent; treatment refusal; withholding treatment

PARENTAL CONSENT *See under*

ABORTION / PARENTAL CONSENT / LEGAL ASPECTS

ABORTION / PARENTAL CONSENT / LEGAL RIGHTS

ABORTION / PARENTAL CONSENT / PLANNED PARENTHOOD OF CENTRAL MISSOURI V. DANFORTH

ABORTION / PARENTAL CONSENT / SUPREME COURT DECISIONS

BEHAVIORAL RESEARCH / PARENTAL CONSENT / SEXUALITY

CONTRACEPTION / PARENTAL CONSENT / LEGAL RIGHTS

HEALTH CARE / PARENTAL CONSENT / LEGAL ASPECTS

HUMAN EXPERIMENTATION / PARENTAL CONSENT

HUMAN EXPERIMENTATION / PARENTAL CONSENT / LEGAL ASPECTS

INVOLUNTARY COMMITMENT / PARENTAL CONSENT

INVOLUNTARY COMMITMENT / PARENTAL CONSENT / LEGAL ASPECTS

PATIENT CARE / PARENTAL CONSENT / CANADA

TRANSPLANTATION / PARENTAL CONSENT / LEGAL ASPECTS

TREATMENT REFUSAL / PARENTAL CONSENT

TREATMENT REFUSAL / PARENTAL CONSENT / RELIGIOUS BELIEFS

TREATMENT REFUSAL / PARENTAL CONSENT / SPINA BIFIDA

VOLUNTARY ADMISSION / PARENTAL CONSENT / PENNSYLVANIA

PARENTS *See under*

WRONGFUL LIFE / PARENTS / COMPENSATION

PASSIVE EUTHANASIA *See* ALLOWING TO DIE

PATENTS *See under*

RECOMBINANT DNA RESEARCH / PATENTS

PATERNALISM *See under*

PHYSICIAN PATIENT RELATIONSHIP / PATERNALISM

PATIENT ACCESS *See also* CONFIDENTIALITY

PATIENT ACCESS / MEDICAL RECORDS / LEGAL ASPECTS

Westin, Alan F. Medical records: should patients have access? *Hastings Center Report* 7(6): 23-28, Dec 1977. 29 fn. BE06156.
   disclosure; hospitals; informed consent; legal rights; *legal aspects; *medical records; malpractice; *patient access; patients; physicians; privacy; property rights; psychiatry

PATIENT ACCESS / MEDICAL RECORDS / LEGAL RIGHTS

Rozovsky, Lorne E. The patients' right to see the record. *Dimensions in Health Service* 54(11): 8-9, Nov 1977. 7 refs. BE06364.
   confidentiality; hospitals; *legal rights; *medical records; *patient access

PATIENT ACCESS / MEDICAL RECORDS / STANDARDS

Melum, Mara M. Balancing information and privacy. *Hospital Progress* 58(7): 68-69+, Jul 1977. 6 fn. BE05730.
   biomedical research; confidentiality; *disclosure; hospitals; *informed consent; institutional policies; *medical records; *patient access; patients; privacy; professional organizations; *standards; statistics; Minnesota; Minnesota Hospital Association

PATIENT CARE *See also* HEALTH CARE

PATIENT CARE / ABORTED FETUSES

Robertson, John A. After *Edelin*: little guidance. *Hastings Center Report* 7(3): 15-17+, Jun 1977. 1 fn. BE05663.
   *aborted fetuses; abortion; judicial action; killing; legal liability; *legal obligations; *patient care; *physicians; viability; *Edelin, Kenneth

PATIENT CARE / ABORTED FETUSES / SUPREME COURT DECISIONS

Ramsey, Paul. The Supreme Court's bicentennial abortion decision: can the 1973 abortion decisions be justly hedged? *In his* Ethics at the Edges of Life: Medical and Legal Intersections. New Haven: Yale University Press, 1978. p. 3-42. 28 fn. ISBN 0-300-02137-2. BE06885.
   *aborted fetuses; *abortion; adolescents; family relationship; fathers; fetuses; legal obligations; *legislation; marital relationship; medical ethics; methods; morality; obstetrics and gynecology; *parental consent; parents; *patient care; physicians; pregnant women; privacy; rights; *spousal consent; state interest; *Supreme Court decisions; *viability; *Missouri; *Planned Parenthood of Central Missouri v. Danforth

---

BE = bioethics accession number   fn. = footnotes   refs. = references

PATIENT CARE / BURN PATIENTS

Imbus, Sharon H.; Zawacki, Bruce E. Autonomy for burned patients when survival is unprecedented. *New England Journal of Medicine* 297(6): 308-311, 11 Aug 1977. 19 refs. BE05759.
*burn patients; *communication; *critically ill; decision making; disclosure; family members; *medical staff; mortality; *patient care; patient participation; prognosis; *self determination; terminally ill

PATIENT CARE / CANCER / ECONOMICS

Drinkwater, C.K.; Roberts, S.H. Cure or care in everyday practice. *Journal of Medical Ethics* 4(1): 12-17, Mar 1978. 20 refs. Case study. BE06920.
alternatives; attitudes; *cancer; case reports; communication; costs and benefits; *decision making; diagnosis; drugs; family members; *economics; *patient care; patient participation; *physicians; prognosis; prolongation of life; quality of life; social adjustment; surgery; *terminal care; terminally ill; treatment refusal

PATIENT CARE / CHILDREN / CANCER

Frankel, Lawrence S.; Damme, Catherine J.; Van Eys, Jan. Childhood cancer and the Jehovah's Witness faith. *Pediatrics* 60(6): 916-921, Dec 1977. 32 refs. BE06853.
blood transfusions; *cancer; case reports; *children; *Jehovah's Witnesses; legal rights; parental consent; *patient care; treatment refusal; withholding treatment

PATIENT CARE / CHILDREN / GENETIC DEFECTS

Stone, Brenda. The 'boy-in-the-bubble' drama. *American Medical News* 20(46): 1+, 21 Nov 1977. BE06366.
*children; genetic counseling; *genetic defects; parents; *patient care; physicians; quality of life; risks and benefits

PATIENT CARE / CHILDREN / LEUKEMIA

Kearney, P.J. Leukaemia in children of Jehovah's Witnesses: issues and priorities in a conflict of care. *Journal of Medical Ethics* 4(1): 32-35, Mar 1978. 6 refs. BE06921.
blood transfusions; cancer; case reports; *children; *decision making; family relationship; informal social control; *Jehovah's Witnesses; *leukemia; parent child relationship; parental consent; parents; *patient care; pediatrics; *physician patient relationship; physicians; religious beliefs; *risks and benefits; *treatment refusal

PATIENT CARE / CHRONICALLY ILL / BRAIN PATHOLOGY

Spudis, Edward V.; Oleck, Howard L. Management of seven stable levels of brain death. *Journal of Legal Medicine* 5(8): 5-11, Aug 1977. 40 refs. BE05767.
*brain death; *brain pathology; case reports; *chronically ill; coma; decision making; determination of death; organ donation; *patient care; physicians; prolongation of life; withholding treatment

PATIENT CARE / COERCION

Faden, Ruth; Faden, Alan. False belief and the refusal of medical treatment. *Journal of Medical Ethics* 3(3): 133-136, Sep 1977. 11 refs. 3 fn. BE05981.
adults; case reports; *coercion; competence; *comprehension; disclosure; informed consent; *moral obligations; paternalism; *patient care; *patients; *physicians; self determination; surgery; *treatment refusal

PATIENT CARE / COSTS AND BENEFITS

Freedman, Benjamin. The case for medical care, inefficient or not. *Hastings Center Report* 7(2): 31-39, Apr 1977. 22 fn. BE05555.
*costs and benefits; *decision making; ethics; health care delivery; moral obligations; *morality; *patient care; patients; physician; physician patient relationship; *preventive medicine; renal dialysis; *resource allocation; *scarcity; selection for treatment; self determination; trust; values

PATIENT CARE / DECEPTION

Cabot, Richard A. The use of truth and falsehood in medicine. *Connecticut Medicine* 42(3): 189-194, Mar 1978. BE06808.
cancer; chronically ill; *deception; *disclosure; *diagnosis; heart diseases; *patient care; *physicians; *placebos; *prognosis; terminally ill

Katz, Jay. Richard C. Cabot—some reflections on deception and placebos in the practice of medicine. *Connecticut Medicine* 42(3): 199-200, Mar 1978. 6 refs. Editorial. BE06809.
communication; *deception; medicine; *patient care; physicians; *placebos

PATIENT CARE / DECISION MAKING

Hunter, Thomas H.; Wenzel, Richard; Fletcher, Joseph. How Does One Determine Acceptable Risks? [Videorecording]. Charlottesville, Va.: University of Virginia Medical Center, Health Sciences Library, 1976. 1 cassette; 60 min.; sound; black and white; 3/4 in. Tape of the Medical Center Hour, 1 Dec 1976. Panel discussion moderated by Thomas H. Hunter. BE07051.
alternatives; *biomedical research; consequences; *decision making; health care; hospitals; *patient care; *physicians; quality of life; recombinant

---

BE = bioethics accession number   fn. = footnotes   refs. = references

DNA research; *risks and benefits; statistics; *surgery; teleological ethics; values

PATIENT CARE / DISCLOSURE

Cabot, Richard A. The use of truth and falsehood in medicine. *Connecticut Medicine* 42(3): 189-194, Mar 1978. BE06808.
cancer; chronically ill; *deception; *disclosure; *diagnosis; heart diseases; *patient care; *physicians; *placebos; *prognosis; terminally ill

Truth and falsehood in medicine. *Journal of the American Medical Association* 239(15): 1554, 14 Apr 1978. Editorial. BE06947.
deception; *disclosure; *patient care; physicians; placebos

PATIENT CARE / DRUGS / RISKS AND BENEFITS

Gifford, Ray W. *Primum non nocere.* *Journal of the American Medical Association* 238(7): 589-590, 15 Aug 1977. 3 refs. BE05758.
*drugs; federal government; government agencies; injuries; *patient care; physician's role; *risks and benefits; Food and Drug Administration

PATIENT CARE / ETHICS COMMITTEES

Levine, Carol. Hospital ethics committees: a guarded prognosis. *Hastings Center Report* 7(3): 25-27, Jun 1977. BE05651.
clergy; *decision making; *ethics committees; family members; hospitals; institutional policies; patient advocacy; *patient care; physicians; prognosis; terminally ill

Veatch, Robert M. Hospital ethics committees: is there a role? *Hastings Center Report* 7(3): 22-25, Jun 1977. BE05674.
counseling; *decision making; *ethics committees; hospitals; institutional policies; *patient care; physician's role; prognosis; terminally ill; values; withholding treatment; In re Quinlan; New Jersey Supreme Court

PATIENT CARE / ETHICS COMMITTEES / INSTITUTIONALIZED PERSONS

Mental health: a model statute to regulate the administration of therapy within mental health facilities. *Minnesota Law Review* 61(5): 841-886, May 1977. 135 fn. BE05931.
competence; *decision making; electrical stimulation of the brain; electroconvulsive therapy; ethical review; *ethics committees; government regulation; informed consent; *institutionalized persons; legislation; *mentally ill; *model legislation; *patient care; psychoactive drugs; psychosurgery; standards; state government

Treadway, Jerry T.; Rossi, Robert B. An ethical review board: its structure, function and province. *Mental Retardation* 15(4): 28-29, Aug 1977. 4 refs. 1 fn. BE05976.
*ethical review; *ethics committees; human experimentation; *institutionalized persons; *mentally retarded; *patient care; Pueblo State Home and Training School

PATIENT CARE / HEALTH PERSONNEL / STRIKES

Baelz, Peter. The right to strike by the caring professions. *Journal of Medical Ethics* 3(3): 150, Sep 1977. Letter. BE05983.
*health personnel; *moral obligations; *patient care; *strikes

PATIENT CARE / INSTITUTIONALIZED PERSONS / MODEL LEGISLATION

Mental health: a model statute to regulate the administration of therapy within mental health facilities. *Minnesota Law Review* 61(5): 841-886, May 1977. 135 fn. BE05931.
competence; *decision making; electrical stimulation of the brain; electroconvulsive therapy; ethical review; *ethics committees; government regulation; informed consent; *institutionalized persons; legislation; *mentally ill; *model legislation; *patient care; psychoactive drugs; psychosurgery; standards; state government

PATIENT CARE / INTENSIVE CARE UNITS

Cohen, Cynthia B. Ethical problems of intensive care. *Anesthesiology* 47(2): 217-227, Aug 1977. 95 fn. BE05836.
*allowing to die; critically ill; *decision making; human rights; extraordinary treatment; *intensive care units; intention; killing; moral obligations; *patient care; *resource allocation; selection for treatment; social worth; *value of life; withholding treatment

PATIENT CARE / JEHOVAH'S WITNESSES / CANCER

Frankel, Lawrence S.; Damme, Catherine J.; Van Eys, Jan. Childhood cancer and the Jehovah's Witness faith. *Pediatrics* 60(6): 916-921, Dec 1977. 32 refs. BE06853.
blood transfusions; *cancer; case reports; *children; *Jehovah's Witnesses; legal rights; parental consent; *patient care; treatment refusal; withholding treatment

PATIENT CARE / LAETRILE / CANCER

Colen, B.D. Dispute over Laetrile focuses attention on rights of patients. *Washington Post,* 29 May 1977, p. A18. BE06024.
attitudes; *cancer; *patient care; patients' rights; physicians; terminally ill; *Laetrile

---

BE = bioethics accession number   fn. = footnotes   refs. = references

Martin, Daniel S. Laetrile—consumer protection is the bottom line. *Medical World News* 18(13): 56, 27 Jun 1977. BE05656.
*cancer; drugs; government regulation; *patient care; public advocacy; self determination; terminally ill; *Laetrile

Saber, Fay A. Laetrile: is it really a matter of free choice? *American Journal of Public Health* 67(9): 871-872, Sep 1977. 3 refs. BE05988.
*cancer; drugs; federal government; government agencies; government regulation; legal aspects; *patient care; terminally ill; Food and Drug Administration; *Laetrile

Stolfi, Julius E. Legalizing human guinea pigs. *New York State Journal of Medicine* 77(9): 1388-1389, Aug 1977. 1 fn. Editorial. BE05975.
*cancer; drugs; law; *patient care; physician's role; terminally ill; *Laetrile

U.S. Congress. Senate. Committee on Human Resources. Subcommittee on Health and Scientific Research. Banning of the Drug Laetrile from Interstate Commerce by FDA. Hearing. Washington: U.S. Government Printing Office, 1977. 420 p. Hearing, 95th Cong., 1st Sess., on evaluation of information [on] which the FDA based its decision to ban the drug Laetrile from interstate commerce, 12 Jul 1977. References included. BE06691.
animal experimentation; attitudes; *cancer; case reports; drugs; federal government; government regulation; nutrition; *patient care; physician's role; *Laetrile

U.S. District Court, W.D. Oklahoma. Rutherford v. United States. 4 Jan 1977. *Federal Supplement* 424: 105-108. 1 fn. BE06277.
*cancer; drugs; federal government; government agencies; government regulation; *patient care; *Food and Drug Administration; *Laetrile

PATIENT CARE / LAETRILE / LEGAL ASPECTS

Thorup, Oscar A., et al. Laetrile: The Right to Choose. [Videorecording]. Charlottesville, Va.: University of Virginia Medical Center, Health Sciences Library, 1977. 1 cassette; 60 min.; sound; black and white; 3/4 in. Tape of the Medical Center Hour, 28 Sep 1977. Panel discussion moderated by Oscar A. Thorup. BE07054.
*cancer; children; control groups; government regulation; historical aspects; human experimentation; *legal aspects; legal liability; legal rights; parental consent; paternalism; *patient care; physicians; *rights; risks and benefits; terminally ill; treatment refusal; *Laetrile

PATIENT CARE / LAETRILE / LEGAL RIGHTS

Annas, George J. Legalizing Laetrile for the terminally ill. *Hastings Center Report* 7(6): 19-20, Dec 1977. BE06131.

*cancer; *drugs; federal government; government agencies; government regulation; *legal rights; *patient care; physician's role; *terminally ill; Food and Drug Administration; *Laetrile

California. Court of Appeal, Fourth District, Division 1. People v. Privitera. 10 Nov 1977. *California Reporter* 141: 764-794. 23 fn. BE07103.
*cancer; *drugs; legal liability; *legal rights; legislation; *patient care; physicians; *privacy; state government; state interest; *California; *Laetrile

PATIENT CARE / LAETRILE / LEGISLATION

California. Superior Court, Appellate Department, Los Angeles County. People v. Privitera. 29 Jan 1976. *California Reporter* 128: 151-160. 12 fn. BE07102.
*cancer; *drugs; equal protection; government regulation; *legislation; *patient care; privacy; *state government; *California; *Laetrile

PATIENT CARE / LAETRILE / NEW YORK ACADEMY OF MEDICINE

New York Academy of Medicine. Statement on amygdalin (Laetrile). *Bulletin of the New York Academy of Medicine* 53(9): 843-846, Nov 1977. Approved by the Committee on Public Health of the New York Academy of Medicine, 14 Jul 1977. 1 fn. BE06361.
cancer; drugs; legal rights; legislation; *organizational policies; *patient care; physicians; state government; *Laetrile; *New York Academy of Medicine

PATIENT CARE / LAETRILE / RIGHTS

Thorup, Oscar A., et al. Laetrile: The Right to Choose. [Videorecording]. Charlottesville, Va.: University of Virginia Medical Center, Health Sciences Library, 1977. 1 cassette; 60 min.; sound; black and white; 3/4 in. Tape of the Medical Center Hour, 28 Sep 1977. Panel discussion moderated by Oscar A. Thorup. BE07054.
*cancer; children; control groups; government regulation; historical aspects; human experimentation; *legal aspects; legal liability; legal rights; parental consent; paternalism; *patient care; physicians; *rights; risks and benefits; terminally ill; treatment refusal; *Laetrile

PATIENT CARE / LAETRILE / RISKS AND BENEFITS

Martin, Daniel S. Laetrile—a dangerous drug. *CA-A Cancer Journal for Clinicians* 27(5): 301-304, Sep/Oct 1977. BE06583.
*drugs; evaluation; federal government; government regulation; *legal aspects; *patient care; *risks and benefits; terminally ill; *Laetrile

PATIENT CARE / LEGAL OBLIGATIONS

Annas, George J. Beyond the Good Samaritan: should

BE = bioethics accession number   fn. = footnotes   refs. = references

doctors be required to provide essential services? *Hastings Center Report* 8(2): 16-17, Apr 1978. 3 fn. BE06835.
*emergency care; *legal obligations; moral obligations; *patient care; physician patient relationship; *physicians

PATIENT CARE / LEUKEMIA / RISKS AND BENEFITS

Kearney, P.J. Leukaemia in children of Jehovah's Witnesses: issues and priorities in a conflict of care. *Journal of Medical Ethics* 4(1): 32-35, Mar 1978. 6 refs. BE06921.
blood transfusions; cancer; case reports; *children; *decision making; family relationship; informal social control; *Jehovah's Witnesses; *leukemia; parent child relationship; parental consent; parents; *patient care; pediatrics; *physician patient relationship; physicians; religious beliefs; *risks and benefits; *treatment refusal

PATIENT CARE / MENTALLY RETARDED / WILLOWBROOK STATE SCHOOL

Hansen, Christopher A. Willowbrook. *In* Wortis, Joseph, ed. Mental Retardation and Developmental Disabilities: An Annual Review—IX. New York: Brunner/Mazel, 1977. p. 6-45. 31 fn. BE06515.
*institutionalized persons; *legal aspects; mental institutions; *mentally retarded; *patient care; review committees; right to treatment; *standards; *Willowbrook State School

PATIENT CARE / MORAL OBLIGATIONS

Childress, James F. Citizen and physician: harmonious or conflicting responsibilities? *Journal of Medicine and Philosophy* 2(4): 401-409, Dec 1977. 3 refs. 3 fn. BE06135.
justice; *moral obligations; obligations to society; *patient care; *physicians; resource allocation; utilitarianism

Jonsen, Albert R.; Jameton, Andrew L. Social and political responsibilities of physicians. *Journal of Medicine and Philosophy* 2(4): 376-400, Dec 1977. 35 refs. 2 fn. BE06140.
diagnosis; health care; justice; *moral obligations; *obligations to society; *patient care; *physicians; political activity; resource allocation

PATIENT CARE / NEWBORNS / CONGENITAL DEFECTS

Ramsey, Paul. The benign neglect of defective infants. *In his* Ethics at the Edges of Life: Medical and Legal Intersections. New Haven: Yale University Press, 1978. p. 189-227. 54 fn. ISBN 0-300-02137-2. BE06889.
active euthanasia; *allowing to die; Christian ethics; *congenital defects; ethics; extraordinary treatment; genetic defects; human characteristics;

human experimentation; involuntary euthanasia; justice; killing; moral obligations; medical ethics; *newborns; pain; parents; *patient care; personhood; physicians; prognosis; quality of life; selection for treatment; self concept; social interaction; socioeconomic factors; spina bifida; *terminally ill; theology; third party consent; *value of life; Jonas, Hans; McCormick, Richard A.

PATIENT CARE / NURSES / LEGAL ASPECTS

Thomas, Barbara. Legal implications. *Nursing Mirror* 145(3): 24-26, 21 Jul 1977. BE05818.
confidentiality; critically ill; hospitals; informed consent; institutional policies; law enforcement; *legal aspects; legal obligations; negligence; *nurses; parental consent; *patient care; professional patient relationship; religious beliefs; technical expertise; treatment refusal

PATIENT CARE / NURSING ETHICS

Doona, Mary E. The ethical dimension in 'ordinary nursing care'. *Linacre Quarterly* 44(4): 320-327, Nov 1977. 15 fn. BE06101.
moral obligations; nurses; *nursing ethics; *patient care; psychological stress

Levine, Myra E. Nursing ethics and the ethical nurse. *American Journal of Nursing* 77(5): 845-849, May 1977. 6 refs. BE05591.
dehumanization; moral obligations; morality; *nurses; *nursing ethics; *patient care; professional patient relationship; *values

PATIENT CARE / PARENTAL CONSENT / CANADA

Sharpe, Gilbert. Valid consent: determining the minor's ability to make decisions. *Canadian Medical Association Journal* 117(8): 934+, 22 Oct 1977. BE06498.
*adolescents; children; emergency care; human experimentation; informed consent; *legal aspects; legislation; *parental consent; *patient care; physician's role; *Canada

PATIENT CARE / PATIENT PARTICIPATION

Bury, Judith. Some thoughts on the doctor/patient relationship. *Midwife, Health Visitor and Community Nurse* 13(7): 202+, Jul 1977. BE06316.
alternatives; *authoritarianism; disclosure; paternalism; *patient care; *patient participation; *physician patient relationship

Newton, Lisa H. The healing of the person. *Connecticut Medicine* 41(10): 641-646, Oct 1977. 22 fn. BE06091.
decision making; dehumanization; paternalism; *health; patient advocacy; *patient care; *patient participation; *patients' rights; *physician patient relationship; physician's role; self determination

Slack, Warner V. The patient's right to decide. *Lancet* 2(8031): 240, 30 Jul 1977. 8 refs. BE05735.

BE = bioethics accession number   fn. = footnotes   refs. = references

communication; *decision making; *patient care; paternalism; *patient participation; patients; physicians; self determination; values

PATIENT CARE / PATIENT PARTICIPATION / PHYSICIAN PATIENT RELATIONSHIP

Hayes, Donald M. Between Doctor and Patient. Valley Forge: Judson Press, 1977. 176 p. 67 fn. ISBN 0-8170-0742-3. BE06184.
abortion; allowing to die; cancer; case reports; community services; death; decision making; dehumanization; drug abuse; drugs; economics; employment; health care delivery; hospitals; leukemia; medical ethics; *patient care; *patient participation; *physician patient relationship; psychosurgery; quality of life; suicide; *terminal care; terminally ill; *values

PATIENT CARE / PEDIATRICS / ATTITUDES

Todres, I. David, et al. Pediatricians' attitudes affecting decision-making in defective newborns. *Pediatrics* 60(2): 197-201, Aug 1977. 13 refs. BE06328.
age; *allowing to die; *attitudes; case reports; *congenital defects; decision making; Down's syndrome; *newborns; parental consent; *patient care; *pediatrics; religion; Roman Catholicism; spina bifida; surgery; treatment refusal; withholding treatment

PATIENT CARE / PHYSICIANS

Jonsen, Albert R. Do no harm: axiom of medical ethics. *In* Spicker, Stuart F.; Engelhardt, H. Tristram, eds. Philosophical Medical Ethics: Its Nature and Significance. Proceedings. Boston: D. Reidel, 1977. p. 27-46. Proceedings of the Third Trans-Disciplinary Symposium on Philosophy and Medicine, Farmington, Connecticut, 11-13 Dec 1975. Discussion by Louis Lasagna on p. 43-46. 25 refs. 1 fn. ISBN 90-277-0772-3. BE05861.
codes of ethics; double effect; injuries; medical ethics; medicine; *moral obligations; *patient care; physician's role; *physicians; risks and benefits; standards

PATIENT CARE / PHYSICIANS / JEWISH ETHICS

Bleich, J. David. Theological considerations in the care of defective newborns. *In* Swinyard, Chester A., ed. Decision Making and the Defective Newborn. Proceedings of a Conference on Spina Bifida and Ethics. Springfield, Ill.: Charles C. Thomas, 1978. p. 512-561. 106 fn. ISBN 0-398-03662-4. BE06963.
adults; contracts; *extraordinary treatment; financial support; informed consent; *Jewish ethics; judicial action; legal liability; legal obligations; *moral obligations; *newborns; pain; *patient care; physician patient relationship; *physicians; *prolongation of life; Roman Catholic ethics;

*scriptural interpretation; *spina bifida; terminally ill; theology; treatment refusal; *value of life; withholding treatment

PATIENT CARE / PHYSICIANS / LEGAL ASPECTS

Grad, Frank P. Medical ethics and the law. *In* Barber, Bernard, ed. Medical Ethics and Social Change. Philadelphia: American Academy of Political and Social Science, 1978. p. 19-36. Annals of the American Academy of Political and Social Science, Vol. 437, May 1978. 69 fn. ISBN 0-87761-226-9. BE06997.
allowing to die; decision making; disclosure; genetic intervention; government regulation; human experimentation; informed consent; *legal aspects; medicine; *patient care; *physicians; withholding treatment

PATIENT CARE / PHYSICIANS / LEGAL LIABILITY

Hirsh, Harold L. Physician's legal liability to third parties who are not patients. *Medical Trial Technique Quarterly* 23(4): 388-400, Spring 1977. 35 refs. BE05950.
communicable diseases; dangerousness; *disclosure; hospitals; injuries; *legal liability; legal obligations; negligence; *patient care; *physicians; psychiatry; torts

PATIENT CARE / PHYSICIANS / RISKS AND BENEFITS

Moore, Rue. Toward an ethical risk-benefit calculus. *American Journal of Digestive Diseases* 22(6): 566-567, Jun 1977. BE05657.
human experimentation; *patient care; physician patient relationship; *physicians; *risks and benefits; values

PATIENT CARE / PLACEBOS

Bishop, Jerry E. Placebos are harmless, but they work, posing problems for medicine. *Wall Street Journal,* 25 Aug 1977, p. 1+. BE06063.
biomedical research; cancer; deception; drugs; evaluation; pain; *patient care; *placebos

Kadlec, James F.; Dinno, Nuhad D. Placebo: a place in medical therapy? *Journal of the Kentucky Medical Association* 75(11): 538-541, Nov 1977. 27 refs. BE06359.
attitudes; *patient care; physicians; *placebos; risks and benefits; Laetrile

Katz, Jay. Richard C. Cabot—some reflections on deception and placebos in the practice of medicine. *Connecticut Medicine* 42(3): 199-200, Mar 1978. 6 refs. Editorial. BE06809.
communication; *deception; medicine; *patient care; physicians; *placebos

---

BE = bioethics accession number   fn. = footnotes   refs. = references

PATIENT CARE / PLACEBOS / ATTITUDES

Comaroff, Jean. A bitter pill to swallow: placebo therapy in general practice. *Sociological Review* 24(1): 79-96, Feb 1976. 37 fn. BE07093.
  *attitudes; drugs; *patient care; physician patient relationship; *physicians; *placebos; psychological stress

PATIENT CARE / PLACEBOS / EVALUATION

Berg, Alfred O. Placebos: a brief review for family physicians. *Journal of Family Practice* 5(1): 97-100, Jul 1977. 24 refs. BE06314.
  attitudes; deception; *evaluation; *patient care; physician patient relationship; *placebos; psychotherapy; trust

PATIENT CARE / PSRO'S

Burnum, John F. The physician as a double agent. *New England Journal of Medicine* 297(5): 278-279, 4 Aug 1977. 19 refs. BE05750.
  conflict of interest; *costs and benefits; *decision making; economics; patient advocacy; *patient care; *physician patient relationship; resource allocation; *PSRO's

PATIENT CARE / PSYCHIATRY / HAWAII DECLARATION

World Psychiatric Association. General Assembly. Declaration of Hawaii. *British Medical Journal* 2(6096): 1204-1205, 5 Nov 1977. Declaration adopted unanimously by the...World Psychiatric Association...6th World Congress...1977. Also appeared in the Journal of the American Medical Association, 238(25): 2732, 19 Dec 1977. BE06367.
  *codes of ethics; coercion; confidentiality; informed consent; mentally ill; *patient care; *psychiatry; *Hawaii Declaration

PATIENT CARE / PSYCHOACTIVE DRUGS / CANCER

Schmeck, Harold M. Addictive drugs as a way of easing death. *New York Times,* 30 Nov 1977, p. B2. BE06413.
  *cancer; human experimentation; pain; *patient care; *psychoactive drugs; suffering; *terminal care; *terminally ill

PATIENT CARE / PSYCHOACTIVE DRUGS / INSTITUTIONALIZED PERSONS

Armstrong, Barbara. The use of psychotropic drugs in state hospitals: a legal or medical decision? *Hospital and Community Psychiatry* 29(2): 118-121, Feb 1978. BE06770.
  decision making; electroconvulsive therapy; *government regulation; *institutionalized persons; legal aspects; legislation; *mentally ill; *patient care;

psychiatry; *psychoactive drugs; self regulation; *state government; *California

PATIENT CARE / RESOURCE ALLOCATION

Freedman, Benjamin. The case for medical care, inefficient or not. *Hastings Center Report* 7(2): 31-39, Apr 1977. 22 fn. BE05555.
  *costs and benefits; *decision making; ethics; health care delivery; moral obligations; *morality; *patient care; patients; physicians; physician patient relationship; *preventive medicine; renal dialysis; *resource allocation; *scarcity; selection for treatment; self determination; trust; values

PATIENT CARE / SPINA BIFIDA / DECISION MAKING

Heymann, Philip B.; Holtz, Sara. The severely defective newborn: the dilemma and the decision process. *In* Swinyard, Chester A., ed. Decision Making and the Defective Newborn. Proceedings of a Conference on Spina Bifida and Ethics. Springfield, Ill.: Charles C. Thomas, 1978. p. 396-434. 11 refs. 13 fn. ISBN 0-398-03662-4. BE06959.
  age; *allowing to die; *congenital defects; *costs and benefits; *decision making; economics; extraordinary treatment; family; handicapped; human development; judicial action; law; *legal aspects; legal obligations; *newborns; parents; *patient care; personhood; physicians; prolongation of life; public policy; *quality of life; resource allocation; society; *spina bifida; standards; suicide; value of life; *withholding treatment

PATIENT CARE / SPINA BIFIDA / JEWISH ETHICS

Bleich, J. David. Theological considerations in the care of defective newborns. *In* Swinyard, Chester A., ed. Decision Making and the Defective Newborn. Proceedings of a Conference on Spina Bifida and Ethics. Springfield, Ill.: Charles C. Thomas, 1978. p. 512-561. 106 fn. ISBN 0-398-03662-4. BE06963.
  adults; contracts; *extraordinary treatment; financial support; informed consent; *Jewish ethics; judicial action; legal liability; legal obligations; *moral obligations; *newborns; pain; *patient care; physician patient relationship; *physicians; *prolongation of life; Roman Catholic ethics; *scriptural interpretation; *spina bifida; terminally ill; theology; treatment refusal; *value of life; withholding treatment

PATIENT CARE / SPINA BIFIDA / LEGAL ASPECTS

Heymann, Philip B.; Holtz, Sara. The severely defective newborn: the dilemma and the decision process. *In* Swinyard, Chester A., ed. Decision Making and the Defective Newborn. Proceedings of a Conference on Spina Bifida and Ethics. Springfield, Ill.: Charles

---

BE = bioethics accession number   fn. = footnotes   refs. = references

C. Thomas, 1978. p. 396-434. 11 refs. 13 fn. ISBN 0-398-03662-4. BE06959.
 age; *allowing to die; *congenital defects; *costs and benefits; *decision making; economics; extraordinary treatment; family; handicapped; human development; judicial action; law; *legal aspects; legal obligations; *newborns; parents; *patient care; personhood; physicians; prolongation of life; public policy; *quality of life; resource allocation; society; *spina bifida; standards; suicide; value of life; *withholding treatment

PATIENT CARE / SPINA BIFIDA / PHYSICIAN'S ROLE

Duff, Raymond S. A physician's role in the decision-making process: a physician's experience. *In* Swinyard, Chester A., ed. Decision Making and the Defective Newborn. Proceedings of a Conference on Spina Bifida and Ethics. Springfield, Ill.: Charles C. Thomas, 1978. p. 194-219. 18 refs. Case study. ISBN 0-398-03662-4. BE06954.
 *allowing to die; alternatives; biomedical technologies; *decision making; disclosure; ethics; extraordinary treatment; family members; family problems; *handicapped; hospitals; informed consent; institutional policies; killing; legal aspects; medical staff; medicine; *newborns; parents; *physician's role; patient care; prognosis; physicians; psychological stress; public opinion; quality of life; risks and benefits; social impact; *spina bifida; terminally ill; withholding treatment

PATIENT CARE / TERMINALLY ILL

Ramsey, Paul. The benign neglect of defective infants. *In his* Ethics at the Edges of Life: Medical and Legal Intersections. New Haven: Yale University Press, 1978. p. 189-227. 54 fn. ISBN 0-300-02137-2. BE06889.
 active euthanasia; *allowing to die; Christian ethics; *congenital defects; ethics; extraordinary treatment; genetic defects; human characteristics; human experimentation; involuntary euthanasia; justice; killing; moral obligations; medical ethics; *newborns; pain; parents; *patient care; personhood; physicians; prognosis; quality of life; selection for treatment; self concept; social interaction; socioeconomic factors; spina bifida; *terminally ill; theology; third party consent; *value of life; Jonas, Hans; McCormick, Richard A.

PATIENT CARE   See under

OPERANT CONDITIONING / PATIENT CARE / CONTRACTS

PATIENT PARTICIPATION   See under

PATIENT CARE / PATIENT PARTICIPATION

PATIENT CARE / PATIENT PARTICIPATION / PHYSICIAN PATIENT RELATIONSHIP

PATIENTS   See under

PRIVACY / PATIENTS / PSYCHIATRY

TREATMENT REFUSAL / PATIENTS / RIGHTS

PATIENTS' RIGHTS

Bandman, Elsie; Bandman, Bertram. There is nothing automatic about rights. *American Journal of Nursing* 77(5): 867-872, May 1977. 20 fn. BE05594.
 confidentiality; dehumanization; disclosure; hospitals; human experimentation; informed consent; institutional obligations; medical records; patient care; *patients' rights; privacy; risks and benefits; treatment refusal

PATIENTS' RIGHTS / PREGNANT WOMEN

The pregnant patient's bill of rights. *MCN: American Journal of Maternal Child Nursing* 2(2): 137-138, Mar/Apr 1977. BE06222.
 drugs; medical records; patient care; *patients' rights; *pregnant women; risks and benefits

PATIENTS' RIGHTS   See under

CHILDREN / PATIENTS' RIGHTS

ELECTROCONVULSIVE THERAPY / PATIENTS' RIGHTS

ELECTROCONVULSIVE THERAPY / TREATMENT REFUSAL / PATIENTS' RIGHTS

PHYSICIAN PATIENT RELATIONSHIP / PATIENTS' RIGHTS

TERMINALLY ILL / PATIENTS' RIGHTS

PEDIATRICS / DECISION MAKING

Vaughan, Victor C. Difficult decisions in pediatrics. *In* Vaughan, Victor C.; McKay, R. James; Nelson, Waldo E., eds. Nelson Textbook of Pediatrics. Tenth Edition. Philadelphia: W.B. Saunders, 1975. p. 144-145. ISBN 0-7216-9018-1. BE07079.
 *decision making; family members; *pediatrics; *physician's role; values

PEDIATRICS   See under

PATIENT CARE / PEDIATRICS / ATTITUDES

PENNSYLVANIA   See under

ABORTION / FINANCIAL SUPPORT / PENNSYLVANIA

COERCION / INSTITUTIONALIZED PERSONS / PENNSYLVANIA

INVOLUNTARY COMMITMENT / STANDARDS / PENNSYLVANIA

RIGHT TO TREATMENT / INSTITUTIONALIZED PERSONS / PENNSYLVANIA

BE = bioethics accession number   fn. = footnotes   refs. = references

PENNSYLVANIA *See under (cont'd.)*

VOLUNTARY ADMISSION / PARENTAL CONSENT / PENNSYLVANIA

PERINATOLOGY

Jonsen, Albert R.; Lister, George. Newborn intensive care: the ethical problems. *Hastings Center Report* 8(1): 15-18, Feb 1978. 9 fn. BE06781.
    etiology; human experimentation; injuries; intensive care units; morbidity; newborns; patient care; *perinatology; prognosis; risks and benefits

PERSONHOOD *See also* BEGINNING OF LIFE

PERSONHOOD

Engelhardt, H. Tristram. Some persons are humans, some humans are persons, and the world is what we persons make of it. *In* Spicker, Stuart F.; Engelhardt, H. Tristram, eds. Philosophical Medical Ethics: Its Nature and Significance. Proceedings. Boston: D. Reidel, 1977. p. 183-194. Proceedings of the Third Trans-Disciplinary Symposium on Philosophy and Medicine, Farmington, Connecticut, 11-13 Dec 1975. 8 refs. 2 fn. ISBN 90-277-0772-3. BE05870.
    evolution; fetuses; human characteristics; moral obligations; natural law; *personhood; self concept

Pastrana, Gabriel. Personhood and the beginning of human life. *Thomist* 41(2): 247-294, Apr 1977. 75 fn. BE05574.
    abortion; *beginning of life; brain; fetal development; fetuses; *personhood; moral policy; philosophy; self concept; social interaction; twinning; viability

PERSONHOOD / FETAL DEVELOPMENT

Paoletti, Robert A. Developmental-genetic and psycho-social positions regarding the ontological status of the fetus. *Linacre Quarterly* 44(3): 243-261, Aug 1977. 80 fn. BE05838.
    beginning of life; *fetal development; *fetuses; *human characteristics; infants; *personhood; *social interaction; viability

PERSONHOOD / FETAL DEVELOPMENT / HUMAN RIGHTS

Warren, Mary A. Do potential people have moral rights? *Canadian Journal of Philosophy* 7(2): 275-289, Jun 1977. 10 fn. BE06306.
    *abortion; consequences; contraception; ethics; *fetal development; future generations; human development; *human rights; moral obligations; *personhood; population growth; quality of life; reproduction; *self concept; value of life; Golden Rule; *Hare, R.M.

PERSONHOOD / FETUSES

Barry, Robert. Personhood: the conditions of identification and description. *Linacre Quarterly* 45(1): 64-81, Feb 1978. 28 fn. BE06771.
    fetal development; *fetuses; human characteristics; human development; *personhood; philosophy; value of life

Tiefel, Hans O. The unborn: human values and responsibilities. *Journal of the American Medical Association* 239(21): 2263-2267, 26 May 1978. 10 refs. BE06970.
    *fetuses; maternal life; moral obligations; obligations of society; *personhood; religious beliefs; resource allocation; rights; *value of life

PERSONHOOD / FETUSES / LEGAL ASPECTS

Tennessee. Supreme Court. Hamby v. McDaniel. 12 Dec 1977. *South Western Reporter, 2d Series,* 559: 774-783. 5 fn. BE07126.
    *fetuses; judicial action; *legal aspects; legislation; mother fetus relationship; *personhood; prenatal injuries; state courts; Supreme Court decisions; viability; *wrongful death; *Tennessee Wrongful Death Statute; United States

PERSONHOOD / FETUSES / RIGHTS

Kluge, E.H.W. The right to life of potential persons. *Dalhousie Law Journal* 3(3): 837-848, Jan 1977. 13 fn. BE06289.
    fetal development; *fetuses; human characteristics; *personhood; prenatal injuries; *rights; selective abortion; self concept; value of life

PERSONHOOD / HUMAN RIGHTS / HARE, R.M.

Warren, Mary A. Do potential people have moral rights? *Canadian Journal of Philosophy* 7(2): 275-289, Jun 1977. 10 fn. BE06306.
    *abortion; consequences; contraception; ethics; *fetal development; future generations; human development; *human rights; moral obligations; *personhood; population growth; quality of life; reproduction; *self concept; value of life; Golden Rule; *Hare, R.M.

PERSONHOOD / SOCIAL INTERACTION

Paoletti, Robert A. Developmental-genetic and psycho-social positions regarding the ontological status of the fetus. *Linacre Quarterly* 44(3): 243-261, Aug 1977. 80 fn. BE05838.
    beginning of life; *fetal development; *fetuses; *human characteristics; infants; *personhood; *social interaction; viability

PERSONHOOD *See under*

ABORTION / PERSONHOOD

DETERMINATION OF DEATH / PERSONHOOD /

BE = bioethics accession number   fn. = footnotes   refs. = references

PERSONHOOD *See under (cont'd.)*
SELF CONCEPT
ORGAN TRANSPLANTATION / PERSONHOOD / BRAIN
VALUE OF LIFE / PERSONHOOD

PHENYLKETONURIA *See under*
GENETIC COUNSELING / PHENYLKETONURIA
GENETIC SCREENING / NEWBORNS / PHENYL-KETONURIA

PHILOSOPHY *See under*
MEDICAL ETHICS / PHILOSOPHY
PHYSICIANS / PHILOSOPHY

PHYSICAL CONTAINMENT *See under*
RECOMBINANT DNA RESEARCH / PHYSICAL CONTAINMENT
RECOMBINANT DNA RESEARCH / PHYSICAL CONTAINMENT / FORT DETRICK

PHYSICIAN PATIENT RELATIONSHIP *See also* PROFESSIONAL PATIENT RELATIONSHIP

PHYSICIAN PATIENT RELATIONSHIP

Dougherty, Charles J. Moral directionality in the doctor-patient relationship. *Linacre Quarterly* 44(4): 361-367, Nov 1977. 5 refs. BE06104.
*deontological ethics; moral obligations; paternalism; patient participation; *physician patient relationship; physician's role; *utilitarianism

PHYSICIAN PATIENT RELATIONSHIP / AUTHORITARIANISM

Bury, Judith. Some thoughts on the doctor/patient relationship. *Midwife, Health Visitor and Community Nurse* 13(7): 202+, Jul 1977. BE06316.
alternatives; *authoritarianism; disclosure; paternalism; *patient care; *patient participation; *physician patient relationship

MacIntyre, Alasdair. Patients as agents. *In* Spicker, Stuart F.; Engelhardt, H. Tristram, eds. Philosophical Medical Ethics: Its Nature and Significance. Proceedings. Boston: D. Reidel, 1977. p. 197-212. Proceedings of the Third Trans-Disciplinary Symposium on Philosophy and Medicine, Farmington, Connecticut, 11-13 Dec 1975. ISBN 90-277-0772-3. BE05871.
*authoritarianism; *cultural pluralism; decision making; medical ethics; *physician patient relationship; physicians; self determination; values

PHYSICIAN PATIENT RELATIONSHIP / HISTORICAL ASPECTS

McCullough, Laurence B. Historical perspectives on the ethical dimensions of the patient-physician rela-

tionship: the medical ethics of Dr. John Gregory. *Ethics in Science and Medicine* 5(1): 47-53, 1978. 33 fn. BE06907.
confidentiality; disclosure; *historical aspects; medical etiquette; moral obligations; philosophy; *physician patient relationship; physicians; *Gregory, John

PHYSICIAN PATIENT RELATIONSHIP / LEGAL ASPECTS

Sadoff, Robert L. Medical-legal aspects of the doctor-patient relationship. *Pennsylvania Medicine* 81(3): 24-27, Mar 1978. BE06825.
battery; confidentiality; criminal law; informed consent; involuntary commitment; *legal aspects; malpractice; medical records; patient access; patients' rights; *physician patient relationship; privileged communication; right to treatment; sexuality; Pennsylvania

PHYSICIAN PATIENT RELATIONSHIP / MALPRACTICE

Harney, David M. Relationship between patient and physician or hospital. *In his* Medical Malpractice: 1977 Pocket Supplement. Indianapolis: Allen Smith, 1977. p. 1-10. Footnotes included. BE05880.
disclosure; *legal liability; legal obligations; *malpractice; negligence; *physician patient relationship

PHYSICIAN PATIENT RELATIONSHIP / PATERNALISM

Lasagna, Louis; Veatch, Robert M.; Kurtz, Paul. Medical Ethics. [Videorecording]. Amherst,N.Y.: American Humanist Association, 1977. 1 cassette; 30 min.; sound; color; 3/4 in. Ethics in America Series presented by the American Humanist Association. Produced at WNED-TV, Buffalo, New York. BE07041.
decision making; disclosure; informed consent; *medical ethics; *paternalism; *patient care; *patient participation; *physician patient relationship; physicians; risks and benefits; self determination; values

Marsh, Frank H. An ethical approach to paternalism in the physician-patient relationship. *Ethics in Science and Medicine* 4(3-4): 135-138, 1977. 7 fn. BE06068.
decision making; informed consent; *paternalism; patient participation; *physician patient relationship; self determination

PHYSICIAN PATIENT RELATIONSHIP / PATIENTS' RIGHTS

Newton, Lisa H. The healing of the person. *Connecticut Medicine* 41(10): 641-646, Oct 1977. 22 fn. BE06091.

BE = bioethics accession number   fn. = footnotes   refs. = references

decision making; dehumanization; paternalism; *health; patient advocacy; *patient care; *patient participation; *patients' rights; *physician patient relationship; physician's role; self determination

PHYSICIAN PATIENT RELATIONSHIP / PSYCHOTHERAPY / SEXUALITY

Redlich, Fritz. The ethics of sex therapy. *In* Masters, William H.; Johnson, Virginia E.; Kolodny, Robert C., eds. Ethical Issues in Sex Therapy and Research. Boston: Little, Brown, 1977. p. 143-181. Includes discussion by Jay Katz and others on p. 161-181. 20 refs. ISBN 0-316-549-835. BE06164.
    patient care; *physician patient relationship; professional ethics; *psychotherapy; risks and benefits; *sexuality; *values

PHYSICIAN PATIENT RELATIONSHIP / VALUES

Kimball, Chase P. The ethics of personal medicine. *Medical Clinics of North America* 61(4): 867-877, Jul 1977. 11 refs. BE05961.
    communication; confidentiality; informed consent; medical ethics; patient care; *physician patient relationship; physicians; *values

PHYSICIAN PATIENT RELATIONSHIP   *See under*

CONFIDENTIALITY / PHYSICIAN PATIENT RELATIONSHIP

CONFIDENTIALITY / PHYSICIAN PATIENT RELATIONSHIP / INSURANCE

DISCLOSURE / PHYSICIAN PATIENT RELATIONSHIP

INFORMED CONSENT / PHYSICIAN PATIENT RELATIONSHIP

INFORMED CONSENT / PHYSICIAN PATIENT RELATIONSHIP / LEGAL ASPECTS

MORAL OBLIGATIONS / PHYSICIAN PATIENT RELATIONSHIP

MORAL OBLIGATIONS / PHYSICIAN PATIENT RELATIONSHIP / CONFLICT OF INTEREST

MORAL OBLIGATIONS / PHYSICIAN PATIENT RELATIONSHIP / SEXUALITY

PATIENT CARE / PATIENT PARTICIPATION / PHYSICIAN PATIENT RELATIONSHIP

SELF DETERMINATION / PHYSICIAN PATIENT RELATIONSHIP

TERMINAL CARE / PHYSICIAN PATIENT RELATIONSHIP

PHYSICIAN'S ROLE   *See under*

ALLOWING TO DIE / PHYSICIAN'S ROLE

BEHAVIOR CONTROL / PSYCHIATRY / PHYSICIAN'S ROLE

CHILD NEGLECT / PHYSICIAN'S ROLE

DEATH / PHYSICIAN'S ROLE

PHYSICIAN'S ROLE   *See under (cont'd.)*

DISCLOSURE / CANCER / PHYSICIAN'S ROLE

HEALTH CARE DELIVERY / PHYSICIAN'S ROLE

NEWBORNS / CONGENITAL DEFECTS / PHYSICIAN'S ROLE

NEWBORNS / SPINA BIFIDA / PHYSICIAN'S ROLE

ORGAN TRANSPLANTATION / PHYSICIAN'S ROLE

PATIENT CARE / SPINA BIFIDA / PHYSICIAN'S ROLE

PROLONGATION OF LIFE / BIOMEDICAL TECHNOLOGIES / PHYSICIAN'S ROLE

TERMINAL CARE / CHILDREN / PHYSICIAN'S ROLE

TERMINAL CARE / PHYSICIAN'S ROLE

TORTURE / MORAL OBLIGATIONS / PHYSICIAN'S ROLE

PHYSICIANS / CONGENITAL DEFECTS / LEGAL LIABILITY

Shaw, Margery W. Procreation and the population problem. *North Carolina Law Review* 55(6): 1165-1185, Sep 1977. 149 fn. BE05989.
    compensation; *congenital defects; contraception; human rights; involuntary sterilization; *legal liability; legal rights; mentally retarded; *newborns; parents; population control; *physicians; quality of life; *reproduction; *rights; state interest; *unwanted children; voluntary sterilization

PHYSICIANS / JUSTICE / STRIKES

Daniels, Norman. On the picket line: are doctors' strikes ethical? *Hastings Center Report* 8(1): 24-29, Feb 1978. 17 fn. BE06775.
    economics; goals; health care delivery; *justice; *physicians; *remuneration; risks and benefits; *strikes

PHYSICIANS / PHILOSOPHY

Stephens, G. Gayle; Baker, Sherwood. The ethical dimensions of behavior. *In* Conn, Robert E.; Rakel, Robert E.; Johnson, Thomas W., eds. Family Practice. Philadelphia: W.B. Saunders, 1973. p. 219-230. 12 refs. ISBN 0-7216-2665-3. BE07058.
    counseling; *death; decision making; determination of death; ethics; *philosophy; *physician's role; *physicians; *psychological stress

PHYSICIANS   *See under*

ABORTION / PHYSICIANS / ATTITUDES

ABORTION / PHYSICIANS / NEGLIGENCE

ALLOWING TO DIE / PHYSICIANS / ATTITUDES

ALLOWING TO DIE / PHYSICIANS / LEGAL ASPECTS

BIOETHICAL ISSUES / PHYSICIANS

DECEPTION / PHYSICIANS / TERMINALLY ILL

BE = bioethics accession number   fn. = footnotes   refs. = references

PHYSICIANS *See under (cont'd.)*

EUTHANASIA / PHYSICIANS

FAMILY PLANNING / PHYSICIANS / ATTITUDES

IN VITRO FERTILIZATION / PHYSICIANS / LEGAL LIABILITY

INFORMED CONSENT / PHYSICIANS

INFORMED CONSENT / PHYSICIANS / LEGAL ASPECTS

INFORMED CONSENT / PHYSICIANS / LEGAL LIABILITY

LEGAL OBLIGATIONS / PHYSICIANS / ABORTED FETUSES

LEGAL OBLIGATIONS / PHYSICIANS / DEINSTITUTIONALIZED PERSONS

MEDICAL ETHICS / PHYSICIANS / EDUCATION

MORAL OBLIGATIONS / PHYSICIANS

MORAL OBLIGATIONS / PHYSICIANS / STRIKES

OBLIGATIONS TO SOCIETY / PHYSICIANS

PATIENT CARE / PHYSICIANS

PATIENT CARE / PHYSICIANS / JEWISH ETHICS

PATIENT CARE / PHYSICIANS / LEGAL ASPECTS

PATIENT CARE / PHYSICIANS / LEGAL LIABILITY

PATIENT CARE / PHYSICIANS / RISKS AND BENEFITS

POPULATION CONTROL / PHYSICIANS / ATTITUDES

PRIVILEGED COMMUNICATION / PHYSICIANS / NEW ZEALAND

PSYCHOSURGERY / PHYSICIANS / VALUES

STERILIZATION / PHYSICIANS / LEGAL RIGHTS

TERMINAL CARE / PHYSICIANS / ATTITUDES

TREATMENT REFUSAL / PHYSICIANS / ATTITUDES

WRONGFUL LIFE / PHYSICIANS / LEGAL LIABILITY

PLACEBOS *See under*

HUMAN EXPERIMENTATION / PLACEBOS

PATIENT CARE / PLACEBOS

PATIENT CARE / PLACEBOS / ATTITUDES

PATIENT CARE / PLACEBOS / EVALUATION

PLANNED PARENTHOOD OF CENTRAL MISSOURI V. DANFORTH *See under*

ABORTION / GOVERNMENT REGULATION / PLANNED PARENTHOOD OF CENTRAL MISSOURI V. DANFORTH

ABORTION / PARENTAL CONSENT / PLANNED PARENTHOOD OF CENTRAL MISSOURI V. DANFORTH

ABORTION / SPOUSAL CONSENT / PLANNED PARENTHOOD OF CENTRAL MISSOURI V. DANFORTH

POLIOMYELITIS *See under*

IMMUNIZATION / POLIOMYELITIS / RISKS AND BENEFITS

IMMUNIZATION / POLIOMYELITIS / TORTS

IMMUNIZATION / PUBLIC POLICY / POLIOMYELITIS

POLITICAL ACTIVITY *See under*

ABORTION / POLITICAL ACTIVITY

ABORTION / ROMAN CATHOLICISM / POLITICAL ACTIVITY

RECOMBINANT DNA RESEARCH / INVESTIGATORS / POLITICAL ACTIVITY

RECOMBINANT DNA RESEARCH / LEGISLATION / POLITICAL ACTIVITY

POLITICAL SYSTEMS *See under*

EUGENICS / POLITICAL SYSTEMS / VALUES

POLITICS *See under*

ABORTION / FINANCIAL SUPPORT / POLITICS

ABORTION / POLITICS

ABORTION / POLITICS / ATTITUDES

ABORTION / POLITICS / GREAT BRITAIN

BEHAVIORAL GENETICS / POLITICS

HUMAN EXPERIMENTATION / POLITICS / HISTORICAL ASPECTS

INVOLUNTARY COMMITMENT / POLITICS

MEDICINE / POLITICS

PSYCHIATRY / POLITICS / USSR

RECOMBINANT DNA RESEARCH / GOVERNMENT REGULATION / POLITICS

SCIENCE / POLITICS

SOCIOBIOLOGY / POLITICS

POOR PERSONS *See* INDIGENTS

POPULATION CONTROL / BLACKS / VALUES

Murray, Robert F. The ethical and moral values of black Americans and population policy. *In* Veatch, Robert M., ed. Population Policy and Ethics: The American Experience. New York: Irvington Publishers, distributed by Halsted Press, 1977. p. 197-209. A volume in the Irvington Population and Demography Series. A project of the Research Group on Ethics and Population, Institute of Society, Ethics and the Life Sciences. 25 refs. ISBN 0-470-15170-6. BE06868.

attitudes; *blacks; common good; disclosure; family; justice; *population control; *public policy; self determination; survival; *values; voluntary programs

POPULATION CONTROL / DEVELOPING COUNTRIES / LEGAL ASPECTS

Mukerjee, Bhupen N. A legal approach to the

---

population problem: a comparative view. *Journal of the Indian Law Institute* 16(4): 649-659, Oct-Dec 1974. 32 fn. BE07063.

> *developing countries; ecology; employment; family planning; *legal aspects; *population control; *population growth; public policy; *social impact; social problems; India

POPULATION CONTROL / DEVELOPING COUNTRIES / SOCIOECONOMIC FACTORS

Taylor, Carl E. Economic triage of the poor and population control. *American Journal of Public Health* 67(7): 660-663, Jul 1977. 17 refs. BE05824.

> community services; *developing countries; education; fertility; health care; mortality; nutrition; *population control; population growth; resource allocation; social impact; *socioeconomic factors

POPULATION CONTROL / ECOLOGY

Potter, Ralph B. The simple structure of the population debate: the logic of the ecology movement. *In* Veatch, Robert M., ed. Population Policy and Ethics: The American Experience. New York: Irvington Publishers, distributed by Halsted Press, 1977. p. 347-363. A volume in the Irvington Population and Demography Series. A project of the Research Group on Ethics and Population, Institute of Society, Ethics and the Life Sciences. 6 refs. ISBN 0-470-15170-6. BE06869.

> coercion; *ecology; goals; justice; natural resources; *population control; population growth; public policy; *resource allocation; scarcity; values; voluntary programs

POPULATION CONTROL / GENETICS

Murray, Robert F. The perspective of the population geneticist. *In* Veatch, Robert M., ed. Population Policy and Ethics: The American Experience. New York: Irvington Publishers, distributed by Halsted Press, 1977. p. 365-375. A volume in the Irvington Population and Demography Series. A project of the Research Group on Ethics and Population, Institute of Society, Ethics and the Life Sciences. 21 refs. ISBN 0-470-15170-6. BE06870.

> ecology; gene pool; genetic diversity; *genetics; negative eugenics; *population control; survival

POPULATION CONTROL / INCENTIVES

Wishik, Samuel M. The use of incentives for fertility reduction. *American Journal of Public Health* 68(2): 113-114, Feb 1978. 4 refs. Editorial. BE06803.

> children; contraception; developing countries; family planning; government regulation; *incentives; informal social control; international aspects; morality; *population control; remuneration; values

POPULATION CONTROL / INDIA

Borders, William. India will moderate birth-curb program. *New York Times,* 3 Apr 1977, sect. 1, p. 1+. BE06009.

> incentives; involuntary sterilization; *population control; public policy; *India

Datta-Ray, Sunanda. Setback seen for population control in India. *Washington Post,* 23 Apr 1977, p. A16. BE06017.

> coercion; politics; *population control; public policy; *India

Jorapur, Pandurang B. Law and population in India. *Indian Journal of Social Work* 38(1): 61-69, Apr 1977. 10 refs. BE05805.

> abortion; birth rate; family planning; incentives; fertility; involuntary sterilization; legislation; mandatory programs; mortality; *population control; population distribution; population growth; public policy; social impact; socioeconomic factors; *India

POPULATION CONTROL / JUSTICE

Veatch, Robert M. Justice. *In* Veatch, Robert M., ed. Population Policy and Ethics: The American Experience. New York: Irvington Publishers, distributed by Halsted Press, 1977. p. 31-39. A volume in the Irvington Population and Demography Series. A project of the Research Group on Ethics and Population, Institute of Society, Ethics and the Life Sciences. 5 fn. ISBN 0-470-15170-6. BE06863.

> equal protection; human rights; *justice; obligations of society; obligations to society; *population control; reproduction; resource allocation

POPULATION CONTROL / JUSTICE / INCENTIVES

Veatch, Robert M. Governmental population incentives: ethical issues at stake. *Studies in Family Planning* 8(4): 100-108, Apr 1977. 40 refs. 7 fn. BE05585.

> coercion; contraception; disclosure; discrimination; economics; family planning; *freedom; *incentives; indigents; *justice; obligations of society; *population control; public policy; self determination; social impact; socioeconomic factors; utilitarianism; voluntary sterilization

POPULATION CONTROL / LEGAL ASPECTS

Edgar, Harold; Greenawalt, Kent. The legal tradition. *In* Veatch, Robert M., ed. Population Policy and Ethics: The American Experience. New York: Irvington Publishers, distributed by Halsted Press, 1977. p. 127-166. A volume in the Irvington Population and Demography Series. A project of the Research Group on Ethics and Population, Institute of Society, Ethics and the Life Sciences. ISBN 0-470-15170-6. BE06867.

BE = bioethics accession number   fn. = footnotes   refs. = references

coercion; common good; economics; equal protection; family planning; freedom; *incentives; justice; *legal aspects; *population control; *public policy; reproduction; *rights; self determination; socioeconomic factors; survival; voluntary programs

POPULATION CONTROL / PHYSICIANS / ATTITUDES

Veatch, Robert M.; Draper, Thomas. The values of physicians. *In* Veatch, Robert M., ed. Population Policy and Ethics: The American Experience. New York: Irvington Publishers, distributed by Halsted Press, 1977. p. 377-408. A volume in the Irvington Population and Demography Series. A project of the Research Group on Ethics and Population, Institute of Society, Ethics and the Life Sciences. 42 refs. 5 fn. ISBN 0-470-15170-6. BE06871.
   abortion; *attitudes; confidentiality; *family planning; justice; organizational policies; *physicians; *population control; professional organizations; public policy; self determination; sterilization; survival; values

POPULATION CONTROL / PUBLIC POLICY

Dyck, Arthur J. Alternative views of moral priorities in population policy. *BioScience* 27(4): 272-276, Apr 1977. 34 refs. 4 fn. BE05552.
   birth rate; coercion; ecology; family planning; justice; natural resources; *population control; population growth; poverty; *public policy; resource allocation; scarcity; socioeconomic factors; voluntary programs

Veatch, Robert M. An ethical analysis of population policy proposals. *In* Veatch, Robert M., ed. Population Policy and Ethics: The American Experience. New York: Irvington Publishers, distributed by Halsted Press, 1977. p. 445-475. A volume in the Irvington Population and Demography Series. A project of the Research Group on Ethics and Population, Institute of Society, Ethics and the Life Sciences. 43 refs. ISBN 0-470-15170-6. BE06872.
   abortion; economics; coercion; education; family planning; *incentives; *mandatory programs; *population control; population distribution; *public policy; social impact; socioeconomic factors; *voluntary programs

POPULATION CONTROL / PUBLIC POLICY / COMMON GOOD

Brown, Peter G. The general welfare. *In* Veatch, Robert M., ed. Population Policy and Ethics: The American Experience. New York: Irvington Publishers, distributed by Halsted Press, 1977. p. 41-46. A volume in the Irvington Population and Demography Series. A project of the Research Group on Ethics and Population, Institute of Society, Ethics and the Life Sciences. 4 fn. ISBN 0-470-15170-6. BE06864.

*common good; *population control; *public policy

POPULATION CONTROL / PUBLIC POLICY / HISTORICAL ASPECTS

Brown, Peter G.; Corfman, Eunice. Moral-political values: an historical analysis. *In* Veatch, Robert M., ed. Population Policy and Ethics: The American Experience. New York: Irvington Publishers, distributed by Halsted Press, 1977. p. 55-126. A volume in the Irvington Population and Demography Series. A project of the Research Group on Ethics and Population, Institute of Society, Ethics and the Life Sciences. 56 refs. 139 fn. ISBN 0-470-15170-6. BE06866.
   *common good; ecology; equal protection; federal government; *freedom; government regulation; *historical aspects; involuntary sterilization; *justice; *legal aspects; legislation; negative eugenics; population distribution; *population control; population growth; *public policy; reproduction; rights; socioeconomic factors; *values

POPULATION CONTROL / PUBLIC POLICY / INCENTIVES

Edgar, Harold; Greenawalt, Kent. The legal tradition. *In* Veatch, Robert M., ed. Population Policy and Ethics: The American Experience. New York: Irvington Publishers, distributed by Halsted Press, 1977. p. 127-166. A volume in the Irvington Population and Demography Series. A project of the Research Group on Ethics and Population, Institute of Society, Ethics and the Life Sciences. ISBN 0-470-15170-6. BE06867.
   coercion; common good; economics; equal protection; family planning; freedom; *incentives; justice; *legal aspects; *population control; *public policy; reproduction; *rights; self determination; socioeconomic factors; survival; voluntary programs

POPULATION CONTROL / PUBLIC POLICY / INDIA

Thomas, John M.; Ryniker, Barbara M.; Kaplan, Milton. Indian abortion law revision and population policy: an overview. *Journal of the Indian Law Institute* 16(4): 513-534, Oct-Dec 1974. 46 fn. BE07060.
   *abortion; attitudes; family planning; government regulation; *international aspects; law; *legislation; physicians; *population control; *public policy; religion; socioeconomic factors; therapeutic abortion; *India; *Medical Termination of Pregnancy Act

POPULATION CONTROL / PUBLIC POLICY / LEGAL ASPECTS

Brown, Peter G.; Corfman, Eunice. Moral-political values: an historical analysis. *In* Veatch, Robert M.,

---

BE = bioethics accession number   fn. = footnotes   refs. = references

ed. Population Policy and Ethics: The American Experience. New York: Irvington Publishers, distributed by Halsted Press, 1977. p. 55-126. A volume in the Irvington Population and Demography Series. A project of the Research Group on Ethics and Population, Institute of Society, Ethics and the Life Sciences. 56 refs. 139 fn. ISBN 0-470-15170-6. BE06866.

*common good; ecology; equal protection; federal government; *freedom; government regulation; *historical aspects; involuntary sterilization; *justice; *legal aspects; legislation; negative eugenics; population distribution; *population control; population growth; *public policy; reproduction; rights; socioeconomic factors; *values

POPULATION CONTROL / PUBLIC POLICY / VALUES

Brown, Peter G.; Corfman, Eunice. Moral-political values: an historical analysis. *In* Veatch, Robert M., ed. Population Policy and Ethics: The American Experience. New York: Irvington Publishers, distributed by Halsted Press, 1977. p. 55-126. A volume in the Irvington Population and Demography Series. A project of the Research Group on Ethics and Population, Institute of Society, Ethics and the Life Sciences. 56 refs. 139 fn. ISBN 0-470-15170-6. BE06866.

*common good; ecology; equal protection; federal government; *freedom; government regulation; *historical aspects; involuntary sterilization; *justice; *legal aspects; legislation; negative eugenics; population distribution; *population control; population growth; *public policy; reproduction; rights; socioeconomic factors; *values

Veatch, Robert M., ed. Population Policy and Ethics: The American Experience. New York: Irvington Publishers, distributed by Halsted Press, 1977. 501 p. A volume in the Irvington Population and Demography Series. A project of the Research Group on Ethics and Population, Institute of Society, Ethics and the Life Sciences. References and footnotes included. ISBN 0-470-15170-6. BE06861.

abortion; blacks; common good; family planning; government regulation; historical aspects; justice; legal aspects; minority groups; physicians; *population control; population growth; professional organizations; *public policy; religion; self determination; socioeconomic factors; survival; *values

POPULATION CONTROL / RIGHTS

Edgar, Harold; Greenawalt, Kent. The legal tradition. *In* Veatch, Robert M., ed. Population Policy and Ethics: The American Experience. New York: Irvington Publishers, distributed by Halsted Press, 1977. p. 127-166. A volume in the Irvington Population and Demography Series. A project of the Re-

search Group on Ethics and Population, Institute of Society, Ethics and the Life Sciences. ISBN 0-470-15170-6. BE06867.

coercion; common good; economics; equal protection; family planning; freedom; *incentives; justice; *legal aspects; *population control; *public policy; reproduction; *rights; self determination; socioeconomic factors; survival; voluntary programs

POPULATION CONTROL / SELF DETERMINATION

Warwick, Donald P. Freedom. *In* Veatch, Robert M., ed. Population Policy and Ethics: The American Experience. New York: Irvington Publishers, distributed by Halsted Press, 1977. p. 17-29. A volume in the Irvington Population and Demography Series. A project of the Research Group on Ethics and Population, Institute of Society, Ethics and the Life Sciences. 7 refs. ISBN 0-470-15170-6. BE06862.

coercion; family planning; *freedom; *population control; psychological stress; *self determination; social impact

POPULATION CONTROL / SURVIVAL

Golding, Martin P. Security/survival. *In* Veatch, Robert M., ed. Population Policy and Ethics: The American Experience. New York: Irvington Publishers, distributed by Halsted Press, 1977. p. 47-52. A volume in the Irvington Population and Demography Series. A project of the Research Group on Ethics and Population, Institute of Society, Ethics and the Life Sciences. 1 ref. ISBN 0-470-15170-6. BE06865.

*population control; *survival; value of life; values

POPULATION GROWTH / DEVELOPING COUNTRIES / SOCIAL IMPACT

Mukerjee, Bhupen N. A legal approach to the population problem: a comparative view. *Journal of the Indian Law Institute* 16(4): 649-659, Oct-Dec 1974. 32 fn. BE07063.

*developing countries; ecology; employment; family planning; *legal aspects; *population control; *population growth; public policy; *social impact; social problems; India

PORTUGAL *See under*

ORGAN TRANSPLANTATION / CADAVERS / PORTUGAL

PRECONCEPTION INJURIES / LEGAL ASPECTS

Ey, Douglas W. Torts—negligence—child has cause of action for preconception medical malpractice. *Vanderbilt Law Review* 31(1): 218-226, Jan 1978. 64 fn. BE06729.

blood transfusions; congenital defects; fetuses; hos-

BE = bioethics accession number   fn. = footnotes   refs. = references

pitals; *legal aspects; *legal obligations; legal rights; malpractice; mothers; negligence; *physicians; *preconception injuries; *prenatal injuries; public policy; *torts; viability

PRECONCEPTION INJURIES / TORTS

Illinois. Appellate Court, Fourth District. Renslow v. Mennonite Hospital. 10 Jun 1976. *North Eastern Reporter, 2d Series,* 351: 870-874. BE07106.
 blood transfusions; congenital defects; hospitals; legal liability; newborns; physicians; *preconception injuries; prenatal injuries; *torts

Illinois. Supreme Court. Renslow v. Mennonite Hospital. 8 Aug 1977. *North Eastern Reporter, 2d Series,* 367: 1250-1266. BE07107.
 blood transfusions; congenital defects; fetuses; hospitals; legal liability; legal obligations; negligence; newborns; physicians; *preconception injuries; prenatal injuries; *torts; viability

Steefel, David S. Preconception torts: foreseeing the unconceived. *University of Colorado Law Review* 48(4): 621-639, Summer 1977. 96 fn. BE06311.
 fetuses; legal aspects; legal liability; legal rights; negligence; parents; *preconception injuries; privacy; reproduction; *torts; wrongful death; wrongful life; *Renslow v. Mennonite Hospital

PREGNANT WOMEN *See under*

 ABORTION / PREGNANT WOMEN / LEGAL RIGHTS

 ALLOWING TO DIE / PREGNANT WOMEN / COMA

 MASS SCREENING / PREGNANT WOMEN / SPINA BIFIDA

 PATIENTS' RIGHTS / PREGNANT WOMEN

 PROLONGATION OF LIFE / PREGNANT WOMEN / BRAIN DEATH

 PROLONGATION OF LIFE / PREGNANT WOMEN / COMA

 TREATMENT REFUSAL / PREGNANT WOMEN

PRENATAL DIAGNOSIS *See also* AMNIOCENTESIS

PRENATAL DIAGNOSIS / MORALITY / LEGAL ASPECTS

Milunsky, Aubrey. Ethics, morality, the law, and prenatal diagnosis. *In his* Know Your Genes. Boston: Houghton Mifflin, 1977. p. 184-195. ISBN 0-395-25374-8. BE05886.
 amniocentesis; costs and benefits; disclosure; family members; fetuses; genetic defects; genetic screening; government regulation; *legal aspects; legal rights; legislation; *morality; personhood; *prenatal diagnosis; selective abortion; sex determination; wrongful death; wrongful life

PRENATAL DIAGNOSIS / SELECTIVE ABORTION

Reid, Robert. The moral dilemmas of the biological revolution. *Times (London),* 28 Jun 1977, p. 14. BE06041.
 amniocentesis; *biomedical technologies; genetic defects; *prenatal diagnosis; *selective abortion; sex determination; *social impact

Stumpf, Samuel E. Genetics and the control of human development: some ethical considerations. *In* Hollis, Harry N., comp. A Matter of Life and Death: Christian Perspectives. Nashville: Broadman Press, 1977. p. 114-128. 18 fn. ISBN 0-8054-6118-3. BE06534.
 Christian ethics; cloning; genetic intervention; in vitro fertilization; *prenatal diagnosis; *reproductive technologies; *selective abortion

PRENATAL DIAGNOSIS / SELECTIVE ABORTION / GENETIC DEFECTS

Thorup, Oscar A., et al. Genetically Transmitted Disease. [Videorecording]. Charlottesville, Va.: University of Virginia Medical Center, Health Sciences Library, 1977. 1 cassette; 60 min.; sound; black and white; 3/4 in. Tape of the Medical Center Hour, 21 Sep 1977. Panel discussion moderated by Oscar A. Thorup. BE07053.
 amniocentesis; case reports; coercion; decision making; economics; future generations; gene pool; *genetic counseling; *genetic defects; health care; goals; human rights; mandatory programs; moral obligations; obligations to society; parents; physicians; *prenatal diagnosis; public policy; reproduction; *selective abortion; self determination; sex determination; socioeconomic factors

PRENATAL DIAGNOSIS / SPINA BIFIDA

Neural-tube defects. *Lancet* 1(8059): 312-313, 11 Feb 1978. 4 fn. Editorial. BE06789.
 decision making; mass screening; newborns; *prenatal diagnosis; selective abortion; *spina bifida; withholding treatment; Great Britain

Prevention of spina bifida—the parents' choice. *Nature* 271(5646): 595, 16 Feb 1978. Editorial. BE06792.
 allowing to die; decision making; handicapped; incidence; *mass screening; methods; parents; pregnant women; *prenatal diagnosis; selective abortion; *spina bifida; Great Britain

Wald, N.J., et al. Maternal serum-alpha-fetoprotein measurement in antenatal screening for anencephaly and spina bifida in early pregnancy. *Lancet* 1(8026): 1323-1332, 25 Jun 1977. Report of U.K. Collaborative Study on Alpha-fetoprotein in Relation to Neural-tube Defects. 23 refs. BE06228.
 amniocentesis; epidemiology; fetal development; genetic screening; pregnant women; *prenatal diagnosis; research design; *spina bifida

---

BE = bioethics accession number fn. = footnotes refs. = references

PRENATAL DIAGNOSIS / SPINA BIFIDA / GREAT BRITAIN

Altman, Lawrence K. U.S. researchers are beginning use of British testing for birth defects. *New York Times,* 11 Sep 1977, sect. 1, p. 26. BE06397.
amniocentesis; mass screening; pregnant women; *prenatal diagnosis; *spina bifida; *Great Britain

PRENATAL INJURIES / COMPENSATION / LEGAL RIGHTS

Connecticut. Superior Court, Hartford County. Simon v. Mullin. 2 Nov 1977. *Atlantic Reporter, 2d Series,* 380: 1353-1357. BE07119.
*compensation; congenital defects; fetuses; *legal rights; newborns; *prenatal injuries; viability

PRENATAL INJURIES / LEGAL ASPECTS

Chabon, Robert S. The legal status of the unborn child. *Journal of Legal Medicine* 5(5): 22-24, May 1977. 20 refs. BE05607.
*abortion; congenital defects; *fetuses; *legal aspects; *legal rights; personhood; *prenatal injuries; viability; wrongful death; wrongful life

Pace, P.J. Civil liability for pre-natal injuries. *Modern Law Review* 40(2): 141-158, Mar 1977. 82 fn. BE06294.
fathers; fetuses; insurance; *legal aspects; *legal liability; legislation; mothers; negligence; parents; personhood; *prenatal injuries; torts

PRESIDENT'S COMMISSION FOR PROTECTION OF HUMAN SUBJECTS

U.S. Congress. Senate. A bill to amend the Public Health Service Act to establish the President's Commission for the Protection of Human Subjects of Biomedical and Behavioral Research, and for other purposes. S. 1893, 95th Cong., 1st Sess., 19 Jul 1977. 33 p. By Edward Kennedy, et al. Referred to the Committee on Human Resources. BE06471.
behavioral research; bioethical issues; confidentiality; ethical review; *ethics committees; evaluation; federal government; financial support; goals; government agencies; government regulation; health care delivery; *human experimentation; *informed consent; legal aspects; *program descriptions; records; research subjects; *President's Commission for Protection of Human Subjects

PRESIDENT'S COMMISSION FOR PROTECTION OF HUMAN SUBJECTS *See under*

BEHAVIORAL RESEARCH / PRESIDENT'S COMMISSION FOR PROTECTION OF HUMAN SUBJECTS

HUMAN EXPERIMENTATION / PRESIDENT'S COMMISSION FOR PROTECTION OF HUMAN SUBJECTS

PREVENTIVE MEDICINE / AMERICAN HEART ASSOCIATION

American Heart Association. Committee on Ethics. The American Heart Association and community cardiovascular health: an ethical comment. Report of the Committee on Ethics of the American Heart Association. *Circulation* 57(3): 645A-649A, Mar 1978. BE06814.
biomedical research; coercion; community medicine; *community services; disadvantaged; *health care; health care delivery; heart diseases; human experimentation; informal social control; obligations to society; *organizational policies; *preventive medicine; *privacy; public advocacy; public participation; self induced illness; *American Heart Association

PREVENTIVE MEDICINE / COSTS AND BENEFITS

Freedman, Benjamin. The case for medical care, inefficient or not. *Hastings Center Report* 7(2): 31-39, Apr 1977. 22 fn. BE05555.
*costs and benefits; *decision making; ethics; health care delivery; moral obligations; *morality; *patient care; patients; physicians; physician patient relationship; *preventive medicine; renal dialysis; *resource allocation; *scarcity; selection for treatment; self determination; trust; values

PREVENTIVE MEDICINE *See under*

SOCIAL CONTROL / PREVENTIVE MEDICINE / COSTS AND BENEFITS

PRISONERS / PSYCHIATRY

Lundy, Phyllis J.; Breggin, Peter R. Psychiatric oppression of prisoners. *Psychiatric Opinion* 11(3): 30-37, Jun 1974. 9 fn. BE07059.
behavior control; *prisoners; psychiatric diagnosis; *psychiatry; psychosurgery; psychotherapy; punishment

PRISONERS *See under*

BEHAVIOR CONTROL / PRISONERS / LEGAL RIGHTS

BEHAVIOR CONTROL / PRISONERS / PROJECT START

BEHAVIORAL RESEARCH / PRISONERS

FORCE FEEDING / PRISONERS / GREAT BRITAIN

HEALTH CARE / PRISONERS / MINNESOTA

HUMAN EXPERIMENTATION / PRISONERS

HUMAN EXPERIMENTATION / PRISONERS / CODES OF ETHICS

INFORMED CONSENT / PRISONERS

INFORMED CONSENT / PRISONERS / CALIFORNIA

OPERANT CONDITIONING / PRISONERS

BE = bioethics accession number   fn. = footnotes   refs. = references

PRISONERS *See under (cont'd.)*

OPERANT CONDITIONING / PRISONERS / LEGAL ASPECTS

PSYCHOACTIVE DRUGS / PRISONERS / CALIFORNIA

PSYCHOSURGERY / PRISONERS / LEGAL RIGHTS

RIGHT TO TREATMENT / PRISONERS / PSYCHOTHERAPY

PRIVACY *See also* CONFIDENTIALITY, PRIVILEGED COMMUNICATION

PRIVACY PROTECTION STUDY COMMISSION *See under*

CONFIDENTIALITY / MEDICAL RECORDS / PRIVACY PROTECTION STUDY COMMISSION

PRIVACY / LAETRILE / LEGAL RIGHTS

U.S. District Court, W.D. Oklahoma. Rutherford v. United States. 5 Dec 1977. *Federal Supplement* 438: 1287-1302. 28 fn. BE07125.
 cancer; constitutional law; drugs; federal government; *government regulation; judicial action; law; *legal aspects; *legal rights; legislation; *patients; *privacy; toxicity; *Food and Drug Administration; Food, Drug, and Cosmetic Act; *Laetrile

PRIVACY / PATIENTS / PSYCHIATRY

New York. Supreme Court, New York County. Doe v. Roe. 21 Nov 1977. *New York Supplement, 2d Series,* 400: 668-680. 9 fn. BE07122.
 compensation; *confidentiality; constitutional law; contracts; *disclosure; insurance; legal obligations; legal rights; legislation; *literature; mentally ill; *patients; physician patient relationship; physicians; *privacy; *psychiatry; psychotherapy; public policy; state government; torts

PRIVACY *See under*

ABORTION / PRIVACY / SUPREME COURT DECISIONS

BEHAVIORAL RESEARCH / PRIVACY

BEHAVIORAL RESEARCH / PRIVACY / DIGNITY

CONFIDENTIALITY / PRIVACY / CONSTITUTIONAL LAW

CONFIDENTIALITY / PRIVACY / LEGAL ASPECTS

HUMAN EXPERIMENTATION / PRIVACY

PRIVATE HOSPITALS *See under*

ABORTION / PRIVATE HOSPITALS / CONSCIENCE

PRIVILEGED COMMUNICATION *See also* CONFIDENTIALITY, PRIVACY

PRIVILEGED COMMUNICATION / LEGISLATION / CALIFORNIA

Olander, Alexander J. Discovery of psychotherapist-patient communications after *Tarasoff. San Diego Law Review* 15(2): 265-285, Mar 1978. 101 fn. BE06820.
 confidentiality; *dangerousness; disclosure; *legal aspects; legal obligations; *legislation; malpractice; obligations to society; patients; physician patient relationship; *privileged communication; *professional patient relationship; psychiatric diagnosis; *psychotherapy; *California; In re Lifschutz; People v. Hopkins; *Tarasoff v. Regents

PRIVILEGED COMMUNICATION / PHYSICIANS / NEW ZEALAND

Fairgray, Ross. Medical privilege. *New Zealand Nursing Journal* 70(2): 20-22, Feb 1977. BE05907.
 disclosure; drug abuse; informed consent; legislation; patients; *physicians; *privileged communication; psychiatry; *New Zealand

PRIVILEGED COMMUNICATION / PSYCHIATRY / COMMON GOOD

Rozovsky, Lorne E. Patient communications are not secret. *Dimensions in Health Service* 54(7): 48-49, Jul 1977. BE05733.
 *common good; confidentiality; judicial action; legal aspects; legal obligations; nurses; physician patient relationship; physicians; *privileged communication; *psychiatry; social workers; society; Canada

PRIVILEGED COMMUNICATION / PSYCHOTHERAPY

Bellamy, William A. Privileged communication and confidentiality in forensic psychiatry. *In* Wolman, Benjamin B., ed. International Encyclopedia of Psychiatry, Psychology, Psychoanalysis, and Neurology. Volume 9. New York: Van Nostrand Reinhold, 1977. p. 74-77. 7 refs. ISBN 0-918228-01-8. BE06568.
 confidentiality; legal aspects; legislation; physician patient relationship; *privileged communication; psychiatry; *psychotherapy

Christensen, Eric L. Constitutional law—right of privacy—evidence law of privileges—the patient-litigant exception to the psychotherapist-patient privilege. *Loyola of Los Angeles Law Review* 10(3): 695-708, Jun 1977. 71 fn. BE06303.
 confidentiality; constitutional law; disclosure; *legal aspects; legislation; patients; physician patient relationship; privacy; *privileged communication; psychiatry; *psychotherapy; rights; state courts;

BE = bioethics accession number   fn. = footnotes   refs. = references

state interest; Caesar v. Mountanos; California; In re Lifschutz

PRIVILEGED COMMUNICATION / PSYCHOTHERAPY / DANGEROUSNESS

Olander, Alexander J. Discovery of psychotherapist-patient communications after *Tarasoff*. *San Diego Law Review* 15(2): 265-285, Mar 1978. 101 fn. BE06820.

confidentiality; *dangerousness; disclosure; *legal aspects; legal obligations; *legislation; malpractice; obligations to society; patients; physician patient relationship; *privileged communication; *professional patient relationship; psychiatric diagnosis; *psychotherapy; *California; In re Lifschutz; People v. Hopkins; *Tarasoff v. Regents

PRIVILEGED COMMUNICATION / TARASOFF V. REGENTS / LEGAL ASPECTS

Olander, Alexander J. Discovery of psychotherapist-patient communications after *Tarasoff*. *San Diego Law Review* 15(2): 265-285, Mar 1978. 101 fn. BE06820.

confidentiality; *dangerousness; disclosure; *legal aspects; legal obligations; *legislation; malpractice; obligations to society; patients; physician patient relationship; *privileged communication; *professional patient relationship; psychiatric diagnosis; *psychotherapy; *California; In re Lifschutz; People v. Hopkins; *Tarasoff v. Regents

PROFESSIONAL ETHICS / AMERICAN MEDICAL ASSOCIATION

Meyer, Lawrence. FTC begins hearings on AMA code of ethics. *Washington Post*, 8 Sep 1977, p. A11. BE06396.

codes of ethics; medical etiquette; *professional ethics; professional organizations; *American Medical Association

PROFESSIONAL ETHICS / PSYCHIATRY

Moore, Robert A. Ethics in the practice of psychiatry—origins, functions, models, and enforcement. *American Journal of Psychiatry* 135(2): 157-163, Feb 1978. 21 refs. BE06788.

codes of ethics; ethical review; peer review; *professional ethics; *psychiatry; self regulation

PROFESSIONAL ETHICS / PSYCHOLOGY

Gurel, Brenda D. Ethical problems in psychology. *In* Wolman, Benjamin B., ed. International Encyclopedia of Psychiatry, Psychology, Psychoanalysis, and Neurology. Volume 4. New York: Van Nostrand Reinhold, 1977. p. 377-380. 9 refs. ISBN 0-918228-01-8. BE06565.

codes of ethics; peer review; education; *professional ethics; professional organizations; *program descriptions; *psychology; *American Psychological Association

PROFESSIONAL ETHICS   *See under*

BEHAVIORAL RESEARCH / PROFESSIONAL ETHICS / EVALUATION

MEDICINE / PROFESSIONAL ETHICS

PSYCHOTHERAPY / PROFESSIONAL ETHICS

SCIENCE / PROFESSIONAL ETHICS

TORTURE / PROFESSIONAL ETHICS

PROFESSIONAL ORGANIZATIONS / JUDICIAL ACTION / STANDARDS

Kaplin, William A. Professional power and judicial review: the health professions. *George Washington Law Review* 44(5): 710-753, Aug 1976. 241 fn. BE06641.

due process; *health personnel; *judicial action; organizational policies; legal liability; physicians; professional competence; *professional organizations; *standards; state action; technical expertise; torts; trust

PROGNOSIS   *See under*

DISCLOSURE / CANCER / PROGNOSIS

PROGRAM DESCRIPTIONS   *See under*

RECOMBINANT DNA RESEARCH / GOVERNMENT REGULATION / PROGRAM DESCRIPTIONS

REVIEW COMMITTEES / INSTITUTIONALIZED PERSONS / PROGRAM DESCRIPTIONS

PROJECT START   *See under*

BEHAVIOR CONTROL / PRISONERS / PROJECT START

OPERANT CONDITIONING / PROJECT START

PROLONGATION OF LIFE   *See also* ALLOWING TO DIE, EUTHANASIA, EXTRAORDINARY TREATMENT, LIFE EXTENSION

PROLONGATION OF LIFE

Lang, Joan A.; Seltzer, Marc M. Review of two books on death and dying. [Title provided]. *DePaul Law Review* 26(4): 891-901, Summer 1977. 24 fn. BE06490.

*allowing to die; biomedical technologies; coma; death; legal aspects; legal rights; *prolongation of life; self determination; treatment refusal; *Quinlan, Karen

Wakin, Edward. Is the right-to-die wrong? *U.S. Catholic* 43(3): 6-12, Mar 1978. BE06831.

active euthanasia; *allowing to die; case reports; coma; determination of death; extraordinary treat-

---

BE = bioethics accession number   fn. = footnotes   refs. = references

ment; living wills; *prolongation of life; religious ethics; terminally ill; withholding treatment

PROLONGATION OF LIFE / BIOMEDICAL TECHNOLOGIES / PHYSICIAN'S ROLE

Van den Berg, Jan H. Medical Power and Medical Ethics. New York: W.W. Norton, 1978. 91 p. 2 fn. ISBN 0-393-06428-X. BE06984.
   active euthanasia; aged; *allowing to die; *biomedical technologies; congenital defects; disclosure; historical aspects; medical ethics; newborns; patients; *physician's role; *prolongation of life; self determination; surgery; terminally ill; withholding treatment

PROLONGATION OF LIFE / CHILDREN / GENETIC DEFECTS

Fletcher, Joseph; Wilson, Raphael; Hendley, L. Owen. The Isolated Immune-Deficient Infant: What About the Future? [Videorecording]. Charlottesville, Va.: University of Virginia Medical Center, Health Sciences Library, 1975. 1 cassette; 60 min.; sound; black and white; 3/4 in. Tape of the Medicine and Society Conference, 2 Dec 1975. Panel discussion moderated by Joseph Fletcher. Case study. BE07057.
   *case reports; *children; communicable diseases; *extraordinary treatment; family relationship; financial support; *genetic defects; human development; human experimentation; infants; informed consent; prenatal diagnosis; *prolongation of life; quality of life; reproduction; selective abortion; sex linked defects; social interaction

PROLONGATION OF LIFE / CHRISTIAN ETHICS

Coggan, Donald. On dying and dying well: extracts from the Edwin Stevens lecture. *Journal of Medical Ethics* 3(2): 57-60, Jun 1977. 13 fn. BE05643.
   allowing to die; *Christian ethics; drugs; economics; extraordinary treatment; hospices; intention; *moral obligations; pain; physician patient relationship; physicians; *prolongation of life; *resource allocation; state medicine; *terminal care; terminally ill

PROLONGATION OF LIFE / EMERGENCY CARE

Tait, Karen M.; Winslow, Gerald. Beyond consent —the ethics of decision-making in emergency medicine. *Western Journal of Medicine* 126(2): 156-159, Feb 1977. 19 refs. BE06292.
   decision making; *emergency care; *informed consent; *paternalism; physicians; *prolongation of life; resuscitation; self determination

PROLONGATION OF LIFE / EXTRAORDINARY TREATMENT

McCormick, Richard A. The quality of life, the sanctity of life. *Hastings Center Report* 8(1): 30-36, Feb 1978. 37 fn. BE06784.
   congenital defects; *extraordinary treatment; newborns; *prolongation of life; *quality of life; terminally ill; treatment refusal; value of life; withholding treatment

O'Neil, Richard. In defense of the "ordinary"/"extraordinary" distinction. *Linacre Quarterly* 45(1): 37-40, Feb 1978. 2 refs. BE06790.
   competence; *extraordinary treatment; *human rights; *moral obligations; *patients; *physicians; *prolongation of life; Roman Catholic ethics; standards; terminally ill; third party consent; *treatment refusal; *Veatch, Robert M.

PROLONGATION OF LIFE / EXTRAORDINARY TREATMENT / GENETIC DEFECTS

Fletcher, Joseph; Wilson, Raphael; Hendley, L. Owen. The Isolated Immune-Deficient Infant: What About the Future? [Videorecording]. Charlottesville, Va.: University of Virginia Medical Center, Health Sciences Library, 1975. 1 cassette; 60 min.; sound; black and white; 3/4 in. Tape of the Medicine and Society Conference, 2 Dec 1975. Panel discussion moderated by Joseph Fletcher. Case study. BE07057.
   *case reports; *children; communicable diseases; *extraordinary treatment; family relationship; financial support; *genetic defects; human development; human experimentation; infants; informed consent; prenatal diagnosis; *prolongation of life; quality of life; reproduction; selective abortion; sex linked defects; social interaction

PROLONGATION OF LIFE / MORAL OBLIGATIONS

O'Neil, Richard. In defense of the "ordinary"/"extraordinary" distinction. *Linacre Quarterly* 45(1): 37-40, Feb 1978. 2 refs. BE06790.
   competence; *extraordinary treatment; *human rights; *moral obligations; *patients; *physicians; *prolongation of life; Roman Catholic ethics; standards; terminally ill; third party consent; *treatment refusal; *Veatch, Robert M.

PROLONGATION OF LIFE / PREGNANT WOMEN / BRAIN DEATH

Meyer, Lawrence. 'Two lives involved': 'brain dead' mother kept alive in effort to save 4-month fetus. *Washington Post,* 2 Dec 1977, p. A1+. BE06414.
   *brain death; fetuses; *pregnant women; prognosis; *prolongation of life

PROLONGATION OF LIFE / PREGNANT WOMEN / COMA

Colen, B.D. Ethics questions raised by pregnancy of comatose woman. *Washington Post,* 5 Dec 1977, p. A3. BE06417.

BE = bioethics accession number   fn. = footnotes   refs. = references

brain death; cadavers; *coma; organ donation; *pregnant women; *prolongation of life

PROLONGATION OF LIFE / QUALITY OF LIFE

McCormick, Richard A. The quality of life, the sanctity of life. *Hastings Center Report* 8(1): 30-36, Feb 1978. 37 fn. BE06784.
congenital defects; *extraordinary treatment; newborns; *prolongation of life; *quality of life; terminally ill; treatment refusal; value of life; withholding treatment

PROLONGATION OF LIFE / VALUE OF LIFE

Needleman, Lionel. Valuing other people's lives. *Manchester School of Economics and Social Studies* 44(4): 309-342, Dec 1976. 57 refs. 14 fn. BE06208.
economics; *family members; kidneys; motivation; *organ donation; organ transplantation; parents; *prolongation of life; *risks and benefits; statistics; treatment refusal; *value of life

PROPERTY RIGHTS   *See under*

ABORTION / VALUE OF LIFE / PROPERTY RIGHTS

PROTESTANT ETHICS   *See under*

EUTHANASIA / PROTESTANT ETHICS / RAMSEY, PAUL

PSRO'S   *See under*

CONFIDENTIALITY / PSRO'S / LEGAL ASPECTS
PATIENT CARE / PSRO'S

PSYCHIATRIC DIAGNOSIS / USSR

Psychiatric terror. *Wall Street Journal,* 22 Sep 1977, p. 20. BE06064.
dissent; involuntary commitment; politics; professional organizations; *psychiatric diagnosis; *psychiatry; schizophrenia; *USSR

PSYCHIATRIC DIAGNOSIS   *See under*

INVOLUNTARY COMMITMENT / PSYCHIATRIC DIAGNOSIS / USSR

PSYCHIATRY / CONFLICT OF INTEREST / VALUES

In the service of the state: the psychiatrist as double agent. A conference on conflicting loyalties, 24-26 Mar 1977. *Hastings Center Report* 8(2): Special Suppl. 1-24, Apr 1978. Cosponsored by the American Psychiatric Association and The Hastings Center. Transcript condensed and edited for readability by Margaret O. Steinfels and Carol Levine. BE06846.
confidentiality; *conflict of interest; institutionalized persons; medical etiquette; involuntary commitment; *mental institutions; *military personnel;

moral obligations; *peer review; *physician's role; *prisoners; professional ethics; *psychiatry; psychology; right to treatment; rights; treatment refusal; *values

PSYCHIATRY / DEINSTITUTIONALIZED PERSONS / LEGAL LIABILITY

Psychotherapists' liability for the release of mentally ill offenders: a proposed expansion of the theory of strict liability. *University of Pennsylvania Law Review* 126(1): 204-240, Nov 1977. 141 fn. BE06363.
dangerousness; decision making; *deinstitutionalized persons; judicial action; *legal liability; *mentally ill; *negligence; *psychiatry; risks and benefits; torts

PSYCHIATRY / DISSENT / USSR

Soviet 'no' to ethics standard in psychiatry. *Times (London),* 31 Aug 1977, p. 6. BE06456.
codes of ethics; *dissent; involuntary commitment; organizational policies; professional organizations; *psychiatry; *USSR; *World Psychiatric Association

Weiss, Ted. Human rights and psychiatry in the Soviet Union. *Congressional Record (Daily Edition)* 123(197): E7487-E7489, 15 Dec 1977. BE06591.
*dissent; involuntary commitment; politics; professional organizations; *psychiatry; public policy; *USSR; World Psychiatric Association

PSYCHIATRY / HISTORICAL ASPECTS / GREAT BRITAIN

Carstairs, G.M. Revolutions and the rights of man. *American Journal of Psychiatry* 134(9): 979-983, Sep 1977. 4 refs. BE06243.
attitudes; dissent; freedom; *historical aspects; human rights; institutionalized persons; involuntary commitment; mental institutions; mentally ill; *methods; physicians; politics; psychiatric diagnosis; *psychiatry; psychotherapy; social control; Eighteenth Century; *Great Britain; Nineteenth Century; Twentieth Century; USSR

PSYCHIATRY / LAW

Coleman, Lee. Law and psychiatry—it's a bad affair. *Law and Liberty* 3(4): 1+, Spring/Summer 1977. Editorial. BE06224.
coercion; *law; mentally ill; patient care; psychiatric diagnosis; *psychiatry; voluntary admission

PSYCHIATRY / LEGAL ASPECTS

Sadoff, Robert L. Changing laws and ethics in psychiatry. *Bulletin of the American Academy of Psychiatry and the Law* 5(1): 34-40, 1977. 21 refs. BE05894.
confidentiality; dangerousness; deinstitutionalized persons; disclosure; informed consent; institution-

---

BE = bioethics accession number   fn. = footnotes   refs. = references

alized persons; *legal aspects; mentally ill; *psychiatry; treatment refusal

PSYCHIATRY / LEGAL RIGHTS

Dunham, H. Warren. Psychiatry, sociology and civil liberties. *Man and Medicine: Journal of Values and Ethics in Health Care* 2(4): 263-278, Summer 1977. Commentary by Martin D. Hanlon, p.275-278. 3 refs. 18 fn. BE06254.
 behavior disorders; behavioral research; coercion; common good; conflict of interest; health care; human rights; institutionalized persons; involuntary commitment; *legal rights; mentally ill; physician's role; political systems; psychiatric diagnosis; *psychiatry; self determination; social impact; social problems; *social sciences; sociology of medicine; stigmatization

PSYCHIATRY / METHODS / HISTORICAL ASPECTS

Carstairs, G.M. Revolutions and the rights of man. *American Journal of Psychiatry* 134(9): 979-983, Sep 1977. 4 refs. BE06243.
 attitudes; dissent; freedom; *historical aspects; human rights; institutionalized persons; involuntary commitment; mental institutions; mentally ill; *methods; physicians; politics; psychiatric diagnosis; *psychiatry; psychotherapy; social control; Eighteenth Century; *Great Britain; Nineteenth Century; Twentieth Century; USSR

PSYCHIATRY / POLITICS / USSR

Reddaway, Peter; Bloch, Sidney. Curbing psychiatry's political misuse. *Washington Post,* 15 Nov 1977, p. A23. BE06411.
 dissent; mass media; *organizational policies; *politics; professional organizations; *psychiatry; *USSR; *World Psychiatric Association

PSYCHIATRY *See under*

 ADOLESCENTS / PSYCHIATRY / CONTRACTS
 BEHAVIOR CONTROL / PSYCHIATRY / PHYSICIAN'S ROLE
 CODES OF ETHICS / PSYCHIATRY
 CONFIDENTIALITY / MEDICAL RECORDS / PSYCHIATRY
 CONFIDENTIALITY / PSYCHIATRY
 CONFIDENTIALITY / PSYCHIATRY / LEGAL ASPECTS
 CONFIDENTIALITY / PSYCHIATRY / LITERATURE
 DISSENT / PSYCHIATRY / USSR
 HUMAN EXPERIMENTATION / PSYCHIATRY
 HUMAN EXPERIMENTATION / PSYCHIATRY / FREEDOM
 INFORMED CONSENT / PSYCHIATRY / LEGAL ASPECTS
 MEDICAL ETHICS / PSYCHIATRY

PSYCHIATRY *See under (cont'd.)*

 NURSING ETHICS / PSYCHIATRY
 PATIENT CARE / PSYCHIATRY / HAWAII DECLARATION
 PRISONERS / PSYCHIATRY
 PRIVACY / PATIENTS / PSYCHIATRY
 PRIVILEGED COMMUNICATION / PSYCHIATRY / COMMON GOOD
 PROFESSIONAL ETHICS / PSYCHIATRY
 SUICIDE / PSYCHIATRY / LEGAL LIABILITY
 WORLD PSYCHIATRIC ASSOCIATION / PSYCHIATRY / USSR

PSYCHOACTIVE DRUGS / ADOLESCENTS / CALIFORNIA

California. Legislature. Assembly. An act...relating to administration of psychotropic drugs. Assembly Bill No. 1221, 1977-78 Regular Session, 30 Mar 1977. 6 p. By Assemblyman Alatorre, et al. Referred to Committee on Criminal Justice; died 1977-78 Session. BE06596.
 *adolescents; coercion; informed consent; *institutionalized persons; *psychoactive drugs; *California

PSYCHOACTIVE DRUGS / DECISION MAKING

Veatch, Robert M. Value foundations for drug use. *Journal of Drug Issues* 7(3): 253-262, Summer 1977. 24 refs. BE05938.
 *decision making; *ethics; Protestant ethics; *psychoactive drugs; psychological stress; self determination; values

PSYCHOACTIVE DRUGS / PRISONERS / CALIFORNIA

California. Legislature. Assembly. An act to add Chapter 4 (commencing with Section 4360) to Title 4 of Part 3 of the Penal Code, relating to jails. Assembly Bill No. 1627, 1977-78 Regular Session, 13 Apr 1977. 7 p. By Assemblyman Alatorre. Referred to Committee on Criminal Justice; died 1977-78 Session. BE06598.
 coercion; informed consent; mentally ill; *prisoners; *psychoactive drugs; *California

California. Legislature. Assembly. An act...relating to drug therapy for prisoners. Assembly Bill No. 16, 1977-78 Regular Session, 7 Dec 1976. 13 p. By Assemblyman Alatorre, et al. Vetoed by Governor Edmund G. Brown, Jr., 28 Sep 1977. BE06594.
 *informed consent; mentally ill; *prisoners; psychiatry; *psychoactive drugs; psychosurgery; *California

PSYCHOACTIVE DRUGS *See under*

 BEHAVIOR CONTROL / PSYCHOACTIVE DRUGS / LEGAL ASPECTS
 GOVERNMENT REGULATION / PSYCHOACTIVE

PSYCHOACTIVE DRUGS *See under (cont'd.)*
DRUGS / CALIFORNIA

HUMAN EXPERIMENTATION / PSYCHOACTIVE DRUGS

HUMAN EXPERIMENTATION / PSYCHOACTIVE DRUGS / GOVERNMENT AGENCIES

INFORMED CONSENT / PSYCHOACTIVE DRUGS / RISKS AND BENEFITS

PATIENT CARE / PSYCHOACTIVE DRUGS / CANCER

PATIENT CARE / PSYCHOACTIVE DRUGS / INSTITUTIONALIZED PERSONS

TERMINAL CARE / PSYCHOACTIVE DRUGS

TREATMENT REFUSAL / COERCION / PSYCHOACTIVE DRUGS

PSYCHOLOGICAL STRESS *See under*

INFORMED CONSENT / PSYCHOLOGICAL STRESS / LEGAL ASPECTS

ORGAN DONATION / PSYCHOLOGICAL STRESS

TERMINALLY ILL / PSYCHOLOGICAL STRESS

PSYCHOLOGY / PUBLIC ADVOCACY

Simon, Gottlieb C. Ethical and social issues in professional psychology. *In* Wolman, Benjamin B., ed. International Encyclopedia of Psychiatry, Psychology, Psychoanalysis, and Neurology. Volume 4. New York: Van Nostrand Reinhold, 1977. p. 380-383. 12 refs. ISBN 0-918228-01-8. BE06566.
community services; employment; obligations to society; patient advocacy; *psychology; *public advocacy; social problems

PSYCHOLOGY *See under*

CODES OF ETHICS / PSYCHOLOGY

PROFESSIONAL ETHICS / PSYCHOLOGY

SELF DETERMINATION / PSYCHOLOGY

PSYCHOSURGERY

Bailey, H.R.; Dowling, J.L.; Davies, E. The ethics of psychiatric surgery. *In* Sweet, William H.; Obrador, Sixto; Martin-Rodriguez, José G., eds. Neurosurgical Treatment in Psychiatry, Pain, and Epilepsy. Baltimore: University Park Press, 1977. p. 497-503. 12 refs. ISBN 0-8391-0881-8. BE05693.
attitudes; behavior disorders; evaluation; normality; physician patient relationship; *psychosurgery; suffering

Dagi, T.F. Psychiatric surgery and the ethics of uncertainty. *In* Sweet, William H.; Obrador, Sixto; Martin-Rodriguez, José G., eds. Neurosurgical Treatment in Psychiatry, Pain, and Epilepsy. Baltimore: University Park Press, 1977. p. 513-523. 3 refs. 22 fn. ISBN 0-8391-0881-8. BE05695.
behavior disorders; decision making; diagnosis; in-

tention; legal aspects; mental health; mentally ill; patient care; *psychosurgery; risks and benefits

De Lange, S.A. Some ethical implications of psychiatric surgery. *In* Sweet, William H.; Obrador, Sixto; Martin-Rodriguez, José G., eds. Neurosurgical Treatment in Psychiatry, Pain, and Epilepsy. Baltimore: University Park Press, 1977. p. 489-496. 9 refs. ISBN 0-8391-0881-8. BE05692.
evaluation; mentally ill; pain; personality; *psychosurgery

Donnelly, John. Psychosurgery. *In* Wolman, Benjamin B., ed. International Encyclopedia of Psychiatry, Psychology, Psychoanalysis, and Neurology. Volume 9. New York: Van Nostrand Reinhold, 1977. p. 305-310. 7 refs. ISBN 0-918228-01-8. BE06569.
government regulation; historical aspects; incidence; institutionalized persons; legal aspects; mentally ill; *psychosurgery; social control; violence

Laitinen, L.V. Ethical aspects of psychiatric surgery. *In* Sweet, William H.; Obrador, Sixto; Martin-Rodriguez, José G., eds. Neurosurgical Treatment in Psychiatry, Pain, and Epilepsy. Baltimore: University Park Press, 1977. p. 483-488. 11 refs. ISBN 0-8391-0881-8. BE05691.
control groups; electrical stimulation of the brain; evaluation; human experimentation; informed consent; *psychosurgery

Older, Jules. Psychosurgery: a dilemma of medicine and society. *In* Diesendorf, Mark, ed. The Magic Bullet: Social Implications and Limitations of Modern Medicine—An Environmental Approach. O'Connor, Australia: Society for Social Responsibility in Science (A.C.T.), 1976. p. 103-111. 10 refs. BE06663.
behavior control; children; evaluation; historical aspects; informed consent; prisoners; *psychosurgery

Shuman, Samuel I. Psychosurgery and the Medical Control of Violence: Autonomy and Deviance. Detroit: Wayne State University Press, 1977. 360 p. 281 fn. ISBN 0-8143-1579-8. BE06001.
attitudes; behavior control; behavior disorders; brain pathology; decision making; etiology; health; human rights; informed consent; institutionalized persons; legal aspects; mentally ill; physicians; *psychosurgery; science; self determination; social control; violence; *Kaimowitz v. Department of Mental Health

U.S. National Commission for the Protection of Human Subjects of Biomedical and Behavioral Research. National Commission for the Protection of Human Subjects of Biomedical and Behavioral Research. Transcript of the Meeting Proceedings (26th Meeting), 7-8 Jan 1977. Springfield, Va.: National Technical Information Service, 1977. 345 p. Report

---

BE = bioethics accession number   fn. = footnotes   refs. = references

No. NCPHS/M-77/06. NTIS No. PB-263 280.
BE06647.
*children; *ethics committees; federal government;
financial support; government regulation; *human
experimentation; informed consent; nontherapeutic research; *parental consent; *psychosurgery; risks and benefits; therapeutic research;
treatment refusal

Valenstein, Elliot S. Brain control: scientific, ethical
and political considerations. *In* Ellison, Craig W.,
ed. Modifying Man: Implications and Ethics.
Washington: University Press of America, 1978. p.
143-168. ISBN 0-8191-0302-0. BE06670.
aggression; *behavior control; brain; brain pathology; *electrical stimulation of the brain; ethics
committees; evaluation; government regulation;
human experimentation; mentally ill; political activity; *psychosurgery; social determinants; *violence

Whitlock, F.A. The ethics of psychosurgery. *In*
Smith, J. Sydney; Kiloh, L.G., eds. Psychosurgery
and Society. New York: Pergamon Press, 1977. p.
129-135. A Symposium Organized by the Neuropsychiatric Institute, Sydney, 26-27 Sep 1974. 17 refs.
ISBN 0-08-021836-9. BE06169.
*behavior control; brain pathology; informed con-
sent; law enforcement; mental health; mentally ill;
morality; *physician's role; prisoners; *psychiatry;
*psychosurgery; self determination; violence

PSYCHOSURGERY COMMISSION

U.S. Congress. House. A bill to prohibit psychosurgery in federally connected health care facilities.
H.R. 7371, 95th Cong., 1st Sess., 23 May 1977. 10
p. By Louis Stokes. Referred to the Committees on
Interstate and Foreign Commerce and Ways and
Means. BE06470.
*federal government; financial support; *govern-
ment regulation; hospitals; *psychosurgery; *Psychosurgery Commission

PSYCHOSURGERY / DECISION MAKING

Andy, O.J. Decision-making in psychiatric surgery.
In Sweet, William H.; Obrador, Sixto; Martin-Rodriguez, José G., eds. Neurosurgical Treatment in Psychiatry, Pain, and Epilepsy. Baltimore: University
Park Press, 1977. p. 505-511. 9 refs. ISBN 0-
8391-0881-8. BE05694.
*decision making; ethics committees; human ex-
perimentation; legal aspects; physicians; *psychosurgery; therapeutic research

PSYCHOSURGERY / ETHICAL REVIEW

**Annas, George J.; Glantz, Leonard H.; Katz, Barbara
F.** Psychosurgery: the regulation of surgical innovation. *In their* Informed Consent to Human Experimentation: The Subject's Dilemma. Cambridge,
Mass.: Ballinger Publishing Company, 1977. p.

215-255. 152 fn. ISBN 0-88410-147-9. BE06859.
*ethical review; ethics committees; evaluation;
*government regulation; informed consent; legisla-
tion; negligence; *psychosurgery; risks and benefits; state government; National Commission for
Protection of Human Subjects

PSYCHOSURGERY / ETHICS COMMITTEES / NEW SOUTH WALES

Kiloh, L.G. Commentary on the report of the Committee of Inquiry into Psychosurgery. *Medical Journal
of Australia* 2(9): 296-301, 27 Aug 1977. BE06324.
electrical stimulation of the brain; *ethics committees; *evaluation; government regulation; human
experimentation; informed consent; *psychosurgery; *New South Wales

PSYCHOSURGERY / EVALUATION

Greenberg, Joel. Psychosurgery at the crossroads. *Science News* 111(20): 314-315+, 14 May 1977.
BE05616.
attitudes; behavior control; *evaluation; federal
government; government regulation; mentally ill;
*psychosurgery; National Commission for Protec-
tion of Human Subjects

PSYCHOSURGERY / EVALUATION / NEW SOUTH WALES

Kiloh, L.G. Commentary on the report of the Committee of Inquiry into Psychosurgery. *Medical Journal
of Australia* 2(9): 296-301, 27 Aug 1977. BE06324.
electrical stimulation of the brain; *ethics committees; *evaluation; government regulation; human
experimentation; informed consent; *psychosurgery; *New South Wales

PSYCHOSURGERY / GOVERNMENT REGULA-TION

**Annas, George J.; Glantz, Leonard H.; Katz, Barbara
F.** Psychosurgery: the regulation of surgical innovation. *In their* Informed Consent to Human Experimentation: The Subject's Dilemma. Cambridge,
Mass.: Ballinger Publishing Company, 1977. p.
215-255. 152 fn. ISBN 0-88410-147-9. BE06859.
*ethical review; ethics committees; evaluation;
*government regulation; informed consent; legisla-
tion; negligence; *psychosurgery; risks and benefits; state government; National Commission for
Protection of Human Subjects

Kiloh, L.G. Psychosurgery. *Medical Journal of Australia* 1(10): 349-350, 5 Mar 1977. BE06223.
*government regulation; mentally ill; *psychosur-
gery; public hospitals; New South Wales

U.S. Congress. House. A bill to prohibit psychosurgery in federally connected health care facilities.
H.R. 7371, 95th Cong., 1st Sess., 23 May 1977. 10
p. By Louis Stokes. Referred to the Committees on

BE = bioethics accession number   fn. = footnotes   refs. = references

Interstate and Foreign Commerce and Ways and Means. BE06470.
> *federal government; financial support; *government regulation; hospitals; *psychosurgery; *Psychosurgery Commission

PSYCHOSURGERY / LEGAL ASPECTS

Greenblatt, Steven J. The ethics and legality of psychosurgery. *New York Law School Law Review* 22(4): 961-980, 1977. 101 fn. BE06071.
> human experimentation; informed consent; institutionalized persons; *legal aspects; legislation; mentally ill; *psychosurgery; state government; third party consent

PSYCHOSURGERY / LEGAL ASPECTS / NEW SOUTH WALES

Maddison, J.C. The legal aspects of psychosurgery in New South Wales. *In* Smith, J. Sydney; Kiloh, L.G., eds. Psychosurgery and Society. New York: Pergamon Press, 1977. p. 85-89. A Symposium Organized by the Neuropsychiatric Institute, Sydney, 26-27 Sep 1974. ISBN 0-08-021836-9. BE06165.
> human experimentation; informed consent; *legal aspects; legislation; *psychosurgery; *New South Wales

PSYCHOSURGERY / LEGAL ASPECTS / UNITED STATES

Bromberger, B. Psychosurgery and the law in the United States of America. *In* Smith, J. Sydney; Kiloh, L.G., eds. Psychosurgery and Society. New York: Pergamon Press, 1977. p. 137-150. A Symposium Organized by the Neuropsychiatric Institute, Sydney, 26-27 Sep 1974. ISBN 0-08-021836-9. BE06170.
> codes of ethics; ethics committees; human experimentation; informed consent; institutionalized persons; *legal aspects; legislation; mentally ill; *psychosurgery; risks and benefits; standards; state government; *United States

PSYCHOSURGERY / LEGISLATION / NEW SOUTH WALES

Edwards, G. Mental health legislation and psychosurgery in New South Wales—a proposed amendment. *In* Smith, J. Sydney; Kiloh, L.G., eds. Psychosurgery and Society. New York: Pergamon Press, 1977. p. 151-153. A Symposium Organized by the Neuropsychiatric Institute, Sydney, 26-27 Sep 1974. ISBN 0-08-021836-9. BE06171.
> ethics committees; institutionalized persons; *legislation; mentally ill; *psychosurgery; *New South Wales

PSYCHOSURGERY / NATIONAL COMMISSION FOR PROTECTION OF HUMAN SUBJECTS

Annas, George J. Psychosurgery: procedural safeguards. *Hastings Center Report* 7(2): 11-13, Apr 1977. BE05542.
> behavior control; confidentiality; ethics committees; evaluation; informed consent; pain; privacy; *psychosurgery; risks and benefits; *National Commission for Protection of Human Subjects

Ryan, Kenneth J.; Yesley, Michael S.; Annas, George J. Psychosurgery: clarifications from the National Commission. *Hastings Center Report* 7(5): 4+, Oct 1977. Letter. BE06073.
> children; ethics committees; federal government; financial support; privileged communication; *psychosurgery; *National Commission for Protection of Human Subjects

PSYCHOSURGERY / NATIONAL COMMISSION FOR PROTECTION OF HUMAN SUBJECTS / EVALUATION

Annas, George J. The attempted revival of psychosurgery. *Medicolegal News* 5(3): 3+, Summer 1977. 3 refs. Editorial. BE05715.
> adults; behavior control; children; ethics committees; *evaluation; human experimentation; informed consent; pain; placebos; *psychosurgery; third party consent; *National Commission for Protection of Human Subjects

PSYCHOSURGERY / PHYSICIANS / VALUES

McCaughey, J.D. Civil liberties and patient rights. *In* Smith, J. Sydney; Kiloh, L.G., eds. Psychosurgery and Society. New York: Pergamon Press, 1977. p. 103-108. A Symposium Organized by the Neuropsychiatric Institute, Sydney, 26-27 Sep 1974. 4 refs. ISBN 0-08-021836-9. BE06166.
> *decision making; human rights; informed consent; law enforcement; mentally ill; *physicians; *psychosurgery; *values

PSYCHOSURGERY / PRISONERS / LEGAL RIGHTS

Kelley, Paul M. Prisoner access to psychosurgery: a constitutional perspective. *Pacific Law Journal* 9(1): 249-280, Jan 1978. 203 fn. BE06739.
> decision making; equal protection; government regulation; informed consent; *legal rights; legislation; patient access; *prisoners; privacy; *psychosurgery; self determination; state government; state interest; treatment refusal; Aden v. Younger; California; First Amendment; Kaimowitz v. Department of Mental Health

PSYCHOSURGERY / RISKS AND BENEFITS

Black, Peter M. The rationale for psychosurgery. *Humanist* 37(4): 6+, Jul-Aug 1977. 8 refs. BE05827.

BE = bioethics accession number   fn. = footnotes   refs. = references

behavior control; evaluation; patients; *psychosurgery; *risks and benefits; self determination

PSYCHOSURGERY / SELF DETERMINATION

Szasz, Thomas S. Aborting unwanted behavior: the controversy on psychosurgery. *Humanist* 37(4): 7+, Jul-Aug 1977. 8 fn. BE05828.
attitudes; *coercion; patients; *psychosurgery; *self determination

PSYCHOSURGERY / SOCIAL CONTROL

Crawford, A. Medico-political considerations of psychosurgery. *In* Smith, J. Sydney; Kiloh, L.G., eds. Psychosurgery and Society. New York: Pergamon Press, 1977. p. 109-113. A Symposium Organized by the Neuropsychiatric Institute, Sydney, 26-27 Sep 1974. 6 refs. ISBN 0-08-021836-9. BE06167.
human experimentation; legal aspects; organ transplantation; peer review; politics; *psychosurgery; *social control

PSYCHOSURGERY / UNITED STATES

Ervin, F.R. The American experience. *In* Smith, J. Sydney; Kiloh, L.G., eds. Psychosurgery and Society. New York: Pergamon Press, 1977. p. 125-128. A Symposium Organized by the Neuropsychiatric Institute, Sydney, 26-27 Sep 1974. ISBN 0-08-021836-9. BE06168.
ethics committees; federal government; government regulation; informed consent; institutionalized persons; mentally ill; prisoners; psychoactive drugs; *psychosurgery; state government; *United States

PSYCHOSURGERY / VIOLENCE

Chavkin, Samuel. The Mind Stealers: Psychosurgery and Mind Control. Boston: Houghton Mifflin, 1978. 228 p. 400 fn. ISBN 0-395-26381-6. BE06977.
*adolescents; aggression; *behavior control; behavioral genetics; brain pathology; children; *electrical stimulation of the brain; evaluation; human experimentation; imprisonment; law enforcement; institutionalized persons; legal aspects; legal rights; *operant conditioning; *prisoners; prognosis; psychoactive drugs; *psychosurgery; socioeconomic factors; *violence; Project START

PSYCHOTHERAPY

Goldberg, Carl. Ethical concerns in the conduct of psychotherapy. *In his* Therapeutic Partnership: Ethical Concerns in Psychotherapy. New York: Springer Publishing Company, 1977. p. 1-14. ISBN 0-8261-2350-3. BE06514.
behavior control; confidentiality; medical records; physician patient relationship; *psychotherapy; social control

PSYCHOTHERAPY / ADOLESCENTS

Miller, Derek. The ethics of practice in adolescent psychiatry. *American Journal of Psychiatry* 134(4): 420-424, Apr 1977. 13 refs. BE05569.
*adolescents; authoritarianism; behavior control; behavior disorders; confidentiality; informed consent; physician patient relationship; privacy; professional competence; *psychiatry; *psychotherapy

PSYCHOTHERAPY / INFORMED CONSENT / CONSENT FORMS

Schwarz, Eitan D. The use of a checklist in obtaining informed consent for treatment with medication. *Hospital and Community Psychiatry* 29(2): 97+, Feb 1978. BE06795.
communication; *consent forms; disclosure; *informed consent; medical records; mentally ill; patients; program descriptions; psychiatry; *psychoactive drugs; psychological stress; *psychotherapy; risks and benefits; *Informed Consent Checklist

PSYCHOTHERAPY / PROFESSIONAL ETHICS

Van Hoose, William H.; Kottler, Jeffrey A. Ethical and Legal Issues in Counseling and Psychotherapy. San Francisco: Jossey-Bass, 1977. 224 p. Bibliography: p. 195-217. ISBN 0-87589-317-1. BE05714.
behavior control; confidentiality; counseling; ethics; group therapy; legal aspects; malpractice; privileged communication; morality; professional competence; *professional ethics; professional organizations; professional patient relationship; psychiatric diagnosis; psychiatry; psychology; *psychotherapy; sexuality; standards; values

PSYCHOTHERAPY   *See under*

BEHAVIOR CONTROL / PSYCHOTHERAPY / RELIGIOUS BELIEFS

CONFIDENTIALITY / DANGEROUSNESS / PSYCHOTHERAPY

CONFIDENTIALITY / PSYCHOTHERAPY / LEGAL ASPECTS

CONFIDENTIALITY / PSYCHOTHERAPY / LEGAL LIABILITY

PHYSICIAN PATIENT RELATIONSHIP / PSYCHOTHERAPY / SEXUALITY

PRIVILEGED COMMUNICATION / PSYCHOTHERAPY

PRIVILEGED COMMUNICATION / PSYCHOTHERAPY / DANGEROUSNESS

RIGHT TO TREATMENT / PRISONERS / PSYCHOTHERAPY

PUBLIC ADVOCACY   *See under*

PSYCHOLOGY / PUBLIC ADVOCACY

BE = bioethics accession number   fn. = footnotes   refs. = references

PUBLIC ADVOCACY *See under (cont'd.)*

RECOMBINANT DNA RESEARCH / PUBLIC ADVOCACY / CAMBRIDGE

PUBLIC HEALTH *See under*

RECOMBINANT DNA RESEARCH / PUBLIC POLICY / PUBLIC HEALTH

PUBLIC HOSPITALS *See under*

ABORTION / PUBLIC HOSPITALS

ABORTION / PUBLIC HOSPITALS / LEGAL ASPECTS

PUBLIC OPINION *See under*

ABORTION / PUBLIC OPINION

ABORTION / SUPREME COURT DECISIONS / PUBLIC OPINION

BEHAVIORAL RESEARCH / ETHICS / PUBLIC OPINION

BIOETHICAL ISSUES / PUBLIC POLICY / PUBLIC OPINION

RECOMBINANT DNA RESEARCH / PUBLIC OPINION

PUBLIC PARTICIPATION / SCIENCE

Culliton, Barbara J. Science's restive public. *Daedalus* 107(2): 147-156, Spring 1978. 11 fn. BE06938.
federal government; *government regulation; fetuses; *human experimentation; investigators; *public participation; *recombinant DNA research; *science; self regulation

Nelkin, Dorothy. Threats and promises: negotiating the control of research. *Daedalus* 107(2): 191-209, Spring 1978. 58 fn. BE06939.
decision making; ethics committees; federal government; *government regulation; municipal government; investigators; *public participation; *recombinant DNA research; risks and benefits; *science; self determination; *self regulation; *social control; social impact

PUBLIC PARTICIPATION *See under*

BEHAVIORAL RESEARCH / PUBLIC PARTICIPATION / DRUG ABUSE

BIOMEDICAL RESEARCH / PUBLIC PARTICIPATION

RECOMBINANT DNA RESEARCH / PUBLIC PARTICIPATION

RECOMBINANT DNA RESEARCH / PUBLIC PARTICIPATION / CAMBRIDGE

PUBLIC POLICY *See under*

BIOETHICAL ISSUES / PUBLIC POLICY / PUBLIC OPINION

FAMILY PLANNING / PUBLIC POLICY / INDIA

GENETIC SCREENING / PUBLIC POLICY / SICKLE CELL ANEMIA

PUBLIC POLICY *See under (cont'd.)*

HEALTH CARE / PUBLIC POLICY

HEALTH CARE / PUBLIC POLICY / ECONOMICS

IMMUNIZATION / PUBLIC POLICY

IMMUNIZATION / PUBLIC POLICY / POLIOMYELITIS

INVOLUNTARY STERILIZATION / PUBLIC POLICY / INDIA

POPULATION CONTROL / PUBLIC POLICY

POPULATION CONTROL / PUBLIC POLICY / COMMON GOOD

POPULATION CONTROL / PUBLIC POLICY / HISTORICAL ASPECTS

POPULATION CONTROL / PUBLIC POLICY / INCENTIVES

POPULATION CONTROL / PUBLIC POLICY / INDIA

POPULATION CONTROL / PUBLIC POLICY / LEGAL ASPECTS

POPULATION CONTROL / PUBLIC POLICY / VALUES

RECOMBINANT DNA RESEARCH / PUBLIC POLICY

RECOMBINANT DNA RESEARCH / PUBLIC POLICY / CONTAINMENT

RECOMBINANT DNA RESEARCH / PUBLIC POLICY / PUBLIC HEALTH

RECOMBINANT DNA RESEARCH / PUBLIC POLICY / RISKS AND BENEFITS

PUNISHMENT *See under*

NEGATIVE REINFORCEMENT / PUNISHMENT

QUALITY OF LIFE *See also* VALUE OF LIFE

QUALITY OF LIFE / EVALUATION

Shaw, Anthony. Defining the quality of life. *Hastings Center Report* 7(5): 11, Oct 1977. BE06075.
decision making; *evaluation; newborns; patients; *quality of life

QUALITY OF LIFE / TERMINALLY ILL

Wilkes, Eric. Quality of life: effects of the knowledge of diagnosis in terminal illness. *Nursing Times* 73(39): 1506-1507, 29 Sep 1977. 4 refs. BE06340.
*cancer; diagnosis; *disclosure; evaluation; nurses; *quality of life; *terminally ill

QUALITY OF LIFE *See under*

ABORTION / QUALITY OF LIFE

ALLOWING TO DIE / QUALITY OF LIFE

ALLOWING TO DIE / QUALITY OF LIFE / JUDICIAL ACTION

PROLONGATION OF LIFE / QUALITY OF LIFE

BE = bioethics accession number   fn. = footnotes   refs. = references

QUALITY OF LIFE See under (cont'd.)

SELECTION FOR TREATMENT / QUALITY OF
LIFE / SPINA BIFIDA

QUINLAN, KAREN

Gold, Jay A. The Quinlan case: a review of two books.
[Title provided]. *American Journal of Law and Med-*
icine 3(1): 89-94, Spring 1977. BE05919.
*allowing to die; brain death; *determination of
death; personhood; state courts; terminally ill;
*withholding treatment; *In re Quinlan; New Jer-
sey Supreme Court; *Quinlan, Karen

Lang, Joan A.; Seltzer, Marc M. Review of two books
on death and dying. [Title provided]. *DePaul Law*
Review 26(4): 891-901, Summer 1977. 24 fn.
BE06490.
*allowing to die; biomedical technologies; coma;
death; legal aspects; legal rights; *prolongation of
life; self determination; treatment refusal; *Quinlan, Karen

Quinlan, Joseph; Quinlan, Julia; Battelle, Phyllis. Karen Ann: The Quinlans Tell Their Story. Garden
City, N.Y.: Doubleday, 1977. 343 p. ISBN 0-
385-12666-2. BE06188.
allowing to die; coma; decision making; *extraordinary treatment; family members; hospitals; legal
aspects; nursing homes; physicians; prolongation
of life; Roman Catholic ethics; *terminally ill;
*withholding treatment; *Quinlan, Karen

Sklar, Zachary; Coburn, Daniel R. Life against death: an
interview with Karen Ann Quinlan's guardian. *Juris*
Doctor 7(11): 27-31, Dec 1977. BE06369.
*allowing to die; fathers; *decision making; law;
*legal guardians; mass media; physicians; termi-
nally ill; *Quinlan, Karen

RADIOLOGY See under

INFORMED CONSENT / RADIOLOGY

INFORMED CONSENT / RADIOLOGY / RISKS
AND BENEFITS

RAMSEY, PAUL See under

EUTHANASIA / PROTESTANT ETHICS / RAMSEY,
PAUL

RANDOM SELECTION See under

HUMAN EXPERIMENTATION / RANDOM SELEC-
TION

ORGAN TRANSPLANTATION / RANDOM SELEC-
TION / RISKS AND BENEFITS

SELECTION FOR TREATMENT / RENAL DIALY-
SIS / RANDOM SELECTION

RECESSIVE GENETIC CONDITIONS See under

GENETIC COUNSELING / RECESSIVE GENETIC
CONDITIONS

RECOMBINANT DNA RESEARCH

Anderson, E.S. One fine day.... *New Scientist*
77(1086): 148-150, 19 Jan 1978. BE06722.
*artificial genes; biological containment; govern-
ment agencies; government regulation; health hazards; hybrids; mass media; *recombinant DNA
research; social control; standards; Asilomar Conference; Berg, Paul; Great Britain; United States;
Williams Report

Cohn, Victor. Scientists duplicate rat insulin gene.
Washington Post, 24 May 1977, p. A1+. BE06022.
insulin; *recombinant DNA research; risks and
benefits

Cooke, Robert. Improving on Nature: The Brave New
World of Genetic Engineering. New York: Quadran-
gle/The New York Times Book Company, 1977.
248 p. ISBN 0-8129-0667-5. BE05699.
amniocentesis; *behavioral genetics; artificial
genes; biological containment; biology; cancer;
cloning; DNA therapy; embryo transfer; evolution; federal government; *food; genetic defects;
*genetic intervention; genetic screening; genetics;
government regulation; health hazards; in vitro fertilization; mutation; physical containment; prenatal diagnosis; *recombinant DNA research; risks
and benefits; social impact

Dubos, René. Genetic engineering. *New York Times,*
21 Apr 1977, p. A25. BE06015.
*recombinant DNA research; risks and benefits

Jackson, M. Virginia. Genetic Engineering and the Recombinant DNA Research Controversy—a Historical Bibliography: 1973-1977. Unpublished document.
Aug 1977. 8 p. Bibliographic Series No. 1. Available
from the Mervyn H. Sterne Library, University of
Alabama, Birmingham. Bibliography. BE05801.
historical aspects; *recombinant DNA research

Lear, John. Recombinant DNA: The Untold Story.
New York: Crown Publishers, 1978. 280 p. ISBN
0-517-53165-8. BE06978.
attitudes; decision making; federal government;
government regulation; health hazards; investigators; municipal government; obligations to society;
*recombinant DNA research; risks and benefits;
science; self regulation

May, Robert M. The recombinant DNA debate. *Sci-*
ence 198(4322): 1144-1146, 16 Dec 1977. 13 fn.
BE06144.
biomedical research; federal government; government regulation; health hazards; public participation; public policy; *recombinant DNA research

Powledge, Tabitha M. The genetic engineers still await
guidelines. *New York Times,* 15 Feb 1976, sect. 4, p.

BE = bioethics accession number   fn. = footnotes   refs. = references

8. BE06611.
food; industry; medicine; *recombinant DNA research; risks and benefits; social control

Rogers, Michael. Biohazard. New York: Alfred A. Knopf, 1977. 209 p. ISBN 0-394-40128-X. BE05890.
attitudes; biological containment; carcinogens; federal government; genetics; *government regulation; *health hazards; historical aspects; *investigators; physical containment; public participation; *recombinant DNA research; *self regulation; social impact; *Asilomar Conference; NIH Guidelines

Schmeck, Harold M. Scientists report using bacteria to produce the gene for insulin. *New York Times,* 24 May 1977, p. 1+. BE06023.
insulin; *recombinant DNA research; risks and benefits

Sinsheimer, Robert L. Genetic engineering and gene therapy: some implications. *In* Cohen, Bernice H.; Lilienfeld, Abraham M.; Huang, P.C., eds. Genetic Issues in Public Health and Medicine. Springfield, Ill.: Charles C. Thomas, 1978. p. 439-461. 30 refs. ISBN 0-398-03659-4. BE06986.
DNA therapy; *recombinant DNA research; risks and benefits

Sinsheimer, Robert L. Genetic research: the importance of maximum safety and forethought. *New York Times,* 30 May 1977, p. 14. Letter. BE06025.
health hazards; *recombinant DNA research

Sullivan, Daniel J. Gene-splicing: the eighth day of creation. *America* 137(20): 440-443, 17 Dec 1977. Article reviews three books on recombinant DNA. BE06590.
attitudes; government regulation; health hazards; investigators; *recombinant DNA research; self regulation

U.S. National Library of Medicine. Recombinant DNA. May 1976 through November 1977. Bethesda, Md.: National Library of Medicine, 1977. 12 p. Literature Search No. 77-15, prepared by P.E. Pothier. 162 citations. Updates L.S. 76-25. BE06695.
*methods; *recombinant DNA research; social impact

Wade, Nicholas. The Ultimate Experiment: Man-Made Evolution. New York: Walker, 1977. 162 p. 95 fn. ISBN 0-8027-0572-3. BE05891.
attitudes; federal government; biological containment; government agencies; *government regulation; *health hazards; *investigators; methods; municipal government; physical containment; public opinion; public participation; *recombinant DNA research; risks and benefits; science; *self regulation; social impact; *Asilomar Conference; NIH Guidelines

RECOMBINANT DNA RESEARCH / AMERICAN FEDERATION FOR CLINICAL RESEARCH

American Federation for Clinical Research. Position statement of the AFCR on the issue of legislation of recombinant DNA research. *Clinical Research* 26(3): 124, Apr 1978. BE06833.
federal government; *government regulation; *legislation; *organizational policies; professional organizations; *recombinant DNA research; *American Federation for Clinical Research

Short, Elizabeth M.; Kohler, Peter O. Position paper of the AFCR Public Policy Committee: recombinant DNA research. *Clinical Research* 26(3): 125-131, Apr 1978. 43 refs. BE06834.
federal government; government regulation; health hazards; investigators; *legislation; *organizational policies; *recombinant DNA research; science; social control; *American Federation for Clinical Research

RECOMBINANT DNA RESEARCH / AMERICAN MEDICAL ASSOCIATION

American Medical Association. Council on Scientific Affairs. Guidelines for recombinant DNA research. *Connecticut Medicine* 41(9): 596, Sep 1977. Statement adopted by AMA House of Delegates, Annual Meeting, 1977. BE05771.
government regulation; *organizational policies; *professional organizations; *recombinant DNA research; review committees; *American Medical Association; NIH Guidelines

RECOMBINANT DNA RESEARCH / ATTITUDES

Lewin, Roger. US changes tack on genetic engineering. *New Scientist* 76(1072): 3, 6 Oct 1977. BE06090.
*attitudes; *federal government; government regulation; *investigators; *legislation; politics; *recombinant DNA research; risks and benefits

Singer, Maxine F. The recombinant DNA debate. *Science* 196(4286): 127, 8 Apr 1977. 1 fn. Editorial. BE05579.
*attitudes; *recombinant DNA research; *risks and benefits; social control

RECOMBINANT DNA RESEARCH / CARCINOGENS

Silcock, Bryan. Scientists to test super-germ. *Sunday Times (London),* 29 May 1977, p. 6. BE06443.
*carcinogens; *recombinant DNA research; risks and benefits; Great Britain

RECOMBINANT DNA RESEARCH / CONTAINMENT

Kukin, Marrick. Research with recombinant DNA: potential risks, potential benefits, and ethical consider-

---

BE = bioethics accession number   fn. = footnotes   refs. = references

ations. *New York State Journal of Medicine* 78(2): 226-235, Feb 1978. 35 refs. BE06916.
*biological containment; *containment; DNA therapy; drug industry; food; freedom; genetic intervention; genetics; government agencies; government regulation; health hazards; hybrids; insulin; investigators; methods; natural selection; *physical containment; public participation; *recombinant DNA research; *risks and benefits; science; standards; universities; *NIH Guidelines; Princeton

RECOMBINANT DNA RESEARCH / CONTAINMENT / HEALTH HAZARDS

Novick, Richard P. Present controls are just a start. *Bulletin of the Atomic Scientists* 33(5): 16+, May 1977. BE05598.
biological containment; *containment; federal government; government regulation; *health hazards; physical containment; *recombinant DNA research; NIH Guidelines

RECOMBINANT DNA RESEARCH / CONTAINMENT / STANDARDS

U.S. National Institutes of Health. Recombinant DNA research: proposed revised guidelines. *Federal Register* 42(187): 49596-49609, 27 Sep 1977. 28 refs. BE05991.
*biological containment; *containment; investigators; legal obligations; *physical containment; *recombinant DNA research; review committees; *standards

RECOMBINANT DNA RESEARCH / DISCLOSURE

Dixon, Bernard. Secret sharers? *Sciences* 18(4): 30-31, Apr 1978. BE06841.
*confidentiality; *disclosure; industry; *recombinant DNA research; *review committees; *Genetic Manipulation Advisory Group; Great Britain

RECOMBINANT DNA RESEARCH / ECOLOGY / SOCIAL IMPACT

Leeper, E.M. NEPA and basic research: DNA debate prompts review of environmental impacts. *BioScience* 27(8): 515-517, Aug 1977. BE05972.
*ecology; federal government; government agencies; health hazards; *recombinant DNA research; *social impact; National Environmental Policy Act; National Institutes of Health; National Science Foundation

RECOMBINANT DNA RESEARCH / EUROPE

Dickson, David. Warning over bacterial research; other countries urged to adopt British guidelines. *Times (London)*, 29 Oct 1976, p. 4. BE06617.
containment; government regulation; *recombi-

nant DNA research; standards; *Europe; Great Britain

RECOMBINANT DNA RESEARCH / EUROPEAN SCIENCE FOUNDATION

Sherwell, Chris. Harmony of practice. *Nature* 270(5633): 94, 10 Nov 1977. BE06121.
legislation; *recombinant DNA research; review committees; social control; *European Science Foundation

RECOMBINANT DNA RESEARCH / FORT DETRICK

Dickson, David. Recombinant DNA risk-assessment studies to begin at Fort Detrick. *Nature* 272(5653): 488, 6 Apr 1978. BE06840.
*evaluation; physical containment; *recombinant DNA research; *risks and benefits; *Fort Detrick; Maryland

RECOMBINANT DNA RESEARCH / GENETIC MANIPULATION ADVISORY GROUP

Great Britain. Genetic Manipulation Advisory Group. First Report of the Genetic Manipulation Advisory Group. London: Her Majesty's Stationery Office, May 1978. 77 p. ISBN 0-10-172150-1. BE06883.
biological containment; *containment; confidentiality; education; goals; government agencies; government regulation; health care; health hazards; international aspects; legal aspects; methods; physical containment; program descriptions; *recombinant DNA research; research personnel; review committees; social control; standards; universities; *Genetic Manipulation Advisory Group; *Great Britain; Williams Report

Walgate, Robert. GMAG wants to stay on. *Nature* 273(5660): 259, 25 May 1978. BE06971.
containment; health hazards; *recombinant DNA research; review committees; social control; *Genetic Manipulation Advisory Group; Great Britain

RECOMBINANT DNA RESEARCH / GOVERNMENT AGENCIES / CONFLICT OF INTEREST

Simring, Francine R. The double helix of self-interest. *Sciences* 17(3): 10-13+, May-Jun 1977. BE05629.
*conflict of interest; *decision making; ecology; federal government; financial support; *government agencies; government regulation; health hazards; industry; public participation; public policy; *recombinant DNA research; risks and benefits; social control; social impact; universities; *National Institutes of Health; NIH Guidelines

RECOMBINANT DNA RESEARCH / GOVERNMENT REGULATION

Adelberg, Edward A. Recombinant DNA research. *Connecticut Medicine* 42(2): 134-135, Feb 1978. BE06768.

BE = bioethics accession number   fn. = footnotes   refs. = references

federal government; *government regulation; *recombinant DNA research; risks and benefits

Baltimore, David, et al. A threat to scientific research. *Bulletin of the Atomic Scientists* 33(10): 9, Dec 1977. BE06368.
  federal government; *government regulation; legislation; municipal government; *recombinant DNA research; state government

Becker, Frank. Law vs. science: legal control of genetic research. *Kentucky Law Journal* 65(4): 880-894, 1977. 97 fn. BE06630.
  containment; federal government; freedom; *government regulation; municipal government; public participation; *recombinant DNA research; risks and benefits; science; self regulation; state government; NIH Guidelines

Bengelsdorf, Irving S. Controlling creation: what's at stake in genetic engineering. *American Medical News* 20(47): Impact 3-4, 28 Nov 1977. BE06247.
  biological containment; communicable diseases; biological warfare; drug industry; evolution; *government regulation; *health hazards; hybrids; legal liability; physical containment; *recombinant DNA research; social impact; technology; NIH Guidelines

Biologists oppose DNA research bills. *Science News* 112(3): 36-37, 16 Jul 1977. BE05721.
  attitudes; *government regulation; health hazards; investigators; legislation; *recombinant DNA research; risks and benefits

Bronson, Gail. Control of all genetic research facilities by HEW proposed by interagency panel. *Wall Street Journal (Eastern Edition),* 14 Mar 1977, p. 20. BE06436.
  federal government; *government regulation; *industry; *recombinant DNA research; risks and benefits

Bronson, Gail. Controlling gene transplants. *Wall Street Journal (Eastern Edition),* 11 May 1977, p. 16. BE06440.
  *federal government; *government regulation; industry; *municipal government; *recombinant DNA research; review committees; risks and benefits

Cohn, Victor. Amid latest gene research flap, easing of rules is eyed. *Washington Post,* 16 Dec 1977, p. A2. BE06425.
  federal government; *government regulation; investigators; *recombinant DNA research; universities

Cohn, Victor. Califano asks Hill to restrict genetic research. *Washington Post,* 7 Apr 1977, p. A1+. BE06011.
  federal government; *government regulation; *recombinant DNA research

Cohn, Victor. DNA research control dims. *Washington Post,* 28 Sep 1977, p. A1+. BE06400.

attitudes; federal government; *government regulation; investigators; legislation; *recombinant DNA research

Cohn, Victor. Gene-study control law lacking. *Washington Post,* 20 Feb 1977, p. A3. BE06215.
  federal government; *government regulation; legislation; *recombinant DNA research

Cohn, Victor. Scientists now downplay risks of genetic research. *Washington Post,* 18 Jul 1977, p. A1+. BE06054.
  *attitudes; federal government; *government regulation; *investigators; legislation; *recombinant DNA research; *risks and benefits

Coughlin, Ellen K. Outside control called threat to scientific research. *Chronicle of Higher Education* 14(2): 6, 12 Sep 1977. BE05776.
  decision making; federal government; *government regulation; municipal government; *recombinant DNA research; science; self regulation; Ball, George W.

Cowen, Robert C. Should biologists redesign organic life? *Christian Science Monitor (Eastern Edition),* 3 Feb 1977, p. 12-13. BE06434.
  attitudes; containment; federal government; *government regulation; investigators; *recombinant DNA research; risks and benefits

Dickson, David. NIH may loosen recombinant DNA research guidelines. *Nature* 273(5659): 179, 18 May 1978. BE06968.
  *federal government; *government regulation; legislation; *recombinant DNA research; review committees; National Institutes of Health

Dismukes, Key. Recombinant DNA: a proposal for regulation. *Hastings Center Report* 7(2): 25-30, Apr 1977. 2 fn. BE05551.
  attitudes; federal government; government agencies; *government regulation; investigators; public policy; *recombinant DNA research; risks and benefits; self regulation; social control; NIH Guidelines; Public Health Service

DNA research: no federal regulation now. *Chemical and Engineering News* 55(47): 22, 21 Nov 1977. BE06353.
  federal government; *government regulation; industry; legislation; *recombinant DNA research; risks and benefits

Genetic engineering. *Christian Science Monitor (Eastern Edition),* 7 Jan 1977, p. E. Editorial. BE06432.
  federal government; *government regulation; *recombinant DNA research; risks and benefits

Gilbert, Walter. Recombinant DNA research: government regulation. *Science* 197(4300): 208, 15 Jul 1977. Letter. BE05724.
  federal government; *government regulation; legislation; *recombinant DNA research; social im-

---

BE = bioethics accession number   fn. = footnotes   refs. = references

pact; Gordon Research Conference on Nucleic Acids 1977

Greenberg, Daniel S. 'Dangerous knowledge': should it be controlled? *Washington Post,* 11 Oct 1977, p. A19. BE06406.
*government regulation; investigators; *recombinant DNA research; risks and benefits

How to regulate basic research. *Washington Post,* 12 Apr 1977, p. A18. Editorial. BE06013.
federal government; *government regulation; legislation; *recombinant DNA research

Johnson, Albert W. Recombinant DNA controversy. *Science* 198(4313): 123-124, 14 Oct 1977. Letter. BE06089.
containment; *government regulation; *municipal government; *recombinant DNA research; risks and benefits; California; *San Diego

Kennedy, Edward M. Legislation to regulate and control recombinant DNA research. [Title provided]. *Congressional Record (Daily Edition)* 123(58): S5335-S5337, 1 Apr 1977. BE06226.
employment; federal government; *government regulation; *legislation; *recombinant DNA research; risks and benefits; *standards

Lewin, Roger. Controls seem inevitable for 'genetic engineering'. *Christian Science Monitor (Eastern Edition),* 14 Mar 1977, p. 1+. BE06437.
attitudes; federal government; *government regulation; public participation; *recombinant DNA research; risks and benefits

Metzenbaum, Howard. Recombinant DNA Research Standards Act of 1977. *Congressional Record (Daily Edition)* 123(40): S3699-S3704, 8 Mar 1977. BE05914.
*federal government; *government regulation; health hazards; investigators; *recombinant DNA research; risks and benefits; self regulation

Nelkin, Dorothy. Threats and promises: negotiating the control of research. *Daedalus* 107(2): 191-209, Spring 1978. 58 fn. BE06939.
decision making; ethics committees; federal government; *government regulation; municipal government; investigators; *public participation; *recombinant DNA research; risks and benefits; *science; self determination; *self regulation; *social control; social impact

Policing the gene-splicers. *New York Times,* 10 Oct 1977, p. 28. Editorial. BE06405.
federal government; *government regulation; health hazards; *recombinant DNA research

Randal, Judith. Who will oversee the gene jugglers? *Change* 9(5): 54-55, May 1977. BE06486.
*federal government; *government regulation; public participation; *recombinant DNA research; risks and benefits

Recombinant DNA research faces federal regulation. *Medical World News* 18(7): 24, 4 Apr 1977.

BE05577.
*federal government; *government regulation; legislation; *recombinant DNA research; risks and benefits

Regulating DNA research: the debate goes on. *Change* 9(6): 53, Jun 1977. BE06488.
drugs; federal government; *government regulation; legislation; *recombinant DNA research

Regulation of recombinant DNA research. [Title provided]. *Medical World News* 18(26): 7, 26 Dec 1977. Editorial. BE06375.
federal government; *government regulation; industry; legislation; *recombinant DNA research

Restricting DNA experiments. *Washington Post,* 19 Dec 1977, p. A26. Editorial. BE06429.
attitudes; *government regulation; investigators; *recombinant DNA research

Rogers, Paul G.; Leeper, E.M. Rogers lists advantages of his DNA bill. *BioScience* 27(9): 591-593, Sep 1977. BE05986.
attitudes; *federal government; *government regulation; investigators; *recombinant DNA research; review committees; NIH Guidelines

Schmeck, Harold M. 'Gene splicing' faces new debate in Congress. *New York Times,* 15 Dec 1977, p. D13. BE06424.
federal government; *government regulation; legislation; municipal government; *recombinant DNA research; state government

Schmeck, Harold M. U.S. agency bids lab halt work on genes. *New York Times,* 16 Dec 1977, p. A11. BE06426.
*federal government; *government regulation; investigators; *recombinant DNA research; *universities; Harvard University; NIH Guidelines

Seidman, Aaron. The U.S. Senate and recombinant DNA research. *Newsletter on Science, Technology, and Human Values* No. 22: 30-32, Jan 1978. BE06755.
*federal government; *government regulation; *legislation; *recombinant DNA research

Stetten, DeWitt, et al. The social implications of research on genetic manipulation: panel discussion. *In* Schultz, J.; Brada, Z., eds. Genetic Manipulation as It Affects the Cancer Problem. Proceedings. New York: Academic Press, 1977. p. 217-260. Proceedings of the Miami Winter Symposia, Jan 1977, sponsored by the Papanicolaou Cancer Research Institute, Miami, Florida. Miami Winter Symposia—Vol. 14. BE06684.
containment; ecology; evolution; *government regulation; government agencies; health hazards; hybrids; international aspects; physical containment; *recombinant DNA research; *risks and benefits; science; social control; social impact; standards; Europe; Great Britain; National Environmental

---

BE = bioethics accession number   fn. = footnotes   refs. = references

Policy Act; NIH Guidelines; United States; Williams Report

Stevenson, Adlai E. Recombinant DNA legislation. *Congressional Record (Daily Edition)* 123(148): S15410-S15413, 22 Sep 1977. BE07034.
 attitudes; federal government; *government regulation; investigators; legislation; *recombinant DNA research; risks and benefits

U.S. Congress. House. A bill to amend the Public Health Service Act to regulate activities involving recombinant DNA, and for other purposes. H.R. 7897, 95th Cong., 1st Sess., 20 Jun 1977. 50 p. By Paul G. Rogers, et al. Referred to the Committee on Interstate and Foreign Commerce. BE07036.
 federal government; *government regulation; *recombinant DNA research; review committees; standards

U.S. Congress. House. A bill to amend the Public Health Service Act to regulate research projects involving recombinant DNA. H.R. 4759, 95th Cong., 1st Sess., 9 Mar 1977. 15 p. By Paul G. Rogers, et al. Referred to the Committee on Interstate and Foreign Commerce. BE07035.
 containment; federal government; *government regulation; *recombinant DNA research; standards

U.S. Congress. House. Recombinant DNA Act. H.R. 11192, 95th Cong., 2d Sess., 28 Feb 1978. 19 p. By Harley O. Staggers and Paul G. Rogers. Referred to the Committee on Interstate and Foreign Commerce. BE06717.
 evaluation; government agencies; *government regulation; health hazards; *law enforcement; *program descriptions; *recombinant DNA research; social impact; *Commission—Research Involving Genetic Manipulation; Department of Health, Education, and Welfare; NIH Guidelines

U.S. Congress. House. Recombinant DNA Safety Assurance Act. H.R. 10453, 95th Cong., 2d Sess., 19 Jan 1978. 11 p. By Harley O. Staggers. Referred to the Committee on Interstate and Foreign Commerce. BE06716.
 evaluation; *federal government; government agencies; *government regulation; health hazards; *program descriptions; *recombinant DNA research; state government; Commission for the Study of Recombinant DNA Activities; Department of Health, Education, and Welfare; NIH Guidelines

U.S. Congress. Senate. Committee on Human Resources. Subcommittee on Health and Scientific Research. Recombinant DNA Regulation Act, 1977. Hearing. Washington: U.S. Government Printing Office, 1977. 472 p. Hearing, 95th Cong., 1st Sess., on S. 1217, to regulate activities involving recombinant deoxyribonucleic acid and related bills, 6 Apr 1977. References and footnotes included. BE06692.

federal government; *government regulation; health hazards; legislation; *recombinant DNA research; state government; NIH Guidelines

U.S. Federal Interagency Committee on Recombinant DNA Research. Interim report of the Federal Interagency Committee on Recombinant DNA Research: suggested elements for legislation. 15 Mar 1977. *In* U.S. National Institutes of Health. Recombinant DNA Research: Volume 2. Documents Relating to "NIH Guidelines for Research...". Washington: U.S. Government Printing Office, Mar 1978. p. 279-345. DHEW Pub. No. (NIH) 78-1139. 16 fn. BE06949.
 disclosure; *federal government; government agencies; *government regulation; health hazards; *legislation; *recombinant DNA research; standards; NIH Guidelines

U.S. National Institutes of Health. Recombinant DNA Research: Volume 2. Documents Relating to "NIH Guidelines for Research Involving Recombinant DNA Molecules"—Jun 1976-Nov 1977. Washington: U.S. Government Printing Office, Mar 1978. 902 p. DHEW Pub. No. (NIH) 78-1139. References and footnotes included. BE06948.
 attitudes; confidentiality; disclosure; *federal government; *government regulation; international aspects; legislation; *patents; public policy; *recombinant DNA research; review committees; risks and benefits; standards

Wahl, Diana H. 'Minimum standards' for DNA research. *Washington Post,* 26 Apr 1977, p. A18. Letter. BE06019.
 federal government; *government regulation; *recombinant DNA research; state government

Watson, J.D. Trying to bury Asilomar. *Clinical Research* 26(3): 113-115, Apr 1978. Editorial. BE06832.
 attitudes; federal government; *government regulation; investigators; *legislation; *recombinant DNA research; self regulation

Zegel, Vikki A.; McCullough, James M. DNA Recombinant Molecule Research. Issue Brief Number IB77024. Washington: Library of Congress, Congressional Research Service, 8 Apr 1977. 10 p. Congressional Research Service Major Issues System. Originally published 31 Mar 1977; updated 8 Apr 1977. Available only through Members of Congress. 54 refs. BE06573.
 federal government; *government regulation; health hazards; legislation; politics; *recombinant DNA research; self regulation

RECOMBINANT DNA RESEARCH / GOVERNMENT REGULATION / ATTITUDES

Cohn, Victor. Discoverer would lift curbs on DNA. *Washington Post,* 17 Dec 1977, p. A1+. BE06427.
 *attitudes; *government regulation; *investiga-

tors; *recombinant DNA research; *risks and benefits; Watson, James

Schmeck, Harold M. Rules on DNA studies viewed as too strict. *New York Times,* 18 Dec 1977, sect. 1, p. 19. BE06428.
> *attitudes; federal government; *government regulation; *investigators; *recombinant DNA research; risks and benefits

RECOMBINANT DNA RESEARCH / GOVERNMENT REGULATION / CALIFORNIA

California. Legislature. Assembly. California Recombinant DNA Research Safety Act. Assembly Bill No. 757, 1977-78 Regular Session, 3 Mar 1977. 15 p. By Assemblyman Keene, et al. Referred to Committee on Health; died 1977-78 Session. BE06595.
> confidentiality; *government regulation; health hazards; *recombinant DNA research; *review committees; state government; *California

RECOMBINANT DNA RESEARCH / GOVERNMENT REGULATION / CAMBRIDGE

McBride, Stewart D. Genetics research guidelines set up in Cambridge. *Christian Science Monitor (Eastern Edition),* 17 Jan 1977, p. 13. BE06433.
> *government regulation; *municipal government; *recombinant DNA research; *review committees; *Cambridge; Massachusetts

Schumacher, Edward. College community votes curb on genetic research. *Washington Post,* 8 Feb 1977, p. A2. BE06214.
> containment; *government regulation; *municipal government; *recombinant DNA research; universities; *Cambridge; Massachusetts

RECOMBINANT DNA RESEARCH / GOVERNMENT REGULATION / CONTAINMENT

U.S. National Institutes of Health. Recombinant DNA research: proposed revised guidelines. *Federal Register* 43(146): 33042-33178, 28 Jul 1978. References and footnotes included. BE06976.
> *biological containment; *containment; *federal government; *government regulation; legal obligations; *physical containment; *recombinant DNA research; review committees; risks and benefits; standards; NIH Guidelines

RECOMBINANT DNA RESEARCH / GOVERNMENT REGULATION / DENMARK

Godtfredsen, Sven. Denmark follows UK on DNA guidelines. *Nature* 271(5647): 700-701, 23 Feb 1978. BE06767.
> *government regulation; international aspects; *recombinant DNA research; *Denmark

RECOMBINANT DNA RESEARCH / GOVERNMENT REGULATION / EUROPE

Tooze, John. Emerging attitudes and policies in Europe. *In* Beers, Roland F.; Bassett, Edward G., eds. Recombinant Molecules: Impact on Science and Society. New York: Raven Press, 1977. p. 455-470. Miles International Symposium Series No. 10. ISBN 0-89004-131-8. BE05855.
> containment; *government regulation; international aspects; *recombinant DNA research; review committees; Ashby Report; Eastern Europe; *Europe; European Molecular Biology Organization; NIH Guidelines; Williams Report

RECOMBINANT DNA RESEARCH / GOVERNMENT REGULATION / GREAT BRITAIN

Simpson, Bill. Genetic manipulation. *Times (London),* 23 Oct 1976, p. 13. Letter. BE06616.
> government agencies; *government regulation; *recombinant DNA research; *Great Britain; Health and Safety Commission

Vickers, T. Flexible DNA regulation: the British model. *Bulletin of the Atomic Scientists* 34(1): 4-5, Jan 1978. 4 refs. BE06760.
> *government regulation; *recombinant DNA research; Genetic Manipulation Advisory Group; *Great Britain; Health and Safety at Work Act

RECOMBINANT DNA RESEARCH / GOVERNMENT REGULATION / HEALTH HAZARDS

Powledge, Tabitha M. You shall be as gods. *Worldview* 20(9): 45-47, Sep 1977. BE05783.
> biological containment; biological warfare; evolution; federal government; government agencies; *government regulation; *health hazards; international aspects; legislation; physical containment; *recombinant DNA research; self regulation; social impact; state government; Europe; NIH Guidelines; United States

RECOMBINANT DNA RESEARCH / GOVERNMENT REGULATION / LEGISLATION

Recombinant DNA and technology assessment. *Georgia Law Review* 11(4): 785-878, Summer 1977. 457 fn. BE06309.
> communicable diseases; constitutional law; containment; decision making; ecology; *federal government; financial support; government agencies; *government regulation; health hazards; injuries; judicial action; *law; legal aspects; *legal liability; *legislation; negligence; public health; public participation; *recombinant DNA research; research institutes; research personnel; review committees; risks and benefits; social impact; state government; *technology assessment; *torts; DNA Research Act; National Environmental Policy Act; National Institutes of Health; NIH Guidelines; Public Health Service Act; Toxic Substances Control Act

---

BE = bioethics accession number   fn. = footnotes   refs. = references

RECOMBINANT DNA RESEARCH / GOVERNMENT REGULATION / NATIONAL ACADEMY OF SCIENCES

Academy opposes local DNA research rules. *Science and Government Report* 7(9): 8, 15 May 1977. BE05602.
>   federal government; *government regulation; legislation; *organizational policies; *professional organizations; *recombinant DNA research; *National Academy of Sciences

RECOMBINANT DNA RESEARCH / GOVERNMENT REGULATION / NIH GUIDELINES

Holtzman, Eric. Recombinant DNA: triumph or Trojan horse? *Man and Medicine: The Journal of Values and Ethics in Health Care* 2(2): 83-102, Winter 1977. 17 fn. BE05898.
>   attitudes; *federal government; *government regulation; health hazards; industry; investigators; peer review; public participation; *public policy; *recombinant DNA research; *risks and benefits; science; self regulation; social impact; technology; *NIH Guidelines

Simring, Francine R. Folio for folly: N.I.H. Guidelines for recombinant DNA research. *Man and Medicine: The Journal of Values and Ethics in Health Care* 2(2): 110-119, Winter 1977. 7 fn. BE05900.
>   conflict of interest; *federal government; *government regulation; industry; public participation; *recombinant DNA research; risks and benefits; *NIH Guidelines

Zimmerman, Burke K. Self-discipline or self-deception? *Man and Medicine: The Journal of Values and Ethics in Health Care* 2(2): 120-132, Winter 1977. 9 fn. BE05901.
>   carcinogens; *federal government; *government regulation; health hazards; investigators; public participation; *recombinant DNA research; *NIH Guidelines

RECOMBINANT DNA RESEARCH / GOVERNMENT REGULATION / POLITICS

Culliton, Barbara J. Recombinant DNA bills derailed: Congress still trying to pass a law. *Science* 199(4326): 274-277, 20 Jan 1978. 1 fn. BE06726.
>   biology; federal government; *government regulation; *investigators; *legislation; *political activity; *politics; professional organizations; public participation; *recombinant DNA research; risks and benefits; *U.S. Congress

Wade, Nicholas. Senate passes back gene-splice cup. *Science* 200(4348): 1368, 23 Jun 1978. BE06975.
>   federal government; *government regulation; legislation; *politics; *recombinant DNA research

RECOMBINANT DNA RESEARCH / GOVERNMENT REGULATION / PROGRAM DESCRIPTIONS

Jacobs, Leon. The role of the National Institutes of Health in rulemaking. *In* Beers, Roland F.; Bassett, Edward G., eds. Recombinant Molecules: Impact on Science and Society. New York: Raven Press, 1977. p. 445-454. Miles International Symposium Series No. 10. ISBN 0-89004-131-8. BE05854.
>   decision making; federal government; government agencies; *government regulation; investigators; *program descriptions; public participation; *recombinant DNA research; science; *National Institutes of Health

RECOMBINANT DNA RESEARCH / GOVERNMENT REGULATION / RISKS AND BENEFITS

Grobstein, Clifford. The recombinant-DNA debate. *Scientific American* 237(1): 22-33, Jul 1977. BE05725.
>   attitudes; *containment; biological containment; ethics committees; evaluation; evolution; genetics; government agencies; *government regulation; health hazards; industry; investigators; legislation; methods; physical containment; politics; *public policy; *recombinant DNA research; *risks and benefits; science; technology; values; National Institutes of Health; NIH Guidelines

U.S. Congress. Senate. Committee on Commerce, Science, and Transportation. Subcommittee on Science, Technology, and Space. Regulation of Recombinant DNA Research. Hearings. Washington: U.S. Government Printing Office, 1978. 432 p. Hearings, 95th Cong., 1st Sess., on regulation of recombinant DNA research, 2, 8, 10 Nov 1977. Serial No. 95-52. BE06983.
>   attitudes; biological containment; *federal government; *government regulation; health hazards; investigators; physical containment; *recombinant DNA research; *risks and benefits; universities

Wade, Nicholas. Gene splicing: Congress starts framing law for research. *Science* 196(4285): 39-40, 1 Apr 1977. BE05587.
>   *federal government; *government regulation; health hazards; *recombinant DNA research; *risks and benefits; NIH Guidelines

Williamson, R. Genetic engineering. *Nursing Mirror* 145(9): 23-24, 1 Sep 1977. BE05845.
>   containment; *government regulation; *recombinant DNA research; *risks and benefits; Genetic Manipulation Advisory Group; Great Britain

RECOMBINANT DNA RESEARCH / GOVERNMENT REGULATION / WEST GERMANY

Germany moves to control DNA research. *New Scientist* 74(1053): 443, 26 May 1977. BE05613.
>   *government regulation; *recombinant DNA research; *West Germany

---

BE = bioethics accession number   fn. = footnotes   refs. = references

RECOMBINANT DNA RESEARCH / GREAT BRITAIN

Great Britain. Genetic Manipulation Advisory Group. First Report of the Genetic Manipulation Advisory Group. London: Her Majesty's Stationery Office, May 1978. 77 p. ISBN 0-10-172150-1. BE06883.
biological containment; *containment; confidentiality; education; goals; government agencies; government regulation; health care; health hazards; international aspects; legal aspects; methods; physical containment; program descriptions; *recombinant DNA research; research personnel; review committees; social control; standards; universities; *Genetic Manipulation Advisory Group; *Great Britain; Williams Report

RECOMBINANT DNA RESEARCH / HARVARD UNIVERSITY

Binyon, Michael. How science has raised the ghost of Frankenstein in an American city. *Times (London),* 6 Nov 1976, p. 14. BE06618.
containment; government regulation; health hazards; *recombinant DNA research; risks and benefits; Cambridge; *Harvard University; Massachusetts; Velucci, Al; Wald, George

RECOMBINANT DNA RESEARCH / HEALTH HAZARDS

Ryan, Allan J. The recombinant bogeyman. *Postgraduate Medicine* 63(1): 17+, Jan 1978. 6 refs. Editorial. BE06753.
federal government; government regulation; *health hazards; *recombinant DNA research

RECOMBINANT DNA RESEARCH / INDUSTRY

Bronson, Gail. Control of all genetic research facilities by HEW proposed by interagency panel. *Wall Street Journal (Eastern Edition),* 14 Mar 1977, p. 20. BE06436.
federal government; *government regulation; *industry; *recombinant DNA research; risks and benefits

Lewin, Roger. GMAG falls foul of privacy constraints. *New Scientist* 76(1082): 683, 15 Dec 1977. BE06256.
confidentiality; *industry; *recombinant DNA research; review committees; Genetic Manipulation Advisory Group

Rawls, Rebecca L. Drug firms becoming active in DNA research. *Chemical and Engineering News* 55(29): 19-20, 18 Jul 1977. BE05968.
containment; drugs; federal government; government regulation; *industry; *recombinant DNA research

RECOMBINANT DNA RESEARCH / INDUSTRY / LEGAL ASPECTS

Cohn, Victor. Genetic patent ruling stirs concern. *Washington Post,* 6 Feb 1977, p. A3. BE06213.
federal government; government regulation; *industry; *legal aspects; *recombinant DNA research

RECOMBINANT DNA RESEARCH / INVESTIGATORS / ATTITUDES

Abelson, Philip H. Recombinant DNA. *Science* 197(4305): 721, 19 Aug 1977. Editorial. BE05749.
*attitudes; federal government; health hazards; *investigators; legislation; *recombinant DNA research

Cohn, Victor. Scientists now downplay risks of genetic research. *Washington Post,* 18 Jul 1977, p. A1+. BE06054.
*attitudes; federal government; *government regulation; *investigators; legislation; *recombinant DNA research; *risks and benefits

Leeper, E.M. Recombinant DNA forum—stellar cast; gripping plot; but no new message. *BioScience* 27(5): 317-319, May 1977. BE05622.
*attitudes; government regulation; federal government; health hazards; *investigators; legislation; public participation; *recombinant DNA research; *social control

Randal, Judith. If the gene splicers win their battle, will they lose the war? *Change* 9(10): 48-49, Oct 1977. BE06497.
*attitudes; federal government; government regulation; *investigators; public participation; *recombinant DNA research; *risks and benefits

RECOMBINANT DNA RESEARCH / INVESTIGATORS / POLITICAL ACTIVITY

Culliton, Barbara J. Recombinant DNA bills derailed: Congress still trying to pass a law. *Science* 199(4326): 274-277, 20 Jan 1978. 1 fn. BE06726.
biology; federal government; *government regulation; *investigators; *legislation; *political activity; *politics; professional organizations; public participation; *recombinant DNA research; risks and benefits; *U.S. Congress

RECOMBINANT DNA RESEARCH / LEGISLATION

Culliton, Barbara J. Recombinant DNA bills derailed: Congress still trying to pass a law. *Science* 199(4326): 274-277, 20 Jan 1978. 1 fn. BE06726.
biology; federal government; *government regulation; *investigators; *legislation; *political activity; *politics; professional organizations; public participation; *recombinant DNA research; risks and benefits; *U.S. Congress

Fields, Cheryl M. Scientists fight Senate DNA bill. *Chronicle of Higher Education* 14(21): 13, 1 Aug 1977. BE05756.
attitudes; *government regulation; *federal government; investigators; *legislation; *recombinant DNA research; review committees; science

Grimes, A.J. An update on DNA research regulation. *BioScience* 27(11): 720, Nov 1977. BE06355.

BE = bioethics accession number   fn. = footnotes   refs. = references

*federal government; government regulation; *legislation; *recombinant DNA research

Halvorson, Harlyn O. Recombinant DNA legislation—what next? *Science* 198(4315): 357, 28 Oct 1977. 1 fn. Editorial. BE06087.
    *federal government; government regulation; investigators; *legislation; *recombinant DNA research

Leeper, E.M. Continuing saga of recombinant DNA: Senate hears proposals for regulation without legislation. *BioScience* 27(12): 775-776, Dec 1977. BE06372.
    *attitudes; federal government; government regulation; investigators; *legislation; *recombinant DNA research; risks and benefits

Lewin, Roger. Scientists' backlash against US legislation on DNA. *New Scientist* 74(1057): 692, 23 Jun 1977. BE05652.
    attitudes; *federal government; *government regulation; investigators; *legislation; *recombinant DNA research

Lewin, Roger. US changes tack on genetic engineering. *New Scientist* 76(1072): 3, 6 Oct 1977. BE06090.
    *attitudes; *federal government; government regulation; *investigators; *legislation; politics; *recombinant DNA research; risks and benefits

Norman, Colin. Committees rushing to write DNA regulations. *Science and Government Report* 7(6): 4-5, 1 Apr 1977. BE05572.
    *federal government; *government regulation; *legislation; municipal government; politics; *recombinant DNA research; state government; NIH Guidelines

Norman, Colin. Full circle. *Nature* 267(5614): 748, 30 Jun 1977. BE05660.
    attitudes; *federal government; *government regulation; investigators; *legislation; *recombinant DNA research; risks and benefits

Norman, Colin. Recombinant DNA pioneers lower the alarm. *Science and Government Report* 7(12): 1-3, 1 Jul 1977. BE05734.
    *federal government; government agencies; *government regulation; health hazards; investigators; *legislation; political activity; *recombinant DNA research; review committees; standards; NIH Guidelines

Recombinant DNA debate three years on. *Nature* 268(5617): 185, 21 Jul 1977. Editorial. BE05969.
    *federal government; *government regulation; *legislation; *recombinant DNA research

Recombinant DNA: local or federal regulation? *Nature* 267(5611): 475, 9 Jun 1977. Editorial. BE05661.
    federal government; *government regulation; *legislation; municipal government; *recombinant DNA research; state government

Rogers, Paul G. Recombinant DNA bill. *Chemical*

and *Engineering News* 55(31): 3, 1 Aug 1977. Letter. BE06327.
    *federal government; government regulation; *legislation; *recombinant DNA research; standards

Schmeck, Harold M. Congress is likely to delay until at least next year DNA research regulations once thought critical. *New York Times,* 25 Oct 1977, p. 25. BE06408.
    attitudes; federal government; government regulation; investigators; *legislation; *recombinant DNA research

Schmeck, Harold M. Scientists seek to influence legislation on gene research. *New York Times,* 6 Jul 1977, p. A15. BE06050.
    *attitudes; federal government; government regulation; *investigators; *legislation; professional organizations; *recombinant DNA research; risks and benefits

Wade, Nicholas. Gene splicing preemption rejected. *Science* 196(4288): 406, 22 Apr 1977. BE05588.
    *federal government; *government regulation; *legislation; public participation; *recombinant DNA research; state government

Wade, Nicholas. Gene splicing: Senate bill draws charges of Lysenkoism. *Science* 197(4301): 348+, 22 Jul 1977. BE05738.
    federal government; government agencies; *government regulation; health hazards; investigators; *legislation; political activity; professional organizations; *recombinant DNA research; review committees; NIH Guidelines

RECOMBINANT DNA RESEARCH / LEGISLATION / POLITICAL ACTIVITY

Dickson, David. Friends of DNA fight back. *Nature* 272(5655): 664-665, 20 Apr 1978. BE06839.
    attitudes; federal government; *government regulation; *legislation; municipal government; *political activity; *recombinant DNA research; science; state government; universities; U.S. Congress

RECOMBINANT DNA RESEARCH / LEGISLATION / U.S. CONGRESS

Wade, Nicholas. Congress set to grapple again with gene splicing. *Science* 199(4335): 1319-1322, 24 Mar 1978. BE06830.
    federal government; government regulation; *legislation; politics; *recombinant DNA research; *U.S. Congress

RECOMBINANT DNA RESEARCH / MARYLAND

Maryland. Recombinant DNA research. *Maryland Annotated Code,* Article 43, Sects. 898-910. Effective 1 Jul 1977. 4 p. BE06657.
    federal government; government regulation; *re-

BE = bioethics accession number   fn. = footnotes   refs. = references

combinant DNA research; review committees; state government; *Maryland

RECOMBINANT DNA RESEARCH / MEDICAL RESEARCH COUNCIL

Mitchell, Mary; Kaplan, J. Gordin. Medical Research Council committee draws up guidelines for research into recombinant DNA. *Canadian Medical Association Journal* 116(7): 802-804, 9 Apr 1977. 7 refs. BE05532.
   biological containment; government regulation; health hazards; investigators; *physical containment; *recombinant DNA research; risks and benefits; *self regulation; *Canada; *Medical Research Council

RECOMBINANT DNA RESEARCH / MORAL OBLIGATIONS

Singer, Maxine. Scientists and the control of science. *New Scientist* 74(1056): 631-634, 16 Jun 1977. BE05668.
   containment; *federal government; *government regulation; health hazards; *investigators; *moral obligations; public participation; *recombinant DNA research; science; *social impact; *NIH Guidelines

RECOMBINANT DNA RESEARCH / NATIONAL INSTITUTES OF HEALTH

Jacobs, Leon. The role of the National Institutes of Health in rulemaking. *In* Beers, Roland F.; Bassett, Edward G., eds. Recombinant Molecules: Impact on Science and Society. New York: Raven Press, 1977. p. 445-454. Miles International Symposium Series No. 10. ISBN 0-89004-131-8. BE05854.
   decision making; federal government; government agencies; *government regulation; investigators; *program descriptions; public participation; *recombinant DNA research; science; *National Institutes of Health

RECOMBINANT DNA RESEARCH / NETHERLANDS

Schuuring, Casper. Dutch go ahead on DNA. *Nature* 266(5604): 671, 21 Apr 1977. BE05578.
   containment; government regulation; *recombinant DNA research; review committees; *Netherlands

Schuuring, Casper. Limited progress. *Nature* 270(5632): 5-6, 3 Nov 1977. BE06120.
   government regulation; industry; *recombinant DNA research; universities; *Netherlands

RECOMBINANT DNA RESEARCH / NIH GUIDELINES

Fox, Jeffrey L. Revisions to DNA research guidelines debated. *Chemical and Engineering News* 56(2): 25-26+, 9 Jan 1978. BE06734.
   attitudes; biological containment; federal government; government regulation; health hazards; industry; investigators; physical containment; *recombinant DNA research; universities; *NIH Guidelines

Kukin, Marrick. Research with recombinant DNA: potential risks, potential benefits, and ethical considerations. *New York State Journal of Medicine* 78(2): 226-235, Feb 1978. 35 refs. BE06916.
   *biological containment; *containment; DNA therapy; drug industry; food; freedom; genetic intervention; genetics; government agencies; government regulation; health hazards; hybrids; insulin; investigators; methods; natural selection; *physical containment; public participation; *recombinant DNA research; *risks and benefits; science; standards; universities; *NIH Guidelines; Princeton

Singer, Maxine F. Scientists and the control of science. *New Scientist* 74(1056): 631-634, 16 Jun 1977. BE05668.
   containment; *federal government; *government regulation; health hazards; *investigators; *moral obligations; public participation; *recombinant DNA research; science; *social impact; *NIH Guidelines

Wade, Nicholas. Gene-splicing rules: another round of debate. *Science* 199(4324): 30-31+, 6 Jan 1978. BE06761.
   *attitudes; federal government; *government regulation; investigators; *recombinant DNA research; *NIH Guidelines

RECOMBINANT DNA RESEARCH / NUCLEAR ENERGY

Federow, Harold L. Recombinant DNA and nuclear energy. *Bulletin of the Atomic Scientists* 34(2): 6-7, Feb 1978. BE06776.
   government regulation; *nuclear energy; *recombinant DNA research; review committees; risks and benefits

RECOMBINANT DNA RESEARCH / ORGANIZATIONAL POLICIES / LEGISLATION

Short, Elizabeth M.; Kohler, Peter O. Position paper of the AFCR Public Policy Committee: recombinant DNA research. *Clinical Research* 26(3): 125-131, Apr 1978. 43 refs. BE06834.
   federal government; government regulation; health hazards; investigators; *legislation; *organizational policies; *recombinant DNA research; science; social control; *American Federation for Clinical Research

RECOMBINANT DNA RESEARCH / PATENTS

Dickson, David. Universities can patent recombinant DNA results. *Nature* 272(5650): 199, 16 Mar 1978. BE06813.

BE = bioethics accession number   fn. = footnotes   refs. = references

federal government; financial support; government regulation; *patents; *recombinant DNA research; *universities; Frederickson, Donald; National Institutes of Health; United States

Patenting biology: DNA and oil-eaters. *Science News* 113(11): 167+, 18 Mar 1978. BE06821.
    federal government; financial support; government agencies; government regulation; industry; legal aspects; *patents; public policy; *recombinant DNA research; universities

Richards, Bill. Life forms can be patented. *Washington Post,* 7 Oct 1977, p. A1+. BE06403.
    industry; *patents; *recombinant DNA research

U.S. National Institutes of Health. Recombinant DNA Research: Volume 2. Documents Relating to "NIH Guidelines for Research Involving Recombinant DNA Molecules"—Jun 1976-Nov 1977. Washington: U.S. Government Printing Office, Mar 1978. 902 p. DHEW Pub. No. (NIH) 78-1139. References and footnotes included. BE06948.
    attitudes; confidentiality; disclosure; *federal government; *government regulation; international aspects; legislation; *patents; public policy; *recombinant DNA research; review committees; risks and benefits; standards

RECOMBINANT DNA RESEARCH / PHYSICAL CONTAINMENT

Rowe, Wallace P. Recombinant DNA guidelines: scientific and political questions. *Science* 198(4317): 563+, 11 Nov 1977. Letter. BE06118.
    decision making; government regulation; *physical containment; public policy; *recombinant DNA research

RECOMBINANT DNA RESEARCH / PHYSICAL CONTAINMENT / FORT DETRICK

Marx, Jean L. The new P4 laboratories: containing recombinant DNA. *Science* 197(4311): 1350-1352, 30 Sep 1977. BE05843.
    health hazards; *physical containment; *recombinant DNA research; *Fort Detrick; Maryland

RECOMBINANT DNA RESEARCH / PUBLIC ADVOCACY / CAMBRIDGE

Cambridge Experimentation Review Board. The Cambridge Experimentation Review Board. *Bulletin of the Atomic Scientists* 33(5): 22-27, May 1977. Recommendations approved by the City Council, with some further restrictions, on February 7, 1977. BE05599.
    federal government; government regulation; institutional obligations; municipal government; *public advocacy; public participation; *recombinant DNA research; *review committees; *social control; standards; universities; *Cambridge; *Cam-

bridge Experimentation Review Board; NIH Guidelines

RECOMBINANT DNA RESEARCH / PUBLIC OPINION

Gaylin, Willard. The Frankenstein factor. *New England Journal of Medicine* 297(12): 665-667, 22 Sep 1977. BE05777.
    attitudes; behavior control; biomedical technologies; health hazards; investigators; political activity; psychosurgery; public participation; *public opinion; *recombinant DNA research; risks and benefits; social control; social impact; Asilomar Conference

RECOMBINANT DNA RESEARCH / PUBLIC PARTICIPATION

Budrys, Algis. The politics of deoxyribonucleic acid. *New Republic* 176(16): 18-21, 16 Apr 1977. BE05543.
    federal government; government regulation; health hazards; investigators; municipal government; *public participation; *recombinant DNA research; review committees; risks and benefits; self regulation; *social control; universities; *Cambridge; Massachusetts; People's Business Commission

Callahan, Daniel. The public and recombinant DNA research. *Connecticut Medicine* 41(9): 569-573, Sep 1977. BE05775.
    *decision making; federal government; government regulation; historical aspects; legislation; obligations of society; political activity; politics; *public participation; *public policy; *recombinant DNA research; review committees; science; state government; National Institutes of Health; NIH Guidelines

Callahan, Daniel. Recombinant DNA: science and the public. *Hastings Center Report* 7(2): 20-23, Apr 1977. BE05544.
    federal government; government regulation; investigators; *public participation; *recombinant DNA research; self regulation; *social control; *NIH Guidelines

Park, Bob; Thacher, Scott. Dealing with experts: the recombinant DNA debate. *Science for the People* 9(5): 28-35, Sep-Oct 1977. 15 refs. 10 fn. BE05848.
    federal government; government regulation; industry; legislation; municipal government; *public participation; *public policy; *recombinant DNA research; risks and benefits; *social control; universities; *Cambridge Experimentation Review Board

RECOMBINANT DNA RESEARCH / PUBLIC PARTICIPATION / CAMBRIDGE

King, Jonathan. A science for the people. *New Scientist* 74(1056): 634-636, 16 Jun 1977. BE05650.

BE = bioethics accession number   fn. = footnotes   refs. = references

decision making; federal government; government regulation; health hazards; investigators; *municipal government; *public participation; *recombinant DNA research; *social control; socioeconomic factors; *Cambridge; Massachusetts

RECOMBINANT DNA RESEARCH / PUBLIC POLICY

Callahan, Daniel. The public and recombinant DNA research. *Connecticut Medicine* 41(9): 569-573, Sep 1977. BE05775.
*decision making; federal government; government regulation; historical aspects; legislation; obligations of society; political activity; politics; *public participation; *public policy; *recombinant DNA research; review committees; science; state government; National Institutes of Health; NIH Guidelines

National Conference of Catholic Bishops. Bishops' Committee for Human Values. Statement on Recombinant DNA research. Washington: United States Catholic Conference, 2 May 1977. 8 p. BE05802.
freedom; health hazards; moral obligations; public participation; *public policy; *recombinant DNA research; risks and benefits; science; social control; *values

U.S. Congress. House. Committee on Science and Technology. Subcommittee on Science, Research and Technology. Science Policy Implications of DNA Recombinant Molecule Research. Hearings. Washington: U.S. Government Printing Office, 1977. 1293 p. Hearings...95th Cong., 1st Sess., 29-31 Mar, 27-28 Apr, 3-5 and 25-26 May, 7-8 Sep 1977. References and footnotes included. BE06196.
biological containment; federal government; government regulation; health hazards; industry; legal aspects; physical containment; public participation; *public policy; *recombinant DNA research; risks and benefits; *science; social impact

RECOMBINANT DNA RESEARCH / PUBLIC POLICY / CONTAINMENT

Grobstein, Clifford. The recombinant-DNA debate. *Scientific American* 237(1): 22-33, Jul 1977. BE05725.
attitudes; *containment; biological containment; ethics committees; evaluation; evolution; genetics; government agencies; government regulation; health hazards; industry; investigators; legislation; methods; physical containment; politics; *public policy; *recombinant DNA research; *risks and benefits; science; technology; values; National Institutes of Health; NIH Guidelines

RECOMBINANT DNA RESEARCH / PUBLIC POLICY / PUBLIC HEALTH

Goldstein, Richard. Public-health policy and recombinant DNA. *New England Journal of Medicine* 296(21): 1226-1228, 26 May 1977. 3 refs. 1 fn. Editorial. BE05614.
containment; health hazards; *public health; *public policy; *recombinant DNA research; *risks and benefits; NIH Guidelines

RECOMBINANT DNA RESEARCH / PUBLIC POLICY / RISKS AND BENEFITS

Holtzman, Eric. Recombinant DNA: triumph or Trojan horse? *Man and Medicine: The Journal of Values and Ethics in Health Care* 2(2): 83-102, Winter 1977. 17 fn. BE05898.
attitudes; *federal government; *government regulation; health hazards; industry; investigators; peer review; public participation; *public policy; *recombinant DNA research; *risks and benefits; science; self regulation; social impact; technology; *NIH Guidelines

RECOMBINANT DNA RESEARCH / REVIEW COMMITTEES / CALIFORNIA

California. Legislature. Assembly. California Recombinant DNA Research Safety Act. Assembly Bill No. 757, 1977-78 Regular Session, 3 Mar 1977. 15 p. By Assemblyman Keene, et al. Referred to Committee on Health; died 1977-78 Session. BE06595.
confidentiality; *government regulation; health hazards; *recombinant DNA research; *review committees; state government; *California

RECOMBINANT DNA RESEARCH / REVIEW COMMITTEES / LEGISLATION

Norman, Colin. Feeling the draft. *Nature* 267(5613): 658-659, 23 Jun 1977. BE05659.
*federal government; *government regulation; *legislation; *recombinant DNA research; *review committees

RECOMBINANT DNA RESEARCH / RISKS AND BENEFITS

Adelberg, Edward A. Recombinant DNA research: benefits and risks. *Connecticut Medicine* 41(7): 397-404, Jul 1977. 16 refs. BE05954.
federal government; government regulation; health hazards; *recombinant DNA research; *risks and benefits

Cohen, Stanley N. Recombinant DNA: fact and fiction. *Science* 195(4279): 654-657, 18 Feb 1977. 5 fn. BE06220.
evolution; government regulation; *health hazards; investigators; physical containment; *recombinant DNA research; *risks and benefits; self regulation

---

BE = bioethics accession number   fn. = footnotes   refs. = references

Cohn, Victor. Genetic experiment raises questions of a federal loophole. *Washington Post,* 12 Jun 1977, p. C8. BE06027.
government regulation; *recombinant DNA research; *risks and benefits

Cohn, Victor. NIH sees a 'small' risk in gene work. *Washington Post,* 18 Nov 1977, p. A8. BE06412.
federal government; government regulation; *recombinant DNA research; *risks and benefits

Cowen, Robert C. Congress waits to review guidelines on DNA research. *Christian Science Monitor (Eastern Edition),* 20 Dec 1977, p. 3. BE06460.
federal government; *government regulation; *recombinant DNA research; risks and benefits

Griffin, Bryan. Genetic engineering: the moral challenge. *Human Life Review* 3(3): 30-39, Summer 1977. BE05682.
biomedical research; *decision making; health hazards; government regulation; public participation; *recombinant DNA research; *risks and benefits; science; self regulation; social impact

Kukin, Marrick. Research with recombinant DNA: potential risks, potential benefits, and ethical considerations. *New York State Journal of Medicine* 78(2): 226-235, Feb 1978. 35 refs. BE06916.
*biological containment; *containment; DNA therapy; drug industry; food; freedom; genetic intervention; genetics; government agencies; government regulation; health hazards; hybrids; insulin; investigators; methods; natural selection; *physical containment; public participation; *recombinant DNA research; *risks and benefits; science; standards; universities; *NIH Guidelines; Princeton

Lappé, Marc. Regulating recombinant DNA research: pulling back from the apocalypse. *Man and Medicine: The Journal of Values and Ethics in Health Care* 2(2): 103-109, Winter 1977. 3 fn. BE05899.
federal government; government regulation; health hazards; *recombinant DNA research; *risks and benefits; NIH Guidelines

Lederberg, Seymour. The least hazardous course: recombinant DNA technology as an option for human genetic, viral, and cancer therapy. *In* Beers, Roland F.; Bassett, Edward G., eds. Recombinant Molecules: Impact on Science and Society. New York: Raven Press, 1977. p. 485-494. 36 refs. ISBN 0-89004-131-8. BE05856.
biological containment; cancer; health hazards; physical containment; *recombinant DNA research; *risks and benefits

Macklin, Ruth. On the ethics of *not* doing scientific research. *Hastings Center Report* 7(6): 11-13, Dec 1977. BE06143.
biomedical research; consequences; decision making; health hazards; *recombinant DNA research; *risks and benefits; social control; values

Newmark, Peter. WHO looks for benefits from genetic engineering. *Nature* 272(5655): 663-664, 20 Apr 1978. BE06943.
attitudes; containment; government regulation; *recombinant DNA research; *risks and benefits

Pandora's box of genes. *Economist* 262(6966): 82-83, 5 Mar 1977. BE05945.
federal government; government regulation; health hazards; industry; *recombinant DNA research; *risks and benefits

Rowe, Wallace P. Guidelines that do the job. *Bulletin of the Atomic Scientists* 33(5): 14-15, May 1977. 1 fn. BE05597.
federal government; government regulation; health hazards; *recombinant DNA research; *risks and benefits; NIH Guidelines

Schmeck, Harold M. Advance in DNA research. *New York Times,* 12 Nov 1977, p. 14. BE06409.
artificial genes; federal government; government regulation; *recombinant DNA research; *risks and benefits

Stetten, DeWitt, et al. The social implications of research on genetic manipulation: panel discussion. *In* Schultz, J.; Brada, Z., eds. Genetic Manipulation as It Affects the Cancer Problem. Proceedings. New York: Academic Press, 1977. p. 217-260. Proceedings of the Miami Winter Symposia, Jan 1977, sponsored by the Papanicolaou Cancer Research Institute, Miami, Florida. Miami Winter Symposia—Vol. 14. BE06684.
containment; ecology; evolution; *government regulation; government agencies; health hazards; hybrids; international aspects; physical containment; *recombinant DNA research; *risks and benefits; science; social control; social impact; standards; Europe; Great Britain; National Environmental Policy Act; NIH Guidelines; United States; Williams Report

Watson, James D. An imaginary monster. *Bulletin of the Atomic Scientists* 33(5): 12-13, May 1977. BE05596.
carcinogens; federal government; government regulation; investigators; *recombinant DNA research; *risks and benefits; self regulation

RECOMBINANT DNA RESEARCH / RISKS AND BENEFITS / SOCIAL IMPACT

Warshaw, Frances R. Gene implantation: proceed with caution—reservations concerning research in recombinant DNA. *In* Beers, Roland F.; Bassett, Edward G., eds. Recombinant Molecules: Impact on Science and Society. New York: Raven Press, 1977. p. 501-514. 44 refs. 6 fn. ISBN 0-89004-131-8. BE05857.
cancer; genetic intervention; health hazards; investigators; public participation; *recombinant DNA research; *risks and benefits; *social impact

---

BE = bioethics accession number   fn. = footnotes   refs. = references

RECOMBINANT DNA RESEARCH / SELF REGULATION

Tinkering with life. *Time* 109(16): 32-34+, 18 Apr 1977. BE05583.
  attitudes; federal government; *government regulation; health hazards; *investigators; physical containment; *recombinant DNA research; *risks and benefits; *self regulation; social control

RECOMBINANT DNA RESEARCH / SOCIAL CONTROL

Budrys, Algis. The politics of deoxyribonucleic acid. *New Republic* 176(16): 18-21, 16 Apr 1977. BE05543.
  federal government; government regulation; health hazards; investigators; municipal government; *public participation; *recombinant DNA research; review committees; risks and benefits; self regulation; *social control; universities; *Cambridge; Massachusetts; People's Business Commission

Callahan, Daniel. Recombinant DNA: science and the public. *Hastings Center Report* 7(2): 20-23, Apr 1977. BE05544.
  federal government; government regulation; investigators; *public participation; *recombinant DNA research; self regulation; *social control; *NIH Guidelines

Chargaff, Erwin. A few remarks regarding research on recombinant DNA. *Man and Medicine: The Journal of Values and Ethics in Health Care* 2(2): 78-82, Winter 1977. BE05897.
  health hazards; politics; *recombinant DNA research; science; *social control; social impact

Fabro, Fred. DNA recombinant research. *Connecticut Medicine* 41(7): 441-443, Jul 1977. Editorial. BE05955.
  attitudes; investigators; professional organizations; *recombinant DNA research; *social control; American Medical Association

Greenberg, Daniel S. Lessons of the DNA controversy. *New England Journal of Medicine* 297(21): 1187-1188, 24 Nov 1977. BE06113.
  attitudes; federal government; government regulation; investigators; legislation; politics; public participation; *recombinant DNA research; risks and benefits; *social control

Grimwade, Sandy. Pieces of action. *Nature* 268(5620): 480, 11 Aug 1977. BE05971.
  ecology; federal government; government regulation; industry; *recombinant DNA research; *social control

Maddocks, Melvin. DNA and the habit of saying yes to whatever's next. *Christian Science Monitor (Eastern Edition),* 24 Mar 1977, p. 26. BE06439.

attitudes; *recombinant DNA research; risks and benefits; *social control

Powledge, Tabitha M. Recombinant DNA: backing off on legislation. *Hastings Center Report* 7(6): 8-10, Dec 1977. BE06146.
  attitudes; federal government; government regulation; health hazards; investigators; legislation; municipal government; *recombinant DNA research; *social control

Powledge, Tabitha M. Recombinant DNA: the argument shifts. *Hastings Center Report* 7(2): 18-19, Apr 1977. BE05575.
  federal government; government regulation; industry; international aspects; legislation; *recombinant DNA research; risks and benefits; *social control; universities; NIH Guidelines

Rowe, Wallace P. Recombinant DNA: what happened? *New England Journal of Medicine* 297(21): 1176-1177, 24 Nov 1977. 1 fn. Editorial. BE06119.
  health hazards; investigators; mass media; public participation; *recombinant DNA research; self regulation; *social control

Shephard, David A. Genetic intervention: the modern chimera. *Canadian Medical Association Journal* 116(7): 705-707, 9 Apr 1977. 26 fn. BE05531.
  cancer; genetic defects; health hazards; hybrids; physician's role; public participation; *recombinant DNA research; risks and benefits; *social control; *social impact

RECOMBINANT DNA RESEARCH / SOCIAL CONTROL / ATTITUDES

Leeper, E.M. Recombinant DNA forum—stellar cast; gripping plot; but no new message. *BioScience* 27(5): 317-319, May 1977. BE05622.
  *attitudes; government regulation; federal government; health hazards; *investigators; legislation; public participation; *recombinant DNA research; *social control

RECOMBINANT DNA RESEARCH / SOCIAL CONTROL / CAMBRIDGE

Cambridge Experimentation Review Board. The Cambridge Experimentation Review Board. *Bulletin of the Atomic Scientists* 33(5): 22-27, May 1977. Recommendations approved by the City Council, with some further restrictions, on February 7, 1977. BE05599.
  federal government; government regulation; institutional obligations; municipal government; *public advocacy; public participation; *recombinant DNA research; *review committees; *social control; standards; universities; *Cambridge; *Cambridge Experimentation Review Board; NIH Guidelines

King, Jonathan. A science for the people. *New Scientist* 74(1056): 634-636, 16 Jun 1977. BE05650.

BE = bioethics accession number   fn. = footnotes   refs. = references

decision making; federal government; government regulation; health hazards; investigators; *munici-pal government; *public participation; *recombi-nant DNA research; *social control; socioeco-nomic factors; *Cambridge; Massachusetts

RECOMBINANT DNA RESEARCH / SOCIAL CONTROL / GREAT BRITAIN

Pirt, S.J. Genetic manipulation. *Times (London)*, 19 Oct 1976, p. 13. Letter. BE06614.
government regulation; *recombinant DNA re-search; self regulation; *social control; *Great Britain

RECOMBINANT DNA RESEARCH / SOCIAL CONTROL / HEALTH HAZARDS

Philipson, Lennart; Tiollais, Pierre. Rational contain-ment on recombinant DNA. *Nature* 268(5616): 90-91, 14 Jul 1977. 21 refs. BE05967.
containment; *health hazards; *recombinant DNA research; *social control

RECOMBINANT DNA RESEARCH / SOCIAL CONTROL / INTERNATIONAL ASPECTS

Fields, Cheryl M. Foreign countries weigh uniform rules on DNA studies. *Chronicle of Higher Educa-tion* 14(7): 3, 11 Apr 1977. BE05554.
health hazards; *international aspects; *recombi-nant DNA research; *social control; Great Britain; NIH Guidelines

Ioirysh, A.I. Law and the new potentials of biology. *Soviet Law and Government* 16(1): 40-54, Summer 1977. 12 fn. BE05678.
biological warfare; codes of ethics; *human experi-mentation; *international aspects; investigators; le-gal aspects; *recombinant DNA research; *risks and benefits; self regulation; *social control; thera-peutic research; Nuremberg Code

U.S. Federal Interagency Committee on Recombinant DNA Research. Report of the Federal Interagency Committee on Recombinant DNA Research: interna-tional activities. Nov 1977. *In* U.S. National In-stitutes of Health. Recombinant DNA Research: Volume 2. Documents Relating to "NIH Guidelines for Research...". Washington: U.S. Government Printing Office, Mar 1978. p. 381-500. DHEW Pub. No. (NIH) 78-1139. References and footnotes in-cluded. BE06950.
containment; *international aspects; professional organizations; *recombinant DNA research; *so-cial control; standards

RECOMBINANT DNA RESEARCH / SOCIAL CONTROL / RISKS AND BENEFITS

Cohen, Carl. When may research be stopped? *New En-gland Journal of Medicine* 296(21): 1203-1210, 26

May 1977. 2 refs. 1 fn. BE05608.
biomedical research; consequences; costs and bene-fits; injuries; *recombinant DNA research; *risks and benefits; *social control; social impact

Goodfield, June. Playing God: Genetic Engineering and the Manipulation of Life. New York: Random House, 1977. 218 p. ISBN 0-394-40692-3. BE05879.
DNA therapy; decision making; federal govern-ment; genetic defects; *genetic intervention; gov-ernment regulation; health hazards; historical aspects; international aspects; investigators; legal liability; morality; public opinion; public participa-tion; *recombinant DNA research; reproductive technologies; *risks and benefits; science; self regu-lation; *social control; values; Asilomar Confer-ence; NIH Guidelines

Price, Susan L. The little beasts. *Pharos* 41(1): 2-9, Jan 1978. 13 refs. BE06748.
attitudes; biomedical research; containment; gov-ernment regulation; investigators; legislation; obli-gations to society; peer review; *recombinant DNA research; rights; *risks and benefits; science; self regulation; *social control

Research with Recombinant DNA: An Academy Fo-rum, March 7-9, 1977. Washington: National Acad-emy of Sciences, 1977. 295 p. 121 refs. ISBN 0-309-02641-5. BE06517.
ecology; employment; epidemiology; evolution; federal government; genetic intervention; govern-ment regulation; health hazards; industry; investi-gators; politics; *public participation; public policy; *recombinant DNA research; *risks and benefits; science; *social control

Tinkering with life. *Time* 109(16): 32-34+, 18 Apr 1977. BE05583.
attitudes; federal government; *government regu-lation; health hazards; *investigators; physical containment; *recombinant DNA research; *risks and benefits; *self regulation; *social control

RECOMBINANT DNA RESEARCH / SOCIAL IMPACT

Beckwith, Jon. Recombinant DNA: does the fault lie within our genes? *Science for the People* 9(3): 14-17, May-Jun 1977. 19 refs. 1 fn. BE05604.
aggression; behavioral genetics; ecology; *genetic intervention; genetic screening; health hazards; hy-perkinesis; industry; minority groups; *recombi-nant DNA research; *risks and benefits; *social impact; social problems

RECOMBINANT DNA RESEARCH / SWEDEN

Barnaby, Wendy. Sweden debates gene-splicing. *Na-ture* 270(5639): 653, 22/29 Dec 1977. BE06249.
government regulation; physical containment; *re-combinant DNA research; social control; *Sweden

---

BE = bioethics accession number   fn. = footnotes   refs. = references

RECOMBINANT DNA RESEARCH / SWITZER-LAND

Waldner, Rosmarie. Switzerland—guidelines emerge. *Nature* 267(5608): 199, 19 May 1977. BE05630.
*investigators; *recombinant DNA research; *self regulation; NIH Guidelines; *Switzerland

RECOMBINANT DNA RESEARCH / TECHNOLOGY ASSESSMENT / LAW

Recombinant DNA and technology assessment. *Georgia Law Review* 11(4): 785-878, Summer 1977. 457 fn. BE06309.
communicable diseases; constitutional law; containment; decision making; ecology; *federal government; financial support; government agencies; *government regulation; health hazards; injuries; judicial action; *law; legal aspects; *legal liability; *legislation; negligence; public health; public participation; *recombinant DNA research; research institutes; research personnel; review committees; risks and benefits; social impact; state government; *technology assessment; *torts; DNA Research Act; National Environmental Policy Act; National Institutes of Health; NIH Guidelines; Public Health Service Act; Toxic Substances Control Act

RECOMBINANT DNA RESEARCH / TORTS / LEGAL LIABILITY

Recombinant DNA and technology assessment. *Georgia Law Review* 11(4): 785-878, Summer 1977. 457 fn. BE06309.
communicable diseases; constitutional law; containment; decision making; ecology; *federal government; financial support; government agencies; *government regulation; health hazards; injuries; judicial action; *law; legal aspects; *legal liability; *legislation; negligence; public health; public participation; *recombinant DNA research; research institutes; research personnel; review committees; risks and benefits; social impact; state government; *technology assessment; *torts; DNA Research Act; National Environmental Policy Act; National Institutes of Health; NIH Guidelines; Public Health Service Act; Toxic Substances Control Act

RECOMBINANT DNA RESEARCH / UNIVERSITIES

Schmeck, Harold M. U.S. agency bids lab halt work on genes. *New York Times,* 16 Dec 1977, p. A11. BE06426.
*federal government; *government regulation; investigators; *recombinant DNA research; *universities; Harvard University; NIH Guidelines

RECOMBINANT DNA RESEARCH / UNIVERSITY OF CALIFORNIA, SAN FRANCISCO

Insulin gene researchers admit breach of rules. *Science News* 112(14): 212, 1 Oct 1977. BE05850.
communication; federal government; government regulation; insulin; investigators; *recombinant DNA research; *universities; *University of California, San Francisco

Wade, Nicholas. Recombinant DNA: NIH rules broken in insulin gene project. *Science* 197(4311): 1342-1345, 30 Sep 1977. BE05844.
federal government; government regulation; insulin; investigators; *recombinant DNA research; review committees; *universities; NIH Guidelines; *University of California, San Francisco

REFUSAL OF TREATMENT *See* TREATMENT REFUSAL

RELIGION *See under*
    ABORTION / RELIGION / ATTITUDES
    HUMAN EXPERIMENTATION / RELIGION

RELIGIOUS BELIEFS *See under*
    BEHAVIOR CONTROL / PSYCHOTHERAPY / RELIGIOUS BELIEFS
    IN VITRO FERTILIZATION / RELIGIOUS BELIEFS
    TREATMENT REFUSAL / BLOOD TRANSFUSIONS / RELIGIOUS BELIEFS
    TREATMENT REFUSAL / PARENTAL CONSENT / RELIGIOUS BELIEFS

RELIGIOUS ETHICS *See under*
    ABORTION / RELIGIOUS ETHICS

RELIGIOUS HOSPITALS *See under*
    VOLUNTARY STERILIZATION / RELIGIOUS HOSPITALS / ROMAN CATHOLICISM

REMUNERATION *See under*
    BLOOD DONATION / REMUNERATION
    ORGAN DONATION / REMUNERATION

RENAL DIALYSIS / COSTS AND BENEFITS

Colen, B.D. The life-and-death cost of health. *Washington Post,* 22 Jul 1977, p. A27. BE06058.
*costs and benefits; *federal government; *financial support; health care; *renal dialysis; resource allocation

RENAL DIALYSIS / ECONOMICS

Rennie, Drummond. Home dialysis and the costs of

BE = bioethics accession number   fn. = footnotes   refs. = references

uremia. *New England Journal of Medicine* 298(7): 399-400, 16 Feb 1978. 7 refs. Editorial. BE06794.
*economics; federal government; financial support; health facilities; incentives; *legislation; medical fees; organ transplantation; politics; *renal dialysis; selection for treatment; statistics

RENAL DIALYSIS / LEGISLATION

Rennie, Drummond. Home dialysis and the costs of uremia. *New England Journal of Medicine* 298(7): 399-400, 16 Feb 1978. 7 refs. Editorial. BE06794.
*economics; federal government; financial support; health facilities; incentives; *legislation; medical fees; organ transplantation; politics; *renal dialysis; selection for treatment; statistics

RENAL DIALYSIS / RISKS AND BENEFITS

Levine, Carol. Dialysis or transplant: values and choices. *Hastings Center Report* 8(2): 8-10, Apr 1978. BE06848.
*decision making; economics; *kidneys; morbidity; mortality; *organ transplantation; psychological stress; *renal dialysis; *risks and benefits

RENAL DIALYSIS *See under*
SELECTION FOR TREATMENT / RENAL DIALYSIS / RANDOM SELECTION

REPRODUCTION / GENETIC DEFECTS / ATTITUDES

Margolin, C.R. Attitudes toward control and elimination of genetic defects. *Social Biology* 25(1): 33-37, Spring 1978. 9 refs. 2 fn. BE06923.
amniocentesis; *attitudes; contraception; *decision making; *genetic defects; *reproduction; selective abortion; students

REPRODUCTION / HUMAN RIGHTS

Reilly, Philip. Genetics, Law, and Social Policy. Cambridge, Mass.: Harvard University Press, 1977. 275 p. 297 refs. ISBN 0-674-34657-2. BE06519.
AID; embryo transfer; eugenics; federal government; *genetic counseling; genetic defects; genetic fathers; *genetic screening; government regulation; *human rights; in vitro fertilization; involuntary sterilization; *legal aspects; *legislation; mandatory programs; medical records; mentally handicapped; *reproduction; *reproductive technologies; sickle cell anemia; *state government; Tay Sachs disease; voluntary programs; wrongful life; XYY karyotype

REPRODUCTION / INDUSTRIAL MEDICINE / HEALTH HAZARDS

Hyatt, James C. Protection for unborn? Work-safety issue isn't as simple as it sounds. *Wall Street Journal,* 2 Aug 1977, p. 1+. BE06062.
congenital defects; employment; females; fetuses;

*health hazards; *industrial medicine; morbidity; pregnant women; *reproduction; women's rights

REPRODUCTION / RIGHTS

Shaw, Margery W. Procreation and the population problem. *North Carolina Law Review* 55(6): 1165-1185, Sep 1977. 149 fn. BE05989.
compensation; *congenital defects; contraception; human rights; involuntary sterilization; *legal liability; legal rights; mentally retarded; *newborns; parents; population control; *physicians; quality of life; *reproduction; *rights; state interest; *unwanted children; voluntary sterilization

REPRODUCTION / WOMEN'S RIGHTS

Swan, George S. State-mandated abortion: has 1984 arrived seven years early? *Law and Liberty* 3(4): 4-6, Spring/Summer 1977. BE06225.
*abortion; financial support; *mandatory programs; *reproduction; self determination; state interest; Supreme Court decisions; *women's rights

REPRODUCTIVE TECHNOLOGIES *See also*
CLONING, EMBRYO TRANSFER, GENETIC INTERVENTION, IN VITRO FERTILIZATION, SEX PRESELECTION

REPRODUCTIVE TECHNOLOGIES

Reilly, Philip. Genetics, Law, and Social Policy. Cambridge, Mass.: Harvard University Press, 1977. 275 p. 297 refs. ISBN 0-674-34657-2. BE06519.
AID; embryo transfer; eugenics; federal government; *genetic counseling; genetic defects; genetic fathers; *genetic screening; government regulation; *human rights; in vitro fertilization; involuntary sterilization; *legal aspects; *legislation; mandatory programs; medical records; mentally handicapped; *reproduction; *reproductive technologies; sickle cell anemia; *state government; Tay Sachs disease; voluntary programs; wrongful life; XYY karyotype

Smith, George P. Uncertainties on the spiral staircase: metaethics and the new biology. *Pharos* 41(1): 10-12, Jan 1978. 13 fn. BE06757.
embryo transfer; eugenics; human experimentation; in vitro fertilization; metaethics; *reproductive technologies; utilitarianism

Stumpf, Samuel E. Genetics and the control of human development: some ethical considerations. *In* Hollis, Harry N., comp. A Matter of Life and Death: Christian Perspectives. Nashville: Broadman Press, 1977. p. 114-128. 18 fn. ISBN 0-8054-6118-3. BE06534.
Christian ethics; cloning; genetic intervention; in vitro fertilization; *prenatal diagnosis; *reproductive technologies; *selective abortion

---

BE = bioethics accession number   fn. = footnotes   refs. = references

REPRODUCTIVE TECHNOLOGIES / MORALITY

Guentert, Kenneth. Will your grandchild be a test-tube baby? *U.S. Catholic* 42(6): 6-11, Jun 1977. BE05648.
AID; AIH; artificial insemination; artificial organs; embryo transfer; cloning; frozen semen; host mothers; in vitro fertilization; *morality; placentas; *reproductive technologies; semen donors; tissue banking

Hammes, John A. Shall we play God? An ethical analysis of biogenetics. *Homiletic and Pastoral Review* 77(7): 59-62, Apr 1977. 14 refs. BE06298.
genetic intervention; *morality; normative ethics; *reproductive technologies

REPRODUCTIVE TECHNOLOGIES / SOCIAL IMPACT

Howard, Ted; Rifkin, Jeremy. Who Should Play God? The Artificial Creation of Life and What It Means for the Future of the Human Race. New York: Dell Publishing Company, 1977. 272 p. 526 fn. ISBN 0-440-19504-7. BE05882.
AID; behavioral genetics; biomedical technologies; cloning; DNA therapy; drugs; embryo transfer; *eugenics; genetic screening; *genetic intervention; government regulation; health hazards; historical aspects; in vitro fertilization; industry; involuntary sterilization; population control; recombinant DNA research; *reproductive technologies; risks and benefits; selective abortion; social control; *social impact

RESEARCH *See* BEHAVIORAL RESEARCH, BIOMEDICAL RESEARCH, HUMAN EXPERIMENTATION, RECOMBINANT DNA RESEARCH

RESEARCH DESIGN *See under*

HUMAN EXPERIMENTATION / RESEARCH DESIGN

RESEARCH SUBJECTS *See under*

HUMAN EXPERIMENTATION / RESEARCH SUBJECTS / COMPENSATION

HUMAN EXPERIMENTATION / RESEARCH SUBJECTS / LEGAL RIGHTS

RESOURCE ALLOCATION

Wojcik, Jan. Muted Consent: A Casebook in Modern Medical Ethics. West Lafayette, Ind.: Purdue University, 1978. 164 p. 237 fn. Case study. ISBN 911198-50-4. BE06704.
*abortion; *behavior control; *case reports; *determination of death; disclosure; drugs; electrical stimulation of the brain; ethics; eugenics; euthanasia; extraordinary treatment; fetuses; *genetic counseling; genetic defects; *genetic intervention; *genetic screening; health care; *human experi-

mentation; *informed consent; involuntary commitment; mentally ill; operant conditioning; pregnant women; psychosurgery; quality of life; reproductive technologies; *resource allocation; rights; selective abortion; value of life; withholding treatment

RESOURCE ALLOCATION / AGED / NURSING HOMES

Lavelle, Michael J. An economic analysis of resource allocation for care of the aged. *Hospital Progress* 58(9): 99-103, Sep 1977. BE05782.
*administrators; *aged; attitudes; children; decision making; *economics; federal government; financial support; law; *moral obligations; morality; *nursing homes; patient care; personhood; philosophy; renal dialysis; *resource allocation; selection for treatment; terminal care

RESOURCE ALLOCATION / HEALTH CARE

Beauchamp, Tom L.; Walters, LeRoy, eds. Contemporary Issues in Bioethics. Encino, Calif.: Dickenson Publishing Company, 1978. 612 p. References and footnotes included. ISBN 0-8221-0200-5. BE06700.
*abortion; allowing to die; *behavior control; *bioethics; biomedical technologies; brain death; children; codes of ethics; confidentiality; decision making; deontological ethics; *determination of death; *euthanasia; fetuses; genetic intervention; *health; *health care; *human experimentation; *informed consent; involuntary commitment; justice; legal aspects; *medical ethics; natural law; *normative ethics; operant conditioning; patients' rights; personhood; psychosurgery; *resource allocation; rights; selection for treatment; treatment refusal; utilitarianism; Kaimowitz v. Department of Mental Health; Quinlan, Karen

RESOURCE ALLOCATION / HYPERTENSION / COSTS AND BENEFITS

Weinstein, Milton C.; Stason, William B. Allocating resources: the case of hypertension. *Hastings Center Report* 7(5): 24-29, Oct 1977. 2 fn. BE06079.
*costs and benefits; decision making; discrimination; economics; *health care; *hypertension; mass screening; public policy; *resource allocation

RESOURCE ALLOCATION / KIDNEY DISEASES

Some kidney patients over 45 'not treated'. *Times (London)*, 19 May 1977, p. 4. BE06441.
*allowing to die; *kidney diseases; renal dialysis; *resource allocation; scarcity; *withholding treatment

RESOURCE ALLOCATION / MEDICINE / COSTS AND BENEFITS

Hellegers, André E. Biologic origins of bioethical

---

problems. *In* Wynn, Ralph M., ed. Obstetrics and Gynecology Annual. Volume 6, 1977. New York: Appleton-Century-Crofts, 1977. p. 1-9. ISBN 0-8385-7181-6. BE06570.
  *costs and benefits; deontological ethics; *health; health care delivery; *medicine; obligations to society; physician patient relationship; physicians; psychiatry; politics; quality of life; *resource allocation; *social control; teleological ethics; value of life

RESOURCE ALLOCATION / MORAL POLICY / SCARCITY

Taurek, John M. Should the numbers count? *Philosophy and Public Affairs* 6(4): 293-316, Summer 1977. 5 fn. BE05937.
  *decision making; moral obligations; *moral policy; philosophy; *resource allocation; *scarcity

RESOURCE ALLOCATION / TERMINALLY ILL

Schiffer, R.B.; Freedman, Benjamin. The last bed in the ICU. *Hastings Center Report* 7(6): 21-22, Dec 1977. Case study. BE06149.
  *decision making; family members; intensive care units; physicians; prolongation of life; *resource allocation; *resuscitation; *terminally ill

RESOURCE ALLOCATION   *See under*

  HEALTH CARE DELIVERY / RESOURCE ALLOCATION

  HEALTH CARE DELIVERY / RESOURCE ALLOCATION / SWEDEN

  HEALTH CARE / RESOURCE ALLOCATION

  HEALTH CARE / RESOURCE ALLOCATION / COSTS AND BENEFITS

  HEALTH CARE / RESOURCE ALLOCATION / ECONOMICS

  PATIENT CARE / RESOURCE ALLOCATION

RESUSCITATION / INTENSIVE CARE UNITS / COSTS AND BENEFITS

Phillips, G.D. Life support systems in intensive care: a review of history, ethics, cost, benefit and rational use. *Anaesthesia and Intensive Care* 5(3): 251-257, Aug 1977. 67 refs. BE06326.
  artificial organs; *costs and benefits; determination of death; economics; historical aspects; *intensive care units; mortality; prolongation of life; renal dialysis; *resuscitation

RESUSCITATION / LEGAL LIABILITY

Huber, Robert. Legal considerations of cardiopulmonary resuscitation. *In* Safar, P., ed. Advances in Cardiopulmonary Resuscitation. New York: Springer-Verlag, 1977. p. 246-249. 2 refs. ISBN 0-387-90234-1. BE06574.
  *allowing to die; *legal aspects; informed consent;

*legal liability; medical staff; physicians; *resuscitation; surgery; withholding treatment

RESUSCITATION / NURSES

Sampson, Chris. The student's dilemma. *Nursing Mirror* 145(3): 16-17, 21 Jul 1977. BE05814.
  *decision making; emergency care; *nurses; *resuscitation; *students

RESUSCITATION / NURSES / CHRONICALLY ILL

Flaherty, M. Josephine; Smith, James M. The nurse and orders not to resuscitate. *Hastings Center Report* 7(4): 27-28, Aug 1977. Case study. BE05757.
  *chronically ill; emergency care; *moral obligations; *nurses; patients; physicians; professional ethics; *resuscitation; self determination

RESUSCITATION / TERMINALLY ILL

Gendrop, Sylvia C. The order: *No Code*. *Linacre Quarterly* 44(4): 312-319, Nov 1977. 15 fn. BE06100.
  *allowing to die; communication; family members; *nurses; physician's role; *resuscitation; *terminal care; *terminally ill; *withholding treatment

Schiffer, R.B.; Freedman, Benjamin. The last bed in the ICU. *Hastings Center Report* 7(6): 21-22, Dec 1977. Case study. BE06149.
  *decision making; family members; intensive care units; physicians; prolongation of life; *resource allocation; *resuscitation; *terminally ill

RETARDED   *See* MENTALLY RETARDED

REVIEW COMMITTEES / INSTITUTIONALIZED PERSONS / PROGRAM DESCRIPTIONS

Crane, Lansing; Zonana, Howard; Wizner, Stephen. Implications of the Donaldson decision: a model for periodic review of committed patients. *Hospital and Community Psychiatry* 28(11): 827-833, Nov 1977. 3 fn. BE06500.
  alternatives; dangerousness; deinstitutionalized persons; *evaluation; *institutionalized persons; interdisciplinary communication; involuntary commitment; mental institutions; mentally ill; patient care; *program descriptions; *review committees; standards; voluntary programs; O'Connor v. Donaldson

REVIEW COMMITTEES   *See under*

  RECOMBINANT DNA RESEARCH / REVIEW COMMITTEES / CALIFORNIA

  RECOMBINANT DNA RESEARCH / REVIEW COMMITTEES / LEGISLATION

BE = bioethics accession number   fn. = footnotes   refs. = references

RIGHT TO TREATMENT / INSTITUTIONALIZED PERSONS

Blocker, Webster; Dowben, Carla. Right to treatment: issues and implications. *Texas Medicine* 73(10): 76-81, Oct 1977. 32 fn. BE06084.
due process; economics; *institutionalized persons; involuntary commitment; legal aspects; legal liability; mentally ill; psychiatry; *right to treatment

RIGHT TO TREATMENT / INSTITUTIONALIZED PERSONS / MENTALLY ILL

Lundeen, Paul A. Pennsylvania's new Mental Health Procedures Act: due process and the right to treatment for the mentally ill. *Dickinson Law Review* 81(3): 627-647, Spring 1977. 141 fn. BE05924.
adolescents; *coercion; competence; confidentiality; dangerousness; *due process; *institutionalized persons; involuntary commitment; legal rights; legislation; medical records; *mentally ill; *patient care; prisoners; *right to treatment; state government; voluntary admission; *Mental Health Procedures Act; *Pennsylvania

Martin, Andrew. A right to treatment. *Nursing Times* 73(42): 1620-1621, 20 Oct 1977. BE06349.
informed consent; *institutionalized persons; *involuntary commitment; *legal aspects; legal rights; *mentally ill; *right to treatment

McGough, Lucy S.; Carmichael, William C. The right to treatment and the right to refuse treatment. *American Journal of Orthopsychiatry* 47(2): 307-320, Apr 1977. 30 refs. BE05567.
constitutional law; dangerousness; due process; *institutionalized persons; involuntary commitment; *legal rights; *mentally ill; peer review; *right to treatment; standards; *treatment refusal

Roth, Loren H. Involuntary civil commitment: the right to treatment and the right to refuse treatment. *Psychiatric Annals* 7(5): 50-51+, May 1977. 55 refs. BE05809.
coercion; dangerousness; competence; drugs; due process; economics; informed consent; *institutionalized persons; *involuntary commitment; judicial action; *legal aspects; legal rights; legislation; mental institutions; *mentally ill; mentally retarded; operant conditioning; patients; psychiatry; review committees; *right to treatment; standards; state government; *treatment refusal

RIGHT TO TREATMENT / INSTITUTIONALIZED PERSONS / O'CONNOR V. DONALDSON

Szasz, Thomas S. Psychiatric Slavery. New York: Free Press, 1977. 159 p. 240 fn. Case study. ISBN 0-02-931600-6. BE05709.
dangerousness; *institutionalized persons; *involuntary commitment; law; *legal aspects; *mentally ill; professional organizations; psychiatric diagnosis; psychiatry; *right to treatment; Supreme Court

decisions; treatment refusal; American Civil Liberties Union; American Psychiatric Association; *Donaldson, Kenneth; *Mental Health Law Project; *O'Connor v. Donaldson

RIGHT TO TREATMENT / INSTITUTIONALIZED PERSONS / PENNSYLVANIA

Lundeen, Paul A. Pennsylvania's new Mental Health Procedures Act: due process and the right to treatment for the mentally ill. *Dickinson Law Review* 81(3): 627-647, Spring 1977. 141 fn. BE05924.
adolescents; *coercion; competence; confidentiality; dangerousness; *due process; *institutionalized persons; involuntary commitment; legal rights; legislation; medical records; *mentally ill; *patient care; prisoners; *right to treatment; state government; voluntary admission; *Mental Health Procedures Act; *Pennsylvania

RIGHT TO TREATMENT / MENTALLY HANDICAPPED

Epstein, Joseph M. Mental disabilities law issues. *Colorado Lawyer* 6(12): 2163-2168, Dec 1977. 36 fn. BE06370.
dangerousness; due process; equal protection; involuntary commitment; *legal aspects; legislation; *mentally handicapped; *right to treatment; state government; *Colorado

RIGHT TO TREATMENT / MENTALLY ILL / NURSING HOMES

Barnett, Cynthia F. Treatment rights of mentally ill nursing home residents. *University of Pennsylvania Law Review* 126(3): 578-629, Jan 1978. 227 fn. BE06723.
*aged; constitutional law; contracts; deinstitutionalized persons; federal government; government regulation; informed consent; involuntary commitment; judicial action; legal aspects; legal rights; mental institutions; mentally handicapped; *nursing homes; *mentally ill; patient care; patients; psychotherapy; *right to treatment; social impact; state action; state government; voluntary admission; Alabama

RIGHT TO TREATMENT / PRISONERS / PSYCHOTHERAPY

Hoard, Steven L. Prisoners' rights—*Bowring v. Godwin*: the limited right of state prisoners to psychological and psychiatric treatment. *North Carolina Law Review* 56(3): 612-621, Apr 1978. 80 fn. BE06844.
legal aspects; *mentally ill; *prisoners; *psychotherapy; *right to treatment; state government

RIGHTS *See also* HUMAN RIGHTS, LEGAL RIGHTS, PATIENTS' RIGHTS

BE = bioethics accession number   fn. = footnotes   refs. = references

RIGHTS *See under*

ABORTION / FETUSES / RIGHTS

ABORTION / RIGHTS

BEHAVIOR CONTROL / RIGHTS

BEHAVIORAL RESEARCH / DRUG ABUSE / RIGHTS

CONFIDENTIALITY / RIGHTS

EUTHANASIA / MORAL OBLIGATIONS / RIGHTS

EUTHANASIA / RIGHTS

HEALTH CARE / RIGHTS

HUMAN EXPERIMENTATION / RIGHTS

KILLING / MORAL OBLIGATIONS / RIGHTS

PATIENT CARE / LAETRILE / RIGHTS

PERSONHOOD / FETUSES / RIGHTS

POPULATION CONTROL / RIGHTS

REPRODUCTION / RIGHTS

TREATMENT REFUSAL / PATIENTS / RIGHTS

VOLUNTARY EUTHANASIA / RIGHTS

RISKS AND BENEFITS *See under*

AMNIOCENTESIS / RISKS AND BENEFITS

BEHAVIORAL RESEARCH / INFORMED CONSENT / RISKS AND BENEFITS

BIOMEDICAL RESEARCH / SOCIAL CONTROL / RISKS AND BENEFITS

DISCLOSURE / ANESTHESIA / RISKS AND BENEFITS

ELECTROCONVULSIVE THERAPY / RISKS AND BENEFITS

GENETIC COUNSELING / HUNTINGTON'S CHOREA / RISKS AND BENEFITS

GENETIC INTERVENTION / RISKS AND BENEFITS

GENETIC SCREENING / XYY KARYOTYPE / RISKS AND BENEFITS

HUMAN EXPERIMENTATION / ALCOHOL ABUSE / RISKS AND BENEFITS

HUMAN EXPERIMENTATION / CHILDREN / RISKS AND BENEFITS

HUMAN EXPERIMENTATION / FETUSES / RISKS AND BENEFITS

HUMAN EXPERIMENTATION / GOVERNMENT REGULATION / RISKS AND BENEFITS

HUMAN EXPERIMENTATION / INFORMED CONSENT / RISKS AND BENEFITS

HUMAN EXPERIMENTATION / MORAL OBLIGATIONS / RISKS AND BENEFITS

HUMAN EXPERIMENTATION / RISKS AND BENEFITS / HISTORICAL ASPECTS

IMMUNIZATION / DISCLOSURE / RISKS AND BENEFITS

IMMUNIZATION / POLIOMYELITIS / RISKS AND BENEFITS

IMMUNIZATION / RISKS AND BENEFITS

INFORMED CONSENT / PSYCHOACTIVE DRUGS / RISKS AND BENEFITS

RISKS AND BENEFITS *See under (cont'd.)*

INFORMED CONSENT / RADIOLOGY / RISKS AND BENEFITS

INFORMED CONSENT / RISKS AND BENEFITS / LEGAL ASPECTS

INFORMED CONSENT / SURGERY / RISKS AND BENEFITS

ORGAN TRANSPLANTATION / KIDNEYS / RISKS AND BENEFITS

ORGAN TRANSPLANTATION / RANDOM SELECTION / RISKS AND BENEFITS

PATIENT CARE / DRUGS / RISKS AND BENEFITS

PATIENT CARE / LAETRILE / RISKS AND BENEFITS

PATIENT CARE / LEUKEMIA / RISKS AND BENEFITS

PATIENT CARE / PHYSICIANS / RISKS AND BENEFITS

PSYCHOSURGERY / RISKS AND BENEFITS

RECOMBINANT DNA RESEARCH / GOVERNMENT REGULATION / RISKS AND BENEFITS

RECOMBINANT DNA RESEARCH / PUBLIC POLICY / RISKS AND BENEFITS

RECOMBINANT DNA RESEARCH / RISKS AND BENEFITS

RECOMBINANT DNA RESEARCH / RISKS AND BENEFITS / SOCIAL IMPACT

RECOMBINANT DNA RESEARCH / SOCIAL CONTROL / RISKS AND BENEFITS

RENAL DIALYSIS / RISKS AND BENEFITS

SEX PRESELECTION / RISKS AND BENEFITS

SOCIAL CONTROL / SCIENCE / RISKS AND BENEFITS

TRANSPLANTATION / CHILDREN / RISKS AND BENEFITS

ROMAN CATHOLIC ETHICS *See under*

ABORTION / ROMAN CATHOLIC ETHICS / DOUBLE EFFECT

ABORTION / ROMAN CATHOLIC ETHICS / HISTORICAL ASPECTS

ABORTION / VALUE OF LIFE / ROMAN CATHOLIC ETHICS

ARTIFICIAL INSEMINATION / ROMAN CATHOLIC ETHICS

EUTHANASIA / ROMAN CATHOLIC ETHICS / NOLAN, KIERAN

INVOLUNTARY STERILIZATION / ROMAN CATHOLIC ETHICS / INDIA

VOLUNTARY STERILIZATION / ROMAN CATHOLIC ETHICS

ROMAN CATHOLICISM *See also* ROMAN CATHOLIC ETHICS

ROMAN CATHOLICISM *See under*

ABORTION / INFORMAL SOCIAL CONTROL / RO-

BE = bioethics accession number   fn. = footnotes   refs. = references

ROMAN CATHOLICISM *See under (cont'd.)*
MAN CATHOLICISM

ABORTION / ROMAN CATHOLICISM / POLITI-
CAL ACTIVITY

VOLUNTARY STERILIZATION / RELIGIOUS HOS-
PITALS / ROMAN CATHOLICISM

RORVIK, DAVID *See under*

CLONING / RORVIK, DAVID

ROYAL COLLEGE OF PSYCHIATRISTS *See
under*

ELECTROCONVULSIVE THERAPY / ROYAL COL-
LEGE OF PSYCHIATRISTS

SAIKEWICZ, JOSEPH

Annas, George J. The incompetent's right to die: the
case of Joseph Saikewicz. *Hastings Center Report*
8(1): 21-23, Feb 1978. 1 fn. BE06769.
*allowing to die; competence; ethics committees;
*judicial action; *legal rights; leukemia; mentally
retarded; *self determination; state interest; terminally ill; *treatment refusal; *withholding treatment; *Saikewicz, Joseph

Ayd, Frank J. Treatment for the terminally ill incompetent: who decides—courts or physicians? *Medical-
Moral Newsletter* 15(4): 13-16, Apr 1978. BE06836.
aged; *allowing to die; competence; *decision making; drugs; judicial action; *legal aspects; leukemia;
*mentally retarded; *terminally ill; treatment re-
fusal; withholding treatment; *Saikewicz, Joseph

Curran, William J. The Saikewicz decision. *New England Journal of Medicine* 298(9): 499-500, 2 Mar
1978. 5 refs. BE06965.
*allowing to die; *decision making; ethics commit-
tees; family members; *judicial action; mentally
handicapped; physicians; *state courts; *terminally ill; withholding treatment; *Massachusetts
Supreme Judicial Court; *Saikewicz, Joseph

Kindregan, Charles P. The court as forum for life and
death decisions: reflections on procedures for substituted consent. *Suffolk University Law Review* 11(4):
919-935, Spring 1977. 78 fn. BE06232.
*allowing to die; competence; institutionalized per-
sons; *judicial action; legal aspects; leukemia;
mentally retarded; organ donation; parental consent; patient care; *third party consent; *withholding treatment; *Saikewicz, Joseph

Massachusetts. Supreme Judicial Court. Jones v. Saikewicz. 9 Jul 1976. Unpublished court decision. 3
p. Docket No. SJC-711, Appeals Court No. 76-369,
Hampshire Probate No. 45596. BE06209.
*allowing to die; *institutionalized persons; *men-
tally retarded; *withholding treatment; *Saikewicz, Joseph

Massachusetts. Supreme Judicial Court, Hampshire.
Superintendent of Belchertown v. Saikewicz. 28 Nov

1977. *North Eastern Reporter, 2d Series,* 370:
417-435. 20 fn. BE06210.
*allowing to die; competence; drugs; *institution-
alized persons; *legal rights; leukemia; *mentally
retarded; privacy; prolongation of life; state interest; terminally ill; treatment refusal; *withholding
treatment; *Saikewicz, Joseph

McCormick, Richard A.; Hellegers, André E. The
specter of Joseph Saikewicz: mental incompetence
and the law. *America* 138(12): 257-260, 1 Apr
1978. BE06849.
aged; *allowing to die; *competence; *decision
making; drugs; family members; *judicial action;
leukemia; *mentally retarded; physicians; prolongation of life; terminally ill; *withholding treatment; *Saikewicz, Joseph

Ramsey, Paul. The strange case of Joseph Saikewicz.
In his Ethics at the Edges of Life: Medical and Legal
Intersections. New Haven: Yale University Press,
1978. p. 300-317. 9 fn. Case study. ISBN 0-
300-02137-2. BE06892.
aged; *allowing to die; *competence; *decision
making; drugs; *judicial action; legal guardians;
leukemia; *mentally retarded; physicians; prognosis; prolongation of life; *quality of life; standards;
*terminally ill; *third party consent; treatment re-
fusal; *withholding treatment; *In re Quinlan;
*Massachusetts Supreme Judicial Court; *Saikew-
icz, Joseph

Relman, Arnold S. The Saikewicz decision: judges as
physicians. *New England Journal of Medicine*
298(9): 508-509, 2 Mar 1978. 6 refs. BE06966.
*allowing to die; *decision making; family mem-
bers; *judicial action; mentally handicapped; physicians; *state courts; *terminally ill; withholding
treatment; *Massachusetts Supreme Judicial
Court; *Saikewicz, Joseph

SCARCITY *See under*

ORGAN TRANSPLANTATION / KIDNEYS / SCAR-
CITY

RESOURCE ALLOCATION / MORAL POLICY /
SCARCITY

**SCIENCE / CONSTITUTIONAL LAW / FREE-
DOM**

Green, Harold P. The boundaries of scientific freedom. *Newsletter on Science, Technology, and Human
Values* No. 20: 17-21, Jun 1977. 20 fn. BE06235.
biomedical research; *constitutional law; federal
government; financial support; *freedom; government regulation; legal rights; municipal government; public policy; recombinant DNA research;
*science; First Amendment

SCIENCE / FREEDOM

Nagel, Ernest; Kurtz, Paul. Ethics of Scientific Research. [Videorecording]. Amherst, N.Y.: American

---

BE = bioethics accession number   fn. = footnotes   refs. = references

Humanist Association, 1977. 1 cassette; 28 min.; sound; color; 3/4 in. Ethics in America Series presented by the American Humanist Association. Produced at WNED-TV, Buffalo, New York. BE07039.
   financial support; *freedom; genetics; government regulation; human experimentation; intelligence; moral obligations; public opinion; recombinant DNA research; risks and benefits; *science; self regulation; social determinants; social impact; society; standards; technology; *values

SCIENCE / POLITICS

Walgate, Robert. The expert has no clothes. *Nature* 271(5647): 698, 23 Feb 1978. BE06766.
   attitudes; *politics; risks and benefits; *science

SCIENCE / PROFESSIONAL ETHICS

Toulmin, Stephen. The meaning of professionalism: doctors' ethics and biomedical science. *In* Engelhardt, H. Tristram; Callahan, Daniel, eds. Knowledge, Value and Belief. Hastings-on-Hudson, N.Y.: Institute of Society, Ethics and the Life Sciences, 1977. p. 254-278. Foundations of Ethics and Its Relationship to Science: Volume 2. 9 fn. ISBN 0-916558-02-9. BE06177.
   confidentiality; goals; *conflict of interest; human experimentation; government regulation; law; legal obligations; medical ethics; *medicine; *moral obligations; obligations to society; *physician patient relationship; physicians; professional competence; *professional ethics; professional organizations; *science; values

SCIENCE *See under*
   CODES OF ETHICS / SCIENCE
   DEATH / SCIENCE
   GOVERNMENT REGULATION / SCIENCE
   PUBLIC PARTICIPATION / SCIENCE
   SOCIAL CONTROL / SCIENCE
   SOCIAL CONTROL / SCIENCE / RISKS AND BENEFITS

SCRIPTURAL INTERPRETATION *See under*
   ABORTION / SCRIPTURAL INTERPRETATION

SELECTION FOR TREATMENT

Ramsey, Paul. "Euthanasia" and dying well enough. *In his* Ethics at the Edges of Life: Medical and Legal Intersections. New Haven: Yale University Press, 1978. p. 145-188. 45 fn. ISBN 0-300-02137-2. BE06888.
   active euthanasia; *adults; *allowing to die; chronically ill; coma; competence; congenital defects; death; *decision making; diagnosis; drugs; *euthanasia; *extraordinary treatment; human rights; legal aspects; medical ethics; moral obligations;

newborns; pain; physicians; prognosis; quality of life; religious ethics; *selection for treatment; social interaction; spina bifida; *standards; terminal care; *terminally ill; theology; *third party consent; *treatment refusal; value of life; withholding treatment; McCormick, Richard A.; Veatch, Robert M.

SELECTION FOR TREATMENT / MORAL POLICY / SPINA BIFIDA

Fletcher, John C. Spina bifida with myelomeningocele: a case study in attitudes towards defective newborns. *In* Swinyard, Chester A., ed. Decision Making and the Defective Newborn. Proceedings of a Conference on Spina Bifida and Ethics. Springfield, Ill.: Charles C. Thomas, 1978. p. 281-303. 39 refs. ISBN 0-398-03662-4. BE06956.
   active euthanasia; allowing to die; *attitudes; attitudes to death; congenital defects; decision making; extraordinary treatment; family problems; handicapped; *moral policy; *newborns; obligations of society; patient care; parents; physicians; prognosis; *quality of life; resource allocation; *selection for treatment; *spina bifida; *value of life; values

SELECTION FOR TREATMENT / NEWBORNS / INTENSIVE CARE UNITS

Ramsey, Paul. An ingathering of other reasons for neonatal infanticide. *In his* Ethics at the Edges of Life: Medical and Legal Intersections. New Haven: Yale University Press, 1978. p. 228-267. 36 fn. ISBN 0-300-02137-2. BE06890.
   abortion; *allowing to die; *congenital defects; *costs and benefits; ethical analysis; fetal development; fetuses; health care; human equality; *intensive care units; killing; medical ethics; *newborns; normality; personhood; prematurity; prognosis; public policy; *quality of life; *resource allocation; resuscitation; *selection for treatment; socioeconomic factors; Supreme Court decisions; *value of life; withholding treatment; *Roe v. Wade; *Sanoma Conference

SELECTION FOR TREATMENT / NEWBORNS / SPINA BIFIDA

Darling, Rosalyn B. Parents, physicians, and spina bifida. *Hastings Center Report* 7(4): 10-14, Aug 1977. 25 fn. BE05743.
   *attitudes; *children; family problems; *handicapped; *newborns; *parents; *physicians; quality of life; *selection for treatment; social adjustment; *spina bifida; withholding treatment

Fletcher, John C. Spina bifida with myelomeningocele: a case study in attitudes towards defective newborns. *In* Swinyard, Chester A., ed. Decision Making and the Defective Newborn. Proceedings of a Conference on Spina Bifida and Ethics. Springfield, Ill.: Charles C. Thomas, 1978. p. 281-303. 39 refs. ISBN

BE = bioethics accession number   fn. = footnotes   refs. = references

0-398-03662-4. BE06956.
active euthanasia; allowing to die; *attitudes; attitudes to death; congenital defects; decision making; extraordinary treatment; family problems; handicapped; *moral policy; *newborns; obligations of society; patient care; parents; physicians; prognosis; *quality of life; resource allocation; *selection for treatment; *spina bifida; *value of life; values

Fletcher, Joseph. Pediatric euthanasia: the ethics of selective treatment for spina bifida. *In* Swinyard, Chester A., ed. Decision Making and the Defective Newborn. Proceedings of a Conference on Spina Bifida and Ethics. Springfield, Ill.: Charles C. Thomas, 1978. p. 477-488. 23 refs. ISBN 0-398-03662-4. BE06961.
allowing to die; costs and benefits; decision making; *ethics; moral obligations; *newborns; personhood; physicians; *quality of life; Roman Catholic ethics; *selection for treatment; situational ethics; *spina bifida; standards; value of life; values; wedge argument; withholding treatment

Hauerwas, Stanley. Selecting children to live or die: an ethical analysis of the debate between Dr. Lorber and Dr. Freeman on the treatment of meningomyelocele. *In* Horan, Dennis J.; Mall, David, eds. Death, Dying, and Euthanasia. Washington: University Publications of America, 1977. p. 228-249. 20 refs. ISBN 0-89093-139-9. BE05688.
attitudes; costs and benefits; *decision making; ethicists; *newborns; parents; physicians; prognosis; *selection for treatment; *spina bifida; suffering; utilitarianism; value of life; withholding treatment; *Freeman, John; *Lorber, John

Lorber, John. Ethical concepts in the treatment of myelomeningocele. *In* Swinyard, Chester A., ed. Decision Making and the Defective Newborn. Proceedings of a Conference on Spina Bifida and Ethics. Springfield, Ill.: Charles C. Thomas, 1978. p. 59-67. 16 refs. ISBN 0-398-03662-4. BE06952.
*extraordinary treatment; family problems; mentally retarded; morbidity; mortality; *newborns; physically handicapped; prognosis; prolongation of life; quality of life; *selection for treatment; *spina bifida; statistics

Reid, Robert. Spina bifida: the fate of the untreated. *Hastings Center Report* 7(4): 16-19, Aug 1977. BE05745.
*active euthanasia; allowing to die; decision making; *newborns; parents; physicians; quality of life; *selection for treatment; selective abortion; *spina bifida; values; wedge argument; withholding treatment

Swinyard, Chester A., ed. Decision Making and the Defective Newborn. Proceedings of a Conference on Spina Bifida and Ethics. Springfield, Ill.: Charles C. Thomas, 1978. 649 p. References and footnotes included. ISBN 0-398-03662-4. BE06951.
*allowing to die; attitudes; congenital defects; *de-

cision making; ethics; extraordinary treatment; informed consent; legal aspects; medical ethics; *newborns; parental consent; *patient care; physician's role; *quality of life; *selection for treatment; socioeconomic factors; *spina bifida; withholding treatment

Veatch, Robert M. The technical criteria fallacy. *Hastings Center Report* 7(4): 15-16, Aug 1977. 9 fn. BE05744.
decision making; *newborns; prognosis; *selection for treatment; *spina bifida; values

SELECTION FOR TREATMENT / QUALITY OF LIFE / SPINA BIFIDA

Reich, Warren T. Quality of life and defective newborn children: an ethical analysis. *In* Swinyard, Chester A., ed. Decision Making and the Defective Newborn. Proceedings of a Conference on Spina Bifida and Ethics. Springfield, Ill.: Charles C. Thomas, 1978. p. 489-511. 35 refs. 4 fn. ISBN 0-398-03662-4. BE06962.
active euthanasia; allowing to die; consequences; decision making; *deontological ethics; *extraordinary treatment; handicapped; killing; *moral obligations; *newborns; *normative ethics; personhood; prognosis; *quality of life; Roman Catholic ethics; *selection for treatment; social interaction; *spina bifida; teleological ethics; utilitarianism; *value of life; values; Fletcher, Joseph; Freeman, John; *Lorber, John; *McCormick, Richard A.

SELECTION FOR TREATMENT / RENAL DIALYSIS / RANDOM SELECTION

Hellegers, André E. A lottery for lives? *Ob. Gyn. News* 12(7): 20-21, 1 Apr 1977. BE05559.
ethics committees; medical education; physicians; *random selection; *renal dialysis; *selection for treatment; social worth; students; value of life

SELECTION FOR TREATMENT / SPINA BIFIDA / EVALUATION

Guiney, E.J., et al. Surgical closure of myelomeningocele: problems and consequences of the introduction of a policy of selection. *Irish Journal of Medical Science* 146(8): 260-262, Aug 1977. 8 refs. BE06323.
*evaluation; mortality; patient care; prognosis; *selection for treatment; *spina bifida; standards

SELECTION FOR TREATMENT / SPINA BIFIDA / NORMATIVE ETHICS

Reich, Warren T. Quality of life and defective newborn children: an ethical analysis. *In* Swinyard, Chester A., ed. Decision Making and the Defective Newborn. Proceedings of a Conference on Spina Bifida and Ethics. Springfield, Ill.: Charles C. Thomas, 1978. p. 489-511. 35 refs. 4 fn. ISBN 0-398-

---

BE = bioethics accession number   fn. = footnotes   refs. = references

03662-4. BE06962.
active euthanasia; allowing to die; consequences; decision making; *deontological ethics; *extraordinary treatment; handicapped; killing; *moral obligations; *newborns; *normative ethics; personhood; prognosis; *quality of life; Roman Catholic ethics; *selection for treatment; social interaction; *spina bifida; teleological ethics; utilitarianism; *value of life; values; Fletcher, Joseph; Freeman, John; *Lorber, John; *McCormick, Richard A.

SELECTION FOR TREATMENT / VALUE OF LIFE / SPINA BIFIDA

Reich, Warren T. Quality of life and defective newborn children: an ethical analysis. *In* Swinyard, Chester A., ed. Decision Making and the Defective Newborn. Proceedings of a Conference on Spina Bifida and Ethics. Springfield, Ill.: Charles C. Thomas, 1978. p. 489-511. 35 refs. 4 fn. ISBN 0-398-03662-4. BE06962.
active euthanasia; allowing to die; consequences; decision making; *deontological ethics; *extraordinary treatment; handicapped; killing; *moral obligations; *newborns; *normative ethics; personhood; prognosis; *quality of life; Roman Catholic ethics; *selection for treatment; social interaction; *spina bifida; teleological ethics; utilitarianism; *value of life; values; Fletcher, Joseph; Freeman, John; *Lorber, John; *McCormick, Richard A.

SELECTION FOR TREATMENT *See under*

ORGAN TRANSPLANTATION / SELECTION FOR TREATMENT

SELECTION OF SUBJECTS *See under*

HUMAN EXPERIMENTATION / SELECTION OF SUBJECTS

HUMAN EXPERIMENTATION / SELECTION OF SUBJECTS / ALCOHOL ABUSE

SELECTIVE ABORTION

Callahan, Daniel. Abortion and medical ethics. *In* Barber, Bernard, ed. Medical Ethics and Social Change. Philadelphia: American Academy of Political and Social Science, 1978. p. 116-127. Annals of the American Academy of Political and Social Science, Vol. 437, May 1978. 7 fn. ISBN 0-87761-226-9. BE07004.
*abortion; amniocentesis; fetuses; genetic defects; human experimentation; prenatal diagnosis; rights; *selective abortion; sex determination; Supreme Court decisions; viability

SELECTIVE ABORTION / ETHICS

Smith, David H. The abortion of defective fetuses: some moral considerations. *In* Smith, David H., ed. No

Rush to Judgement: Essays on Medical Ethics. Bloomington, Ind.: Indiana University, Poynter Center, 1977. p. 18-43. 50 fn. BE05794.
*abortion; Christian ethics; conscience; contraception; cultural pluralism; decision making; *ethics; fetal development; fetuses; genetic defects; love; *moral obligations; moral policy; normative ethics; personhood; pregnant women; rights; *selective abortion; society; theology; viability

SELECTIVE ABORTION / MORAL OBLIGATIONS / GENETIC DEFECTS

Doyle, Carolyn T. Amniocentesis and the defective child. *Biological Psychiatry* 12(5): 611-612, Oct 1977. 3 refs. Editorial. BE06346.
amniocentesis; attitudes; *genetic defects; *moral obligations; *parents; reproduction; *selective abortion; social impact

SELECTIVE ABORTION / UNWANTED CHILDREN / TORTS

Gursky, Sandra. Birth despite vasectomy and abortion held not a 'wrong'. *Journal of Legal Medicine* 5(7): 29-31, Jul 1977. 10 refs. BE05726.
children; compensation; genetic defects; legal aspects; legal rights; males; malpractice; negligence; parents; physicians; *selective abortion; *torts; *unwanted children; value of life; *voluntary sterilization; wrongful life; Speck v. Feingold and Schwartz

SELECTIVE ABORTION *See under*

AMNIOCENTESIS / SELECTIVE ABORTION / CHRISTIAN ETHICS

AMNIOCENTESIS / SELECTIVE ABORTION / GENETIC DEFECTS

PRENATAL DIAGNOSIS / SELECTIVE ABORTION

PRENATAL DIAGNOSIS / SELECTIVE ABORTION / GENETIC DEFECTS

SELF CONCEPT *See under*

DETERMINATION OF DEATH / PERSONHOOD / SELF CONCEPT

SELF DETERMINATION / BURN PATIENTS

Imbus, Sharon H.; Zawacki, Bruce E. Autonomy for burned patients when survival is unprecedented. *New England Journal of Medicine* 297(6): 308-311, 11 Aug 1977. 19 refs. BE05759.
*burn patients; *communication; *critically ill; decision making; disclosure; family members; *medical staff; mortality; *patient care; patient participation; prognosis; *self determination; terminally ill

Pratt, David S., et al. Autonomy for severely burned patients. *New England Journal of Medicine* 297(21): 1182-1183, 24 Nov 1977. Letter. BE06097.
*allowing to die; *burn patients; costs and benefits;

BE = bioethics accession number   fn. = footnotes   refs. = references

critically ill; *extraordinary treatment; prognosis; *self determination; *terminally ill; withholding treatment

SELF DETERMINATION / LAETRILE / CANCER

Clinite, Barbara J. Freedom of choice in medical treatment: reconsidering the efficacy requirement of the FDCA. *Loyola University Law Journal* 9(1): 205-226, Fall 1977. 150 fn. BE07033.
*cancer; *drugs; due process; federal government; government agencies; human experimentation; *government regulation; patient care; privacy; *self determination; standards; state interest; *Food and Drug Administration; *Laetrile

SELF DETERMINATION / MENTALLY ILL

Monahan, John. John Stuart Mill on the liberty of the mentally ill: a historical note. *American Journal of Psychiatry* 134(12): 1428-1429, Dec 1977. 4 refs. BE06145.
*involuntary commitment; *mentally ill; *self determination; Mill, John Stuart

SELF DETERMINATION / PHYSICIAN PATIENT RELATIONSHIP

Stalley, R.F. Self-determination. *Journal of Medical Ethics* 4(1): 40-41, Mar 1978. 6 refs. BE06931.
coercion; human experimentation; patient care; *physician patient relationship; *self determination

SELF DETERMINATION / PSYCHOLOGY

Hyatt, I. Ralph. Psychology's concern with freedom. *Intellect* 105(2377): 99-100, Sep-Oct 1976. 4 refs. BE06642.
behavior control; behavioral genetics; genetic intervention; *psychology; *self determination

SELF DETERMINATION / TERMINALLY ILL

Padgett, Jack F. Is there a right to die? *Bioethics Digest* 2(4): 1-4, Aug 1977. 34 fn. BE05763.
*allowing to die; legal rights; patients' rights; physician's role; *self determination; *terminally ill; *treatment refusal

SELF DETERMINATION *See under*

ALLOWING TO DIE / SELF DETERMINATION
BEHAVIOR CONTROL / SELF DETERMINATION
POPULATION CONTROL / SELF DETERMINATION
PSYCHOSURGERY / SELF DETERMINATION
TERMINAL CARE / SELF DETERMINATION
TREATMENT REFUSAL / SELF DETERMINATION
TREATMENT REFUSAL / SELF DETERMINATION / LEGAL RIGHTS

SELF INDUCED ILLNESS *See under*

HEALTH CARE / MORAL OBLIGATIONS / SELF INDUCED ILLNESS

SELF REGULATION *See under*

HUMAN EXPERIMENTATION / SELF REGULATION
RECOMBINANT DNA RESEARCH / SELF REGULATION

SEX PRESELECTION / RISKS AND BENEFITS

Rorvik, David M.; Shettles, Landrum B. But is it moral? The "bioethical" debate. *In their* Choose Your Baby's Sex: The One Sex-Selection Method that Works. New York: Dodd, Mead, 1977. p. 136-149. ISBN 0-396-07356-5. BE05704.
attitudes; females; males; morality; parents; *risks and benefits; selective abortion; sex linked defects; *sex preselection; *sex ratio; *social impact

SEX PRESELECTION / SOCIAL IMPACT

Whelan, Elizabeth M. Sex control methods on the horizon: do we really want them? *In her* Boy or Girl? The Sex Selection Technique That Makes All Others Obsolete. New York: Bobbs-Merrill, 1977. p. 96-104. ISBN 0-672-52276-4. BE06174.
birth order; selective abortion; *sex preselection; sex ratio; *social impact; X bearing sperm; Y bearing sperm

SEXUALITY *See under*

BEHAVIORAL RESEARCH / ADOLESCENTS / SEXUALITY
BEHAVIORAL RESEARCH / PARENTAL CONSENT / SEXUALITY
HUMAN EXPERIMENTATION / CONFIDENTIALITY / SEXUALITY
HUMAN EXPERIMENTATION / INFORMED CONSENT / SEXUALITY
HUMAN EXPERIMENTATION / SEXUALITY
MORAL OBLIGATIONS / PHYSICIAN PATIENT RELATIONSHIP / SEXUALITY
PHYSICIAN PATIENT RELATIONSHIP / PSYCHOTHERAPY / SEXUALITY

SICKLE CELL ANEMIA *See under*

GENETIC SCREENING / NEWBORNS / SICKLE CELL ANEMIA
GENETIC SCREENING / PUBLIC POLICY / SICKLE CELL ANEMIA

SOCIAL ADJUSTMENT *See under*

ORGAN DONORS / KIDNEYS / SOCIAL ADJUSTMENT

SOCIAL CONTROL *See also* GOVERNMENT REGULATION, JUDICIAL ACTION, LEGISLATION,

---

BE = bioethics accession number   fn. = footnotes   refs. = references

SELF REGULATION

SOCIAL CONTROL / MEDICINE

Hellegers, André E. Biologic origins of bioethical problems. *In* Wynn, Ralph M., ed. Obstetrics and Gynecology Annual. Volume 6, 1977. New York: Appleton-Century-Crofts, 1977. p. 1-9. ISBN 0-8385-7181-6. BE06570.
    *costs and benefits; deontological ethics; *health; health care delivery; *medicine; obligations to society; physician patient relationship; physicians; psychiatry; politics; quality of life; *resource allocation; *social control; teleological ethics; value of life

Szasz, Thomas S.; Kurtz, Paul. Ethics of Medicine and Psychiatry. [Videorecording]. Amherst, N.Y.: American Humanist Association, 1977. 1 cassette; 29 min.; sound; color; 3/4 in. Ethics in America Series presented by the American Humanist Association. Produced at WNED-TV, Buffalo, New York. BE07040.
    allowing to die; *economics; drugs; federal government; financial support; government regulation; health; health care; health insurance; hospitals; human rights; involuntary commitment; malpractice; *medicine; mentally ill; patients; physician patient relationship; physician's role; physicians; politics; psychiatry; punishment; self determination; self induced illness; *social control

SOCIAL CONTROL / PREVENTIVE MEDICINE / COSTS AND BENEFITS

McCullough, Laurence B. Some additional bioethical questions related to hepatitis B antigen. *American Journal of Clinical Pathology* 68(3): 340-342, Sep 1977. 2 refs. BE06335.
    carriers; common good; communicable diseases; *costs and benefits; decision making; freedom; health personnel; *hepatitis; *human rights; justice; *preventive medicine; risks and benefits; self determination; *social control; values; Rawls, John

SOCIAL CONTROL / SCIENCE

Graham, Loren R. Concerns about science and attempts to regulate inquiry. *Daedalus* 107(2): 1-21, Spring 1978. 18 fn. 3E06934.
    biomedical technologies; costs and benefits; *government regulation; human experimentation; public opinion; recombinant DNA research; resource allocation; *science; *social control; *social impact; wedge argument

Morison, Robert S. Introduction. *Daedalus* 107(2): vii-xvi, Spring 1978. 4 fn. BE06933.
    costs and benefits; investigators; public opinion; *science; self regulation; *social control; social impact

SOCIAL CONTROL / SCIENCE / RISKS AND BENEFITS

Bok, Sissela. Freedom and risk. *Daedalus* 107(2): 115-127, Spring 1978. 19 fn. BE06936.
    human experimentation; *risks and benefits; *science; self regulation; *social control; standards

SOCIAL CONTROL *See under*
    BIOMEDICAL RESEARCH / SOCIAL CONTROL
    BIOMEDICAL RESEARCH / SOCIAL CONTROL / HEALTH HAZARDS
    BIOMEDICAL RESEARCH / SOCIAL CONTROL / RISKS AND BENEFITS
    BIOMEDICAL RESEARCH / SOCIAL CONTROL / SOCIAL IMPACT
    HUMAN EXPERIMENTATION / SOCIAL CONTROL / INTERNATIONAL ASPECTS
    HUMAN EXPERIMENTATION / SOCIAL CONTROL / SOCIAL IMPACT
    LIFE EXTENSION / SOCIAL CONTROL / SOCIAL IMPACT
    PSYCHOSURGERY / SOCIAL CONTROL
    RECOMBINANT DNA RESEARCH / SOCIAL CONTROL
    RECOMBINANT DNA RESEARCH / SOCIAL CONTROL / ATTITUDES
    RECOMBINANT DNA RESEARCH / SOCIAL CONTROL / CAMBRIDGE
    RECOMBINANT DNA RESEARCH / SOCIAL CONTROL / GREAT BRITAIN
    RECOMBINANT DNA RESEARCH / SOCIAL CONTROL / HEALTH HAZARDS
    RECOMBINANT DNA RESEARCH / SOCIAL CONTROL / INTERNATIONAL ASPECTS
    RECOMBINANT DNA RESEARCH / SOCIAL CONTROL / RISKS AND BENEFITS

SOCIAL IMPACT *See under*
    ABORTION / BLACKS / SOCIAL IMPACT
    ABORTION / FINANCIAL SUPPORT / SOCIAL IMPACT
    ABORTION / LAW ENFORCEMENT / SOCIAL IMPACT
    ABORTION / SOCIAL IMPACT / INDIA
    ABORTION / SUPREME COURT DECISIONS / SOCIAL IMPACT
    ALLOWING TO DIE / LEGISLATION / SOCIAL IMPACT
    BEHAVIORAL RESEARCH / GOVERNMENT REGULATION / SOCIAL IMPACT
    BEHAVIORAL RESEARCH / INFORMED CONSENT / SOCIAL IMPACT
    BIOLOGY / SOCIAL IMPACT
    BIOMEDICAL RESEARCH / SOCIAL CONTROL / SOCIAL IMPACT
    BIOMEDICAL TECHNOLOGIES / INDUSTRY / SOCIAL IMPACT

BE = bioethics accession number   fn. = footnotes   refs. = references

SOCIAL IMPACT *See under (cont'd.)*
  BIOMEDICAL TECHNOLOGIES / SOCIAL IMPACT
  CARCINOGENS / MASS MEDIA / SOCIAL IMPACT
  CONFIDENTIALITY / MEDICAL RECORDS / SOCIAL IMPACT
  DEINSTITUTIONALIZED PERSONS / MENTALLY ILL / SOCIAL IMPACT
  GENETIC INTERVENTION / SOCIAL IMPACT
  GENETIC SCREENING / SOCIAL IMPACT
  HEALTH CARE DELIVERY / GOVERNMENT REGULATION / SOCIAL IMPACT
  HUMAN EXPERIMENTATION / SOCIAL CONTROL / SOCIAL IMPACT
  LIFE EXTENSION / SOCIAL CONTROL / SOCIAL IMPACT
  MEDICAL ETHICS / SOCIAL IMPACT
  MEDICAL RECORDS / FREEDOM OF INFORMATION ACT / SOCIAL IMPACT
  POPULATION GROWTH / DEVELOPING COUNTRIES / SOCIAL IMPACT
  RECOMBINANT DNA RESEARCH / ECOLOGY / SOCIAL IMPACT
  RECOMBINANT DNA RESEARCH / RISKS AND BENEFITS / SOCIAL IMPACT
  RECOMBINANT DNA RESEARCH / SOCIAL IMPACT
  REPRODUCTIVE TECHNOLOGIES / SOCIAL IMPACT
  SEX PRESELECTION / SOCIAL IMPACT

SOCIAL INTERACTION *See under*
  PERSONHOOD / SOCIAL INTERACTION

SOCIAL SCIENCES / LEGAL RIGHTS

Dunham, H. Warren. Psychiatry, sociology and civil liberties. *Man and Medicine: Journal of Values and Ethics in Health Care* 2(4): 263-278, Summer 1977. Commentary by Martin D. Hanlon, p. 275-278. 3 refs. 18 fn. BE06254.
  behavior disorders; behavioral research; coercion; common good; conflict of interest; health care; human rights; institutionalized persons; involuntary commitment; *legal rights; mentally ill; physician's role; political systems; psychiatric diagnosis; *psychiatry; self determination; social impact; social problems; *social sciences; sociology of medicine; stigmatization

SOCIAL SCIENCES *See under*
  ETHICS / SOCIAL SCIENCES / EDUCATION

SOCIAL WORTH *See under*
  TERMINAL CARE / TERMINALLY ILL / SOCIAL WORTH

SOCIETY *See under*
  AGED / SOCIETY / ATTITUDES

SOCIOBIOLOGY

Gould, Stephen J. Biological potential vs. biological determinism. *Natural History* 85(5): 12+, May 1976. BE06204.
  altruism; behavioral genetics; human characteristics; *sociobiology

Holton, Gerald. Sociobiology: the new synthesis? *Newsletter on Science, Technology, and Human Values* No. 21: 28-43, Oct 1977. 31 fn. BE06246.
  ancient history; attitudes; biology; evolution; historical aspects; methods; philosophy; science; social impact; social sciences; *sociobiology; Nineteenth Century; Twentieth Century; Wilson, Edward O.

Science for the People. Ann Arbor Editorial Collective. Biology as a Social Weapon. Minneapolis: Burgess Publishing Company, 1977. 154 p. 339 refs. ISBN 0-8087-4534-4. BE06157.
  aggression; *behavioral genetics; *discrimination; ecology; eugenics; evaluation; females; historical aspects; human characteristics; human equality; *intelligence; males; minority groups; politics; population control; sexuality; social determinants; *sociobiology; violence; *XYY karyotype

Science for the People. Sociobiology Study Group. Sociobiology—a new biological determinism. *In* Science for the People. Ann Arbor Editorial Collective. Biology as a Social Weapon. Minneapolis: Burgess Publishing Company, 1977. p. 133-149. 33 refs. 2 fn. ISBN 0-8087-4534-4. BE06159.
  behavioral genetics; evolution; human characteristics; natural selection; *sociobiology; Wilson, Edward O.

Thorup, Oscar A., et al. Sociobiology: Are There Areas of Forbidden Knowledge? [Videorecording]. Charlottesville, Va.: University of Virginia Medical Center, Health Sciences Library, 1978. 1 cassette; 60 min.; sound; black and white; 3/4 in. Tape of the Medical Center Hour, 1 Mar 1978. Panel discussion moderated by Oscar A. Thorup. BE07056.
  behavioral genetics; *biomedical research; discrimination; ethics committees; government regulation; informal social control; intelligence; legal aspects; risks and benefits; *social control; *sociobiology; Science for the People; Wilson, Edward O.

SOCIOBIOLOGY / ALTRUISM

Stent, Gunther S. You can take the ethics out of altruism but you can't take the altruism out of ethics. *Hastings Center Report* 7(6): 33-36, Dec 1977. BE06151.
  *altruism; evolution; cultural evolution; genetics; natural selection; *sociobiology; values; *Dawkins, Richard

SOCIOBIOLOGY / BEHAVIORAL GENETICS

Currier, Richard. Sociobiology: the new heresy.

BE = bioethics accession number   fn. = footnotes   refs. = references

Human Behavior 5(11): 16-22, Nov 1976. BE06206.
> attitudes; *behavioral genetics; discrimination; intelligence; journalism; *sociobiology; Wilson, Edward O.

SOCIOBIOLOGY / HUMAN CHARACTERISTICS

Simon, Michael A. Sociobiology: the Aesop's fables of science. *Sciences* 18(2): 18-21+, Feb 1978. BE06797.
> behavioral research; genetics; *human characteristics; social determinants; social sciences; *sociobiology

SOCIOBIOLOGY / MORALITY

Lewin, Roger. Biological limits to morality. *New Scientist* 76(1082): 694-696, 15 Dec 1977. BE06255.
> *altruism; biology; evolution; genetics; human characteristics; *morality; natural selection; philosophy; *sociobiology

SOCIOBIOLOGY / POLITICS

Cowen, Robert C. Who oppresses scientists? *Christian Science Monitor (Eastern Edition),* 19 Oct 1977, p. 25. BE06457.
> freedom; *politics; *rights; science; *sociobiology

SOCIOECONOMIC FACTORS *See under*

ABORTION / SOCIOECONOMIC FACTORS
POPULATION CONTROL / DEVELOPING COUNTRIES / SOCIOECONOMIC FACTORS

SOCIOLOGY OF MEDICINE

Hortenstine, John C. Value change and medicine in today's society. *Virginia Medical* 104(3): 169-173, Mar 1977. 11 refs. BE05944.
> goals; health; health personnel; medical education; medicine; patients; professional ethics; social interaction; *sociology of medicine

SOUTH CAROLINA *See under*

ALLOWING TO DIE / LEGISLATION / SOUTH CAROLINA

SOUTHAMPTON UNIVERSITY MEDICAL SCHOOL *See under*

MEDICAL ETHICS / MEDICAL EDUCATION / SOUTHAMPTON UNIVERSITY MEDICAL SCHOOL

SPINA BIFIDA *See under*

ACTIVE EUTHANASIA / NEWBORNS / SPINA BIFIDA
ALLOWING TO DIE / NEWBORNS / SPINA BIFIDA
ALLOWING TO DIE / SPINA BIFIDA / DECISION MAKING

SPINA BIFIDA *See under (cont'd.)*

ALLOWING TO DIE / SPINA BIFIDA / LEGAL LIABILITY
EXTRAORDINARY TREATMENT / NEWBORNS / SPINA BIFIDA
GENETIC SCREENING / SPINA BIFIDA / COSTS AND BENEFITS
MASS SCREENING / PREGNANT WOMEN / SPINA BIFIDA
MASS SCREENING / SPINA BIFIDA
MASS SCREENING / SPINA BIFIDA / COSTS AND BENEFITS
NEWBORNS / SPINA BIFIDA / ATTITUDES
NEWBORNS / SPINA BIFIDA / DECISION MAKING
NEWBORNS / SPINA BIFIDA / PHYSICIAN'S ROLE
PARENTAL CONSENT / SPINA BIFIDA / LEGAL ASPECTS
PATIENT CARE / SPINA BIFIDA / DECISION MAKING
PATIENT CARE / SPINA BIFIDA / JEWISH ETHICS
PATIENT CARE / SPINA BIFIDA / LEGAL ASPECTS
PATIENT CARE / SPINA BIFIDA / PHYSICIAN'S ROLE
PRENATAL DIAGNOSIS / SPINA BIFIDA
PRENATAL DIAGNOSIS / SPINA BIFIDA / GREAT BRITAIN
SELECTION FOR TREATMENT / MORAL POLICY / SPINA BIFIDA
SELECTION FOR TREATMENT / NEWBORNS / SPINA BIFIDA
SELECTION FOR TREATMENT / QUALITY OF LIFE / SPINA BIFIDA
SELECTION FOR TREATMENT / SPINA BIFIDA / EVALUATION
SELECTION FOR TREATMENT / SPINA BIFIDA / NORMATIVE ETHICS
SELECTION FOR TREATMENT / VALUE OF LIFE / SPINA BIFIDA
TREATMENT REFUSAL / PARENTAL CONSENT / SPINA BIFIDA
TREATMENT REFUSAL / SPINA BIFIDA / ALTERNATIVES
TREATMENT REFUSAL / SPINA BIFIDA / LEGAL ASPECTS

SPOUSAL CONSENT *See under*

ABORTION / SPOUSAL CONSENT / LEGAL ASPECTS
ABORTION / SPOUSAL CONSENT / LEGAL RIGHTS
ABORTION / SPOUSAL CONSENT / PLANNED PARENTHOOD OF CENTRAL MISSOURI V. DANFORTH
ABORTION / SPOUSAL CONSENT / SUPREME COURT DECISIONS

BE = bioethics accession number   fn. = footnotes   refs. = references

SPOUSAL CONSENT *See under (cont'd.)*
VOLUNTARY STERILIZATION / SPOUSAL CONSENT / LEGAL ASPECTS

STANDARDS *See under*
BRAIN DEATH / STANDARDS
HUMAN EXPERIMENTATION / CHILDREN / STANDARDS
INFORMED CONSENT / STANDARDS
INVOLUNTARY COMMITMENT / STANDARDS / PENNSYLVANIA
PATIENT ACCESS / MEDICAL RECORDS / STANDARDS
PROFESSIONAL ORGANIZATIONS / JUDICIAL ACTION / STANDARDS
RECOMBINANT DNA RESEARCH / CONTAINMENT / STANDARDS

STATE GOVERNMENT *See under*
GENETIC COUNSELING / STATE GOVERNMENT

STATE INTEREST *See under*
TREATMENT REFUSAL / STATE INTEREST

STATE UNIVERSITY OF NEW YORK AT ALBANY *See under*
HUMAN EXPERIMENTATION / STATE UNIVERSITY OF NEW YORK AT ALBANY

STATISTICS *See under*
ABORTION / STATISTICS
HUMAN EXPERIMENTATION / STATISTICS / METHODS

STERILIZATION *See also* INVOLUNTARY STERILIZATION, VOLUNTARY STERILIZATION

STERILIZATION

Anderson, Norman. Issues of Life and Death: Abortion, Birth Control, Capital Punishment, Euthanasia. Downers Grove, Ill.: InterVarsity Press, 1977. 130 p. 189 fn. ISBN 0-87784-721-5. BE05787.
AID; *abortion; *artificial insemination; *capital punishment; Christian ethics; *contraception; DNA therapy; *euthanasia; genetic intervention; justifiable killing; organ transplantation; personhood; scriptural interpretation; *sterilization; theology; *value of life

STERILIZATION / CHRISTIAN ETHICS

Anderson, Norman. Birth control, sterilisation and abortion. *In his* Issues of Life and Death: Abortion, Birth Control, Capital Punishment, Euthanasia. Downers Grove, Ill.: InterVarsity Press, 1977. p. 58-84. 46 fn. ISBN 0-87784-721-5. BE05790.
*abortion; *Christian ethics; *contraception; eugenics; legal aspects; methods; *personhood; popu-

lation control; reproduction; selective abortion; sexuality; *sterilization; therapeutic abortion; voluntary sterilization

STERILIZATION / COERCION

Largey, Gale. Reversible sterilization. *Society* 14(5): 57-59, Jul/Aug 1977. 3 refs. BE05826.
adolescents; *coercion; human rights; indigents; mentally ill; mentally retarded; *minority groups; prisoners; *sterilization

STERILIZATION / GOVERNMENT REGULATION

DHEW proposes 30-day waiting period for sterilizations; no funds for under 21s, contraceptive hysterectomies. *Family Planning Perspectives* 10(1): 39, Jan/Feb 1978. BE06910.
adolescents; ethics committees; *federal government; *government regulation; *informed consent; mentally handicapped; *sterilization; Department of Health, Education, and Welfare

Right about sterilization. *New York Times,* 4 Dec 1977, sect. 4, p. 20. Editorial. BE06416.
adolescents; federal government; *government regulation; informed consent; mentally retarded; *sterilization

U.S. Public Health Service. Proposed restrictions applicable to sterilizations funded by the Department of Health, Education, and Welfare. *Federal Register* 42(239): 62718-62732, 13 Dec 1977. BE06472.
adolescents; adults; competence; disclosure; ethics committees; family planning; *federal government; financial support; *government regulation; *informed consent; institutionalized persons; involuntary sterilization; *mentally retarded; *sterilization; voluntary sterilization

STERILIZATION / ICELAND

Iceland. Abortion, sterilization, and sex education. *International Digest of Health Legislation* 28(3): 614-620, 1977. Law of 22 May 1975. BE06464.
*abortion; counseling; legal liability; physicians; sexuality; socioeconomic factors; standards; *sterilization; therapeutic abortion; *Iceland

STERILIZATION / INFORMED CONSENT / LEGAL ASPECTS

Sharpe, Gilbert. Consent and sterilization. *Canadian Medical Association Journal* 118(5): 591-593, 4 Mar 1978. BE06930.
adolescents; disclosure; contraception; *legal aspects; *informed consent; mentally handicapped; *spousal consent; *sterilization; *voluntary sterilization

STERILIZATION / LEGAL ASPECTS

Morris, Norman; Arthure, Humphrey. Medico-legal

aspects of sterilization. *In their* Sterilization as a Means of Birth Control in Men and Women. London: Peter Owen, 1976. p. 111-121. 5 refs. ISBN 0-7206-0363-3. BE06662.
informed consent; involuntary sterilization; *legal aspects; married persons; mentally retarded; negative eugenics; *sterilization; unwanted children; voluntary sterilization

STERILIZATION / LEGISLATION / HISTORICAL ASPECTS

Simms, Madeleine. A sterilisation debate. *New Humanist* 92(4): 141-143, Nov-Dec 1976. BE06645.
*historical aspects; involuntary sterilization; *legislation; males; mentally retarded; negative eugenics; *sterilization; voluntary sterilization; *Great Britain

STERILIZATION / MENTALLY RETARDED / LEGAL ASPECTS

Henry, Diane. Parents of 3 retarded girls fight hospital refusal to sterilize them. *New York Times,* 2 Oct 1977, sect. 1, p. 1+. BE06401.
*adolescents; children; hospitals; involuntary sterilization; *legal aspects; legal liability; *mentally retarded; parental consent; *sterilization

STERILIZATION / MINORITY GROUPS

Largey, Gale. Reversible sterilization. *Society* 14(5): 57-59, Jul/Aug 1977. 3 refs. BE05826.
adolescents; *coercion; human rights; indigents; mentally ill; mentally retarded; *minority groups; prisoners; *sterilization

STERILIZATION / NEW ZEALAND

Contraception, sterilisation and abortion in New Zealand. *New Zealand Medical Journal* 85(588): 441-445, 25 May 1977. Recommendations from the Report of the Royal Commission of Inquiry into Contraception, Sterilisation and Abortion (Government Printer, Wellington). BE05601.
*abortion; adolescents; alternatives; community services; conscience; *contraception; counseling; education; employment; ethics committees; family planning; health personnel; information dissemination; *legal aspects; mentally handicapped; pregnant women; *public policy; review committees; sexuality; social workers; *sterilization; therapeutic abortion; *New Zealand

STERILIZATION / PHYSICIANS / LEGAL RIGHTS

Curran, William J. The freedom of medical practice, sterilization, and economic medical philosophy. *New England Journal of Medicine* 298(1): 32-33, 5 Jan 1978. 4 refs. BE06727.
*coercion; discrimination; federal government; freedom; financial support; indigents; informed

consent; *involuntary sterilization; legal aspects; *legal rights; *physicians; pregnant women; socioeconomic factors; *sterilization; Medicaid; *Walker v. Pierce

STRIKES   *See under*

MORAL OBLIGATIONS / PHYSICIANS / STRIKES

NURSES / STRIKES / MORALITY

PATIENT CARE / HEALTH PERSONNEL / STRIKES

PHYSICIANS / JUSTICE / STRIKES

STUDENTS   *See under*

EUTHANASIA / STUDENTS / ATTITUDES

MEDICAL ETHICS / STUDENTS / EVALUATION

SUICIDE

Lebacqz, Karen; Engelhardt, H. Tristram. Suicide. *In* Horan, Dennis J.; Mall, David, eds. Death, Dying, and Euthanasia. Washington: University Publications of America, 1977. p. 669-705. 46 refs. 48 fn. ISBN 0-89093-139-9. BE05689.
Christian ethics; competence; consequences; deontological ethics; family; freedom; injuries; intention; justice; killing; *moral obligations; *self determination; *suicide; teleological ethics; value of life; values; voluntary euthanasia; wedge argument

May, William F. The right to die and the obligation to care: allowing to die, killing for mercy, and suicide. *In* McMullin, Ernan, ed. Death and Decision. Boulder, Colo.: Westview Press, 1978. p. 111-130. American Association for the Advancement of Science Selected Symposium 18. 13 fn. ISBN 0-89158-152-9. BE06710.
*allowing to die; euthanasia; hospitals; institutional policies; *rights; prolongation of life; self determination; *suicide; terminal care; terminally ill; theology; withholding treatment

Weir, Robert F., ed. Ethical Issues in Death and Dying. New York: Columbia University Press, 1977. 405 p. References and footnotes included. ISBN 0-231-04307-4. BE06696.
*allowing to die; brain death; cancer; congenital defects; *determination of death; diagnosis; *disclosure; *euthanasia; legal aspects; newborns; physicians; *suicide; *terminally ill; treatment refusal

SUICIDE / LEGAL ASPECTS

Sandak, Lawrence R. Suicide and the compulsion of lifesaving medical procedures: an analysis of the refusal of treatment cases. *Brooklyn Law Review* 44(2): 285-316, Winter 1978. 140 fn. BE06754.
allowing to die; blood transfusions; intention; Jehovah's Witnesses; judicial action; *legal aspects; *legal rights; privacy; *suicide; Supreme Court de-

---

BE = bioethics accession number   fn. = footnotes   refs. = references

cisions; *treatment refusal; withholding treatment; Quinlan, Karen

SUICIDE / MORALITY

Beauchamp, Tom L.; Perlin, Seymour, eds. Ethical Issues in Death and Dying. Englewood Cliffs, N.J.: Prentice-Hall, 1978. 368 p. References and footnotes included. ISBN 0-13-290114-5. BE06878.
active euthanasia; allowing to die; brain death; coma; decision making; *determination of death; disclosure; *euthanasia; informed consent; killing; legal aspects; legislation; living wills; moral obligations; *morality; newborns; *patients' rights; *suicide; terminal care; *terminally ill; theology; treatment refusal; *value of life; values; withholding treatment; Quinlan, Karen

Grisez, Germain. Suicide and euthanasia. *In* Horan, Dennis J.; Mall, David, eds. Death, Dying, and Euthanasia. Washington: University Publications of America, 1977. p. 742-817. 57 fn. ISBN 0-89093-139-9. BE05690.
abortion; active euthanasia; extraordinary treatment; intention; killing; legal aspects; moral obligations; *morality; physician's role; self determination; *suicide; *teleological ethics; utilitarianism; value of life; *voluntary euthanasia; wedge argument

SUICIDE / PSYCHIATRY / LEGAL LIABILITY

California. Court of Appeal, First District, Division 2. Bellah v. Greenson. 5 Oct 1977. *California Reporter* 141: 92-95. 2 fn. BE07117.
*confidentiality; dangerousness; disclosure; *legal liability; negligence; parents; *psychiatry; *suicide; torts

SUICIDE / TELEOLOGICAL ETHICS

Grisez, Germain. Suicide and euthanasia. *In* Horan, Dennis J.; Mall, David, eds. Death, Dying, and Euthanasia. Washington: University Publications of America, 1977. p. 742-817. 57 fn. ISBN 0-89093-139-9. BE05690.
abortion; active euthanasia; extraordinary treatment; intention; killing; legal aspects; moral obligations; *morality; physician's role; self determination; *suicide; *teleological ethics; utilitarianism; value of life; *voluntary euthanasia; wedge argument

SUPREME COURT DECISIONS   *See under*

ABORTION / EQUAL PROTECTION / SUPREME COURT DECISIONS

ABORTION / FINANCIAL SUPPORT / SUPREME COURT DECISIONS

SUPREME COURT DECISIONS   *See under*
(cont'd.)

ABORTION / INSTITUTIONAL POLICIES / SUPREME COURT DECISIONS

ABORTION / PARENTAL CONSENT / SUPREME COURT DECISIONS

ABORTION / PRIVACY / SUPREME COURT DECISIONS

ABORTION / SPOUSAL CONSENT / SUPREME COURT DECISIONS

ABORTION / SUPREME COURT DECISIONS

ABORTION / SUPREME COURT DECISIONS / PUBLIC OPINION

ABORTION / SUPREME COURT DECISIONS / SOCIAL IMPACT

PATIENT CARE / ABORTED FETUSES / SUPREME COURT DECISIONS

SURGERY   *See under*

INFORMED CONSENT / SURGERY / ALTERNATIVES

INFORMED CONSENT / SURGERY / RISKS AND BENEFITS

TREATMENT REFUSAL / SURGERY / JUDICIAL ACTION

SURVIVAL   *See under*

POPULATION CONTROL / SURVIVAL

SWEDEN   *See under*

HEALTH CARE DELIVERY / RESOURCE ALLOCATION / SWEDEN

RECOMBINANT DNA RESEARCH / SWEDEN

SWITZERLAND   *See under*

ALLOWING TO DIE / SWITZERLAND

ALLOWING TO DIE / TERMINALLY ILL / SWITZERLAND

RECOMBINANT DNA RESEARCH / SWITZERLAND

TARASOFF V. REGENTS

American Academy of Psychiatry and the Law. The Tarasoff Decision. [Sound recording]. San Francisco: Convention Cassettes, 22 Oct 1977. 2 cassettes; 60 min. Tapes No. 5 and 6. Taped during a convention of the American Academy of Psychiatry and the Law, San Francisco, 22-24 Oct 1977. BE07038.
*confidentiality; *dangerousness; decision making; *disclosure; expert testimony; involuntary commitment; judicial action; law; *legal aspects; legal liability; obligations to society; professional patient relationship; prognosis; *psychotherapy; *Tarasoff v. Regents

---

BE = bioethics accession number   fn. = footnotes   refs. = references

TARASOFF V. REGENTS *See under*
PRIVILEGED COMMUNICATION / TARASOFF V. REGENTS / LEGAL ASPECTS

TAY SACHS DISEASE *See under*
GENETIC COUNSELING / TAY SACHS DISEASE
GENETIC SCREENING / TAY SACHS DISEASE
WRONGFUL LIFE / TAY SACHS DISEASE

TEACHING METHODS *See under*
MEDICAL ETHICS / TEACHING METHODS

TECHNOLOGY ASSESSMENT *See under*
RECOMBINANT DNA RESEARCH / TECHNOLOGY ASSESSMENT / LAW

TELEOLOGICAL ETHICS *See under*
BEHAVIORAL RESEARCH / TELEOLOGICAL ETHICS
MEDICINE / TELEOLOGICAL ETHICS
SUICIDE / TELEOLOGICAL ETHICS

TENNESSEE *See under*
ALLOWING TO DIE / LEGISLATION / TENNESSEE
LIVING WILLS / LEGISLATION / TENNESSEE
TREATMENT REFUSAL / LEGISLATION / TENNESSEE

TERMINAL CARE *See also* EUTHANASIA

TERMINAL CARE

Coggan, Donald. On dying and dying well: extracts from the Edwin Stevens lecture. *Journal of Medical Ethics* 3(2): 57-60, Jun 1977. 13 fn. BE05643.
allowing to die; *Christian ethics; drugs; economics; extraordinary treatment; hospices; intention; *moral obligations; pain; physician patient relationship; physicians; *prolongation of life; *resource allocation; state medicine; *terminal care; terminally ill

Hayes, Donald M. Between Doctor and Patient. Valley Forge: Judson Press, 1977. 176 p. 67 fn. ISBN 0-8170-0742-3. BE06184.
abortion; allowing to die; cancer; case reports; community services; death; decision making; dehumanization; drug abuse; drugs; economics; employment; health care delivery; hospitals; leukemia; medical ethics; *patient care; *patient participation; *physician patient relationship; psychosurgery; quality of life; suicide; *terminal care; terminally ill; *values

Kastenbaum, Robert J. Dying—innovations in care. *In his* Death, Society, and Human Experience. St. Louis: C.V. Mosby, 1977. p. 217-240. 14 refs. ISBN 0-8016-2617-X. BE06185.
*communication; *death; *education; family mem-

bers; goals; *hospices; nursing education; pain; patients' rights; standards; suffering; *terminal care; *terminally ill

The last taboo. *Times (London),* 14 Dec 1976, p. 15. Editorial. BE06620.
allowing to die; attitudes to death; decision making; disclosure; pain; prolongation of life; psychoactive drugs; resource allocation; *terminal care; Church of England

May, William F. The right to die and the obligation to care: allowing to die, killing for mercy, and suicide. *In* Smith, David H., ed. No Rush to Judgement: Essays on Medical Ethics. Bloomington, Ind.: Indiana University, Poynter Center, 1977. p. 68-92. 13 fn. BE05796.
*active euthanasia; *allowing to die; Christian ethics; critically ill; *death; decision making; extraordinary treatment; hospitals; informed consent; institutional policies; moral obligations; obligations of society; physicians; prolongation of life; quality of life; resuscitation; suffering; suicide; *terminal care; terminally ill; theology; value of life; withholding treatment

Miller, Albert J.; Acri, Michael J. Medical profession and nursing experiences. *In their* Death: A Bibliographical Guide. Metuchen, N.J.: Scarecrow Press, 1977. p. 95-174. Bibliography. ISBN 0-8108-1025-5. BE06186.
*attitudes to death; cancer; communication; *death; determination of death; disclosure; family members; nurses; physicians; psychological stress; *terminal care; terminally ill

Searle, J.F.; Walter, Nicolas; Ellsworth, L.E. The doctor and the dying patient. *Times (London),* 16 Dec 1976, p. 15. Letter. BE06621.
euthanasia; hospices; public opinion; resource allocation; suffering; *terminal care; Church of England

Sell, Irene L., comp. Dying and Death: An Annotated Bibliography. New York: Tiresias Press, 1977. 144 p. Bibliography. ISBN 0-913292-36-2. BE06685.
audiovisual aids; *death; nurses; *terminal care

TERMINAL CARE / ADOLESCENTS

Hofmann, Adele D.; Becker, R.D.; Gabriel, H. Paul. The dying adolescent. *In their* The Hospitalized Adolescent: A Guide to Managing the Ill and Injured Youth. New York: Free Press, 1976. p. 196-210. ISBN 0-02-914790-5. BE07087.
*adolescents; cancer; diagnosis; goals; disclosure; medical staff; parents; physician's role; prognosis; psychological stress; *terminal care; *terminally ill

TERMINAL CARE / CHILDREN / COMMUNICATION

Vaughan, Victor C. The care of the child with a fatal illness. *In* Vaughan, Victor C.; McKay, R. James;

BE = bioethics accession number   fn. = footnotes   refs. = references

Nelson, Waldo E., eds. Nelson Textbook of Pediatrics. Tenth Edition. Philadelphia: W.B. Saunders, 1975. p. 142-144. 4 refs. ISBN 0-7216-9018-1. BE07078.
   autopsies; *children; *communication; disclosure; *parents; *physician's role; psychological stress; *terminal care; *terminally ill

TERMINAL CARE / CHILDREN / PHYSICIAN'S ROLE

Vaughan, Victor C. The care of the child with a fatal illness. In Vaughan, Victor C.; McKay, R. James; Nelson, Waldo E., eds. Nelson Textbook of Pediatrics. Tenth Edition. Philadelphia: W.B. Saunders, 1975. p. 142-144. 4 refs. ISBN 0-7216-9018-1. BE07078.
   autopsies; *children; *communication; disclosure; *parents; *physician's role; psychological stress; *terminal care; *terminally ill

TERMINAL CARE / CHURCH OF ENGLAND

Longley, Clifford. Wrong to prolong life at any cost, Dr. Coggan says. Times (London), 14 Dec 1976, p. 1. BE06619.
   disclosure; euthanasia; extraordinary treatment; organizational policies; prolongation of life; resource allocation; *terminal care; terminally ill; *Church of England

TERMINAL CARE / COMMUNICATION

Krant, Melvin J. Caring for the terminally ill: expectations of patient and family. Hospital Medical Staff 7(2): 1-6, Feb 1978. 9 refs. BE06782.
   cancer; *communication; disclosure; *family members; hospitals; medical staff; psychological stress; *terminal care; *terminally ill

Le Roux, Rose S. Communicating with the dying person. Nursing Forum 16(2): 144-155, 1977. 9 refs. BE06476.
   *communication; disclosure; family members; nurses; physicians; *terminal care; terminally ill

TERMINAL CARE / DECISION MAKING

Drinkwater, C.K.; Roberts, S.H. Cure or care in everyday practice. Journal of Medical Ethics 4(1): 12-17, Mar 1978. 20 refs. Case study. BE06920.
   alternatives; attitudes; *cancer; case reports; communication; costs and benefits; *decision making; diagnosis; drugs; family members; *economics; *patient care; patient participation; *physicians; prognosis; prolongation of life; quality of life; social adjustment; surgery; *terminal care; terminally ill; treatment refusal

TERMINAL CARE / DIGNITY

Vanderpool, Harold Y. The ethics of terminal care. Journal of the American Medical Association 239(9):

850-852, 27 Feb 1978. 8 refs. BE06800.
   allowing to die; attitudes to death; coma; *dignity; disclosure; extraordinary treatment; hospices; prolongation of life; psychological stress; *self concept; self determination; social interaction; *social worth; *terminal care; *terminally ill; values; voluntary euthanasia

TERMINAL CARE / HOSPICES

Hyer, Marjorie. Hospice care supported for terminally ill patients. Washington Post, 17 Apr 1977, p. B5. BE06014.
   *hospices; *terminal care

Rossman, Parker. Hospice: Creating New Models of Care for the Terminally Ill. New York: Association Press, 1977. 240 p. Bibliographies included. ISBN 0-8096-1926-1. BE06175.
   alternatives; cancer; *case reports; clergy; counseling; dehumanization; economics; family members; *hospices; hospitals; medical staff; nurses; pain; patients' rights; physician's role; program descriptions; psychological stress; self determination; social workers; *terminal care; *terminally ill; *New Haven Hospice; St. Christopher's Hospice

Will, George F. A good death. Newsweek 91(2): 72, 9 Jan 1978. BE06763.
   cancer; *hospices; euthanasia; pain; physician's role; *terminal care; terminally ill

Woodward, Kenneth L., et al. Living with dying. Newsweek 91(18): 52-56+, 1 May 1978. BE06972.
   *attitudes to death; family members; historical aspects; *hospices; psychological stress; *terminal care; terminally ill

TERMINAL CARE / HOSPICES / UNITED STATES

Libman, Joan. Death's door: hospice movement stresses family care for the terminally ill. Wall Street Journal, 27 Mar 1978, p. 1+. BE06818.
   allowing to die; cancer; case reports; economics; *family members; financial support; goals; health personnel; *hospices; hospitals; prolongation of life; *social interaction; *terminal care; terminally ill; *United States

TERMINAL CARE / HOSPITALS

Freedman, Theodore J.; Stuart, Allison J. Caring for the terminally ill: one hospital's self-assessment and solutions. Hospital Medical Staff 7(2): 7-12, Feb 1978. 8 refs. BE06778.
   attitudes; communication; disclosure; *hospitals; nurses; physicians; *terminal care; terminally ill

TERMINAL CARE / MEDICAL EDUCATION / LIVING WILLS

Bursztajn, Harold. The role of a training protocol in formulating patient instructions as to terminal care

choices. *Journal of Medical Education* 52(4): 347-348, Apr 1977. 10 refs. BE05927.
case studies; *living wills; decision making; *medical education; patient participation; *terminal care; values

TERMINAL CARE / NURSES

Gendrop, Sylvia C. The order: *No Code*. *Linacre Quarterly* 44(4): 312-319, Nov 1977. 15 fn. BE06100.
*allowing to die; communication; family members; *nurses; physician's role; *resuscitation; *terminal care; *terminally ill; *withholding treatment

Lamerton, Richard. The care of the dying: a specialty. *Nursing Times* 74(11): 436, 16 Mar 1978. 2 refs. BE06816.
cancer; communication; diagnosis; moral obligations; *nurses; pain; physician nurse relationship; *terminal care; terminally ill

TERMINAL CARE / PHYSICIAN PATIENT RELATIONSHIP

Roberts, Cecilia M. Doctors to the dying. *South Dakota Journal of Medicine* 29(11): 23-28, Nov 1976. 7 refs. 1 fn. BE06207.
*communication; *physician patient relationship; prognosis; *terminal care; terminally ill

TERMINAL CARE / PHYSICIAN'S ROLE

Nicholson, John. Doctor and the dying. *Times (London)*, 30 Dec 1976, p. 11. Letter. BE06623.
euthanasia; *physician's role; *terminal care

TERMINAL CARE / PHYSICIANS / ATTITUDES

Noyes, Russell; Jochimsen, Peter R.; Travis, Terry A. The changing attitudes of physicians toward prolonging life. *Journal of the American Geriatrics Society* 25(10): 470-474, Oct 1977. 5 refs. BE06351.
active euthanasia; *allowing to die; *attitudes; communication; disclosure; medical education; *physicians; students; *terminal care; terminally ill; withholding treatment

Roberts, F.J. The doctors' attitude to the dying patient. *New Zealand Medical Journal* 87(607): 181-184, 8 Mar 1978. 13 refs. BE06824.
*attitudes; attitudes to death; communication; diagnosis; death; *physicians; physician's role; prognosis; psychology; *terminal care; terminally ill

TERMINAL CARE / PSYCHOACTIVE DRUGS

Schmeck, Harold M. Addictive drugs as a way of easing death. *New York Times*, 30 Nov 1977, p. B2. BE06413.
*cancer; human experimentation; pain; *patient care; *psychoactive drugs; suffering; *terminal care; *terminally ill

TERMINAL CARE / SELF DETERMINATION

Cassell, Eric J. Autonomy and ethics in action. *New England Journal of Medicine* 297(6): 333-334, 11 Aug 1977. 4 refs. Editorial. BE05751.
physician patient relationship; *self determination; *terminal care; *terminally ill

TERMINAL CARE / TERMINALLY ILL / SOCIAL WORTH

Vanderpool, Harold Y. The ethics of terminal care. *Journal of the American Medical Association* 239(9): 850-852, 27 Feb 1978. 8 refs. BE06800.
allowing to die; attitudes to death; coma; *dignity; disclosure; extraordinary treatment; hospices; prolongation of life; psychological stress; *self concept; self determination; social interaction; *social worth; *terminal care; *terminally ill; values; voluntary euthanasia

TERMINALLY ILL / PATIENTS' RIGHTS

Beauchamp, Tom L.; Perlin, Seymour, eds. Ethical Issues in Death and Dying. Englewood Cliffs, N.J.: Prentice-Hall, 1978. 368 p. References and footnotes included. ISBN 0-13-290114-5. BE06878.
active euthanasia; allowing to die; brain death; coma; decision making; *determination of death; disclosure; *euthanasia; informed consent; killing; legal aspects; legislation; living wills; moral obligations; *morality; newborns; *patients' rights; *suicide; terminal care; *terminally ill; theology; treatment refusal; *value of life; values; withholding treatment; Quinlan, Karen

TERMINALLY ILL / PSYCHOLOGICAL STRESS

Problem?...to Think of Dying. [Motion picture]. Bloomington, Ind.: Indiana University, Audiovisual Center, 1974. 2 reels; 59 min.; sound; color; 16 mm. Presented by the Minnesota State College System; produced by Rita Shaw and directed by Denny Spence. Case study. BE07037.
*cancer; *case reports; clergy; communication; disclosure; family members; physicians; *psychological stress; *social adjustment; *terminally ill; *Caine, Lynn; *Kelly, Orville

TERMINALLY ILL *See under*

BE = bioethics accession number   fn. = footnotes   refs. = references

TERMINALLY ILL *See under (cont'd.)*

COMMUNICATION / TERMINALLY ILL / CANCER

DECEPTION / PHYSICIANS / TERMINALLY ILL

DISCLOSURE / TERMINALLY ILL

DISCLOSURE / TERMINALLY ILL / CANCER

EUTHANASIA / TERMINALLY ILL

EXTRAORDINARY TREATMENT / TERMINALLY ILL

INSTITUTIONAL POLICIES / HOSPITALS / TERMINALLY ILL

PATIENT CARE / TERMINALLY ILL

QUALITY OF LIFE / TERMINALLY ILL

RESOURCE ALLOCATION / TERMINALLY ILL

RESUSCITATION / TERMINALLY ILL

SELF DETERMINATION / TERMINALLY ILL

TERMINAL CARE / TERMINALLY ILL / SOCIAL WORTH

TREATMENT REFUSAL / TERMINALLY ILL

TREATMENT REFUSAL / TERMINALLY ILL / LEGAL RIGHTS

WITHHOLDING TREATMENT / TERMINALLY ILL

TEST TUBE FERTILIZATION *See* IN VITRO FERTILIZATION

TEXAS *See under*

ALLOWING TO DIE / LIVING WILLS / TEXAS

THEOLOGY *See under*

VALUE OF LIFE / THEOLOGY

THIRD PARTY CONSENT / MENTALLY RETARDED

Hellegers, André E. 'Incompetence' and consent. *Ob. Gyn. News* 12(21): 20-21, 1 Nov 1977. BE06114.
*children; competence; informed consent; *mentally ill; *mentally retarded; *parental consent; *third party consent

THIRD PARTY CONSENT / MENTALLY RETARDED / LEGAL RIGHTS

Turnbull, H. Rutherford. Individualizing the law for the mentally handicapped. *North Carolina Journal of Mental Health* 8(6): 1-7, Summer 1977. 3 refs. BE06312.
competence; *due process; *equal protection; involuntary sterilization; *legal rights; *mentally retarded; self determination; *third party consent

THIRD PARTY CONSENT *See under*

ALLOWING TO DIE / THIRD PARTY CONSENT

ALLOWING TO DIE / THIRD PARTY CONSENT / JUDICIAL ACTION

THIRD PARTY CONSENT *See under (cont'd.)*

ELECTROCONVULSIVE THERAPY / THIRD PARTY CONSENT

TORTS *See under*

IMMUNIZATION / POLIOMYELITIS / TORTS

PRECONCEPTION INJURIES / TORTS

RECOMBINANT DNA RESEARCH / TORTS / LEGAL LIABILITY

SELECTIVE ABORTION / UNWANTED CHILDREN / TORTS

VOLUNTARY STERILIZATION / UNWANTED CHILDREN / TORTS

TORTURE

Amnesty International. Danish Medical Group. Evidence of Torture: Studies by the Amnesty International Danish Medical Group. London: Amnesty International Publications, 1977. 39 p. 24 refs. ISBN 0-900058-56-0. BE05696.
*biomedical research; case reports; *diagnosis; human experimentation; injuries; international aspects; medicine; methods; moral obligations; physician's role; *torture; Amnesty International Danish Medical Group

Cooperman, Earl M. Doctors, torture and abuse of the doctor-patient relationship. *Canadian Medical Association Journal* 116(7): 707+, 9 Apr 1977. 21 refs. BE05548.
dissent; government regulation; physician patient relationship; *physician's role; politics; *torture

Keogh, James P.; Spodick, David H. Physicians and torture. *New England Journal of Medicine* 297(12): 675, 22 Sep 1977. Letter. BE05781.
codes of ethics; informal social control; legislation; medicine; *physician's role; physicians; prisoners; professional organizations; *punishment; *torture; Amnesty International; Pakistan; Tokyo Declaration

Meyer-Lie, Arnt K. Torture. *Nature* 268(5618): 294, 28 Jul 1977. 1 ref. Letter. BE05964.
physician's role; political activity; professional organizations; *torture; Amnesty International

TORTURE / BIOMEDICAL RESEARCH

Rich, Vera. Medical profession to investigate torture. *Nature* 272(5652): 394, 30 Mar 1978. BE06807.
*biomedical research; international aspects; dissent; involuntary commitment; medicine; physician's role; social control; *torture

TORTURE / CODES OF ETHICS

Heijder, Alfred; Van Geuns, Herman. Professional Codes of Ethics. London: Amnesty International, 1976. 32 p. 11 refs. ISBN 0-900058-32-3. BE06554.
*codes of ethics; law; law enforcement; medical

BE = bioethics accession number   fn. = footnotes   refs. = references

ethics; *moral obligations; *physician's role; physicians; *professional ethics; *torture

TORTURE / MEDICAL ETHICS

Jonsen, Albert R.; Sagan, Leonard. Torture and the ethics of medicine. *Man and Medicine: The Journal of Values and Ethics in Health Care* 3(1): 33-53, 1978. Article followed by commentaries by Seth Goldsmith and Frank S. Jewett, p. 50-53. 3 refs. 31 fn. BE06906.
international aspects; morality; *medical ethics; physician's role; physicians; prisoners; psychiatry; *torture; utilitarianism

TORTURE / MORAL OBLIGATIONS / PHYSICIAN'S ROLE

Heijder, Alfred; Van Geuns, Herman. Professional Codes of Ethics. London: Amnesty International, 1976. 32 p. 11 refs. ISBN 0-900058-32-3. BE06554.
*codes of ethics; law; law enforcement; medical ethics; *moral obligations; *physician's role; physicians; *professional ethics; *torture

Van Geuns, Herman. The responsibilities of the medical profession in connection with torture. *In* Heijder, Alfred; Van Geuns, Herman. Professional Codes of Ethics. London: Amnesty International, 1976. p. 17-22. 11 refs. ISBN 0-900058-32-3. BE06556.
*moral obligations; *physician's role; physicians; *torture

TORTURE / PROFESSIONAL ETHICS

Heijder, Alfred. Professional codes of ethics against torture. *In* Heijder, Alfred; Van Geuns, Herman. Professional Codes of Ethics. London: Amnesty International, 1976. p. 7-15. ISBN 0-900058-32-3. BE06555.
codes of ethics; law; law enforcement; medical ethics; physicians; *professional ethics; *torture

Heijder, Alfred; Van Geuns, Herman. Professional Codes of Ethics. London: Amnesty International, 1976. 32 p. 11 refs. ISBN 0-900058-32-3. BE06554.
*codes of ethics; law; law enforcement; medical ethics; *moral obligations; *physician's role; physicians; *professional ethics; *torture

TOTALITY *See under*

HEALTH / TOTALITY / HÄRING, BERNARD

TRANSPLANTATION / AUSTRALIA

Law Reform Commission. Human Tissue Transplants. Canberra: Australian Government Publishing Service, 1977. 154 p. Law Reform Commission Report No. 7. 513 fn. Bibliography: p. 144-150. ISBN 0-642-03018-9. BE06172.

autopsies; blood donation; brain death; cadavers; children; determination of death; economics; international aspects; legal liability; legislation; organ donation; organ donors; *organ transplantation; physicians; privacy; public opinion; transplant recipients; *transplantation; *Australia

TRANSPLANTATION / CHILDREN / RISKS AND BENEFITS

Nolan, James G. Anatomical transplants between family members—the problems facing court and counsel. *Family Law Reporter* 1(21): 1-4, 8 Apr 1975. Monograph No. 6. BE07083.
bone marrow; *children; kidneys; *legal aspects; legal guardians; *organ donors; organ transplantation; *parental consent; parents; *risks and benefits; siblings; *transplantation

TRANSPLANTATION / PARENTAL CONSENT / LEGAL ASPECTS

Nolan, James G. Anatomical transplants between family members—the problems facing court and counsel. *Family Law Reporter* 1(21): 1-4, 8 Apr 1975. Monograph No. 6. BE07083.
bone marrow; *children; kidneys; *legal aspects; legal guardians; *organ donors; organ transplantation; *parental consent; parents; *risks and benefits; siblings; *transplantation

TRANSSEXUALISM / LEGAL ASPECTS

Belli, Melvin M. Transsexual surgery: a new tort? *Journal of the American Medical Association* 239(20): 2143-2148, 19 May 1978. 41 refs. BE06967.
battery; informed consent; *legal aspects; legal liability; legal rights; malpractice; physicians; privacy; psychological stress; self concept; *surgery; *torts; *transsexualism

TREATMENT REFUSAL

Cassell, Eric J. What is the function of medicine? *In* McMullin, Ernan, ed. Death and Decision. Boulder, Colo.: Westview Press, 1978. p. 35-44. American Association for the Advancement of Science Selected Symposium 18. 4 refs. ISBN 0-89158-152-9. BE06708.
*allowing to die; chronically ill; critically ill; goals; medicine; physician patient relationship; *self determination; terminally ill; *treatment refusal

Ramsey, Paul. Ethics at the Edges of Life: Medical and Legal Intersections. New Haven: Yale University Press, 1978. 353 p. The Bampton Lectures in America. 302 fn. ISBN 0-300-02137-2. BE06884.
*abortion; adults; *allowing to die; congenital defects; *decision making; *euthanasia; extraordinary treatment; fetuses; law; legal aspects; legislation; medicine; newborns; physician's role; quality of life; *selection for treatment; *Supreme Court decisions; terminal care; *terminally ill;

BE = bioethics accession number   fn. = footnotes   refs. = references

*third party consent; *treatment refusal; *withholding treatment; California Natural Death Act; Commonwealth v. Edelin; *In re Quinlan; Saikewicz, Joseph

TREATMENT REFUSAL / BLOOD TRANSFUSIONS / RELIGIOUS BELIEFS

Pennsylvania. Court of Common Pleas, Allegheny County. In re William J. Dell, II. 8 Jan 1975. *Pennsylvania District and County Reports, 3d Series,* 1: 655-660. BE07081.
　　*blood transfusions; *religious beliefs; state interest; *treatment refusal; value of life

TREATMENT REFUSAL / COERCION / PSYCHOACTIVE DRUGS

New Jersey. Superior Court. Matter of B. 22 Dec 1977. *Atlantic Reporter, 2d Series,* 383: 760-764. BE07127.
　　adults; *coercion; informed consent; *institutionalized persons; involuntary commitment; *legal aspects; legal rights; legislation; *mentally ill; *psychoactive drugs; schizophrenia; *treatment refusal; *New Jersey; *Prolixin

TREATMENT REFUSAL / COMPETENCE

Lipp, Martin R. Competency, refusal of treatment, and committability. *In his* Respectful Treatment: The Human Side of Medical Care. Hagerstown, Md.: Harper and Row, Medical Department, 1977. 61-73. 11 refs. ISBN 0-06-141550-2. BE06679.
　　adults; children; *competence; dangerousness; duration of commitment; decision making; emergency care; *informed consent; *involuntary commitment; legal aspects; legal guardians; *mentally ill; parental consent; patient care; physician patient relationship; psychiatric diagnosis; *treatment refusal

TREATMENT REFUSAL / COMPREHENSION

Faden, Ruth; Faden, Alan. False belief and the refusal of medical treatment. *Journal of Medical Ethics* 3(3): 133-136, Sep 1977. 11 refs. 3 fn. BE05981.
　　adults; case reports; *coercion; competence; *comprehension; disclosure; informed consent; *moral obligations; paternalism; *patient care; *patients; *physicians; self determination; surgery; *treatment refusal

TREATMENT REFUSAL / CRITICALLY ILL

Siegler, Mark. Critical illness: the limits of autonomy. *Hastings Center Report* 7(5): 12-15, Oct 1977. Case study. BE06076.
　　age; competence; *critically ill; *decision making; diagnosis; personality; physicians; prognosis; self determination; *treatment refusal; values

TREATMENT REFUSAL / CRITICALLY ILL / COMPETENCE

Tennessee. Court of Appeals, Middle Section. State Department of Human Services v. Northern. 7 Feb 1978. *South Western Reporter, 2d Series,* 563: 197-215. BE06715.
　　*aged; *competence; comprehension; *critically ill; due process; government agencies; judicial action; legal aspects; *legislation; state government; surgery; third party consent; *treatment refusal; *Tennessee

TREATMENT REFUSAL / CRITICALLY ILL / LEGAL RIGHTS

New Jersey. Morris County Court, Probate Division. Matter of Quackenbush. 13 Jan 1978. *Atlantic Reporter, 2d Series,* 383: 785-790. 2 fn. BE06713.
　　aged; competence; *critically ill; *legal rights; *privacy; prognosis; state interest; *surgery; *treatment refusal

TREATMENT REFUSAL / HUMAN RIGHTS

O'Neil, Richard. In defense of the "ordinary"/"extraordinary" distinction. *Linacre Quarterly* 45(1): 37-40, Feb 1978. 2 refs. BE06790.
　　competence; *extraordinary treatment; *human rights; *moral obligations; *patients; *physicians; *prolongation of life; Roman Catholic ethics; standards; terminally ill; third party consent; *treatment refusal; *Veatch, Robert M.

TREATMENT REFUSAL / INSTITUTIONALIZED PERSONS / CALIFORNIA

California. Legislature. Assembly. An act...relating to mental health. Assembly Bill No. 1365, 1977-78 Regular Session, 4 Apr 1977. 14 p. By Assemblyman Torres, et al. Referred to Committee on Health; died 1977-78 Session. BE06597.
　　competence; electroconvulsive therapy; informed consent; *institutionalized persons; involuntary commitment; legal rights; *mentally ill; physician's role; *psychoactive drugs; *treatment refusal; voluntary admission; *California

TREATMENT REFUSAL / INSTITUTIONALIZED PERSONS / LEGAL RIGHTS

Hester, Deborah W. The right to refuse treatment: a medicolegal paradox. *Advocate: The Suffolk University Law School Journal* 8(2): 28-31, Spring 1977. 36 fn. BE05921.
　　competence; informed consent; *institutionalized persons; *legal rights; *mentally ill; privacy; *treatment refusal

McGough, Lucy S.; Carmichael, William C. The right to treatment and the right to refuse treatment. *American Journal of Orthopsychiatry* 47(2): 307-320, Apr 1977. 30 refs. BE05567.

BE = bioethics accession number　　　fn. = footnotes　　　refs. = references

constitutional law; dangerousness; due process; *institutionalized persons; involuntary commitment; *legal rights; *mentally ill; peer review; *right to treatment; standards; *treatment refusal

New Jersey. Superior Court. Matter of B. 22 Dec 1977. *Atlantic Reporter, 2d Series,* 383: 760-764. BE07127.
    adults; *coercion; informed consent; *institutionalized persons; involuntary commitment; *legal aspects; legal rights; legislation; *mentally ill; *psychoactive drugs; schizophrenia; *treatment refusal; *New Jersey; *Prolixin

Scott, Edward P. The right to treatment, Part II: the right to refuse treatment. *Bioethics Digest* 1(10): 1-7, Feb 1977. 35 fn. BE06221.
    competence; electroconvulsive therapy; ethics committees; informed consent; *institutionalized persons; *legal rights; *mentally handicapped; right to treatment; self determination; *treatment refusal

TREATMENT REFUSAL / INSTITUTIONALIZED PERSONS / MENTALLY ILL

McGough, Lucy S.; Carmichael, William C. The right to treatment and the right to refuse treatment. *American Journal of Orthopsychiatry* 47(2): 307-320, Apr 1977. 30 refs. BE05567.
    constitutional law; dangerousness; due process; *institutionalized persons; involuntary commitment; *legal rights; *mentally ill; peer review; *right to treatment; standards; *treatment refusal

New Jersey. Superior Court. Matter of B. 22 Dec 1977. *Atlantic Reporter, 2d Series,* 383: 760-764. BE07127.
    adults; *coercion; informed consent; *institutionalized persons; involuntary commitment; *legal aspects; legal rights; legislation; *mentally ill; *psychoactive drugs; schizophrenia; *treatment refusal; *New Jersey; *Prolixin

Roth, Loren H. Involuntary civil commitment: the right to treatment and the right to refuse treatment. *Psychiatric Annals* 7(5): 50-51+, May 1977. 55 refs. BE05809.
    coercion; dangerousness; competence; drugs; due process; economics; informed consent; *institutionalized persons; *involuntary commitment; judicial action; *legal aspects; legal rights; legislation; mental institutions; *mentally ill; mentally retarded; operant conditioning; patients; psychiatry; review committees; *right to treatment; standards; state government; *treatment refusal

TREATMENT REFUSAL / JEHOVAH'S WITNESSES / LEGAL RIGHTS

District of Columbia. Superior Court. In the Matter of Bentley. 25 Apr 1974. *Daily Washington Law Reporter* 102(117): 1221+, 17 Jun 1974. BE06605.
    *blood transfusions; fetuses; informed consent;

*Jehovah's Witnesses; *legal rights; privacy; religious beliefs; state interest; *treatment refusal; value of life

TREATMENT REFUSAL / JEHOVAH'S WITNESSES / LEUKEMIA

Kearney, P.J. Leukaemia in children of Jehovah's Witnesses: issues and priorities in a conflict of care. *Journal of Medical Ethics* 4(1): 32-35, Mar 1978. 6 refs. BE06921.
    blood transfusions; cancer; case reports; *children; *decision making; family relationship; informal social control; *Jehovah's Witnesses; *leukemia; parent child relationship; parental consent; parents; *patient care; pediatrics; *physician patient relationship; physicians; religious beliefs; *risks and benefits; *treatment refusal

TREATMENT REFUSAL / JUDICIAL ACTION

Rothenberg, Leslie S. Demands for life and requests for death: the judicial dilemma. *In* McMullin, Ernan, ed. Death and Decision. Boulder, Colo.: Westview Press, 1978. p. 131-154. American Association for the Advancement of Science Selected Symposium 18. 66 fn. ISBN 0-89158-152-9. BE06711.
    *allowing to die; blood transfusions; decision making; Jehovah's Witnesses; *judicial action; legal aspects; surgery; *treatment refusal; withholding treatment

TREATMENT REFUSAL / LEGAL ASPECTS

Creighton, Helen. The right to refuse treatment. *Supervisor Nurse* 8(2): 13-16, Feb 1977. 25 fn. BE05906.
    adults; blood transfusions; children; competence; *disclosure; *informed consent; *legal aspects; legal obligations; patients; physicians; religious beliefs; risks and benefits; self determination; *treatment refusal

Rothman, Daniel A.; Rothman, Nancy L. Death. *In their* The Professional Nurse and the Law. Boston: Little, Brown, 1977. p. 125-141. 32 fn. ISBN 0-316-75768-3. BE06520.
    active euthanasia; allowing to die; brain death; *determination of death; *euthanasia; *legal aspects; legal rights; organ transplantation; religious beliefs; *treatment refusal; withholding treatment

TREATMENT REFUSAL / LEGAL RIGHTS

Sandak, Lawrence R. Suicide and the compulsion of lifesaving medical procedures: an analysis of the refusal of treatment cases. *Brooklyn Law Review* 44(2): 285-316, Winter 1978. 140 fn. BE06754.
    allowing to die; blood transfusions; intention; Jehovah's Witnesses; judicial action; *legal aspects; *legal rights; privacy; *suicide; Supreme Court de-

---

BE = bioethics accession number   fn. = footnotes   refs. = references

cisions; *treatment refusal; withholding treatment;
Quinlan, Karen

TREATMENT REFUSAL / LEGISLATION / TENNESSEE

Tennessee. Court of Appeals, Middle Section. State
Department of Human Services v. Northern. 7 Feb
1978. *South Western Reporter, 2d Series,* 563:
197-215. BE06715.
    *aged; *competence; comprehension; *critically ill;
due process; government agencies; judicial action;
legal aspects; *legislation; state government; surgery; third party consent; *treatment refusal;
*Tennessee

TREATMENT REFUSAL / LIVING WILLS

Veatch, Robert M. Death and dying: the legislative
options. *Hastings Center Report* 7(5): 5-8, Oct
1977. BE06074.
    active euthanasia; *allowing to die; competence;
*decision making; family members; legal guard-
ians; *legislation; *living wills; physicians; model
legislation; state government; *terminally ill;
*treatment refusal; withholding treatment

TREATMENT REFUSAL / MENTALLY ILL / LEGAL ASPECTS

Wing, Kenneth R. The right to refuse treatment: legal
issues. *Bulletin of the American Academy of Psychiatry and the Law* 5(1): 15-19, 1977. 14 fn.
BE05893.
    government regulation; *legal rights; *mentally ill;
privacy; state government; *state interest; *treatment refusal

TREATMENT REFUSAL / MENTALLY ILL / LEGAL RIGHTS

Brooks, Alexander D. The right to refuse treatment.
Administration in Mental Health 4(2): 90-95, Spring
1977. 17 fn. BE05918.
    competence; institutionalized persons; involuntary
commitment; *legal rights; *mentally ill; psychoactive drugs; psychotherapy; *treatment refusal

Cocozza, Joseph J.; Melick, Mary E. A right to refuse
treatment: a broad view. *Bulletin of the American
Academy of Psychiatry and the Law* 5(1): 1-7, 1977.
12 refs. BE05892.
    alternatives; informed consent; institutionalized
persons; involuntary commitment; *legal rights;
*mentally ill; right to treatment; *treatment refusal

Hester, Deborah W. The right to refuse treatment: a
medicolegal paradox. *Advocate: The Suffolk University Law School Journal* 8(2): 28-31, Spring 1977. 36
fn. BE05921.
    competence; informed consent; *institutionalized

persons; *legal rights; *mentally ill; privacy;
*treatment refusal

TREATMENT REFUSAL / MODEL LEGISLATION

Grisez, Germain; Boyle, Joseph M. An alternative to
"death with dignity". *Human Life Review* 4(1):
26-43, Winter 1978. 15 fn. BE06736.
    adults; allowing to die; competence; euthanasia;
informed consent; judicial action; *legal rights; living wills; *model legislation; self determination;
*treatment refusal

TREATMENT REFUSAL / NEWBORNS / CONGENITAL DEFECTS

**Shaw, Anthony; Randolph, Judson G.; Manard, Bar-
bara.** Ethical issues in pediatric surgery: a national
survey of pediatricians and pediatric surgeons. *Pediatrics* 60(4-Part 2): 588-599, Oct 1977. BE06587.
    abortion; *allowing to die; *attitudes; *congenital
defects; *decision making; Down's syndrome; extraordinary treatment; legal aspects; *newborns;
parental consent; parents; *pediatrics; *physicians;
prolongation of life; spina bifida; *surgery; *treatment refusal; *withholding treatment

TREATMENT REFUSAL / PARENTAL CONSENT

Parents prevented from transferring seriously ill child.
Hospitals 51(20): 20, 16 Oct 1977. BE06092.
    children; *legal rights; leukemia; *parental con-
sent; *treatment refusal

TREATMENT REFUSAL / PARENTAL CONSENT / RELIGIOUS BELIEFS

New York. Family Court, Kings County. Matter of
Gregory S. 23 Feb 1976. *New York Supplement, 2d
Series,* 380: 620-623. BE07090.
    children; health care; *parental consent; *religious
beliefs; *treatment refusal

TREATMENT REFUSAL / PARENTAL CONSENT / SPINA BIFIDA

Veatch, Robert M. Abnormal newborns and the physician's role: models of physician decision making. *In*
Swinyard, Chester A., ed. Decision Making and the
Defective Newborn. Proceedings of a Conference on
Spina Bifida and Ethics. Springfield, Ill.: Charles C.
Thomas, 1978. p. 174-193. 37 fn. ISBN 0-
398-03662-4. BE06953.
    allowing to die; blood transfusions; congenital de-
fects; *contracts; *decision making; extraordinary
treatment; judicial action; killing; legal aspects; legal obligations; medical ethics; *moral obligations;
*newborns; *parental consent; physician patient
relationship; *physician's role; physicians; prognosis; *prolongation of life; *quality of life; *selection
for treatment; *spina bifida; technical expertise;

BE = bioethics accession number   fn. = footnotes   refs. = references

*treatment refusal; values; withholding treatment; Hippocratic Oath

TREATMENT REFUSAL / PATIENTS / RIGHTS

Bice, Michael K. Is the right to die a solo decision? *Hospital Medical Staff* 6(7): 1-7, Jul 1977. 5 fn. BE05720.
    allowing to die; burn patients; case reports; competence; decision making; human rights; legal guardians; legal rights; legislation; *patients; paternalism; *physicians; prolongation of life; quality of life; *rights; self determination; terminally ill; *treatment refusal; value of life

TREATMENT REFUSAL / PHYSICIANS / ATTITUDES

Shaw, Anthony; Randolph, Judson G.; Manard, Barbara. Ethical issues in pediatric surgery: a national survey of pediatricians and pediatric surgeons. *Pediatrics* 60(4-Part 2): 588-599, Oct 1977. BE06587.
    abortion; *allowing to die; *attitudes; *congenital defects; *decision making; Down's syndrome; extraordinary treatment; legal aspects; *newborns; parental consent; parents; *pediatrics; *physicians; prolongation of life; spina bifida; *surgery; *treatment refusal; *withholding treatment

TREATMENT REFUSAL / PREGNANT WOMEN

McFayden, I.R. Fetal survival—who decides? *Journal of Medical Ethics* 4(1): 30-31, Mar 1978. BE06922.
    case reports; congenital defects; *decision making; *fetuses; morbidity; mothers; physicians; *pregnant women; *treatment refusal

TREATMENT REFUSAL / SELF DETERMINATION

Cassell, Eric J. The function of medicine. *Hastings Center Report* 7(6): 16-19, Dec 1977. BE06134.
    *allowing to die; *chronically ill; *critically ill; Jehovah's Witnesses; motivation; physician's role; *self determination; suicide; *treatment refusal

TREATMENT REFUSAL / SELF DETERMINATION / LEGAL RIGHTS

Annas, George J. The incompetent's right to die: the case of Joseph Saikewicz. *Hastings Center Report* 8(1): 21-23, Feb 1978. 1 fn. BE06769.
    *allowing to die; competence; ethics committees; *judicial action; *legal rights; leukemia; mentally retarded; *self determination; state interest; terminally ill; *treatment refusal; *withholding treatment; *Saikewicz, Joseph

TREATMENT REFUSAL / SPINA BIFIDA / ALTERNATIVES

Robertson, John A. Legal issues in nontreatment of defective newborns. *In* Swinyard, Chester A., ed. Decision Making and the Defective Newborn. Proceedings of a Conference on Spina Bifida and Ethics. Springfield, Ill.: Charles C. Thomas, 1978. p. 359-383. 8 refs. 9 fn. ISBN 0-398-03662-4. BE06957.
    *allowing to die; *alternatives; child neglect; contracts; *congenital defects; criminal law; decision making; costs and benefits; financial support; killing; hospitals; law enforcement; *legal liability; *legal obligations; legal rights; medical staff; moral obligations; nurses; *newborns; obligations of society; *parental consent; parents; physician patient relationship; *physicians; resource allocation; *spina bifida; *treatment refusal; value of life; *withholding treatment

TREATMENT REFUSAL / SPINA BIFIDA / LEGAL ASPECTS

Robertson, John A. Legal issues in nontreatment of defective newborns. *In* Swinyard, Chester A., ed. Decision Making and the Defective Newborn. Proceedings of a Conference on Spina Bifida and Ethics. Springfield, Ill.: Charles C. Thomas, 1978. p. 359-383. 8 refs. 9 fn. ISBN 0-398-03662-4. BE06957.
    *allowing to die; *alternatives; child neglect; contracts; *congenital defects; criminal law; decision making; costs and benefits; financial support; killing; hospitals; law enforcement; *legal liability; *legal obligations; legal rights; medical staff; moral obligations; nurses; *newborns; obligations of society; *parental consent; parents; physician patient relationship; *physicians; resource allocation; *spina bifida; *treatment refusal; value of life; *withholding treatment

TREATMENT REFUSAL / STATE INTEREST

Wing, Kenneth R. The right to refuse treatment: legal issues. *Bulletin of the American Academy of Psychiatry and the Law* 5(1): 15-19, 1977. 14 fn. BE05893.
    government regulation; *legal rights; *mentally ill; privacy; state government; *state interest; *treatment refusal

TREATMENT REFUSAL / SURGERY / JUDICIAL ACTION

Foster, Henry H. Refusal to consent to amputation. [Title provided]. *Bulletin of the American Academy of Psychiatry and the Law* 5(1): 113-115, 1977. 1 fn. BE05896.
    adults; competence; emergency care; informed consent; *judicial action; psychiatric diagnosis; *surgery; *treatment refusal

TREATMENT REFUSAL / TERMINALLY ILL

Kennedy, Ian. M. The doctor and the dying patient. *Times (London)*, 21 Dec 1976, p. 13. Letter.

BE = bioethics accession number   fn. = footnotes   refs. = references

BE06622.
    informed consent; living wills; socioeconomic fac-
    tors; *terminally ill; *treatment refusal; California
    Natural Death Act

Ramsey, Paul. "Euthanasia" and dying well enough.
    In his Ethics at the Edges of Life: Medical and Legal
    Intersections. New Haven: Yale University Press,
    1978. p. 145-188. 45 fn. ISBN 0-300-02137-2.
    BE06888.
    active euthanasia; *adults; *allowing to die; chron-
    ically ill; coma; competence; congenital defects;
    death; *decision making; diagnosis; drugs; *eutha-
    nasia; *extraordinary treatment; human rights; le-
    gal aspects; medical ethics; moral obligations;
    newborns; pain; physicians; prognosis; quality of
    life; religious ethics; *selection for treatment; social
    interaction; spina bifida; *standards; terminal care;
    *terminally ill; theology; *third party consent;
    *treatment refusal; value of life; withholding treat-
    ment; McCormick, Richard A.; Veatch, Robert
    M.

TREATMENT REFUSAL / TERMINALLY ILL / LEGAL RIGHTS

Randall, Margaret W. The right to die a natural death:
    a discussion of *In re Quinlan* and the California Nat-
    ural Death Act. *University of Cincinnati Law Review*
    46(1): 192-206, 1977. 77 fn. BE05940.
    *allowing to die; legal liability; *legal rights; legis-
    lation; *living wills; physician's role; privacy; self
    determination; state government; state interest;
    *terminally ill; *treatment refusal; withholding
    treatment; *California Natural Death Act; *In re
    Quinlan

TREATMENT REFUSAL   *See under*

ELECTROCONVULSIVE THERAPY / TREATMENT
    REFUSAL
ELECTROCONVULSIVE THERAPY / TREATMENT
    REFUSAL / PATIENTS' RIGHTS

TRUTHTELLING   *See* DISCLOSURE

U.S. CONGRESS   *See under*

RECOMBINANT DNA RESEARCH / LEGISLA-
    TION / U.S. CONGRESS

UNIFORM ANATOMICAL GIFT ACT   *See under*

ORGAN DONATION / UNIFORM ANATOMICAL
    GIFT ACT

UNITED STATES   *See under*

ABORTION / ATTITUDES / UNITED STATES
ABORTION / LEGAL ASPECTS / UNITED STATES
DETERMINATION OF DEATH / LEGISLATION /
    UNITED STATES
GENETIC SCREENING / UNITED STATES

UNITED STATES   *See under (cont'd.)*

INFANTICIDE / HISTORICAL ASPECTS / UNITED
    STATES
PSYCHOSURGERY / LEGAL ASPECTS / UNITED
    STATES
PSYCHOSURGERY / UNITED STATES
TERMINAL CARE / HOSPICES / UNITED STATES

UNIVERSITIES   *See under*

RECOMBINANT DNA RESEARCH / UNIVERSI-
    TIES

UNIVERSITY OF CALIFORNIA, SAN FRAN-CISCO   *See under*

RECOMBINANT DNA RESEARCH / UNIVERSITY
    OF CALIFORNIA, SAN FRANCISCO

UNWANTED CHILDREN   *See under*

SELECTIVE ABORTION / UNWANTED CHIL-
    DREN / TORTS
VOLUNTARY STERILIZATION / UNWANTED
    CHILDREN / LEGAL LIABILITY
VOLUNTARY STERILIZATION / UNWANTED
    CHILDREN / TORTS

USSR   *See under*

DISSENT / PSYCHIATRY / USSR
EUGENICS / HISTORICAL ASPECTS / USSR
HUMAN EXPERIMENTATION / USSR
INVOLUNTARY COMMITMENT / CASE RE-
    PORTS / USSR
INVOLUNTARY COMMITMENT / DISSENT /
    USSR
INVOLUNTARY COMMITMENT / PSYCHIATRIC
    DIAGNOSIS / USSR
INVOLUNTARY COMMITMENT / WORLD PSY-
    CHIATRIC ASSOCIATION / USSR
ORGANIZATIONAL POLICIES / WORLD PSYCHI-
    ATRIC ASSOCIATION / USSR
PSYCHIATRIC DIAGNOSIS / USSR
PSYCHIATRY / DISSENT / USSR
PSYCHIATRY / POLITICS / USSR
WORLD PSYCHIATRIC ASSOCIATION / PSYCHIA-
    TRY / USSR

VALUE OF LIFE   *See also* QUALITY OF LIFE

VALUE OF LIFE

Beauchamp, Tom L.; Perlin, Seymour, eds. Ethical Is-
    sues in Death and Dying. Englewood Cliffs, N.J.:
    Prentice-Hall, 1978. 368 p. References and foot-
    notes included. ISBN 0-13-290114-5. BE06878.
    active euthanasia; allowing to die; brain death;
    coma; decision making; *determination of death;
    disclosure; *euthanasia; informed consent; killing;
    legal aspects; legislation; living wills; moral obliga-
    tions; *morality; newborns; *patients' rights; *sui-

BE = bioethics accession number   fn. = footnotes   refs. = references

cide; terminal care; *terminally ill; theology; treatment refusal; *value of life; values; withholding treatment; Quinlan, Karen

Feinberg, Joel. Voluntary euthanasia and the inalienable right to life. *Philosophy and Public Affairs* 7(2): 93-123, Winter 1978. 21 fn. BE06732.
allowing to die; human rights; killing; *moral obligations; paternalism; philosophy; *rights; suicide; *value of life; *voluntary euthanasia

Frankena, William K. The ethics of respect for life. *In* Temkin, Owsei; Frankena, William K.; Kadish, Sanford H. Respect for Life in Medicine, Philosophy and the Law. Baltimore: Johns Hopkins University Press, 1976. p. 24-62. 49 fn. ISBN 0-8018-1942-3. BE05853.
abortion; biological life; Christian ethics; dignity; ethics; historical aspects; killing; moral obligations; morality; personality; quality of life; religious beliefs; *value of life

Roupas, T.G. The value of life. *Philosophy and Public Affairs* 7(2): 154-183, Winter 1978. 14 fn. BE06751.
*abortion; fetuses; justice; moral obligations; personhood; philosophy; utilitarianism; *value of life

VALUE OF LIFE / BIOLOGICAL LIFE

Engelhardt, H. Tristram. Medicine and the concept of person. *In* Beauchamp, Tom L.; Perlin, Seymour, eds. Ethical Issues in Death and Dying. Englewood Cliffs, N.J.: Prentice-Hall, 1978. p. 271-284. 23 fn. ISBN 0-13-290114-5. BE06882.
adults; *biological life; brain death; children; fetuses; *personhood; self concept; social interaction; *value of life

VALUE OF LIFE / FETUSES

Tiefel, Hans O. The unborn: human values and responsibilities. *Journal of the American Medical Association* 239(21): 2263-2267, 26 May 1978. 10 refs. BE06970.
*fetuses; maternal life; moral obligations; obligations of society; *personhood; religious beliefs; resource allocation; rights; *value of life

VALUE OF LIFE / HANDICAPPED

Hostler, John. The right to life. *Journal of Medical Ethics* 3(3): 143-145, Sep 1977. 5 refs. BE05982.
allowing to die; *handicapped; *justifiable killing; killing; moral obligations; personhood; quality of life; *value of life

VALUE OF LIFE / JEWISH ETHICS

Bleich, J. David. Theological considerations in the care of defective newborns. *In* Swinyard, Chester A., ed. Decision Making and the Defective Newborn. Proceedings of a Conference on Spina Bifida and Ethics. Springfield, Ill.: Charles C. Thomas, 1978. p.

512-561. 106 fn. ISBN 0-398-03662-4. BE06963.
adults; contracts; *extraordinary treatment; financial support; informed consent; *Jewish ethics; judicial action; legal liability; legal obligations; *moral obligations; *newborns; pain; *patient care; physician patient relationship; *physicians; *prolongation of life; Roman Catholic ethics; *scriptural interpretation; *spina bifida; terminally ill; theology; treatment refusal; *value of life; withholding treatment

VALUE OF LIFE / MEDICINE / HISTORICAL ASPECTS

Temkin, Owsei. The idea of respect for life in the history of medicine. *In* Temkin, Owsei; Frankena, William K.; Kadish, Sanford H. Respect for Life in Medicine, Philosophy and the Law. Baltimore: Johns Hopkins University Press, 1976. p. 1-23. 42 fn. ISBN 0-8018-1942-3. BE05852.
codes of ethics; disclosure; euthanasia; health; *historical aspects; *medicine; *physicians; therapeutic abortion; *value of life

VALUE OF LIFE / PERSONHOOD

Engelhardt, H. Tristram. Medicine and the concept of person. *In* Beauchamp, Tom L.; Perlin, Seymour, eds. Ethical Issues in Death and Dying. Englewood Cliffs, N.J.: Prentice-Hall, 1978. p. 271-284. 23 fn. ISBN 0-13-290114-5. BE06882.
adults; *biological life; brain death; children; fetuses; *personhood; self concept; social interaction; *value of life

VALUE OF LIFE / THEOLOGY

Anderson, Norman. The sanctity of human life: some contemporary problems. *In his* Issues of Life and Death: Abortion, Birth Control, Capital Punishment, Euthanasia. Downers Grove, Ill.: InterVarsity Press, 1977. p. 11-33. 61 fn. ISBN 0-87784-721-5. BE05788.
biology; evolution; human characteristics; humanism; scriptural interpretation; *theology; *value of life

VALUE OF LIFE *See under*

BE = bioethics accession number   fn. = footnotes   refs. = references

VALUE OF LIFE *See under (cont'd.)*
EMBRYO TRANSFER / VALUE OF LIFE
EUTHANASIA / VALUE OF LIFE
EUTHANASIA / VALUE OF LIFE / CHRISTIAN
  ETHICS
HEALTH CARE / VALUE OF LIFE
HEALTH CARE / VALUE OF LIFE / ECONOMICS
KILLING / VALUE OF LIFE
PROLONGATION OF LIFE / VALUE OF LIFE
SELECTION FOR TREATMENT / VALUE OF
  LIFE / SPINA BIFIDA
VOLUNTARY EUTHANASIA / VALUE OF LIFE

VALUES *See under*
BEHAVIORAL GENETICS / VALUES
EUGENICS / POLITICAL SYSTEMS / VALUES
FAMILY PLANNING / DEVELOPING
  COUNTRIES / VALUES
HEALTH CARE / MORAL OBLIGATIONS / VAL-
  UES
HUMAN EXPERIMENTATION / CHILDREN /
  VALUES
OPERANT CONDITIONING / VALUES
PHYSICIAN PATIENT RELATIONSHIP / VALUES
POPULATION CONTROL / BLACKS / VALUES
POPULATION CONTROL / PUBLIC POLICY / VAL-
  UES
PSYCHIATRY / CONFLICT OF INTEREST / VAL-
  UES
PSYCHOSURGERY / PHYSICIANS / VALUES

VIABILITY *See under*
ABORTION / VIABILITY / LEGAL ASPECTS
WRONGFUL DEATH / VIABILITY / LEGAL AS-
  PECTS

VIOLENCE *See under*
PSYCHOSURGERY / VIOLENCE

VOLUNTARY ADMISSION / ADOLESCENTS / MENTALLY ILL

U.S. Supreme Court. Kremens v. Bartley. 16 May
1977. *United States Reports* 431: 119-144. 22 fn.
BE07110.
> *adolescents; children; due process; legislation;
> *mentally ill; mentally retarded; *parental con-
> sent; state government; *voluntary admission;
> *Pennsylvania

VOLUNTARY ADMISSION / PARENTAL CONSENT / PENNSYLVANIA

U.S. Supreme Court. Kremens v. Bartley. 16 May
1977. *United States Reports* 431: 119-144. 22 fn.
BE07110.
> *adolescents; children; due process; legislation;
> *mentally ill; mentally retarded; *parental con-

sent; state government; *voluntary admission;
*Pennsylvania

VOLUNTARY EUTHANASIA

Reed, Nicholas. Recent thinking about voluntary euthanasia. *New Humanist* 92(5): 173-174, Jan-Feb
1977. BE06290.
> active euthanasia; *allowing to die; family mem-
> bers; extraordinary treatment; legislation; living
> wills; physicians; terminally ill; treatment refusal;
> *voluntary euthanasia; withholding treatment

Vere, Duncan. Euthanasia. *In* Vale, J.A., ed. Medicine and the Christian Mind. London: Christian
Medical Fellowship Publications, 1975. p. 75-82. 3
refs. ISBN 85111-956-5. BE07071.
> allowing to die; social impact; terminal care; termi-
> nally ill; value of life; *voluntary euthanasia; wedge
> argument

VOLUNTARY EUTHANASIA / LEGAL ASPECTS / AUSTRALIA

Buddin, Terry. The Elusive Right to Voluntary Sterilisation: A Statutory Panacea. Sydney, N.S.W.: Australian Council of Social Service, 1976. 50 p. Law
and Social Welfare: Discussion Paper No. 1. 172 fn.
ISBN 0-85871-070-6. BE07085.
> attitudes; compensation; criminal law; family plan-
> ning; human rights; informed consent; *legal as-
> pects; legal liability; legislation; model legislation;
> physicians; public hospitals; spousal consent; *vol-
> untary sterilization; *Australia

VOLUNTARY EUTHANASIA / MORALITY

Grisez, Germain. Suicide and euthanasia. *In* Horan,
Dennis J.; Mall, David, eds. Death, Dying, and Euthanasia. Washington: University Publications of
America, 1977. p. 742-817. 57 fn. ISBN 0-
89093-139-9. BE05690.
> abortion; active euthanasia; extraordinary treat-
> ment; intention; killing; legal aspects; moral obli-
> gations;   *morality;   physician's   role;   self
> determination; *suicide; *teleological ethics; utili-
> tarianism; value of life; *voluntary euthanasia;
> wedge argument

VOLUNTARY EUTHANASIA / RIGHTS

Feinberg, Joel. Voluntary euthanasia and the inalienable right to life. *Philosophy and Public Affairs* 7(2):
93-123, Winter 1978. 21 fn. BE06732.
> allowing to die; human rights; killing; *moral obli-
> gations; paternalism; philosophy; *rights; suicide;
> *value of life; *voluntary euthanasia

VOLUNTARY EUTHANASIA / VALUE OF LIFE

Kohl, Marvin. Euthanasia and the right to life. *In*
Spicker, Stuart F.; Engelhardt, H. Tristram, eds.
Philosophical Medical Ethics: Its Nature and

BE = bioethics accession number   fn. = footnotes   refs. = references

Significance. Proceedings. Boston: D. Reidel, 1977. p. 73-84. Proceedings of the Third Trans-Disciplinary Symposium on Philosophy and Medicine, Farmington, Connecticut, 11-13 Dec 1975. 9 refs. 6 fn. ISBN 90-277-0772-3. BE05864.
  beneficence; human rights; killing; *value of life; *voluntary euthanasia

VOLUNTARY STERILIZATION / BLACKS

The sinister side of sterilization. *South African Medical Journal* 53(2): 38-39, 14 Jan 1978. Editorial. BE06756.
  attitudes; *blacks; mass media; population control; *voluntary sterilization

VOLUNTARY STERILIZATION / GOVERNMENT REGULATION

U.S. Court of Appeals, District of Columbia Circuit. Relf v. Weinberger. 13 Sep 1977. *Federal Reporter, 2d Series,* 565: 722-727. 6 fn. BE07115.
  *federal government; financial support; *government regulation; sterilization; *voluntary sterilization; Department of Health, Education, and Welfare

VOLUNTARY STERILIZATION / INFORMED CONSENT / LEGAL ASPECTS

Maryland. Court of Appeals. Sard v. Hardy. 9 Nov 1977. *Atlantic Reporter, 2d Series,* 379: 1014-1027. 6 fn. BE07121.
  alternatives; *disclosure; females; *informed consent; law; *legal aspects; legal liability; *legal obligations; legal rights; patients; *physicians; *risks and benefits; *standards; surgery; *voluntary sterilization

VOLUNTARY STERILIZATION / INSTITUTIONAL POLICIES

Bernstein, Arthur H. The changing law on sterilization. *Hospitals* 52(3): 36+, 1 Feb 1978. BE06772.
  *hospitals; *institutional policies; *legal liability; negligence; *physicians; private hospitals; public hospitals; religious hospitals; spousal consent; *unwanted children; *voluntary sterilization

VOLUNTARY STERILIZATION / LEGAL ASPECTS

McKenzie, James F. Contraceptive sterilization: the doctor, the patient, and the United States Constitution. *University of Florida Law Review* 25(2): 327-347, Winter 1973. 201 fn. BE06601.
  battery; hospitals; institutional policies; *legal aspects; legal liability; legislation; negligence; physicians; privacy; spousal consent; state action; state government; unwanted children; *voluntary sterilization

VOLUNTARY STERILIZATION / MEDICAID

U.S. District Court, D. Connecticut. Voe v. Califano. 14 Jul 1977. *Federal Supplement* 434: 1058-1063. BE06275.
  adults; *age; *federal government; females; *financial support; government regulation; informed consent; *voluntary sterilization; *Medicaid

VOLUNTARY STERILIZATION / NEW ZEALAND

New Zealand. Royal Commission of Inquiry. Contraception, Sterilisation, and Abortion in New Zealand: Report of the Royal Commission of Inquiry. Wellington, N.Z.: E.C. Keating, 1977. 455 p. Duncan W. McMullin, Chairman of the Commission. Bibliography: p. 421-445. BE06682.
  *abortion; adolescents; adults; *contraception; family planning; fetuses; historical aspects; human experimentation; illegal abortion; law; legal aspects; *legislation; mentally handicapped; methods; morality; pregnant women; public opinion; public policy; *social impact; socioeconomic factors; statistics; *voluntary sterilization; women's rights; *New Zealand

VOLUNTARY STERILIZATION / RELIGIOUS HOSPITALS / ROMAN CATHOLICISM

Boyle, John P. The Sterilization Controversy: A New Crisis for the Catholic Hospital? New York: Paulist Press, 1977. 101 p. 150 fn. ISBN 0-8091-2016-X. BE06183.
  Christian ethics; contraception; institutional policies; intention; moral obligations; morality; natural law; *religious hospitals; *Roman Catholic ethics; *Roman Catholicism; theology; totality; *voluntary sterilization

VOLUNTARY STERILIZATION / ROMAN CATHOLIC ETHICS

Boyle, John P. The Sterilization Controversy: A New Crisis for the Catholic Hospital? New York: Paulist Press, 1977. 101 p. 150 fn. ISBN 0-8091-2016-X. BE06183.
  Christian ethics; contraception; institutional policies; intention; moral obligations; morality; natural law; *religious hospitals; *Roman Catholic ethics; *Roman Catholicism; theology; totality; *voluntary sterilization

VOLUNTARY STERILIZATION / SPOUSAL CONSENT / LEGAL ASPECTS

Sharpe, Gilbert. Consent and sterilization. *Canadian Medical Association Journal* 118(5): 591-593, 4 Mar 1978. BE06930.
  adolescents; disclosure; contraception; *legal aspects; *informed consent; mentally handicapped;

BE = bioethics accession number   fn. = footnotes   refs. = references

*spousal consent; *sterilization; *voluntary sterilization

VOLUNTARY STERILIZATION / UNWANTED CHILDREN / LEGAL LIABILITY

Bernstein, Arthur H.   The changing law on sterilization. *Hospitals* 52(3): 36+, 1 Feb 1978. BE06772.
  *hospitals; *institutional policies; *legal liability; negligence; *physicians; private hospitals; public hospitals; religious hospitals; spousal consent; *unwanted children; *voluntary sterilization

VOLUNTARY STERILIZATION / UNWANTED CHILDREN / TORTS

Gursky, Sandra.   Birth despite vasectomy and abortion held not a 'wrong'. *Journal of Legal Medicine* 5(7): 29-31, Jul 1977. 10 refs. BE05726.
  children; compensation; genetic defects; legal aspects; legal rights; males; malpractice; negligence; parents; physicians; *selective abortion; *torts; *unwanted children; value of life; *voluntary sterilization; wrongful life; Speck v. Feingold and Schwartz

VOLUNTARY STERILIZATION / YUGOSLAVIA

Socialist Republic of Slovenia (Yugoslavia).   Abortion, sterilization, contraception, and artificial insemination. *International Digest of Health Legislation* 28(4): 1112-1115, 1977. Law of 20 Apr 1977. BE07131.
  *abortion; abortion on demand; age; *artificial insemination; competence; confidentiality; contraception; counseling; decision making; fertility; fetal development; rights; reproduction; semen donors; therapeutic abortion; third party consent; *voluntary sterilization; *Socialist Republic of Slovenia; *Yugoslavia

VOLUNTEERS   See under

HUMAN EXPERIMENTATION / DRUG INDUSTRY / VOLUNTEERS

HUMAN EXPERIMENTATION / VOLUNTEERS / MOTIVATION

WALKER V. PIERCE   See under

INVOLUNTARY STERILIZATION / WALKER V. PIERCE

WEDGE ARGUMENT   See under

ABORTION / VALUE OF LIFE / WEDGE ARGUMENT

ABORTION / WEDGE ARGUMENT

KILLING / WEDGE ARGUMENT

WEST GERMANY   See under

ABORTION / WEST GERMANY

WEST GERMANY   See under (cont'd.)

RECOMBINANT DNA RESEARCH / GOVERNMENT REGULATION / WEST GERMANY

WEST VIRGINIA   See under

DETERMINATION OF DEATH / WEST VIRGINIA

WILLOWBROOK STATE SCHOOL   See under

PATIENT CARE / MENTALLY RETARDED / WILLOWBROOK STATE SCHOOL

WITHHOLDING TREATMENT / CATHOLIC HOSPITAL ASSOCIATION

From the Catholic Hospital Association: a statement on the dying patient. *Linacre Quarterly* 44(2): 155-158, May 1977. BE05611.
  disclosure; health personnel; moral obligations; *organizational policies; patients' rights; *professional organizations; religious hospitals; Roman Catholicism; *terminally ill; treatment refusal; *withholding treatment; *Catholic Hospital Association

WITHHOLDING TREATMENT / NEWBORNS / CONGENITAL DEFECTS

MacMillan, Elizabeth S.   Birth-defective infants: a standard for nontreatment decisions. *Stanford Law Review* 30(3): 599-633, Feb 1978. 167 fn. BE06787.
  child neglect; *congenital defects; *decision making; judicial action; legal aspects; legal obligations; *newborns; parents; physicians; prognosis; quality of life; state interest; *withholding treatment

WITHHOLDING TREATMENT / TERMINALLY ILL

Forse, Max A.   Patient rights and doctors' dilemmas. *Sh'ma* 7(132): 94-96, 15 Apr 1977. BE05536.
  allowing to die; extraordinary treatment; living wills; physician patient relationship; *physician's role; *terminally ill; treatment refusal; *withholding treatment

WOMEN'S RIGHTS   See under

REPRODUCTION / WOMEN'S RIGHTS

WORLD HEALTH ORGANIZATION   See under

MEDICAL ETHICS / WORLD HEALTH ORGANIZATION

WORLD PSYCHIATRIC ASSOCIATION / PSYCHIATRY / USSR

Psychiatrists unite! *Times (London),* 27 Aug 1977, p. 15. Editorial. BE06455.
  dissent; involuntary commitment; *organizational policies; politics; *professional organizations;

BE = bioethics accession number   fn. = footnotes   refs. = references

*psychiatry; *USSR; *World Psychiatric Association

Reddaway, Peter; Bloch, Sidney. Curbing psychiatry's political misuse. *Washington Post,* 15 Nov 1977, p. A23. BE06411.
dissent; mass media; *organizational policies; *politics; professional organizations; *psychiatry; *USSR; *World Psychiatric Association

WORLD PSYCHIATRIC ASSOCIATION *See under*

INVOLUNTARY COMMITMENT / WORLD PSYCHIATRIC ASSOCIATION / USSR

ORGANIZATIONAL POLICIES / WORLD PSYCHIATRIC ASSOCIATION / USSR

WRONGFUL DEATH / FETUSES

California. Supreme Court, In Bank. Justus v. Atchison. 8 Jun 1977. *Pacific Reporter, 2d Series,* 565: 122-137. 15 fn. BE07112.
compensation; fathers; *fetuses; legislation; personhood; psychological stress; state government; *wrongful death; California

Florida. Supreme Court. Stern v. Miller. 9 Jun 1977. *Southern Reporter, 2d Series,* 348: 303-308. 8 fn. BE06261.
*fetuses; legal aspects; *personhood; prenatal injuries; viability; *wrongful death

WRONGFUL DEATH / FETUSES / LEGAL ASPECTS

Tennessee. Supreme Court. Hamby v. McDaniel. 12 Dec 1977. *South Western Reporter, 2d Series,* 559: 774-783. 5 fn. BE07126.
*fetuses; judicial action; *legal aspects; legislation; mother fetus relationship; *personhood; prenatal injuries; state courts; Supreme Court decisions; viability; *wrongful death; *Tennessee Wrongful Death Statute; United States

WRONGFUL DEATH / VIABILITY / LEGAL ASPECTS

Reilly, Timothy P. Torts—wrongful death—a viable fetus is a "person" for purposes of the Rhode Island Wrongful Death Act. *University of Cincinnati Law Review* 46(1): 266-275, 1977. 53 fn. BE05941.
*fetuses; personhood; *legal aspects; prenatal injuries; torts; *viability; *wrongful death

WRONGFUL LIFE / COMPENSATION / LEGAL ASPECTS

Kashi, Joseph S. The case of the unwanted blessing: wrongful life. *University of Miami Law Review* 31(5): 1409-1432, Nov 1977. 66 fn. BE06360.
*compensation; congenital defects; genetic defects; *legal aspects; legal liability; legitimacy; negligence; physicians; selective abortion; torts; un-

wanted children; voluntary sterilization; *wrongful life

Kronenwetter, Patrick. Wrongful birth and emotional distress damages: a suggested approach. *University of Pittsburgh Law Review* 38(3): 550-560, Spring 1977. 51 fn. BE05923.
*compensation; *genetic defects; *legal aspects; *newborns; *parents; prenatal diagnosis; *psychological stress; selective abortion; Tay Sachs disease; *wrongful life; Howard v. Lecher

WRONGFUL LIFE / GENETIC DEFECTS / LEGAL ASPECTS

New York. Supreme Court, Appellate Division, Second Department. Park v. Chessin. 12 Dec 1977. *New York Supplement, 2d Series,* 400: 110-119. BE06525.
abortion; compensation; genetic counseling; *genetic defects; judicial action; *legal aspects; legal obligations; legal rights; legislation; malpractice; negligence; newborns; obstetrics and gynecology; parents; physician patient relationship; physicians; suffering; torts; *wrongful life

WRONGFUL LIFE / LEGAL ASPECTS

Creighton, Helen. Action for wrongful life. *Supervisor Nurse* 8(4): 12-15, Apr 1977. 29 fn. BE05928.
abortion; compensation; congenital defects; *legal aspects; legitimacy; negligence; newborns; public policy; rubella; unwanted children; voluntary sterilization; *wrongful life

Veazey, Lex H. Torts—an action for wrongful life brought on behalf of the wrongfully conceived infant. *Wake Forest Law Review* 13(3): 712-727, Fall 1977. 91 fn. BE06343.
congenital defects; disclosure; genetic defects; infants; *legal aspects; legal liability; malpractice; negligence; pain; parents; physicians; *preconception injuries; suffering; torts; *wrongful life; *Park v. Chessin

WRONGFUL LIFE / PARENTS / COMPENSATION

Minnesota. Supreme Court. Sherlock v. Stillwater Clinic. 14 Oct 1977. *North Western Reporter, 2d Series,* 260: 169-177. 16 fn. BE06265.
*compensation; negligence; *parents; physicians; risks and benefits; unwanted children; voluntary sterilization; *wrongful life

New York. Court of Appeals. Johnson v. Yeshiva University. 12 May 1977. *North Eastern Reporter, 2d Series,* 364: 1340-1341. 1 fn. BE06269.
amniocentesis; children; genetic counseling; *compensation; genetic defects; malpractice; negligence; *parents; physicians; *wrongful life

New York. Supreme Court, Appellate Division, Fourth Department. Karlsons v. Guerinot. 15 Apr 1977.

BE = bioethics accession number   fn. = footnotes   refs. = references

New York Supplement, 2d Series, 394: 933-940. 4 fn. BE06270.

*compensation; *genetic defects; negligence; *parents; physicians; prenatal diagnosis; psychological stress; selective abortion; *wrongful life

WRONGFUL LIFE / PHYSICIANS / LEGAL LIABILITY

Curran, William J. Tay-Sachs disease, wrongful life, and preventive malpractice. *American Journal of Public Health* 67(6): 568-569, Jun 1977. 3 refs. BE06304.

carriers; compensation; diagnosis; fetuses; genetic counseling; genetic defects; legal aspects; *legal liability; obstetrics and gynecology; parents; *physicians; prenatal diagnosis; state courts; *Tay Sachs disease; *wrongful life; Howard v. Lecher; New York

WRONGFUL LIFE / TAY SACHS DISEASE

Curran, William J. Tay-Sachs disease, wrongful life, and preventive malpractice. *American Journal of Public Health* 67(6): 568-569, Jun 1977. 3 refs. BE06304.

carriers; compensation; diagnosis; fetuses; genetic counseling; genetic defects; legal aspects; *legal liability; obstetrics and gynecology; parents; *physi-

cians; prenatal diagnosis; state courts; *Tay Sachs disease; *wrongful life; Howard v. Lecher; New York

XYY KARYOTYPE

Pyeritz, Reed, et al. The XYY male: the making of a myth. *In* Science for the People. Ann Arbor Editorial Collective. Biology as a Social Weapon. Minneapolis: Burgess Publishing Company, 1977. p. 86-100. 74 refs. 4 fn. ISBN 0-8087-4534-4. BE06158.

aggression; behavior disorders; behavioral research; genetic screening; methods; social determinants; *XYY karyotype

XYY KARYOTYPE   *See under*

BEHAVIORAL RESEARCH / XYY KARYOTYPE / AGGRESSION

GENETIC SCREENING / XYY KARYOTYPE

GENETIC SCREENING / XYY KARYOTYPE / RISKS AND BENEFITS

YUGOSLAVIA   *See under*

ABORTION / YUGOSLAVIA

ARTIFICIAL INSEMINATION / YUGOSLAVIA

VOLUNTARY STERILIZATION / YUGOSLAVIA

BE = bioethics accession number   fn. = footnotes   refs. = references

TITLE INDEX

TITLE INDEX

AUTHOR INDEX

AUTHOR INDEX

Annas, George J.
The attempted revival of psychosurgery, 224
Avoiding malpractice suits through the use of informed consent, 167
Beyond the Good Samaritan: should doctors be required to provide essential services, 119, 203
The incompetent's right to die: the case of Joseph Saikewicz, 74, 249, 268
Legalizing Laetrile for the terminally ill, 203
Let them eat cake, 57
Psychosurgery: procedural safeguards, 224

Annas, George J.
See Ryan, Kenneth J.; Yesley, Michael S.; Annas, George J.

Annas, George J.; Glantz, Leonard H.; Katz, Barbara F.
Compensation for harm: an additional protection for human subjects, 144
The current status of the law of informed consent to human experimentation, 151, 167
Fetal research: the limited role of informed consent in protecting the unborn, 148
Informed Consent to Human Experimentation: The Subject's Dilemma, 151
Law of informed consent in human experimentation: institutionalized mentally infirm, 150, 152, 166
Origins of the law of informed consent to human experimentation, 151
Psychosurgery: the regulation of surgical innovation, 223, 223

Annas, George J.; Meisel, Alan; Horne, Rikki.
Legal aspects of informed consent, 168

Appelbaum, Judith C.
See Schultz, Stephen; Swartz, William; Appelbaum, Judith C.

Apple, William.
Ethical problems for subject or experimenter, 90

Appleman, Michael A.
The Legal Issues Involved in the Use of Stimulants on Hyperactive School Children, 86, 88

Arditti, Rita.
Abortion legislation, 57

Argentina.
Removal of human tissues and organs for therapeutic or scientific purposes, 197

Arkansas.
Death with dignity, 76

Armiger, Bernadette.
Ethics of nursing research: profile, principles, perspective, 155

Armstrong, Barbara.
The use of psychotropic drugs in state hospitals: a legal or medical decision, 132, 206

Armstrong, David.
Clinical sense and clinical science, 145

Arnold, John D.
Ethical considerations in selecting subject populations for drug research, 145, 158

Aroskar, Mila A.
Ethical dilemmas and community health nursing, 134
Ethics in the nursing curriculum, 185

Aroskar, Mila A.; Veatch, Robert M.
Ethics teaching in nursing schools, 185

Arthure, Humphrey.
See Morris, Norman; Arthure, Humphrey.

Ashdown-Sharp, Patricia; Darroch, Sandra.
What MPs really think about abortion, 64

Atthowe, John M.
Legal and ethical accountability in everyday practice, 195

Avery, Gordon B.
The morality of drastic intervention, 78

Ayd, Frank J.
Ethical and legal dilemmas posed by tardive dyskinesia, 168
Treatment for the terminally ill incompetent: who decides—courts or physicians, 77, 249

Baelz, Peter.
The right to strike by the caring professions, 202

Bailey, H.R.; Dowling, J.L.; Davies, E.
The ethics of psychiatric surgery, 222

Baker, Sherwood.
See Stephens, G. Gayle; Baker, Sherwood.

Baltimore, David.
Limiting science: a biologist's perspective, 96

Baltimore, David, et al.
A threat to scientific research, 230

Bandman, Bertram.
See Bandman, Elsie; Bandman, Bertram.

Bandman, Elsie; Bandman, Bertram.
There is nothing automatic about rights, 207

Bandow, Doug.
Should society condone abortion as 'convenient', 67

Barber, Bernard.
Perspectives on medical ethics and social change, 185

Barber, Bernard, ed.
Medical Ethics and Social Change, 92, 183

Barnaby, Wendy.
Sweden debates gene-splicing, 242

Barnett, Cynthia F.
Treatment rights of mentally ill nursing home residents, 247

Barry, Robert.
Personhood: the conditions of identification and description, 208

Bartholome, William G.
Central themes in the debate over involvement of infants and children in biomedical research: a critical examination, 141, 155

Bartholome, William G.
See Ramsey, Paul; Bartholome, William G.

Batchelder, Barron M.
See Lankton, James W.; Batchelder, Barron M.; Ominsky, Alan J.

Bates, Richard C.
It's *our* right to pull the plug, 80

Battelle, Phyllis.
See Quinlan, Joseph; Quinlan, Julia; Battelle, Phyllis.

Damme, Catherine J.
Infanticide: the worth of an infant under law, 163, 163, 164

Damme, Catherine J.
See Frankel, Lawrence S.; Damme, Catherine J.; Van Eys, Jan.

Daniels, Norman.
On the picket line: are doctors' strikes ethical, 210

Darling, Rosalyn B.
Parents, physicians, and spina bifida, 250

Darroch, Sandra.
See Ashdown-Sharp, Patricia; Darroch, Sandra.

Datta-Ray, Sunanda.
Setback seen for population control in India, 212
Sterilization backlash in India, 124

Davidson, G.P.; Roberts, F.J.
Relational ethics in medicine, 182

Davies, E.
See Bailey, H.R.; Dowling, J.L.; Davies, E.

Davis, Alan; Horobin, Gordon.
The problem of priorities, 187

Davis, Ann J.
Ethical dilemmas and nursing practice, 194

Davis, Bernard D.
The moralistic fallacy, 88

Davis, Bernard D.
See Hunter, Thomas H.; Davis, Bernard D.; Fletcher, Joseph.

De Lange, S.A.
Some ethical implications of psychiatric surgery, 222

De Wet, B.S.
Clinical trials, children and the law, 150

Delgado, Richard.
Organically induced behavioral change in correctional institutions: release decisions and the "new man" phenomenon, 87

DeMere, McCarthy.
My position is to save lives, not to terminate them, 80

Dennis, K.J.; Hall, M.R.P.
The teaching of medical ethics at Southampton University Medical School, 185

Diamond, Eugene F.
The deformed child's right to life, 79

Diamond, Milton.
See Steinhoff, Patricia G.; Diamond, Milton.

Dickinson, George E.
Death education in U.S. medical schools, 113

Dickson, David.
Friends of DNA fight back, 236
NIH may loosen recombinant DNA research guidelines, 230
Recombinant DNA risk-assessment studies to begin at Fort Detrick, 229
Universities can patent recombinant DNA results, 237
Warning over bacterial research: other countries urged to adopt British guidelines, 229

Diggory, Peter.
See Potts, Malcolm; Diggory, Peter; Peel, John.

DiGiacomo, Robert E.
Behavior modification: toward the understanding and reform of federal policy, 87, 87

Dinno, Nuhad D.
See Kadlec, James F.; Dinno, Nuhad D.

Dinsmore, Janet L., ed.
Death and Dying: An Examination of Legislative and Policy Issues, 112

Dismukes, Key.
Recombinant DNA: a proposal for regulation, 230

District of Columbia.   Superior Court.
In the Matter of Bentley.   25 Apr 1974, 266

Dixon, Bernard.
Secret sharers, 229

Dodge, John A.
See Gairdner, Douglas; Dodge, John A.; Evans, John.

Donagan, Alan.
Informed consent in therapy and experimentation, 150, 167

Donnelly, John.
Psychosurgery, 222

Donnerstein, Edward.
See Wilson, David W.; Donnerstein, Edward.

Doona, Mary E.
The ethical dimension in 'ordinary nursing care', 204

Dorwart, Robert A.
See Mufson, Michael; Dorwart, Robert A.; Eisenberg, Leon.

Dougherty, Charles J.
Moral directionality in the doctor-patient relationship, 209

Dowben, Carla.
See Blocker, Webster; Dowben, Carla.

Dowling, J.L.
See Bailey, H.R.; Dowling, J.L.; Davies, E.

Doyle, Carolyn T.
Amniocentesis and the defective child, 252

Draper, Thomas.
See Veatch, Robert M.; Draper, Thomas.

Drinkwater, C.K.; Roberts, S.H.
Cure or care in everyday practice, 201, 261

Drinkwater, C.K.; Thorne, Susan; Wilson, Michael.
Strive officiously to keep alive, 80

Dua-Agyemang, J.E.
Disclosure of medical notes, 107, 107

Dubos, René.
Genetic engineering, 227

Duff, Raymond S.
A physician's role in the decision-making process: a physician's experience, 81, 192, 207

Duncan, A.S.; Dunstan, G.R.; Welbourn, R.B.
Dictionary of Medical Ethics, 92

Duncan, Gaylen.
See Shires, David B.; Duncan, Gaylen.

Duncan, George E.
Christian ethics and biomedical issues: a physician's perspective, 197

Dunham, H. Warren.
Psychiatry, sociology and civil liberties, 221, 255

Dunstan, G.R.
See Duncan, A.S.; Dunstan, G.R.; Welbourn, R.B.

Hostler, John.
  The right to life, 175, 270
Howard, Ted.
  See Rifkin, Jeremy; Howard, Ted.
Howard, Ted; Rifkin, Jeremy.
  Who Should Play God? The Artificial Creation of Life
  and What It Means for the Future of the Human
  Race, 120, 129, 245
Hoyt, Robert.
  See Bennett, John; Hoyt, Robert.
Huber, Robert.
  Legal considerations of cardiopulmonary resuscitation, 75, 246
Humber, James M.
  Abortion, fetal research, and the law, 52, 148
Hungary.
  Genetic counselling, 127
Hunter, Ian A.
  Abortion: reflections on a protracted debate, 60, 65
Hunter, Thomas H.
  A philosophy of clinical research based on needs, 139
Hunter, Thomas H.; Davis, Bernard D.; Fletcher,
    Joseph.
  Is Behavioral Genetics Taboo? The Neo-Lysenkoism. Videorecording, 88, 89
Hunter, Thomas H.; Marston, Robert; Fletcher,
    Joseph.
  Dilemmas of Clinical Trials. Videorecording, 139
Hunter, Thomas H.; Smith, William.
  Attitudes Toward Death and Dying. Videorecording, 73, 80
Hunter, Thomas H.; Wenzel, Richard; Fletcher,
    Joseph.
  How Does One Determine Acceptable Risks?
  Videorecording, 95, 201
Hunter, Thomas H., et al.
  Fetal Research. Videorecording, 147
  Has the Physician the Right to Strike? Videorecording, 190
  The Health of Public Figures: What Should Be Disclosed? Videorecording, 104
  Medical confidentiality—who protects the victims, 104
  The Physician as Double Agent. Videorecording, 108, 194
  Research Using "Live" Human Fetuses: When Is It
    Justifiable? Videorecording, 147
  What Rights Do Patients Have? Videorecording, 70
Hyatt, I. Ralph.
  Psychology's concern with freedom, 253
Hyatt, James C.
  Protection for unborn? Work-safety issue isn't as simple as it sounds, 244
Hyde, Henry J.
  The heart of the matter, 56, 67
  The humanity of the unborn, 67
Hyer, Marjorie.
  Hospice care supported for terminally ill patients, 261
  Theologians react cautiously to test-tube baby process, 163

Iceland.
  Abortion, sterilization, and sex education, 59, 257
Illinois. Appellate Court, Fourth District.
  Renslow v. Mennonite Hospital. 10 Jun 1976, 215
Illinois. Supreme Court.
  Renslow v. Mennonite Hospital. 8 Aug 1977, 215
Illman, John.
  ECT: therapy or trauma, 118
  Vanishing medical secrets, 108
Imbus, Sharon H.; Zawacki, Bruce E.
  Autonomy for burned patients when survival is unprecedented, 201, 252
Ioirysh, A.I.
  Law and the new potentials of biology, 159, 242
Isaacs, Leonard.
  Death, where is thy distinguishing, 114
Israel.
  Abortion, 59
Italy.
  Removal of organs and tissues for therapeutic purposes, 198

Jackson, Jacquelyne J.
  Informed consent: ethical issues in behavioral research, 89, 150
Jackson, M. Virginia.
  Genetic Engineering and the Recombinant DNA Research Controversy—a Historical Bibliography:
  1973-1977, 227
Jacobs, Leon.
  The role of the National Institutes of Health in rulemaking, 234, 237
Jacobs, Sarah.
  Abortion—the question is whether the government should pay, 57
Jamali, Naseem Z.
  Compulsory sterilization, 173
James, A. Everette; Johson, Burton A.; Hall, Donald J.
  Informed consent: some newer aspects and their relation to the specialty of radiology, 168
Jameton, Andrew L.
  See Jonsen, Albert R.; Jameton, Andrew L.
Jameton, Andrew.
  The nurse: when roles and rules conflict, 194
Jamir, Vinson F.
  Children: health services for minors in Oklahoma
    —capacity to give self-consent to medical care and
    treatment, 135, 135, 164
Jarvis, Peter.
  Nursing and the ethics of withdrawing professional services, 194
  Some comments on the Rcn code of professional conduct, 102
Jellinek, Michael; Parmelee, Dean.
  Is there a role for medical ethics in postgraduate psychiatry courses, 185
Jobson, K.
  Consent to treatment—problems of prisoners, 168
Jochimsen, Peter R.
  See Noyes, Russell; Jochimsen, Peter R.; Travis,
    Terry A.

Macklin, Ruth.
On the ethics of *not* doing scientific research, 240

MacMillan, Elizabeth S.
Birth-defective infants: a standard for nontreatment decisions, 273

MacPherson, Myra.
See Colen, B.D.; MacPherson, Myra.

Maddison, J.C.
The legal aspects of psychosurgery in New South Wales, 224

Maddocks, Melvin.
DNA and the habit of saying yes to whatever's next, 241

Maguire, Daniel C.
Death and the moral domain, 69, 75, 122

Mahoney, Edward J.
The morality of terminating life vs. allowing to die, 69, 82

Mais, Joseph E.
See Franklin, Marc A.; Mais, Joseph E.

Maley, Roger; Hayes, Steven C.
Coercion and control: ethical and legal issues, 87

Mall, David.
See Horan, Dennis J.; Mall, David.

Malter, Susan.
Genetic counseling...a responsibility of health-care professionals, 128

Manard, Barbara.
See Shaw, Anthony; Randolph, Judson G.; Manard, Barbara.

Mancini, Marguerite.
Nursing, minors, and the law, 135

Marcolongo, Francis J.
Moral Choices in Contemporary Society: A Study Guide for Courses by Newspaper, 92

Margolin, C.R.
Attitudes toward control and elimination of genetic defects, 244

Markey, Kathleen.
Federal regulation of fetal research: toward a public policy founded on ethical reasoning, 148, 149

Marks, Cathy L.
Torts—psychiatry and the law—duty to warn potential victim of a homicidal patient, 104

Marsh, Frank H.
An ethical approach to paternalism in the physician-patient relationship, 209

Marshall, John R.; Fellner, Carl H.
Kidney donors revisited, 197

Marston, Robert.
See Hunter, Thomas H.; Marston, Robert; Fletcher, Joseph.

Martin, Andrew J.
Consent to treatment, 167
Law on mercy killing, 122

Martin, Andrew J.
See Leonard, Graham; Martin, Andrew J.

Martin, Andrew.
A right to treatment, 172, 247

Martin, Daniel S.
Laetrile—a dangerous drug, 203
Laetrile—consumer protection is the bottom line, 203

Martindale, David.
The ethics of experimentation, 150

Marty, Martin E.
Physicians leave too much to the bedside encounter; economists stop everything before it starts, 137, 137

Marx, Jean L.
The new P4 laboratories: containing recombinant DNA, 238

Maryland.
Recombinant DNA research, 236

Maryland. Court of Appeals.
Sard v. Hardy. 9 Nov 1977, 167, 272

Massachusetts. Supreme Judicial Court.
Jones v. Saikewicz. 9 Jul 1976, 74, 249

Massachusetts. Supreme Judicial Court, Hampshire.
Superintendent of Belchertown v. Saikewicz. 28 Nov 1977, 74, 249

Massachusetts. Supreme Judicial Court, Norfolk.
Schroeder v. Lawrence. 15 Feb 1977, 169

Massachusetts. Supreme Judicial Court, Suffolk.
Commonwealth v. Golston. 26 Aug 1977, 72
Doe v. Commissioner of Mental Health. 10 May 1977, 108

Masserman, Jules H.
American Psychiatric Association (APA): ethical standards, 102

Masterman, John.
How the right to die could improve the quality of life, 72

Masters, William H.; Johnson, Virginia E.; Kolodny, Robert C., eds.
Ethical Issues in Sex Therapy and Research, 159

Maurice, Shirley; Warrick, Louise.
Ethics and morals in nursing, 194

May, Robert M.
The recombinant DNA debate, 227

May, William F.
The right to die and the obligation to care: allowing to die, killing for mercy, and suicide, 71, 71, 258, 260

McBride, Stewart D.
Genetics research guidelines set up in Cambridge, 233

McCaughey, J.D.
Civil liberties and patient rights, 224

McCormick, Richard A.
Man's moral responsibility for health, 136
The quality of life, the sanctity of life, 219, 220

McCormick, Richard A.
See Hellegers, André E.; McCormick, Richard A.

McCormick, Richard A.; Hellegers, André E.
The specter of Joseph Saikewicz: mental incompetence and the law, 73, 77, 249

McCullough, James M.
See Zegel, Vikki A.; McCullough, James M.

McCullough, Laurence B.
Historical perspectives on the ethical dimensions of the patient-physician relationship: the medical eth-

Stent, Gunther S.
You can take the ethics out of altruism but you can't take the altruism out of ethics, 255

Stephens, G. Gayle; Baker, Sherwood.
The ethical dimensions of behavior, 113, 210

Stetten, DeWitt, et al.
The social implications of research on genetic manipulation: panel discussion, 231, 240

Stevens, Carey; Firth, S.T.
Responses to moral dilemmas in medical students and psychiatric residents, 185

Stevenson, Adlai E.
Recombinant DNA legislation, 232

Stewart, J.H.
The law and transplantation, 198

Stewart, Raymond D.
See Novak, Ervin; Seckman, Clarence E.; Stewart, Raymond D.

Stine, Gerald J.
The future: promise or peril, 129
Genetic counseling, 126
Genetic screening, 130
Genetics, society, and the law, 129, 131

Stokes, Joseph, et al.
Science with a halo or hubris, 96

Stolfi, Julius E.
Legalizing human guinea pigs, 203

Stone, Brenda.
The 'boy-in-the-bubble' drama, 201
Telling the patient he has cancer, 116

Straus, Thomas R.
Planned Parenthood v. Danforth: resolving the antinomy, 52, 63, 66

Stuart, Allison J.
See Freedman, Theodore J.; Stuart, Allison J.

Stumpf, Samuel E.
Genetics and the control of human development: some ethical considerations, 215, 244

Suckiel, Ellen K.
Death and benefit in the permanently unconscious patient; a justification of euthanasia, 122

Sullivan, Daniel J.
Gene-splicing: the eighth day of creation, 228

Sullivan, Thomas D.
Active and passive euthanasia: an impertinent distinction, 68, 72

Sullivan, Walter.
Implants of monkeys may explain success with human embryo, 162
New era in reproduction seen in British laboratory's embryo, 162
Scientists praise British birth as triumph: doctors' success in conception in the laboratory intensifies the debate over reproductive control, 162
Woman gives birth to baby conceived outside the body, 83

Suran, Bernard G.; Lavigne, John V.
Rights of children in pediatric settings: a survey of attitudes, 100

Susser, Mervyn; Stein, Zena; Kline, Jennie.
Ethics in epidemiology, 146

Sutnick, Alton I., et al.
Ethical issues in investigation of screening strategies, 182

Sutton, R.N.P.
Science versus safety: who should judge the balance, 96

Swan, George S.
Abortion, parental burdens, and the right to choose: is a penumbral or Ninth-Amendment right about to arise from the ashes of a common law liberty, 55
State-mandated abortion: has 1984 arrived seven years early, 61, 244

Swartz, William.
See Schultz, Stephen; Swartz, William; Appelbaum, Judith C.

Swazey, Judith P.
Protecting the "animal of necessity": limits to inquiry in clinical investigation, 148, 159

Swinyard, Chester A., ed.
Decision Making and the Defective Newborn. Proceedings of a Conference on Spina Bifida and Ethics, 79, 193, 251

Swiss Academy of Medical Sciences.
Swiss guidelines on care of the dying, 82

Szasz, Thomas S.
Aborting unwanted behavior: the controversy on psychosurgery, 225
Psychiatric Slavery, 172, 247

Szasz, Thomas S.; Kurtz, Paul.
Ethics of Medicine and Psychiatry. Videorecording, 254

Szasz, Thomas.
Patriotic poisoners: psychiatrists, physicians and the CIA drug program, 158

Tait, Karen M.; Winslow, Gerald.
Beyond consent—the ethics of decision-making in emergency medicine, 219

Talbert, Jeffrey T.
The validity of parental consent statutes after *Planned Parenthood*, 52, 63

Tancredi, Laurence R.; Slaby, Andrew E.
Ethical Policy in Mental Health Care: The Goals of Psychiatric Intervention, 136

Tangen, Ottar.
Medical ethics and child abuse, 100

Tannenbaum, Arnold S.; Cooke, Robert A.
Report on the mentally infirm, 91, 152

Taurek, John M.
Should the numbers count, 246

Taylor, Carl E.
Economic triage of the poor and population control, 212

Taylor, Michael A.
Abortion statistics and parental consent: a state-by-state review, 63, 67

Taylor, Nancy K., comp.
Bibliography of Society, Ethics and the Life Sciences: Supplement for 1977-78, 93, 183

Tedeschi, James T.; Gallup, Gordon G.
Human subjects research, 159

Teff, Harvey.
Compensating vaccine-damaged children, 160
Temkin, Owsei.
The idea of respect for life in the history of medicine, 270
Templeton, Allan.
AID—what are the problems, 70
Tendler, Moshe D.
Torah ethics prohibit natural death, 81
Tennessee. Court of Appeals, Middle Section.
State Department of Human Services v. Northern. 7 Feb 1978, 265, 267
Tennessee. Supreme Court.
Hamby v. McDaniel. 12 Dec 1977, 208, 274
Olson v. Molzen. 21 Nov 1977, 64
Texas.
Natural death act, 77
Thacher, Scott.
See Park, Bob; Thacher, Scott.
Thiroux, Jacques P.
Abortion, 52
Euthanasia and allowing someone to die, 69, 77
Thomas, Agnes.
See Affleck, Glenn; Thomas, Agnes.
Thomas, Barbara.
Legal implications, 204
Thomas, Jo.
C.I.A. sought to spray drug on partygoers, 86, 154
Thomas, John M.; Ryniker, Barbara M.; Kaplan, Milton.
Indian abortion law revision and population policy: an overview, 59, 61, 213
Thomas, Marthalee.
See Weimer, William; Thomas, Marthalee.
Thomas, W.D.S.
The Badgley report on the abortion law, 53
Thomson, W.A.R.
See Dunstan, G.R.; Thomson, W.A.R.
Thorne, Susan.
See Drinkwater, C.K.; Thorne, Susan; Wilson, Michael.
Thorup, Oscar A., et al.
Genetically Transmitted Disease. Videorecording, 126, 215
Laetrile: The Right to Choose. Videorecording, 203, 203
Privacy and the Computer: Everything You Know About Yourself but Hoped They'd Never Find Out. Videorecording, 105
Sociobiology: Are There Areas of Forbidden Knowledge? Videorecording, 96, 255
Tiefel, Hans O.
The unborn: human values and responsibilities, 208, 270
Tiollais, Pierre.
See Philipson, Lennart; Tiollais, Pierre.
Tobin, Charles.
A statement on legislation concerning death and dying, 75
Todres, I. David, et al.
Pediatricians' attitudes affecting decision-making in defective newborns, 79, 205

Tolchin, Martin.
Compromise is voted by House and Senate in abortion dispute, 61
Tooze, John.
Emerging attitudes and policies in Europe, 233
Toulmin, Stephen.
The meaning of professionalism: doctors' ethics and biomedical science, 188, 190, 250
Trachtman, Gilbert M.
Ethical standards for school psychologists, 102
Trammell, Richard L.
The presumption against taking life, 68, 72
Travis, Terry A.
See Noyes, Russell; Jochimsen, Peter R.; Travis, Terry A.
Treadway, Jerry T.; Rossi, Robert B.
An ethical review board: its structure, function and province, 202
Treaster, Joseph B.
Researchers say that students were among 200 who took LSD in tests financed by C.I.A. in early '50's, 86, 154
Tripp, John.
Ethical problems in paediatrics, 78, 79, 142
Troland, Mary B.
Involuntary commitment of the mentally ill, 172
Trout, Monroe E.
Immunizations—a societal dilemma, 161
Troyer, John.
Euthanasia, the right to life, and moral structures: a reply to Professor Kohl, 123, 123
Trumbull, Robert.
Anglican report in Canada leans toward euthanasia, 72
Tuck, Christine.
Unwanted pregnancy, 53
Turnage, John R.
See Logan, Daniel L.; Turnage, John R.
Turnbull, H. Rutherford.
Individualizing the law for the mentally handicapped, 263
Turnbull, H.R.
Consent procedures: a conceptual approach to protection without overprotection, 146, 150
Twiss, Sumner B.
The problem of moral responsibility in medicine, 189

U.S. Library of Congress. Congressional Research Service.
See Zegel, Vikki A.; Parratt, Donna, eds.

U.S. Congress. House.
  A bill to amend the Public Health Service Act to
    regulate activities involving recombinant DNA,
    and for other purposes, 232
  A bill to amend the Public Health Service Act to
    regulate research projects involving recombinant
    DNA, 232
  A bill to amend...the Social Security Act to provide
    for the confidentiality of personal medical informa-
    tion created or maintained by medical-care institu-
    tions providing services under...Medicare..., 105
  A bill to limit use of prison inmates in medical re-
    search, 154, 157
  A bill to prohibit psychosurgery in federally con-
    nected health care facilities, 223, 223
  Recombinant DNA Act, 232
  Recombinant DNA Safety Assurance Act, 232

U.S. Congress. House. Committee on Science and
    Technology. Subcommittee on Science, Research
    and Technology.
  Science Policy Implications of DNA Recombinant
    Molecule Research. Hearings, 239

U.S. Congress. House. Committee on the Judiciary.
    Subcommittee on Civil and Constitutional
    Rights.
  Proposed Constitutional Amendments on Abortion.
    Hearings, 54, 54

U.S. Congress. Senate.
  A bill to amend the Public Health Service Act to
    establish the President's Commission for the Pro-
    tection of Human Subjects of Biomedical and Be-
    havioral Research, and for other purposes, 148,
    216
  President's Commission for the Protection of Human
    Subjects of Biomedical and Behavioral Research
    Act of 1978, 91, 156
  Project MKULTRA, the CIA's Program of Research
    in Behavioral Modification. Joint Hearing, 140

U.S. Congress. Senate. Committee on Commerce,
    Science, and Transportation. Subcommittee on
    Science, Technology, and Space.
  Regulation of Recombinant DNA Research. Hear-
    ings, 234

U.S. Congress. Senate. Committee on Finance.
    Subcommittee on Health.
  Confidentiality of Medical Records. Hearing, 105

U.S. Congress. Senate. Committee on Human
    Resources.
  President's Commission for the Protection of Human
    Subjects of Biomedical and Behavioral Research
    Act of 1978. Report, 156

U.S. Congress. Senate. Committee on Human
    Resources. Subcommittee on Health and
    Scientific Research.
  Banning of the Drug Laetrile from Interstate Com-
    merce by FDA. Hearing, 203
  Biological Testing Involving Human Subjects by the
    Department of Defense, 1977. Hearings, 95
  Recombinant DNA Regulation Act, 1977. Hearing,
    232

U.S. Congress. Senate. Committee on the Judiciary.
    Subcommittee on the Constitution.
  Civil Rights of Institutionalized Persons. Hearings,
    169

U.S. Congress. Senate. Committee on the Judiciary.
    Subcommittee to Investigate the Administration of
    the Internal Security Act....
  Humans Used as Guinea Pigs in the Soviet Union.
    Hearing, 160

U.S. Court of Appeals, District of Columbia Circuit.
  Relf v. Weinberger. 13 Sep 1977, 272

U.S. Court of Appeals, Eighth Circuit.
  M.S. v. Wermers. 29 Jun 1977, 110

U.S. Court of Appeals, Fifth Circuit.
  Givens v. Lederle. 8 Aug 1977, 161

U.S. Court of Appeals, Fourth Circuit.
  Semler v. Psychiatric Institute of Washington, D.C.
    17 Feb 1976, 179
  Walker v. Pierce. 26 Jul 1977, 174

U.S. Department of Health, Education, and Welfare.
  Federal financial participation in expenditures for
    abortions funded through various HEW programs,
    56
  Protection of human subjects: proposed regulations
    on research involving prisoners, 157
  Protection of human subjects: research involving
    those institutionalized as mentally infirm; report
    and recommendations for public comment, 153
  Research involving children: report and recommenda-
    tions of the National Commission for the Protec-
    tion of Human Subjects of Biomedical and
    Behavioral Research, 141, 151, 165

U.S. District Court, D. Connecticut.
  Voe v. Califano. 14 Jul 1977, 272

U.S. District Court, D. Minnesota, Fourth Division.
  Hines v. Anderson. 27 May 1977, 137

U.S. District Court, D. South Carolina, Columbia
    Division.
  Floyd v. Anders. 4 Nov 1977, 68

U.S. District Court, E.D. Wisconsin.
  Doe v. Mundy. 2 Sep 1977, 56, 59

U.S. District Court, M.D. Tennessee, Northeastern
    Division.
  Wolfe v. Merrill National Laboratories, Inc. 13 Jun
    1977, 161

U.S. District Court, W.D. Michigan, S.D.
  Doe v. Irwin. 23 Nov 1977, 110, 111

U.S. District Court, W.D. Oklahoma.
  Rutherford v. United States. 4 Jan 1977, 203
  Rutherford v. United States. 5 Dec 1977, 126, 132,
    217

U.S. Energy Research and Development Administration.
  Title 10—Energy, Chapter III, Part 745—Protection
    of human subjects, 148, 148

U.S. Federal Interagency Committee on Recombinant
    DNA Research.
  Interim report of the Federal Interagency Committee
    on Recombinant DNA Research: suggested ele-
    ments for legislation. 15 Mar 1977, 232
  Report of the Federal Interagency Committee on Re-
    combinant DNA Research: international activi-
    ties. Nov 1977, 242

U.S. National Commission for the Protection of
    Human Subjects of Biomedical and Behavioral
    Research.
  Disclosure of Research Information under the Free-
    dom of Information Act. Report and Recommen-

Veatch, Robert M.
Abnormal newborns and the physician's role: models of physician decision making, 193, 193, 267
Death and dying: the legislative options, 76, 267
An ethical analysis of population policy proposals, 213
Governmental population incentives: ethical issues at stake, 212
Hospital ethics committees: is there a role, 202
Justice, 212
Medical ethics, 183
Medicine, biology, and ethics, 184
The technical criteria fallacy, 251
Value foundations for drug use, 221

Veatch, Robert M.
See Lasagna, Louis; Veatch, Robert M.; Kurtz, Paul.
See Callahan, Daniel; Cassell, Eric J.; Veatch, Robert M.
See Aroskar, Mila; Veatch, Robert M.

Veatch, Robert M.; Draper, Thomas.
The values of physicians, 125, 213

Veatch, Robert M., ed.
Population Policy and Ethics: The American Experience, 214

Veazey, Lex H.
Torts—an action for wrongful life brought on behalf of the wrongfully conceived infant, 274

Vecsey, George.
Religious leaders differ on implant, 163

Veith, Frank J., et al.
Brain death. I. A status report of medical and ethical considerations, 98
Brain death. II. A status report of legal considerations, 98

Vere, D.W.
Ethics of clinical trials, 145

Vere, Duncan.
Euthanasia, 271

Vickers, T.
Flexible DNA regulation: the British model, 233

Vrhovac, B.
Placebo and its importance in medicine, 156

Vuori, Hannu.
Privacy, confidentiality and automated health information systems, 106

Wade, Nicholas.
Congress set to grapple again with gene splicing, 236
Gene splicing preemption rejected, 236
Gene splicing: Congress starts framing law for research, 234
Gene splicing: Senate bill draws charges of Lysenkoism, 236
Gene-splicing rules: another round of debate, 237
Recombinant DNA: NIH rules broken in insulin gene project, 243
Senate passes back gene-splice cup, 234
The Ultimate Experiment: Man-Made Evolution, 228

Wahl, Diana H.
'Minimum standards' for DNA research, 232

Wakin, Edward.
Is the right-to-die wrong, 72, 218

Wakin, Edward.
See Hellegers, André E.; Wakin, Edward.

Wald, N.J., et al.
Maternal serum-alpha-fetoprotein measurement in antenatal screening for anencephaly and spina bifida in early pregnancy, 215

Waldner, Rosmarie.
Switzerland—guidelines emerge, 243

Walgate, Robert.
The expert has no clothes, 250
GMAG wants to stay on, 229

Walshe-Brennan, K.S.
Confidentiality in the N.H.S, 108

Walter, Nicolas.
See Searle, J.F.; Walter, Nicolas; Ellsworth, L.E.

Walters, Hugh.
Medical ethics, 184, 188

Walters, LeRoy.
See Beauchamp, Tom L.; Walters, LeRoy.

Wampler, J. Paul.
The physician and his patient: ethical decisions, 93

Wand, Barbara F.
Parental consent abortion statutes: the limits of state power, 52, 63

Warren, Mary A.
Do potential people have moral rights, 64, 208, 208

Warrick, Louise.
See Maurice, Shirley; Warrick, Louise.

Warshaw, Frances R.
Gene implantation: proceed with caution—reservations concerning research in recombinant DNA, 240

Warwick, Donald P.
Freedom, 214
Social sciences and ethics, 90, 119

Washington. Court of Appeals, Division One.
Archer v. Galbraith. 15 Aug 1977, 169

Watson, Charles G.
Can and should the ethics of a cottage industry survive? Ethics for young surgeons, 184

Watson, J.D.
Trying to bury Asilomar, 232

Watson, James D.
An imaginary monster, 240

Wax, Murray L.
Field workers and research subjects: who needs protection, 89

Wecht, Cyril H.
Human experimentation and clinical investigation—legal and ethical considerations, 139

Weimer, William; Thomas, Marthalee.
Toward a hospital policy on drug abuse patients, 106, 107

Weinstein, Milton C.; Stason, William B.
Allocating resources: the case of hypertension, 245

Weintraub, Richard M.
First test-tube baby born in British hospital, 162

Weir, Robert F., ed.
Ethical Issues in Death and Dying, 72, 114, 116, 122, 258

Wyatt, Philip R.
    Who's for amniocentesis, 83

Yanchinski, Stephanie.
    Doctors split on screening pregnant women to detect
        spina bifida babies, 131
Yankauer, Alfred.
    Abortions and public policy, II, 57
Yeaworth, Rosalee C.
    The agonizing decisions in mental retardation, 79, 83
Yesley, Michael S.
    See Ryan, Kenneth J.; Yesley, Michael S.; Annas,
        George J.

Zachary, R.B.
    Life with spina bifida, 80
Zawacki, Bruce E.
    See Imbus, Sharon H.; Zawacki, Bruce E.
Zegel, Vikki A.

Biomedical Ethics: Human Experimentation. Issue
    Brief Number IB74095, 149, 156
Zegel, Vikki A.; McCullough, James M.
    DNA Recombinant Molecule Research. Issue Brief
        Number IB77024, 232
Zegel, Vikki A.; Parratt, Donna, eds.
    CRS Bioethics Workshop for Congress, 16 Jun 1977,
        94
Zellick, Graham.
    The hunger-striking prisoner, 126
Zimmerman, Burke K.
    Self-discipline or self-deception, 234
Zonana, Howard.
    See Crane, Lansing; Zonana, Howard; Wizner,
        Stephen.
Zucker, Karin W.
    Legislatures provide for death with dignity, 76

BIOETHICS THESAURUS

Sample Entry

Main Entry Term

HOSPITALS

Main Entry

SN	Institutions in which sick or injured persons are given medical or surgical treatment
UF	Clinics
	Hospital policies†
BT	Institutions
NT	Mental institutions
	Private hospitals↓
RT	Intensive care units
	Residential facilities↓

Cross-Reference

Clinics
  USE   HOSPITALS

SN = Scope Note. The Scope Note is a definition or a brief statement of the intended usage of a term.

UF = Used For. Terms following the UF notation are synonyms or variant forms of the main entry term.

USE = Use. USE references lead to main entry terms.

BT = Broader Term. This notation designates the larger class of which the main entry term is a member.

NT = Narrower Term. This notation indicates terms which are more specific than the main entry term.

RT = Related Term. Terms following the RT notation do not fall into the broader-narrower relationship but are conceptually related to the main entry term.

↓ = See main entry for narrower terms. If there are narrower terms for either a Narrower Term or a Related Term, the existence of those narrower terms is indicated by an arrow (↓).

† = Consult phrase in alphabetical listing. The dagger (†) indicates that a phrase is represented in the Bioethics Thesaurus by two terms rather than by one term.